ATOMIC, MOLECULAR AND OPTICAL PHYSICS: NEW RESEARCH

ATOMIC, MOLECULAR AND OPTICAL PHYSICS: NEW RESEARCH

L.T. CHEN
EDITOR

Nova Science Publishers, Inc.
New York

LIBRARY OF CONGRESS CATALOGING-IN-PUBLICATION DATA

Atomic, molecular, and optical physics : new research / editor, L.T. Chen.
    p. cm.
  Includes bibliographical references and index.
  ISBN 978-1-60456-907-0 (hardcover : alk. paper)
  1. Lasers. 2. Physical optics. I. Chen, L. T., 1963-
  QC688.A86 2009
  621.36--dc22
                              2008047493

Published by Nova Science Publishers, Inc. ✦ New York

CONTENTS

PREFACE

Atomic, molecular, and optical physics is the study of matter-matter and light-matter interactions on the scale of single atoms or structures containing a few atoms. The three areas are grouped together because of their interrelationships, the similarity of methods used, and the commonality of the energy scales that are relevant. All three areas include both classical and quantum treatments. This new book presents the latest research from around the world.

Chapter 1 revisits, with an in-depth and self-consistent approach, the small and the strong-harmonic expansion methods that handle the dynamic properties of the well-known integro-differential "Maxwell-Bloch" equations, which describe light-matter interactions inside a single-mode inhomogeneously broadened [SMIB] laser. The chapter foremost re-examines the well-established small-side-band routine, outlining the limits of its applicability in the unstable regime of operation. These limits are overcome with the application of a strong harmonic-expansion procedure that allows for the derivation of new analytical information in the control-parameter space within which regular pulsing behavior takes place. In particular, despite the complicated nature of the SMIB laser equations, owing to the presence of an integral over the polarization variable, we show that an extension of the analytical procedure, up to third order in field amplitude, is sufficient to obtain the oscillation frequencies of the asymptotic solutions and an analytical expression that illustrates the dynamic-gain contour during pulse build-up. Furthermore, the evaluation of the long-term operating frequency is shown to bear fundamental importance in the construction of complete analytical solutions, inside the control parameter space that exhibits periodic self-pulsing, through simple iterative extraction of the high-order field amplitudes. Hopefully, these new findings will allow for significant advance and understanding of the physics underneath self-pulsing and time structuring in bad cavity configured lasers.

Chapter 2 extends the strong harmonic expansion method of the preceding report to the much simpler Lorenz-Haken model that describes self-pulsing in single-mode homogeneously broadened lasers. The study focuses on a typical pulse-structuring hierarchy of periodic solutions, which include a period-doubling cascade that builds up with increasing excitation levels. Despite the uncomplicated formalism they display, the Lorenz-Haken equations are shown to carry the same qualitative features as those of the much more complex integro-differential system. Remarkable analogies are found between a few typical solutions of both systems as direct one-to-one comparisons reveal. The obvious advantage of a simpler set of equations lies in the fact that it renders the analytical and numerical handling much easier to carry out. As a consequence, new and ampler information with respect to pulse structuring are more easily extracted. In particular, the iterative harmonic-expansion algorithm is carried out, up to the fifth-order in field amplitude, giving persuasive and explicit credit to the strong-sideband ability to provide an accurate description of pulse structuring in laser-light dynamics.

Chapter 3 gives an update of the study of laser assisted collisions in dense resonantly laser-excited In or Ga homo/heteronuclear vapours. In this way, Rydberg levels, as well as autoionizing levels and collisional ionization, have been evidenced for each element and the relative cross sections have been measured via laser induced fluorescence spectroscopy (LIF). The presence of a large density of excited atoms favours the collision of an atom in the fundamental state and another one in the first excited state to give a molecule directly in an excited electronic state from which it radiates. At the same time, the presence of a rather large density of ions can also favour the onset of molecular ions via collisions, which can radiate after electron/ion recombination and/or eventually dissociate. All of these processes produce molecular signals in the fluorescence spectrum following the resonant atomic excitation. In this way, some unresolved molecular bands and excimer emission have been detected. Special emphasis is dedicated to the description of the most recent LIF experiments performed in a heteronuclear In-Ga vapour by the resonant excitation of either element. Besides the evidence of homonuclear molecules, the presence of the GaIn molecules is inferred from the time-resolved analysis of some fluorescence features. The experiments are carried out in the 900–1100°C temperature range with the vapors confined in fused silica cells. The permanent modifications induced in the silica optical properties by the migration of the In/Ga atoms inside the silica matrix are also described. In this chapter, a section is dedicated to the analysis of how the latter effect, added to all the processes mentioned previously, affects the basic physical properties of laser-excited dense vapors as the radiation trapping, which turns out to be greatly reduced, if not quenched.

Electron-ion collision processes play an important role in the understanding of energy balance in various types of astrophysical plasmas and in controlled thermonuclear fusion plasmas. Electron collisional excitation rates and transition probabilities are important for computing electron temperatures and densities, ionization equilibria and for deriving elemental abundances from emission lines formed in the collisional and photoionized plasmas. Over the last several years our research work has been focused on accurate calculations of electron-ion and atom excitation processes using the close-coupling methods for application to astrophysical and laboratory plasmas. Our effort has been to benchmark theory against experiment for the electron impact excitation of selected atomic systems for transitions of diagnostic importance. Accurate representation of target wave functions that properly account for the important correlation and relaxation effects and inclusion of coupling effects including coupling to the continuum are essential components of a reliable collision calculation. The resonant excitation processes are important in the low energy region. Chapter 4 presents a discussion of the state of our knowledge of the electron impact excitation processes together with some recent results for neutral atoms and singly and multiply charged ions.

Since the discovery of photosensitivity in optical fibers by Hill et al. in 1978 [1], Bragg gratings fabricated in optical fibers and planar waveguides have been extensively investigated over the last three decades and have been widely used in optical communication and sensor applications. In Chapter 5, the properties of birefringence in optical fibers and planar waveguides with Bragg grating structures are investigated experimentally, and the birefringence-induced impairments in communication systems with Bragg-grating–based components are evaluated by simulation.

As an important waveguiding medium, optical fiber plays significant roles in optical communications, optoelectronics, and sensors. A new type of microstructure inscribed in the optical fibers, i.e., fiber Bragg gratings (FBGs), has received considerable attention in recent years. A FBG is a type of distributed Bragg reflector constructed in a short segment of optical fiber that reflects specific wavelengths of light and transmits all the other components. In Chapter 6, optical properties of FBGs will be reviewed first with the underlying physical mechanisms. Different techniques to fabricate FBGs will be illustrated with the comparison of their advantages and drawbacks. For their important sensing applications, FBGs as temperature and humidity sensors will be discussed. The FBG sensors exceed other conventional electric sensors in many aspects, for instance, immunity to electromagnetic interference, compact size, light weight, flexibility, stability, high temperature tolerance, and resistive to harsh environment. A novel approach to realize separate temperature or humidity measurement, and their simultaneous measurement will be demonstrated by the use of FBGs coated with different polymers. The polymer-coated FBGs indicate linear shifts in the Bragg resonance wavelengths of the gratings with the temperature changes. A polyimide-coated FBG is sensitive to humidity due to the unique hygroscopic properties of polyimide while an acrylate-coated FBG shows insensitivity to humidity. The experimental results are in good agreement with the theoretical analysis.

Spatial optical techniques have shown great potential in the field of information security to encode high-security images. Among them, the dual random phase encoding method has received much attention since it was proposed by Réfrégier and Javidi in the middle 1990s. Since then, a number of works in the field were proposed introducing different variations of this technique. On the other hand, the space-time duality refers to the close mathematical analogy that exists between the equations describing the paraxial diffraction of beams in space and the first-order temporal dispersion of optical pulses in dielectric media. It is generally used for extending to the temporal domain well-known properties of spatial optical configurations. In Chapter 7 a new approach is developed for the secure data transmission problem in fiber optic links. We propose the encoding of time-varying optical signals, mainly for short-haul applications, with encryption methods that can be considered the time domain counterparts of the dual random phase encoding process and two of its more frequent variations: the fractional Fourier transform dual random phase encoding and the Fresnel transform dual random phase encoding. Further, as performance is a very relevant subject in fiber optic links, we will analyze mechanisms to produce time limited, as well as bandwidth limited, encoded signals. To this end, the different signal broadenings produced by each stage of the encoding process, in both time and frequency domains, are analyzed by using the Wigner distribution function formalism, and general expressions for the time width, as well as the bandwidth, in every encryption stage is obtained. The numerical simulations show good system performances, and a comparison between the different encryption processes is made. Furthermore, the robustness of the proposed methods is analyzed against the variation of typical parameters of the encryption-decryption setup. Finally, the implementation of this proposal with current photonic technology is discussed.

In Chapter 8, we report electronic and magnetic structure of pure and (As-) doped manganese clusters from density functional theory using generalized gradient approximation for the exchange-correlation energy. Ferromagnetic to ferrimagnetic transition takes place at

$n = 5$ for pure manganese clusters, Mn_n, and remarkable lowering of magnetic moment is found for Mn_{13} and Mn_{19} due to their closed icosahedral growth pattern and results show excellent agreement with experiment. On the other hand, for As-doped manganese clusters, Mn_nAs, ferromagnetic coupling is found only in Mn_2As and Mn_4As and inclusion of a single As stabilizes manganese clusters. Exchange coupling in the Mn_nAs clusters are anomalous and behave quite differently from the Ruderman-Kittel-Kasuya-Yosida like predictions. Finally, possible relevance of the observed magnetic behaviour is discussed in the context of Mn-doped GaAs semiconductor ferromagnetism.

The investigation of the optical properties of porous silica samples is presented. Optical spectroscopy measurements, including Raman scattering, steady state and time resolved photoluminescence, optical absorption and excitation of photoluminescence are reported. Chapter 9 reviews the results of the research we carried out upon the emission features of porous silica in the ultraviolet and visible wavelength range and the characterization of the emission properties of dye-doped sol-gel synthesized silica samples. In particular, the study of the emission band recorded at about 3.7 eV and its correlation with the chemical and physical conditions of the surface is discussed. As regards dye-doped silica samples, the analysis of the spectroscopic features of pre- and post-doped hybrid samples is presented and their potential feasibility as solid state dye laser is proposed.

Noise figure analysis is one of the key topics in optical amplifiers design and analysis. The pump to signal RIN transfer is one of the major causes for the noise in optical fiber amplifiers. If there is intensity modulation to the pump power, the relative intense noise (RIN) will transfer from the pump to the signal wavelength and degrade the system performance. There have been corresponding papers published on this issue with analytical expressions, The existing analysis gives deep insight into the problem; however, the analysis mentioned above focused on Raman amplifiers or Brillouin fiber lasers with single pump and single signal channel, which are not the most general case. For the case of multiple pumps case, there has been no model for the pump to signal RIN transfer. Moreover, the current analyses are based on the temporal model, which is very time-consuming and does not give a clear picture of the frequency response. In Chapter 10, we will propose a novel frequency model to evaluate the pump to signal RIN transfer in optical amplifiers and lasers with arbitrary pumps. Analytical expressions could be derived for one pump specific case based on the model.

Among various candidates considered to implement quantum information processing (QIP), cavity quantum electrodynamics (QED) has attracted much attention over past years due to the availability to demonstrate few-qubit quantum gates experimentally and the possibility to construct future quantum network. In Chapter 11 we review recent work of QIP using cavity QED in weak dissipation. The concrete work we review includes W-state preparation, Toffoli gating, Grover search implementation, and QIP by geometric phase. Under the idea of quantum trajectory, we could present analytical expressions to show the detrimental influence of cavity decay in QIP.

The response of gas atoms or molecules to strong laser fields depends on their internal electronic state. This fact can be exploited to gain insight into bound electron structure, nuclear dynamics, and even electronic dynamics by measuring the emitted photons created by the process of high harmonic generation (HHG). HHG is customarily explained by recombination of a virtually detached electron upon returning to its initial bound state. In this chapter we investigate the emission of high harmonic radiation from molecular systems in excited electronic states.

Specifically we report on numerical results obtained for systems in two types of electronic excited states: (i) states with pronounced net internal angular momentum and (ii) states which are excited by nonlinear plasmon oscillations in highly polarizable molecules by a strong laser field. Our results can be summarized as follows.

If the involved state exhibits pronounced angular momentum both ionization and recombination are influenced and the symmetry of the three-stage process is broken. Chapter 12 shows that this can be used to gain access to the phase of the bound state and that recombination to such a bound state leads to creation of circularly polarized, spatially coherent attosecond X-ray pulses.

By solving the time-dependent Schrödinger equation for a model system containing 4 active electrons using the multi-configuration time-dependent Hartree-Fock approximation, we show that the harmonic spectrum exhibits two cut-offs. The first cut-off is in agreement with the well-established, single active electron cut-off law. The second cut-off presents a signature of multielectron dynamics. Electrons that are ionized from an excited multi-plasmon state and recombine to the ground state gain additional energy, thereby creating the second plateau.

In Chapter 13, a unified theory is given of dynamically modified decay and decoherence of field-driven multipartite systems. When this universal framework is applied to two-level systems (TLS) or qubits experiencing either amplitude or phase noise (AN or PN) due to their coupling to a thermal bath, it results in completely analogous formulae for the modified decoherence rates in both cases. The spectral representation of the modified decoherence rates underscores the main insight of this approach, namely, the decoherence rate is the spectral overlap of the noise and modulation spectra. This allows us to come up with general recipes for modulation schemes for the optimal reduction of decoherence under realistic constraints. An extension of the treatment to multilevel and multipartite systems exploits intra-system symmetries to dynamically protect multipartite entangled states. Another corollary of this treatment is that entanglement, which is very susceptible to noise and can die, i.e., vanish at finite times, can be resuscitated by appropriate modulations prescribed by our universal formalism. This dynamical decoherence control is also shown to be advantageous in quantum computation setups, where control fields are applied concurrently with the gate operations to increase the gate fidelity.

In: Atomic, Molecular and Optical Physics…   ISBN: 978-1-60456-907-0
Editor: L.T. Chen, pp. 1-59   © 2009 Nova Science Publishers, Inc.

Chapter 1

PULSE STRUCTURING IN LASER-LIGHT DYNAMICS: FROM WEAK TO STRONG SIDEBAND ANALYSES; I. THE INTEGRO-DIFFERENTIAL "MAXWELL-BLOCH" SYSTEM

Belkacem Meziane[*]

Université d'Artois, UCCS Artois, UMR CNRS 8181, Rue Jean Souvraz, SP18, 62307,
Lens Cedex, France

Abstract

We revisit, with an in-depth and self-consistent approach, the small and the strong-harmonic expansion methods that handle the dynamic properties of the well-known integro-differential "Maxwell-Bloch" equations, which describe light-matter interactions inside a single-mode inhomogeneously broadened [SMIB] laser. The chapter foremost re-examines the well-established small-side-band routine, outlining the limits of its applicability in the unstable regime of operation. These limits are overcome with the application of a strong harmonic-expansion procedure that allows for the derivation of new analytical information in the control-parameter space within which regular pulsing behavior takes place. In particular, despite the complicated nature of the SMIB laser equations, owing to the presence of an integral over the polarization variable, we show that an extension of the analytical procedure, up to third order in field amplitude, is sufficient to obtain the oscillation frequencies of the asymptotic solutions and an analytical expression that illustrates the dynamic-gain contour during pulse build-up. Furthermore, the evaluation of the long-term operating frequency is shown to bear fundamental importance in the construction of complete analytical solutions, inside the control parameter space that exhibits periodic self-pulsing, through simple iterative extraction of the high-order field amplitudes. Hopefully, these new findings will allow for significant advance and understanding of the physics underneath self-pulsing and time structuring in bad cavity configured lasers.

---

[*] E-mail address: belkacem.meziane@univ-artois.fr

I. Introduction

The dynamic properties of the laser-matter non-linear interactions inside a unidirectional cavity containing an inhomogeneously broadened medium are described in terms of a set of three non-linearly coupled integro-differential equations. These, known as the "Maxwell-Bloch" equations, received considerable attention, and the study of their self-pulsing behavior has attracted a large number of research groups during the past three decades. Their theoretical foundation departs from the self-consistent Lamb picture in which the polarization of the amplifying medium is supposed to adiabatically follow the electric field oscillating inside the laser cavity [1]. Such an adiabatic approximation is indeed valid in many laser systems that emit stable output signals. However, it was soon recognized that the behavior of lasers may vary from one system to another. Regular pulsing, as well as erratic, intensity outputs, challenging the Lamb self-consistent analysis, could not be avoided in some cases [2, 3]. The readjustment of the Lamb self-consistent approach is based on the removal of the adiabatic condition imposed to the polarization variable with respect to the laser field. In such a picture, the laser matter interaction is treated in terms of three non-linearly coupled variables. In addition to the usual laser field and population inversion, the polarization of the medium is taken into account and considered as a third variable, owing to the fact that its relaxation rate may, in some cases, be of the same order of magnitude as that of the population inversion. Such an approach was initiated by Casperson, who demonstrated the efficacy of the integro-differential "Maxwell-Bloch" equations to describe the low instability-threshold discovered, experimentally, in the high gain He-Xe Laser [3].

The first theoretical investigations related to the emergence of unstable solutions in the "Maxwell-Bloch" equations are based on the so-called weak-side-band approach (WSBA) introduced by Casperson [4-8]. According to the Casperson approach, the unstable state occurs under the assumption of one side-mode build-up stemming from side-mode gain and anomalous dispersion generated in the nonlinear polarization by spectral hole-burning in the inhomogeneous profile. The single side-band approach provides a quite simple picture of the physics underlying the emergence of an unstable regime, yet it only yields a qualitative description of the instability problem. A few of its limitations were removed by Hendow and Sargent, who re-examined the procedure by assuming the presence of a pair of weak side modes symmetrically placed about a strong central mode, a situation much closer to the physical reality [9-11]. The whole analysis assumes that the central mode remains arbitrarily strong, whereas the side modes are not allowed to saturate. As shown in this chapter, such a hypothesis constrains the validity of the weak-side-band approach to the transient build-up of the solution that precedes the long-term signals, both in the stable and in the unstable regime of operation, and to a very small range of the instability domain, in the vicinity of the instability threshold inside which the solutions consist of small-amplitude signals oscillating around steady-state.

The instability problem was also investigated along a distinct line of thought by a number of groups [12-18]. Such an approach is based on the application of a standard linear stability analysis (LSA), which consists of the application of small perturbing terms, to the steady state solution and probing for control-parameter values that yield an exponential growth of the initially small perturbations, thus propelling the system away from stable operation. The standard linear stability analysis yields the same results as those given by the small side-band approach at the boundary of the instability domain where the two procedures were shown to be fully equivalent [13].

Both LSA and WSBA give good indications on the two essential criteria for the emergence of single-mode instabilities, i.e., that the field decay rate exceeds the population inversion relaxation rate (bad cavity condition) and that the excitation level be larger than a critical value termed as the second laser threshold. In order to avoid confusion, the first laser threshold refers to the onset of laser action. Both methods also give the oscillation frequencies of the unstable solutions during their transient outgrowth towards the final self-pulsing regime. It is of interest to mention that as compared to the exceedingly high value of the instability threshold of the single-mode homogeneously broadened model (discussed in the following chapter), the SMIB second laser threshold is hardly higher than the first laser threshold. This explains the early experimental observations of a self-pulsing regime in the high-gain He-Xe laser [3], for example, long before any satisfactory theoretical modeling was achieved [4].

It is worth mentioning that the original work of Lamb's group did not deal with the instability problem in laser physics [19], and it took more than a decade to see complete book chapters devoted to laser dynamics [20].

Numerical analysis of the "Maxwell-Bloch" equations must be performed in order to find the details of the long term pulsing comportment. The features of the obtained time traces bear no resemblance with those of the transient regime. In particular, as largely discussed in this chapter, in most of the instability domain the laser field amplitude and the polarization variables undergo perpetual oscillations around zero-mean values while the population inversion variables oscillate around *dc* components distinct from their corresponding stable steady-state values. These oscillations take place with frequencies different from those of the transient relaxation oscillations that precede the permanent temporal traces. Furthermore, apart from the exact localization of the instability threshold and the evaluation of the transient frequencies, no further analytical information can be extracted from both approaches and most of SMIB laser studies, based either on numerical simulations or on analytical aspects, focused mainly on qualitative results. For example, the WSBA, as first studied by Casperson, takes into account one side mode only in the gain profile. On the other hand, the readjustment brought to the method by Hendow and Sargent seemed to be much closer to the physical situation where two side modes coexist. These two modes were assumed to give rise to a beat note that initiates the observed Microwave oscillations. However, as shown in this paper, these readjustments only handle the instability problem at the onset of self-pulsing and cannot be extended to describe any feature of the long-term solutions. Thus, evidently both LSA and WSBA fail to describe the features of the permanent self-pulsing solutions.

Nevertheless, despite these obvious limitations, the small side-band approach constitutes a good ground for the physical mechanism underlying the occurrence of an unstable state in SMIB lasers in a large range of the instability domain. These physical mechanisms have led for the construction of low dimensional models that retrieve much of the dynamic properties of the original infinite-dimensional equations. These models allow for simple and straightforward analytical handling [14, 21-30].

This chapter gives comprehensive analysis of SMIB laser instabilities under the side-band point of view aiming at extending the WSBA to a strong harmonic expansion method that is shown to yield closed form solutions to the integro-differential "Maxwell-Bloch" equations. The study aims at giving full details of the instability issue that go beyond well-established results, and which seemed to become necessary in view of the profusion of new contributions on the subject that have been published for the past few years [31-42].

The presentation focuses on two main points: First, we reconsider the weak side-band analysis in a much simpler framework and carry an in-depth exploration of the SMIB-laser transient regime. The weak side-band method is shown to describe all the features of the laser-matter interaction during the transient oscillations that precede pulse build-up. In particular, we show that the initially Gaussian gain profile undergoes little distortion with respect to the steady-state gain contour while a closed form relation giving the exact values of the transient oscillation frequencies is shown to apply both in the stable and the unstable regimes. The second part presents an original analysis that extends the harmonic expansion method from a weak to a strong side-band method, where the Hendow-Sargent condition that forbids the small side-band components to undergo saturation effects is removed. From this strong harmonic expansion approach, we obtain a closed form expression that gives fairly good values of the long term frequencies and a few analytical solutions in the control parameter space inside which the long term signals consist of regular oscillations are constructed. In addition, the study gives clear evidence that during pulse-build-up, the gain contour undergoes strong distortion that evolves along the instantaneous value of the laser field variable. Such a distortion effect is expressed by a strong lateral saturation phenomenon demonstrated numerically and analytically with the strong harmonic expansion method.

The presentation is organized according to the following hierarchy: Section II is devoted to a quick review of the integro-differential "Maxwell-Bloch" equations. In Section III, we recall the essential features of the steady-state properties, followed by a rapid survey of the fundamental results that characterize the unstable state, as obtained from Linear Stability Analysis, in Sec. IV. Section V inspects typical self-pulsing solutions obtained in the transition region from transient to long-term development as well as typical regular small amplitude oscillations. Section VI reconsiders the weak-side-band method in a simpler framework which allows for the exact determination of the transient properties. The WSBA also yields quantitative information pertaining to the operating frequencies when the solutions consist of small-amplitude orbits in the vicinity of the stable steady-state. Section VII concentrates on a typical hierarchy of regular pulsing solutions obtained with increasing excitation levels. Section VIII presents the first few steps of the strong harmonic expansion method that allow to obtain an expression of the long term operating frequency as well as the first and third order field amplitudes corresponding to the first two terms of the strong harmonic expansion. In Section IX, we show how some of the laser properties are modified under unstable operating conditions. In particular, we show that an anomalous dispersion effect, resulting in dissipation of the electromagnetic wave, lowers the laser output intensity at the onset of instability, resulting in a kink-shaped laser-output versus excitation characteristic, much like the one observed in semiconductor lasers when subject to optical feedback [24]. The gain profiles are also shown to undergo regular oscillations with the presence of a hole-burning effect localized at line center, while the long-term gain profiles undergo strong distortion during pulse build-up. While the small-side band approach is able to describe the first effect, the strong gain modification is shown to be contained only in the strong side-band approach. Analytical solutions corresponding to the typical hierarchy of Sec. VII are constructed following an adapted Fourier series for each interacting variable in Sec. X. The excellent agreement between the numerical solutions and those of the expansion method are indicative of the accurateness of the method. Sec. XI gives further considerations of the dynamic properties that go beyond regular self-pulsing operation. In Sec.XII, we obtain other regular solutions with other control parameters and construct the corresponding analytical solutions. In Sec. XIII, we discuss some pertinent points that still remain open for discussion. For example, we demonstrate with an example obtained in the radiative limit, that the instability of SMIB lasers does not seem to persist for very high excitation levels. Finally, Sec. XIV ends with some conclusions and perspectives towards the construction of analytical solutions in other regions of the control parameter space that are not handled in this paper. For example regions of regular pulsing with non-zero mean values, chaotic solutions, etc.

In addition to the bulk of the report, we devote three appendixes to outline the important steps of the lengthy calculations. A closed form expression that gives the transient frequencies as a function of the decay rates and excitation level is derived in Appendix A, while the expression of the long term frequencies is derived in Appendix B. Also in Appendix B, we derive closed form expression for the first and the third-order terms of the electric-field harmonic expansions. Finally, Appendix C gives the essential steps that lead to delimiting the parameter zone of regular field-pulsing around zero-mean values.

II. Basic Equations

Semi-classical theory of the atom-field interactions inside a unidirectional ring cavity containing an inhomogeneously broadened collection of pumped two-level atoms yields a set of non-linearly coupled equations of motion for the resonant optical field and atomic variables. Within the plane wave and slowly varying envelope approximations, and in the absence of detuning between the centers of the laser line and the spectral profile of the atomic transition, these equations take the form [14-18, 24-30, 39-42]

$$\frac{d}{dt}E(t) = -\kappa \left\{ E(t) + 2C \int_{-\infty}^{+\infty} dw g(w) p(w,t) \right\} \tag{1a}$$

$$\frac{d}{dt}p(w,t) = -(1+iw)p(w,t) + E(t)d(w,t) \tag{1b}$$

$$\frac{d}{dt}d(w,t) = -\wp \left\{ d(w,t) + 1 + \tfrac{1}{2}\left[E*(t)p(w,t) + E(t)p*(w,t) \right] \right\} \tag{1c}$$

and are usually referred to as the integro-differential "Maxwell-Bloch" equations.

In our notations, $E(t)$ is the slowly varying output-field amplitude, scaled to the square root of the saturation intensity, $p(w,t)$ and $d(w,t)$ denote the polarization and population inversion, respectively, of an arbitrary atomic homogeneous packet, positioned at w away from line center in the spectral profile, κ and $\wp$ are, respectively, the cavity decay rate and the relaxation rate of the population inversion, both scaled to the polarization relaxation-rate, $g(w)$ is the atomic spectral distribution, $2C$ is the excitation parameter, and t is a dimensionless time variable representing the product of time and the polarization relaxation-rate. With such scaled variables and parameters, Eqs. (1) are conveniently adapted for numerical simulations.

Because of the polarization integral, the set of equations (1) is a dynamical system of an infinitely high dimension that must be suitably discretized in order to reach a good compromise between accuracy and the speed of execution, as will be discussed in Sec. IV.

For simplifying purposes, but without loss of generality, our presentation focuses on the case of a symmetric gain profile, considering real field-amplitudes. These assumptions are put to use for numerical convenience only, since they allow for the saving of half the numerical time consumption without modifying the solution structure. In this case Eq. (1a) writes

$$\frac{d}{dt}E(t) = -\kappa \left\{ E(t) + 2C \int_0^{+\infty} dw g(w) [p(w,t) + p*(w,t)] \right\} \tag{1a'}$$

Let us recall that Eqs. (1) apply to a single-mode inhomogeneously broadened laser. For example, in a gas laser, the atoms undergo permanent motion. Each atom with a velocity v undergoes a Doppler shift $w = kv$ (where k is the wave vector along the laser axial-direction) with respect to the central frequency of the atomic transition, corresponding to atoms at rest. The Doppler broadening associated with this motion (for which a Maxwellian distribution is assumed) is taken care of by the following Gaussian spectral-distribution

$$g(w) = \frac{1}{\sqrt{2\pi}\sigma_D} \exp\left(-\frac{w^2}{2\sigma_D^2}\right) \tag{2}$$

Where σ_D is the half width, at half maximum, of the atomic spectral profile, also scaled to the polarization relaxation rate.

Before investigating the dynamic properties of Eqs (1), let us first recall some basic features of the stable steady-state which will serve as a reference level for each set of control parameters inside the unstable domain of operation.

III. Steady-State Properties

The steady-state of the atomic components is found when the derivatives appearing in Eqs. (1b) and (1c) are set equal to zero. The following relations are readily derived

$$p_s(w) = -\frac{1 - iw}{1 + w^2 + |E_s|^2} E_s \tag{3a}$$

$$d_s(w) = -\frac{1 + w^2}{1 + w^2 + |E_s|^2} \tag{3b}$$

A state equation which relates the excitation parameter $2C$ to the steady-state field-intensity follows from Eq. (1a) and Eq. (3a)

$$2C \int_{-\infty}^{+\infty} dw \frac{g(w)}{1+w^2+\left|E_s\right|^2} = 1 \tag{4}$$

allowing for representations of the excitation parameter $2C$ with respect to the laser steady-state intensity $\left|E_s\right|^2$, given the value of the atomic spectral-profile half-width σ_D. For numerical convenience, the excitation level may thus be scanned in terms of the field-amplitude modulus $\left|E_s\right|$, instead of $2C$.

The excitation parameter at laser threshold follows from Eq. (4) and writes

$$2C_{th} = \frac{1}{\int_{-\infty}^{+\infty} dw \frac{g(w)}{1+w^2}} \tag{5}$$

The ratio of (4) and (5) defines a normalized excitation parameter, with respect to the pump level at the onset of laser action, in the form

$$r = \frac{\int_{-\infty}^{+\infty} dw \frac{g(w)}{1+w^2}}{\int_{-\infty}^{+\infty} dw \frac{g(w)}{1+w^2+E_s^2}} \tag{6}$$

$r=1$ corresponds to the first laser threshold ($E_s=0$).

Figure 1 gives a representation of the stationary field E_s with respect to the excitation parameter r, indicating how the instability threshold, as obtained numerically following linear scanning of the excitation parameter in Eqs (1), is close to the onset of laser action ($r_{2th}=1.15$).

The unsaturated gain of the medium is given by the Gaussian spectral profile $g(w)$ (Eq. (2)), while the saturated gain profile writes

$$\Gamma(w) = g(w)d_s(w) = -\frac{1+w^2}{1+w^2+E_s^2}g(w) \tag{7}$$

Typical examples of the gain contour, obtained with parameter values corresponding to a bad cavity laser with $\kappa=5$, $\wp=1$, $\sigma_D=5$, and increasing excitation levels, starting from $\left|E_s\right|=0$ to $\left|E_s\right|=1.0, 1.5, 2.0$ respectively, are plotted in Fig. 2. A noteworthy feature is the strong saturation (hole-burning) at line center while the lateral part of the gain curve remains relatively important. The side-band approach attributes the onset of instability in inhomogeneously broadened systems to the presence of this lateral gain. It gives rise to small side band components in the laser spectrum which modulates the output intensity at the frequency of the beat note between these two symmetric side-bands.

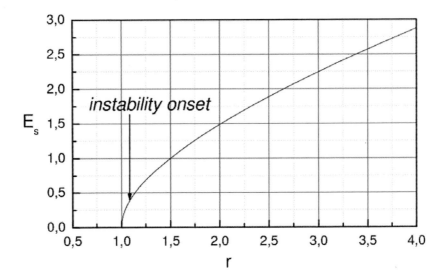

Figure 1. Normalized field-amplitude versus excitation parameter corresponding to stable operation. The downward arrow indicates points of minimum excitations for the onset of an unstable regime. This second threshold depends on the exact values of the field and population inversion decay rates. It comes closer to the threshold for laser action with increasing κ and decreasing γ values. In the case of $\kappa = 100$ and $\gamma = 0.01$ $r_{2th} \approx 1.08$.

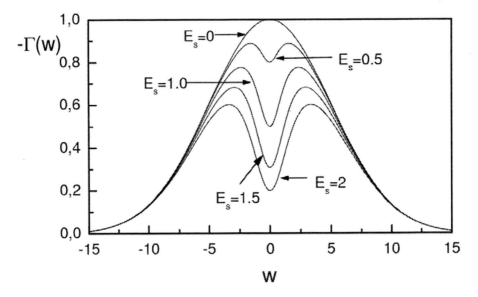

Figure 2. Evolution of the steady-state gain-profile with increasing excitation levels. Note the strong saturation effect at line center.

We will give a full description of the side band method in Sec. VI. It is here worth mentioning that, in the stable regime of operation, the phenomenon of hole-burning is strictly limited to line center, irrespective of the level of excitation, as it clearly transpires from the multiple plots of Fig. 2.

IV. Charcreristics of the Unstable State

Equations (1) enter an unstable regime of operation when the following conditions

$$\kappa > \wp \tag{8a}$$

and

$$r > r_{2th} \tag{8b}$$

are simultaneously satisfied [6-8, 20, 24]

Eq (8a) imposes that the cavity decay rate be larger than the population relaxation rate. It is known as the bad cavity condition. In such a situation, the population inversion evolves at a faster rate with respect to the round trip time between the cavity mirrors, thus preventing the field to establish in a stable steady-state, inside the laser cavity, giving rise to a permanently perturbed output signal, stemming from a lack of coherence between the running waves inside the cavity.

r_{2th} is the second laser threshold. It corresponds to the required level of excitation to initiate unstable behavior.. Its value is obtained with a standard linear stability analysis. However, in the case of the Gaussian lineshape (2), the second laser threshold cannot be evaluated with a closed form expression but requires numerical computations of lengthy combinations of the error function (for details see Refs 6, 7 and 15).

Here, we just mention that r_{2th} shows a dependence on κ and γ and that its value is very close to the first laser threshold, as will be demonstrated in Sec. VI, and already found with numerical simulations in Fig. 1.

The dynamics inherent in Eqs (1) reveals a wealth of properties. These are closely dependent on the numerical values of the control parameters κ and $\wp$ that satisfy the bad cavity condition (8a) and on the excitation levels, above the instability threshold (8b). Two cases of interest are an oscillatory regime, typical of $\wp$ values close to the radiative limit ($\wp =1$) and a self-pulsing regime, characterized by much lower values of $\wp$ [16, 26-28].

The complexity of the simulation procedure, as compared to the much simpler Lorenz-Haken model treated in Chapter 2, arises from the presence of the polarization integral in Eq. (1a) which must be suitably discretized in order to avoid numerical artifacts. The numerical operation consists in simultaneously solving for Eqs. (1a), (1b) and (1c), with scanned values of w, over the spectral distribution (2), adopting a standard fourth order Runge-Kutta routine. A careful division of the spectral profile is required, for a good compromise between accuracy and speed of execution. An accurate match between the results of long-term numerical integration under stable conditions and those predicted by the state equation (4) is obtained with 100 spectral components. However, despite the seemingly converging solutions under stable conditions, the system still possesses unsuspected differences, which are detectable only in the unstable regime of operation. This is particularly true at bifurcation points where the solution undergoes dramatic changes. For example, at the instability threshold of the self pulsing regime, the solution was shown to converge towards the same pulse-train, only when the spectral profile was divided into 120 components (for details see Ref. 28). These elements show how critical the handling of the numerical simulations can be. However, the analytical handling presented in the following sections does not require any particular care, since the dynamic process involves only the population components that are not too far away from line-center.

V. Typical Self-pulsing Solutions

Under conditions (8a) and (8b), the solutions of Eqs 1 remain stable for a small range of excitation levels beyond the first laser threshold. The instability first develops in the form of regular small-amplitude oscillations around steady-state. The amplitude of these regular oscillations grows with increasing excitation levels (supercritical Hopf bifurcation) ultimately yielding long term pulsing solutions.

Here, we shall focus on the parameter zone inside which the long term field solution undergoes perpetual switching around zero mean-values. For other control parameters, the unstable solution develops with exponentially growing amplitude (transient regime) before reaching the final long term pulsing time trace. A typical example of such a solution is represented in Fig. 3a for the field amplitude, in Fig 3b for the polarization, and in Fig 3c for the population inversion components, both obtained at the center of the gain profile. We note that for the field and polarization, as well as for the population inversion variables, the transient part of the signal consists of growing 'relaxation oscillations' around steady-state, while the long term signals oscillate around zero-mean values for the field and polarization and around a *dc* level, distinct from its corresponding steady-state value, for the population inversion. Obviously, any correct treatment of the instability issue should describe these features. We will show, in the following section, that the small side band method of Casperson-Hendow-and-Sargent contains most of the quantitative features related to the transient part of the perturbed equations 'relaxing' towards the long-term solution (either stable or unstable). It also describes the properties of the system for parameter values for which the solutions consist of small amplitude limit cycles. An example of such a solution is represented in Fig. 4 for the three interacting variables. The following comments are worth to be pointed out: First, we note that the mean values of all three variables depart from their steady-state counterparts; Second, the time traces undergo an almost sinusoidal evolution, as indicated by the corresponding frequency spectrum (represented in Fig. 4d) from which one single dominant frequency emerges; Third, the three interacting variables do not show an in-phase evolution but evolve with some time lag, which has previously not been taken into account neither by Casperson nor by Hendow and Sargeant, but must be inferred in order to find valid analytical expressions for the transient relaxation frequencies with respect to excitation level, for fixed decay rates. A straightforward readjustment of the original small-sideband method is proposed in the following section to handle these time lags, consequently giving a more accurate description of the competing nature of light-matter interactions during the transient part of the signals.

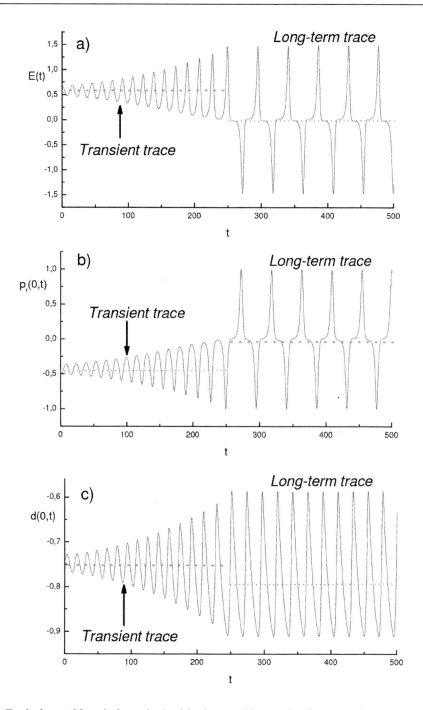

Figure 3. Typical unstable solutions obtained in the transition region from transient to long-term time traces with $\kappa = 4$, $\gamma = 0.1$ and $Es = 0.8$; a) field amplitude, b) polarization and c) population inversion, of the central component ($w = 0$) of the spectral profile. Note the discontinuous jump from an orbiting around a state close to the stationary state to oscillations around zero-mean values for the field and polarization variables while the long term population inversion oscillates around a *dc* component situated away from the corresponding steady state value d_0. This transition region also delimits the passage from a weak-side to a strong-side band approach, developed in this paper.

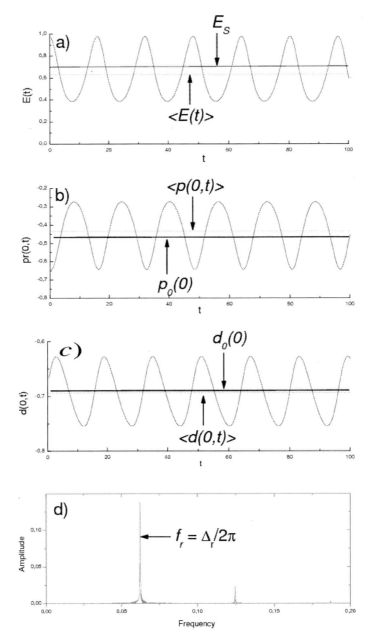

Figure 4. Regular small amplitude oscillations corresponding to the long-term solution of Eqs 1 obtained with $\kappa = 5$, $\gamma = 1$ and $Es = 0.67$; a) field amplitude, b) polarization and c) population inversion, of the central component ($w = 0$) of the spectral profile, d) frequency spectrum of the signals showing the presence of a dominant frequency whose value exactly corresponds to the graphical solution of Figure 5. Note that even in the case of small-amplitude regular oscillations, the *dc* components (dashed lines) slightly deviate from the values of the stationary state (solid lines). A dissipation effect occurs when the system enters the unstable domain of operation. The gap between the steady state and the *dc*-value increases, with increasing excitations, until the field and polarization variables are forced to orbit around zero mean values.

After these obvious, but necessary introductory elements, let us dig into the bulk of our analyses.

VI. Weak-Harmonic Expansion Analysis

a) The foundations of the method

The weak side-band approach is based on a hypothesis that allows for the build-up of two small (not allowed to saturate) side-modes in the laser spectrum, along with the oscillating mode. In the slowly varying amplitude approximation, the output field is written as a superposition of three components (the strong central mode and two small side-bands equally spaced with respect to the frequency of the strong mode). Projected onto a simple analytical framework, the laser output-field-envelope takes the form

$$E(t) = E_s + E_1 \exp(i\Delta t) + E_1^* \exp(-i\Delta t) \tag{9a}$$

Interacting with such a modulated electric-field, both the atomic polarization and the population inversion variables must contain small-amplitude oscillating components superimposed to their stationary values, which also must take the same form as the driving field. Obviously adequate expansions for the matter variables write

$$P(t) = P_s + p_1 \exp(i\Delta t) + p_1^* \exp(-i\Delta t) \tag{9b}$$

$$D(t) = D_s + d_1 \exp(i\Delta t) + d_1^* \exp(-i\Delta t) \tag{9c}$$

Without loss of generality, we consider the case of a real field, so that the projection of the above expansions onto a real framework yields

$$E(t) = E_s + E_1 \cos(\Delta t) \tag{10a}$$

$$p(w,t) = p_s(w) + p_1(w)\cos(\Delta t) + p_2(w)\sin(\Delta t) \tag{10b}$$

$$d(w,t) = d_s(w) + d_1(w)\cos(\Delta t) + d_2(w)\sin(\Delta t) \tag{10c}$$

where E_s, $p_s(w)$, $d_s(w)$ refer to the stable steady-state values obtained from Eqs. (1) and evaluated respectively from Eqs (4), (3a) and (3b), for a given value of the excitation parameter 2C. Now, the time-lag between all three variables is taken care off with the presence of out-of phase components $p_2(w)$, for the polarization and $d_2(w)$ for the population inversion.

Inserting the above relations into Eqs. (1b) and (1c), we find determining equations for the in-phase and out-of-phase components $p_1(w)$, $d_1(w)$ and $p_2(w)$, $d_2(w)$, respectively. When we inject the obtained expressions into Eq. (1a), we find closed form relations for the gain and dispersion experienced by the side-mode component E_1. The following relations are cast (by lengthy but straightforward calculations (see Appendix A for complete step by step derivations) into

$$G_{sb}(\Delta) = 2C \int_{-\infty}^{+\infty} dw g(w) d_s(w) \frac{A_1 B_1 + A_2 B_2}{A_1^2 + A_2^2} = \int_{-\infty}^{+\infty} dw p_1(w, \Delta) = 1 \qquad (11a)$$

and

$$D_{sb}(\Delta) = 2C \int_{-\infty}^{+\infty} dw g(w) d_s(w) \frac{A_1 B_2 - A_2 B_1}{A_1^2 + A_2^2} = \int_{-\infty}^{+\infty} dw p_2(w, \Delta) = -\frac{\Delta}{\kappa} \qquad (11b)$$

where

$$A_1 = (1 + \Delta^2)\left(1 + \frac{\wp^2}{\Delta^2 + \wp^2} E_s^2\right) + w^2 \qquad (11c)$$

$$A_2 = \Delta\left\{(1 + \Delta^2)\left(1 - \frac{\wp}{\Delta^2 + \wp^2} E_s^2\right) - w^2\right\} \qquad (11d)$$

$$B_1 = (1 + \Delta^2)\left(1 - \frac{\wp^2}{\Delta^2 + \wp^2} \frac{1}{1 + w^2} E_s^2\right) \qquad (11e)$$

$$B_2 = \frac{\Delta}{\wp} \frac{1 + \Delta^2}{1 + w^2} \frac{\wp^2}{\Delta^2 + \wp^2} E_s^2 \qquad (11f)$$

b) Frequency-determining relation for the transient regime

For fixed values of the decay rates κ and $\wp$, and for given values of σ_D and excitation parameter $2C$, a determining equation for the unknown frequency Δ follows from the ratio of Eqs (11b) and (11a) and takes the form

$$\frac{\int_{-\infty}^{+\infty} dw g(w) p_2(w, \Delta)}{\int_{-\infty}^{+\infty} dw g(w) p_1(w, \Delta)} = -\frac{\Delta}{\kappa} \qquad (12)$$

A graphical representation of the left and right hand sides of this relation is drawn in Fig. 5, for parameter values corresponding to the harmonic solution of Fig. 4, i.e. κ=5, $\wp$=0.1, $\sigma_D = 5$, and $E_s = 0.67$. As one may quickly check from the figure, the graphical solution $\Delta \approx 0.39$ exactly matches the pulsation frequency of the harmonic signals.

For a given set of parameters, the above relation yields a frequency characteristic with respect to the excitation parameter. We have scanned over large range of values of the parameters, and in each case the graphical solution obtained with Eq. (12) yields a frequency value exactly matching either the small-amplitude stable solutions or the transient relaxation oscillations of the integro-differential equations (1).

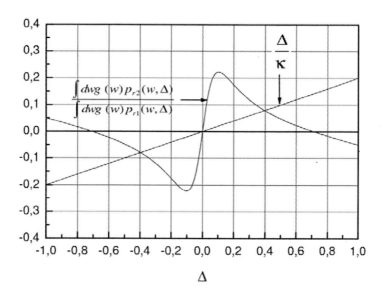

Figure 5. Graphical solution of Eq. (12) giving the pulsation frequency of the oscillations of Figure 4.

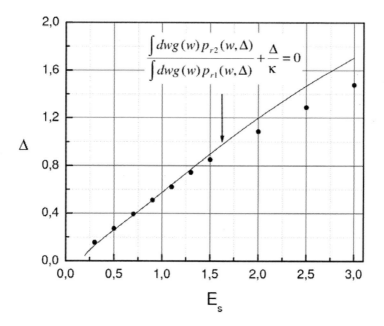

Figure 6. Frequency versus excitation level, corresponding to the solutions of Eq. 12 (solid line), obtained with $\kappa = 4$ and $\gamma = 0.1$. The dotted points were obtained with the transient solutions of the integro-differential equations (1). Note that up to twice threshold ($Es = 1.5$, $r = 2$), the frequencies of the numerical solutions exactly match those of the closed form expression. For higher excitations, indeed, higher order terms participate to the transient dynamics and the first order expression deviates from the actual frequencies of the transient solutions.

Relation (12) is valid both above and below the instability threshold, in the whole domain of stable and unstable operation when the signal relaxes towards constant intensity output or towards a permanent pulsing signal.

Figure 6 represents the variation of the relaxation pulsation with respect to the excitation parameter (solid line) satisfying the above relation along with a few scattered bold points indicating the pulsation values of the relaxation oscillations that are obtained through direct evaluation with the simulated signals of the integro-differential equations, at various excitation levels. For these examples, the decay-rate values are κ=4 and $\wp$=0.1, and the instability threshold is E_s = 0.65. As one may rapidly check from Fig. 6, the simulated pulsations exactly match those obtained with the closed form relation, for low excitation levels, up to E_s = 1.5, corresponding to twice the laser threshold (r=2). Away from this parameter zone, the frequencies of the simulated signals slightly depart from those of the closed form relation. This indicates that for high excitation levels, the relaxation oscillations are no longer driven solely by the first order terms. Higher order terms must be taken into account. However, the presented analysis focuses on the parameter space that harbors regular long term solutions. These occur in the region of excitation levels where a perfect match is obtained between the results of the closed form relation and those of the numerical time traces. Higher excitation levels end in erratic signals, ultimately becoming chaotic. These solutions are not considered in this study.

Now that the features of the transient oscillation relaxation oscillations are identified, with the help of the small side-band approach, let us focus on a few aspects that concern the long-term solutions.

VII. Typical Hierarchy of Regular Pulsing Solutions

The temporal traces of Fig. 3 indicate that when the long term solution is reached, the field and polarization variables undergo a perpetual switching around zero-mean values, while the population inversion oscillates around a *dc* component distinct from its value at steady-state. On the other hand, relations (10) force the harmonic components of the dynamic variables to evolve around the stationary state defined by $E_s, p_s(w)$, and $d_s(w)$. Consequently, it becomes clear that the weak side-band approach of the former section is far from being adapted for the description of the long term solutions represented in Fig. 3. These solutions undergo a perpetual switching between positive and negative values. The field amplitude and each of the polarization components permanently satisfy

$$< E(t> = 0 \tag{13a}$$

$$< p(w,t> = 0 \tag{13b}$$

while the population inversion components evolve according to

$$< d(w,t> = d_{dc} \neq d_s \tag{13c}$$

The recognition of such an evolution constitutes the basis for the construction of the strong-harmonic expansion analysis, presented in the ensuing section.

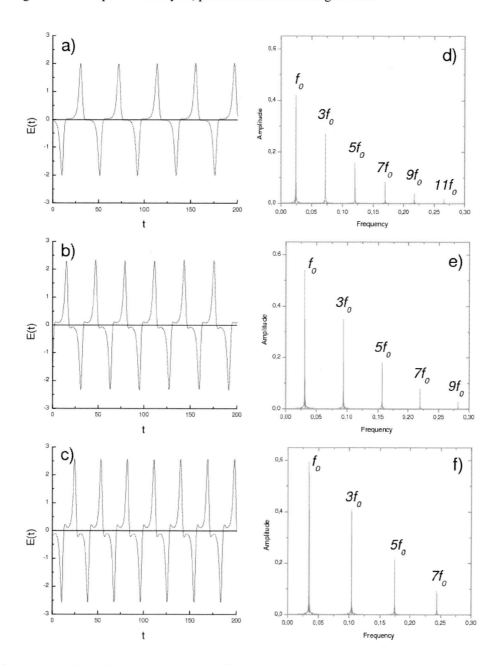

Figure 7. Typical regular solutions obtained with $\kappa = 4$, $\gamma = 0.1$, and increasing excitation levels: a) $Es = 0.8$, b)$Es = 0.95$, and c)$Es = 1.05$. d), e), f) corresponding frequency spectra obtained with a Fast Fourier Algorithm. For each solution the features of the corresponding spectrum will be quite useful in the complete modeling of the corresponding analytical solutions.

Now, let us consider the integro-differential equations (1) and solve for typical examples of parameter values that yield periodic oscillating pulses for the electric-field variable. A classic hierarchy of solutions obtained with κ=4, $\wp$=0.1 and increasing excitation levels is represented in Figs. 7(a), (b), and (c). The corresponding Frequency spectra, as obtained with a Fast Fourier Transform, are represented, along with the temporal signals, in Figs 7(d), (e), and (f), respectively. Note that the pulsation frequency increases, while, as expected, the number of field components appearing in the frequency spectrum decreases, with increasing excitation level. The features of the spectral representations will allow for obtaining the corresponding closed form solutions to a highest possible order. First, we present the first few steps of the strong harmonic expansion method and derive a closed form expression for the permanent long-term frequencies, equivalent to Eq. (12), which describes the transient relaxation oscillations.

VIII. Strong-Harmonic Expansion Analysis

A simple adjustment of the weak side-band method consists in imposing solution tests that satisfy the conditions imposed by Eqs. (13). We start with the hypothesis of harmonic solutions to Eqs. (1) undergoing strong oscillations, around zero-mean values for the field and polarization variables, and around a *dc* component (to be determined) for the population inversion. Therefore, an intrinsic expansion of the field variable consists of the following Fourier series

$$E(t) = \sum_m E_m \cos(m\Delta t), \text{ m} = 1, 2, 3... \tag{14a}$$

while the response of the medium's variables to such a field; is naturally described through

$$p(w,t) = \sum_m \left[p_{ip_m}(w)\cos(m\Delta t) + p_{op_m}(w)\sin(m\Delta t) \right] \tag{14b}$$

$$d(w,t) = d_{dc}(w) + \sum_m \left[d_{ip_m}(w)\cos(m\Delta t) + d_{op_m}(w)\sin(m\Delta t) \right] \tag{14c}$$

where the indexes *ip* and *op* stand for in-phase and out-of phase, respectively.

Inserting these relations into Eqs. (1b) and (1c) yield determining equations for the in-phase and the out-of-phase components $p_{ip1}(w)$, $d_{ip1}(w)$ and $p_{op1}(w)$, $d_{op1}(w)$, respectively, and for the population inversion *dc* component d_{dc}.

When the obtained relations are inserted into Eq (1a), we are left with the following first-order relations (see Appendix B for a complete derivation)

$$2C \int_{-\infty}^{+\infty} dw\, g(w) p_{r1}(w, \Delta, E_1) = -1 \tag{15a}$$

$$2C \int_{-\infty}^{+\infty} dw\, g(w) p_{r2}(w, \Delta, E_1) = \frac{\Delta}{\kappa} \tag{15b}$$

As a first approximation (which reveals to be quite satisfactory, in a first approach), we consider that the main part of the intensity output is carried by the signal through its first term. Such a hypothesis translates into

$$< E(t)^2 > = \frac{E_1^2}{2} \cong E_s^{\;2} \qquad (16)$$

The long term operating frequency, obtained with the ratio of relations (15a) and (15b), satisfies the following relation

$$\int_{-\infty}^{+\infty} dw g(w) p_{r2}(w, \Delta, E_s) = -\frac{\Delta}{\kappa} \int_{-\infty}^{+\infty} dw g(w) p_{r1}(w, \Delta, E_s) \qquad (17)$$

In which, for a given laser system with a fixed cavity decay rate κ, and for a given amplifying medium with an unsaturated gain profile $g(w)$, the operating pulsation is related to the excitation parameter (through E_s).

The dependences of p_{r1} and p_{r2} on w, Δ and E_s are also derived in Appendix B.

The value of the operating frequency is obtained graphically for fixed values of the decay rates κ and γ and fixed excitation levels E_s.. A typical example, representing the left and the right hand sides of relation (17) with respect to Δ is drawn in Fig. 8a. The parameter values are $\kappa = 4$, $\wp = 0.1$ and $E_s = 1$. The obtained long term frequency, in this example, exactly corresponds to the pulsation frequency of the simulated signal obtained with the same parameter values, i.e. $\Delta_{lt} = 0.21$. For a quick comparison, Fig. 8b represents the graphical solution corresponding to the frequency of the transient relaxation oscillation (Eq. 12) preceding the long-term solution. Its value $\Delta_t = 0.58$ is almost three times the value of Δ_{lt}, an indication of the fact that while the transient part of the signal is driven by the first order component of the electric field, the permanent state is driven by the third-order one.

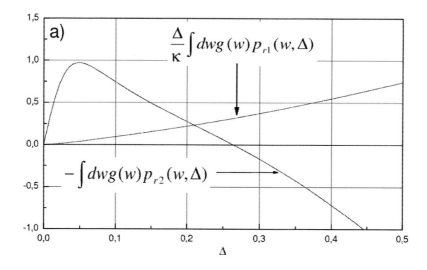

Figure 8. Continued on next page.

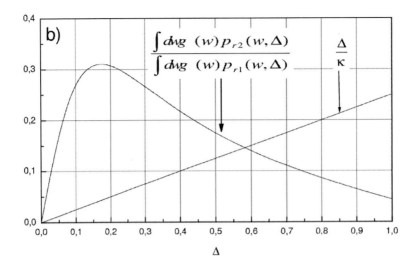

Figure 8. Graphical solution giving the operating frequency of a) the long term solution of Figure 7a and b) the transient solution obtained with the same parameter values. A quick comparison between the values of the two frequencies indicates a ratio $\Delta_t / \Delta_{lt} \approx 3$.

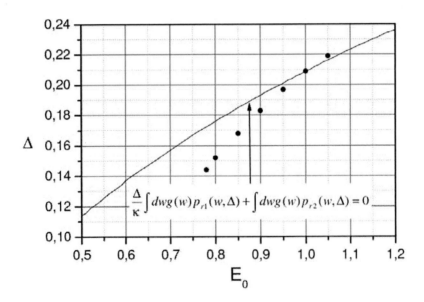

Figure 9. Frequency versus excitation level, corresponding to the solutions of Eq. 17 (solid line), obtained with $\kappa = 4$ and $\gamma = 0.1$. The dotted points were obtained with the permanent solutions of the integro-differential equations (1). Note that the frequencies of the numerical solutions satisfactory match those of the closed form expression. Indeed, as in the case of the transient solutions higher order terms participate to the long-term dynamics and the first order expression deviates from the actual frequencies of the permanent solutions. Contrary to the transient results, these deviations seem to decrease with increasing excitation levels. This unexpected behavior can be understood by the fact that with increasing excitations, the solutions tend towards more regular time traces, less higher order terms arise in the corresponding spectra, and the validity of the first order expression increases with excitation level.

Figure 9 represents the variation of the long term frequency with respect to the excitation parameter (solid line) satisfying relation 17, along with a few scattered bold points indicating the frequencies of the long term oscillations that correspond to the simulated solutions of the integro-differential equations, at various excitation levels. The small deviations between the analytical curve and the simulated points are a signature of the influence of higher order terms. It is worth noticing that the deviations between the numerical pulsations (dotted points) and those of the analytical values evaluated from Eq. (17), are more important at low excitation levels. This can be easily understood, since the signal pulsation takes place at lower repetition rates, and the higher order electric field components become more important. While for higher excitation levels, the signal periodicity is described with a Fourier series with fewer terms. In that case, the third-order analysis is sufficient to obtain a close value of the pulsation frequency. This explains the satisfactory results in the vicinity of $E_0 = 1\text{-}1.1$.

These results suggest that the approximation in Eq. 16 does not excessively deviate from the real physical situation. Thus, in a first step, we may consider that the first order field amplitude is related to the excitation level through

$$E_1 = \sqrt{2}E_s \qquad (18)$$

However, it is clear, both from the long-time trace of Fig. (3) and the permanent signals of Fig. 7, that the actual solution is not a pure harmonic of the form $E_1 \cos(\Delta t)$. Higher order term must be evaluated in order to provide a more accurate analytical description of the pulsing solutions. As a consequence an increased accuracy can only be reached with the iterative inclusion of higher order components.

To third-order, Eqs (14) expand as

$$E(t) = E_1 \cos(\Delta t) + E_3 \cos(3\Delta t) + E_4 \sin(3\Delta t) \qquad (19a)$$

$$p_r(w,t) = p_{r1} \cos(\Delta t) + p_{r2} \sin(\Delta t) + p_{r3} \cos(3\Delta t) + p_{r4} \sin(3\Delta t) \qquad (19b)$$

$$d(w,t) = d_{0d} + d_3 \cos(2\Delta t) + d_4 \sin(2\Delta t) \qquad (19c)$$

An out of phase contribution between the first and the third order field components is introduced in Eq. (19a). Indeed, there is no physical reason why the third component should evolve in-phase with the first one. Indeed, as our investigation will reveal, the out-of phase field-components do participate, to some extent, to the exact modeling of the long-term solutions. These out of phase terms are responsible of the asymmetric nature of the pulse shapes in Figs 7b and 7c. The asymmetry problem will be handled in a much straightforward way in the second Chapter.

Inserting Eqs (19) into Eqs (1b) and (1c) allows for the determination of the third-order polarization components which take the following form

$$p_{r3} = F(w, \Delta, E_s, \wp)\frac{E_1^3}{4} \qquad (20a)$$

$$p_{r4} = G(w, \Delta, E_s, \wp) \frac{E_1^3}{4} \tag{20b}$$

(see Appendix C for the major steps of the derivation)

As expected from the non-linear nature of the interaction between the field and the matter equations, the third order polarization components scale to the third power in first-order field-amplitude.

Inserting relations (20a) and (20b) into Eq. (1a) results in

$$E_3 = 2C \frac{\kappa^2}{\kappa^2 + 9\Delta^2} \left[\frac{3\Delta}{\kappa} \int_{-\infty}^{+\infty} dw g(w) p_{r4}(w, \Delta, E_s, \wp) - \int_{-\infty}^{+\infty} dw g(w) p_{r3}(w, \Delta, E_s, \wp) \right] \tag{21a}$$

and

$$E_4 = -2C \frac{\kappa^2}{\kappa^2 + 9\Delta^2} \left[\frac{3\Delta}{\kappa} \int_{-\infty}^{+\infty} dw g(w) p_{r3}(w, \Delta, E_s, \wp) + \int_{-\infty}^{+\infty} dw g(w) p_{r4}(w, \Delta, E_s, \wp) \right] \tag{21b}$$

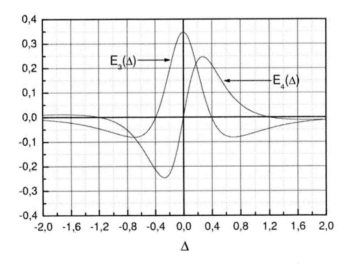

Figure 10. Representation of the in-phase and out-of phase third-order field amplitudes versus frequency for a fixed excitation level ($Es = 1$). These curves indicate that for fixed excitation levels, the out-of phase component increases at the expense of the in-phase part when the operating frequency increases. The physical understanding of this evolution is quite simple: For stable operation all the variables evolve without any dephasing process. At the onset of instability, the system enters a natural evolution (corresponding to the non-linear nature of the laser -matter interaction, represented by the set of Eqs 1 characterized by a strong competing effect between the interacting variables yielding an enhanced out-of phase evolution. The exact values of the in-phase and out-phase components depend on the operating frequency which is related to the excitation level through the first-order closed form expression (17).

Figure 10 represents the evolution of the in-phase and out of phase third-order field components E_3 and E_4 with respect to Δ. These graphs call for a few comments: The curves contain all the features of anomalous dispersion in the neighborhood of a resonance. The in-phase component E_3 is a signature of resonant absorption, while the out-of phase term E_4 is related to an anomalous dispersion effect. It is well known that anomalous dispersion represents dissipation of energy from the electromagnetic wave into the medium. With this in mind, it is expected that the emergence of an out-of phase component in the integro-differential equations (1) will be accompanied by a dissipation effect resulting in a decrease of laser output intensity. This is what is found in the laser characteristic, at the onset of instability. This point will be discussed in some detail in the following section. Let us first analyze the quantitative information extracted from Eqs (21).

The numerical values obtained for E_3 and E_4, with the parameter set of Fig. 8a, i.e. $\kappa = 4$, $\wp = 0.1$, $E_s = 1$, and $\Delta = 0.21$ are : $E_3 = 0.19$ and $E_4 = 0.21$.

So that the analytical solution, to third order, writes

$$E(t) = 1.44\cos(.215t) + 0.19\cos(3 \times 0.215t) + 0.21\sin(3 \times 0.215t) \qquad (22)$$

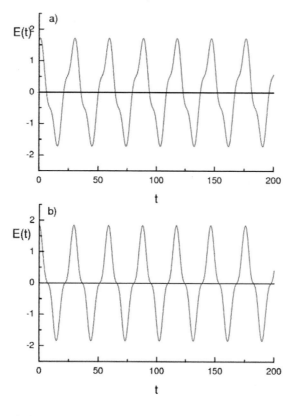

Figure 11. Analytical solution to third-order in field amplitude; a) asymmetric time trace stemming from the influence of the out-of-phase terms, b) symmetric signals, obtained when no out of phase components are included in the field expansion.

It is represented in Fig 11a. Note the asymmetric shape of each pulse stemming from the presence of the out-of-phase component of the third harmonic with respect to the first one. Indeed, the neglect of this out phase term yields symmetrically shaped pulses, as shown in Fig 11b, obtained when the amplitude of the third term in Eq. 22 is solely added to that of the second one.

The solution to third order already shows the dominant features of the corresponding numerical simulations. Indeed, only the inclusion of all higher order terms can yield a correct match with the numerical solutions. Yet, the representations of Figs 11 give clear evidence of the validity of the strong harmonic procedure.

After these few qualitative elements, let us turn back to quantitative considerations. In order to clearly distinguish between the results of the small side-band approach and those of the strong harmonic expansion method, we have represented in Fig. 8a and 8b the graphical solutions that give the long term and the transient operating frequencies for the same parameter values. The frequency obtained with Fig. 8b exactly matches the value of the relaxation frequency while that of Fig. 8a corresponds to the long-term solution.

Fig. 9 represents the frequency-versus-excitation characteristic (Eq.17) in the range where regular oscillations occur. A few values obtained with the numerical time traces are represented as dots along the curve. One can see that the agreement between the numerical values and their analytical counterparts increases with the excitation level and that an exact match is obtained just before an excitation value that corresponds to the critical pulsation

frequency $\Delta_c = \sqrt{\dfrac{\wp}{2}}$. We shall proceed to further consideration of this point in Sec. XI.

Note also that the average intensity output as computed from Eq. 22, satisfies

$$\left\langle E(t)^2 \right\rangle = \frac{E_1^2}{2} + \frac{E_3^2}{2} + \frac{E_4^2}{2} \cong \frac{E_1^2}{2} \cong E_s^2 \qquad (23)$$

within less than 4% of a margin error, giving good credit to the approximation put forward in Eq. 16.

IX. Laser Characteristics and Dynamic-Gain Structuring

When the unstable solution sets-in the system departs from its stationary state. The field and matter variables start to oscillate around *dc* components that keep on moving away from the corresponding steady-state values as the excitation parameter increases, owing to a dissipation effect stemming from anomalous dispersion. The laser field amplitude is ultimately forced to oscillate around a zero-mean value. Such a behavior yields a lowering of the average output intensity and the laser-output versus pump input characteristic deviates from the stationary representation.

Figure 12 depicts a typical average-laser-field-output versus pump-input characteristic, obtained with fine scanning of Eqs. (1) in the numerical simulations. Note the smooth kink that goes along the continuous transition from stable to unstable operation, as compared to the discontinuous transition in the Lorenz-Haken model (see Fig. 5 of the following chapter).

Note also that while the laser output intensity undergoes a dissipation effect in the vicinity of the instability threshold, away from this zone the average intensity of the fluctuating solution is higher than its corresponding steady-state value. This is due to the inhomogeneous nature of the gain profile: At low excitation levels, only the center of the gain contour contributes to the laser field, while, at high excitations, a broader part of the spectral profile participates to the unstable laser emission, ensuing in higher output intensities. This behavior constitutes a significant difference with the Lorenz-Hken equations which will be shown to always deliver lower average intensities with respect to their corresponding stationary values.

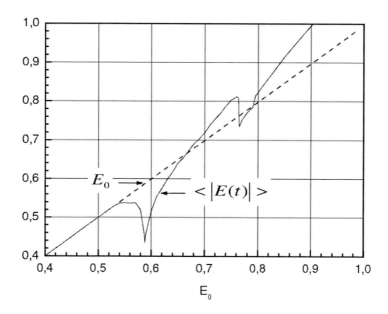

Figure 12. Comparison between the output-field versus excitation-level characteristics corresponding to stable and unstable operation. Note that in the vicinity of the instability threshold, the fluctuating signals result in less average output intensities. The characteristic is kink shaped. For high excitations, however, the lateral part of the gain profile participates much more efficiently in the output pulse modeling, resulting in enhanced average intensities compared to stable operation.

In order to understand the distortion (kink-shaped) of the laser characteristic of Fig. 12, let us first consider the integro-differential equations (1) and solve for parameter values that yield periodic oscillations (in the weak side-band zone, around non-zero mean values). A typical solution is represented in Fig. 13a, while Fig. 13b represents snapshots of the spectral profile obtained along the temporal solution of Fig. 13a. These profiles undergo time changes according to

$$\Gamma(w,t) = g(w)d(w,t) \qquad (24)$$

The curves of Fig. 13b demonstrate that during small amplitude oscillations, only the spectral components close to line center undergo sensible changes. In other words, only these components participate to the dynamic evolution of laser-matter interaction during unstable operation. We also note that the average spectral profile $-<\Gamma(w,t)>$ (dotted curve) exceeds the steady-state gain contour $-\Gamma_0(w)$ (dashed curve). Indeed, such an increase in average population inversion is conform to the decrease in average output field-amplitude that appears in Fig 12. The anomalous dispersion effect results in dissipation of the electromagnetic wave into the medium, ensuing in lesser simulated emission and a higher population inversion.

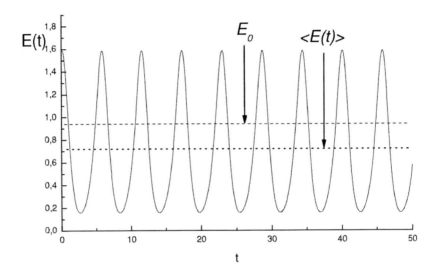

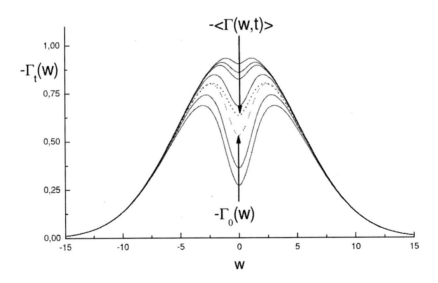

Figure 13. Dynamic gain profiling in the small-amplitude oscillatory regime. a) small amplitude oscillations solution of Eqs 1 with $\kappa = 5$, $\gamma = 1$, and $Es = 0.95$ b) instantaneous snapshots of the spectral profile during the temporal evolution of the oscillations of Figure a). Note that the whole gain profile undergoes an oscillating behavior around a *dc* profile distinct from the steady-state profile. Saturation effects are enhanced only at line center. Note also that the spectral components situated outside the range [-5, 5] do not contribute to the dynamic evolution of the system. These remain clamped to their values at steady-state.

The analytical description of these numerically simulated gain changes are also contained in the small-side band approach of Sec.V. The population-inversion components evolve according to

$$d(w,t) = d_0(w) + d_1(w)\cos(\Delta t) + d_2(w)\sin(\Delta t) \tag{25}$$

(Closed form expressions of the dc and the first-order amplitudes $d_1(w)$ and $d_2(w)$ are evaluated in Appendix B).

The saturated gain of the medium was defined in Eq 7 as

$$\Gamma_0(w) = g(w)d_0(w) \tag{26}$$

During the transient stage of the laser-matter interaction, the population inversion corresponding to each spectral component evolves according to Eq. (25). Thus, the dynamic gain-profile evidently evolves according to

$$\Gamma(w,t) = \Gamma_0(w) + \Gamma_1(w)\cos(\Delta t) + \Gamma_2(w)\sin(\Delta t) \tag{27}$$

that likewise contains a dc component $\Gamma_0(w)$, adding to an oscillating component with an in-phase amplitude

$$\Gamma_1(w) = g(w)d_1(w) \tag{28a}$$

and an out-of-phase term

$$\Gamma_2(w) = g(w)d_2(w) \tag{28b}$$

Representations of $\Gamma_1(w)$ and $\Gamma_2(w)$ show, as expected, that the values of the first order population components rapidly decrease away from line center (Fig. 14). According to Eq. 27, these first order terms undergo harmonic oscillations, ensuing in profile modulation with time.

Instantaneous snapshots of Eq. 27, obtained inside a single oscillation period, are given in Fig. 15. As expected, these show an excellent agreement with the numerical profiles of Fig. 13b.

Now, let us focus on the permanent part of the pulsing regime. Fig. 16b represents a few snapshots of the gain profiles (obtained likewise with numerical simulations of Eqs. (1) along the temporal solution of Fig. (16a). The spectral profile undergoes a much more enhanced distortion along with pulse building. Note the lateral hole-burning effect (dotted curve) stemming from a strong competing effect between the detuned components and the center of the profile. Strong saturation of the gain profile is observed at the peaks of the field-amplitude (lower curve in Fig. 16 relates to the intensity peaks of Fig. (16a).

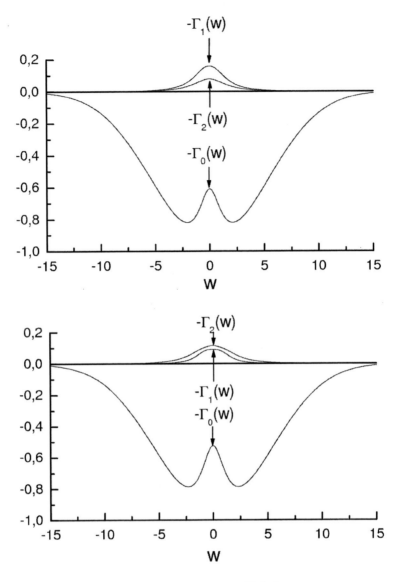

Figure 14. Representations of the stationary spectral profile $\Gamma_0(w)$ along with the contributions of the in-phase $\Gamma_1(w)$ and out-of-phase $\Gamma_2(w)$ components as evaluated from the weak-side band approach; a)with Es = 0.80, the in phase dominates the out-of phase component, while b)an increase of the excitation level (here Es = 0.95) yields an inverted situation, in which the anomalous dispersion effect increases, yielding the out of phase term to predominate.

The analytical description of these strong gain distortions calls for the strong harmonic expansion method outlined in Sec. VIII. To second order, the evolution of the population components follows Eq. (19c)

$$d(w,t) = d_{dc}(w) + d_3(w)\cos(2\Delta t) + d_4(w)\sin(2\Delta t) \tag{29}$$

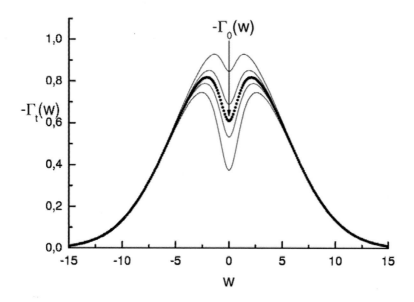

Figure 15. Analytical gain profiling in the weak-side-band approach (Eq. 27). The profile zone close to line center oscillates around the *dc* gain curve $-\Gamma_0(w)$. Compare with the corresponding numerical curves of Figure 13 b.

We define, likewise, a dynamic gain profile corresponding to the long term solution as

$$\Gamma(w,t) = \Gamma_{dc}(w) + \Gamma_3(w)\cos(2\Delta t) + \Gamma_4(w)\sin(2\Delta t) \qquad (30)$$

where

$$\Gamma_{dc}(w) = g(w)d_{dc}(w), \qquad (31a)$$

$$\Gamma_3(w) = g(w)d_3(w), \qquad (31b)$$

and,

$$\Gamma_4(w) = g(w)d_4(w) \qquad (31c)$$

In order to get a qualitative feeling of the contribution of each term, let us represent all the gain components in the same graph.

Snapshots of the solution obtained with $\kappa = 4$, $\Gamma_2(w) = g(w)d_2(w) = 0.1$, and $\Delta = 0.215$ along a single oscillation period are represented in Fig. 17. Note the presence of a lateral hole-burning effect already in $\Gamma_{dc}(w)$. The second order terms show distinct features, forecasting strong competing effects between the lateral and the center components of the gain profile when all the terms are combined together through Eq. 29. This is what is observed in Fig. 18, which shows instantaneous representations of Eq. 29. Every detail of the strong distortion features obtained with the numerical simulations is retrieved, as a direct comparison between Fig. 16 and Fig. 18 reveals.

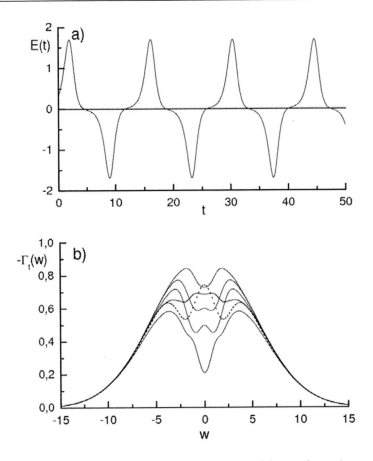

Figure 16. Dynamic gain profiling in the strong-amplitude pulsing regime: a) regular permanent oscillations, solution of Eqs 1 with $\kappa = 4$, $\gamma = 0.1$ and $Es = 1$; b) instantaneous snapshots of the spectral profiles during the temporal evolution of the oscillations of Figure a), as simulated from Eqs 1. Note that the gain profile undergoes a much stronger distortion effect than in Fig 13b). Saturation effects are extended to the lateral part of the profile. Note the lateral hole burning effect during part of the evolution. A competing effect between some localized lateral components and the center of the profile clearly appears. Note, likewise that the spectral components situated outside the range [-5, 5] do not contribute efficiently to the dynamic evolution of the system.

The dynamic evolution of the spectral profiles clarifies the outgrowth of the output characteristic represented in Fig. 12. In the weak-side band zone, the inhomogeneously broadened system evolves with a fixed dominant frequency at line center, thus conferring it all the properties of a homogeneously broadened laser. This results in the lowering of the average output intensity just beyond the instability threshold, as in the case of the Lorenz-Haken model. In other words, in the weak side band zone, only the atomic homogeneous packets situated in the vicinity of the center of the gain profile efficiently seeds the laser signal.

At higher excitation levels, lateral saturation occurs, so that a larger part of the gain profile seeds the laser field. Thus, during unstable operation, the lateral part of the spectral profile serves as an energy reservoir to the laser signal yielding enhanced output intensities as compared to the stationary values.

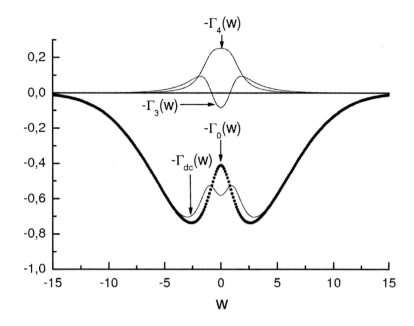

Figure 17. Comparison between the stationary dc (Eq. 31a), the in-phase (Eq. 31b) and the out-of-phase (Eq. 31c) of the second order components of the gain profiles. These representations clearly demonstrate the importance of both the in-phase and the out-of-phase terms. These do not undergo the same dependence with respect to w. Note also that even the dc profile undergoes some distortion, showing some lateral hole burning effect with respect to the stationary profile.

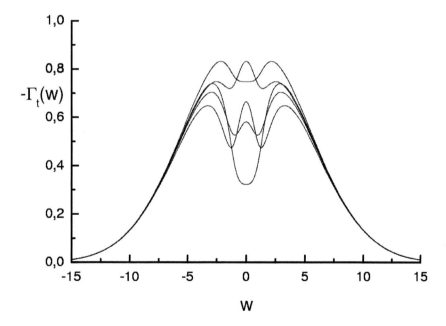

Figure 18. Analytical gain profiling in the strong side-band approach (Eq. 30). All the features of the lateral saturation effects found in the numerical curves of Figure 14b are retrieved. These curves constitute an additional confirmation of the validity of the strong harmonic expansion method.

In other words, the gain profiles undergo small changes when the field undergoes small-amplitude oscillations (both in the unstable zone and in the transient regime) and strong distortion, yielding complete lateral saturation of the profile, in the pulsing regime.

As thoroughly discussed in the preceding sections, in the regular pulsing regime, the field amplitude undergoes a perpetual switching around a zero-mean value, the corresponding average intensity is lower than that obtained at steady state, an increase of the excitation parameter beyond the instability threshold always results in a decrease in average output field amplitude, because of the dissipation effect stemming from anomalous dispersion, implying the following relation

$$\left\langle E(t)^2 \right\rangle < E_s^2 \tag{32}$$

The decreasing process continues until a zero mean value is reached ($\left\langle E(t) \right\rangle = 0$).

However, intensity lowering occurs in a limited (but large) range of excitations only, beyond the second laser threshold. At higher excitation levels the dissipation effect is compensated with the participation to the dynamic process of lateral components in the gain profile, enhancing the field output.

Let us now take advantage of these fruitful results to turn to analytical considerations of a typical self-pulsing structure obtained with a fixed decay rate value and increasing excitation levels.

X. Typical Hierarchy of Analytical Solutions

Now that the features of the long-term regular pulsing solutions have been analyzed in some detail, let us focus on the typical hierarchy of Fig. 7. The third order solution corresponding to the same parameter values as those of Fig. 7a, represented in Fig. 11, indicates that the time trace still requires further adjustments in order to reach a better fit with the numerical simulation. The frequency spectra of Fig 7 suggest that the contributions of the high order cannot be ignored. However, the very lengthy and quite tedious algebra involved in the analytical evaluation is of no encouragement towards further attempts in obtaining analytical expressions for the high-order field amplitudes. Instead, we call for the spectral information of Fig. 7 to evaluate these components. Since these spectra do not contain any phase information, we first consider a phase-locked situation between the high and the first order components. In this case, the analytical solutions may be taken in the form

$$E(t) = \sum_{n=0}^{5} E_{2n+1} \cos\left((2n+1)\Delta t\right) \tag{33}$$

The amplitude values of each of the components, evaluated from Fig. 7, are given in Table I, and the locked solutions corresponding to the hierarchy of Fig. 7 a,b,c are represented, respectively in Fig. 19 a, b,c. Note that both the frequencies and peak-heights match those of Fig. 7. However, as one may verify, the solutions do not contain the asymmetric features of the numerical counterparts. These features are retreived with the inclusion of minor phase corrections to the above symmetric expansion. The final solutions thus write

$$E(t) = \sum_{n=0}^{5} E_{2n+1} \cos\big((2n+1)\Delta t + \varphi_{2n+1}\big) \tag{34}$$

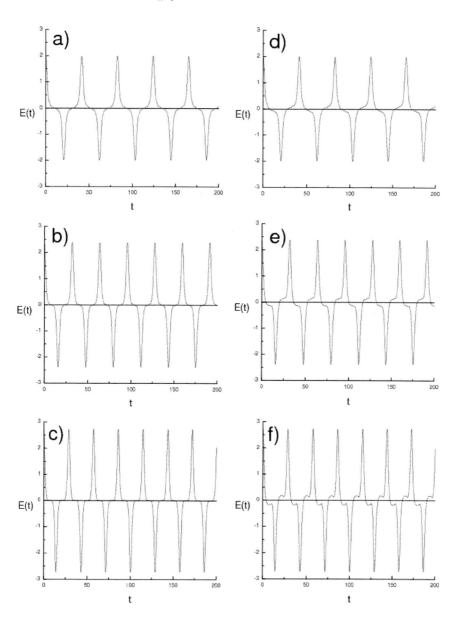

Figure 19. Hierarchy of analytical solutions obtained with the same parameter values as those of Figure 7. If we limit the solution to a symmetric harmonic expansion of the form Eq. 33, we obtain regular and symmetric solutions represented in a) for $Es = 0.8$, b) $Es = 0.95$ and c) $Es = 1.05$. When out-of-phase terms are introduced, as in Eq. 34, the small asymmetries of Figure 7 are retrieved, as indicated in the solutions of d), f) and e) corresponding, respectively to the solutions of a), b) and c).

It takes little effort to scan over the phase terms in order to find the best possible match with the numerical time traces. When these are included, all the details of the numerical solutions show up in their numerical counterparts, as a one to one comparison between Figs 7 a, b, c and Figs 19 d, e, f reveals.

XI. Further Considerations of the Dynamic Properties beyond Regular Self-pulsing

The regular amplitude field switching is typical of any value of the decay rates that satisfy the bad cavity condition. However, as the excitation parameter is increased, this regular behavior only persists for a small range of excitation values before abruptly yielding irregular oscillations. Figure 20 represents such a transition. The regular pulsing behavior obtained in Fig. 20a with $E_s = 1.06$, along with a well resolved spectrum (Fig. 20d), transforms into a completely irregular time trace at $E_s = 1.07$ (Fig. 20b), as demonstrated in the corresponding spectrum of Fig. 20e.

The onset of irregular long term solutions always occurs in the vicinity (at the exact value for adiabatic field evolution) of the following critical frequency value (see also the comments at the end of Appendix C)

$$\Delta_c = \sqrt{\frac{\wp}{2}} \tag{35}$$

This frequency is linked to the behavior of the second order population components as clarified in the following.

The second order population components are given by

$$d_3 = \Gamma_4 \left\{ -M_1(\Delta, w) + \frac{2\Delta}{\wp} M_2(\Delta, w) \right\} d_{0d} \frac{E_1^2}{2} \tag{36a}$$

$$d_4 = -\Gamma_4 \left\{ M_2(\Delta, w) + \frac{2\Delta}{\wp} M_1(\Delta, w) \right\} d_{0d} \frac{E_1^2}{2} \tag{36b}$$

These show dependence on Δ and on w. Let us represent the variations of d_3 and d_4 versus Δ, for the central tuned velocity component $w = 0$. It transpires from Fig.21 that the out-of-phase component is always positive, while the in-phase component changes its sign at the point where

$$-M_1(\Delta, 0) + \frac{2\Delta}{\wp} M_2(\Delta, 0) = 0 \tag{37}$$

satisfied for $\Delta = \sqrt{\dfrac{\wp}{2}}$ (i.e. Eq. 35)

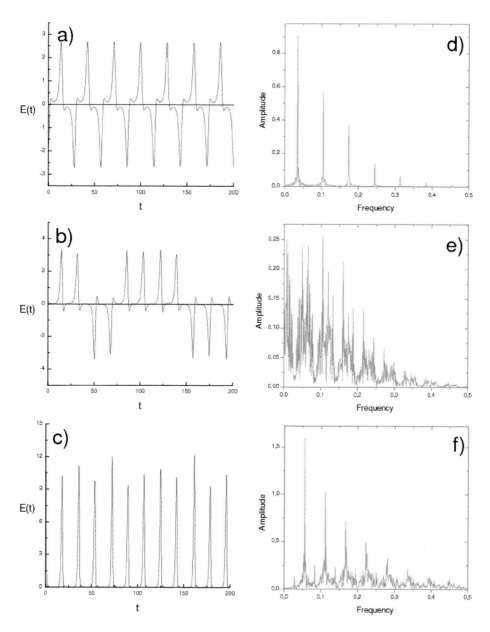

Figure 20. Numerical solutions following the hierarchy of Figure 7 a) for $Es = 1.06$, the solution is still a period-one regular oscillation, b) for $Es = 1.07$, the field-amplitude undergoes what seems to be a chaotic evolution as confirmed by the corresponding frequency spectrum of Figure e) The physical origin of such a dramatic change in behavior is explained in the text and in the Figure 21 caption; c) representation of the field-intensity of the trace b) and f) the corresponding spectrum shows the presence of a sub-harmonic frequencies at $\Delta/2$, $3\Delta/2$ etc..

For $\Delta^2 < \dfrac{\wp}{2}$, the in-phase and out-of-phase of the population inversion harmonic components are both characterized by positive amplitudes. Their effect is a tendency towards an increase of the average population inversion for the central tuned velocity component. For $\Delta^2 > \dfrac{\wp}{2}$, the out-of phase component still remains positive while the in-phase part becomes negative. Thus a competition effect between the tendency of the out-of phase component to increase the population inversion and the opposite inclination of the in-phase component, takes place in the vicinity of $\Delta = \sqrt{\dfrac{\wp}{2}}$. This competition yields completely irregular long-term solutions, as that of Fig. 20b. An increase of the excitation parameter may result in situations where the two effects exactly compensate and the solution becomes regular again. Here a symmetry-breaking effect occurs, forbidding oscillations around zero mean-value for the field and polarization.

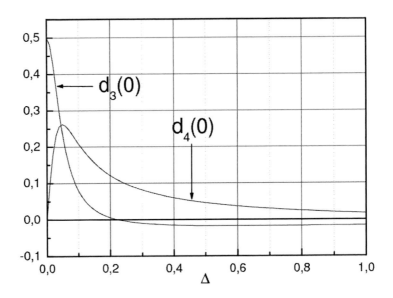

Figure 21. Amplitude of the in-phase and out-of phase second-order population components versus frequency. The curves show the same evolution as those of the in-phase and out-of phase components of the field amplitude represented in Figure 10. While the in-phase component rapidly decreases with increasing frequencies, the out-of phase component increases and rapidly overcomes the in-phase component. Note that while d_4 always remains positive, the in phase component becomes negative above the critical value $\Delta = \sqrt{\dfrac{\wp}{2}}$, and a competition effect takes place between d_3 and d_4 yielding the irregular time trace of Figure 20 b.

These last results imply the following comments. The central velocity components play a driving role in the dynamic evolution of the integro-differential equations. When the in-phase and the out-of-phase harmonic components of the population inversion evolve with positive amplitude values (we refer to this situation as normal behavior), the field-matter interaction takes place with a self-consistent nature: that is, an increase of the excitation parameter yields an increase of the field amplitudes and a decrease of the population inversion components, thus conferring to the solutions regular features. An increase of the excitation parameter increases the oscillation frequency of the signal and the regular behavior persists up to the above described situation where the in-phase and out of phase population components enter into competition, yielding irregular time traces.

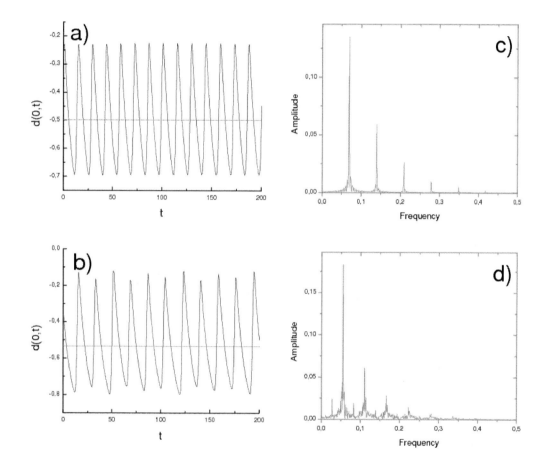

Figure 22. Evolution of the population inversion of the center-line component along with the field traces of Figure 20; a) regular period-one evolution and b) regular period-two oscillations. The corresponding frequency spectra represented respectively in c) shows perfectly localized and strong even components at 2Δ, 4Δ, 6Δ etc, while d) shows small odd components at Δ, 3Δ, 5Δ, etc., a signature of the electric field-beat note superimposed to the population inversion variable during light-matter interactions.

Let us focus on the intensity time-trace corresponding to the field variations of Fig. 20b. Opposed to the irregular evolution of Fig. 20b, the intensity time trace of Fig 20c shows less irregular variations, with clear sub-harmonic components at $\Delta/2$, $3\Delta/2$, $5\Delta/2$, $7\Delta/2$, etc., that appear in the corresponding frequency spectrum of Fig. 20f. These features also emerge in the population inversion time traces, represented in Fig. 22. This demonstrates that the population inversion ignores the field variations but is sensitive rather to intensity variations. In other words, the laser medium interacts not with the laser field amplitude but rather with the laser-field intensity. Indeed, in the self-pulsing regime, the population inversion components evolve at twice the frequency of the field-amplitude, i.e. at the frequency of the laser field intensity. This evolution constitutes the natural property of the laser-matter interaction. And, one may say that when the field and population inversion evolve with the same frequency (as in the case of the weak-side band approach) it constitutes a violation of that natural evolution. When the system departs from oscillating around the stationary state, this constitutes a natural tendency of the system to retrieve the natural laser-matter interaction property, which is contained in the integro-differential set (1).

XII. Further Solutions with Extreme Control Parameter Values

In this section, we focus on other examples obtained with extremely different control parameter values, and demonstrate that the strong harmonic expansion method applies whenever regular oscillations characterize the long-term solutions irrespective of neither the decay rate values nor the excitation level. Figure 23 represent two solutions along with their analytical counterparts, obtained with far apart control parameters. The two sets of parameters corresponding to Figs 23a and 23d are, $\kappa = 100$, $\wp = 1$, $E_s = 0.75$, and $\kappa = 5$, $\wp = 1$, and $E_s = 13$, respectively.

In both cases, the corresponding spectra show a limited number of components (Figs. 23b, 23c). The evaluation of the field-amplitudes, according to the analytical expressions of Sec. VIII, yields the following solutions

$$E(t) = 2 \times \left(0.55\cos(0.86t) + 0.13\cos(3 \times 0.86t - 0.2) + 0.05\cos(5 \times 0.86t)\right) \quad (38a)$$

for the case of Fig. 23a, and is represented in Fig 23c, and

$$E(t) = 2 \times \left(8.5\cos(7.64t) + 2.2\cos(3 \times 7.64t) + 0.025\cos(5 \times 7.64t)\right) \quad (38b)$$

for the case of Fig. 23d, as represented in Fig. 23f. Note that, in this case, an almost sinusoidal trace is obtained, and the signal shows symmetric oscillations, as expected from the absence of any out-phase term in Eq. 38b. The asymmetric nature of the time trace in Fig. 23a is contained in the out-phase term appearing in Eq. 38a.

Now, let us consider a case that contains a double peak structure. The signal represented in Fig. 24a was obtained with $\kappa = 100$, $\wp = 1$, and $E_s = 0.795$. The corresponding spectrum, represented in Fig. 24b, shows that, in this case, the third-order field amplitude dominates the first-order term. The third order amplitude is higher that the first-order one, and higher order terms (up to the 13th) appear in the spectrum. The analytical solution writes

$$E(t) = 2\left\{ \begin{array}{l} 0.4\cos(0.4t) - 0.41\cos(3 \times 0.4t + 0.25) - 0.1\cos(5 \times 0.4t - 0.65) + 0.1\cos(7 \times 0.4t + 0.1) \\ \\ + 0.05\cos(9 \times 0.4t - 0.4) - 0.018\cos(11 \times 0.4t) - 0.015\cos(13 \times 0.4t) \end{array} \right\}$$

$$(39)$$

and is represented in Fig. 24c. Again, an excellent agreement between the numerical and analytical solution is found.

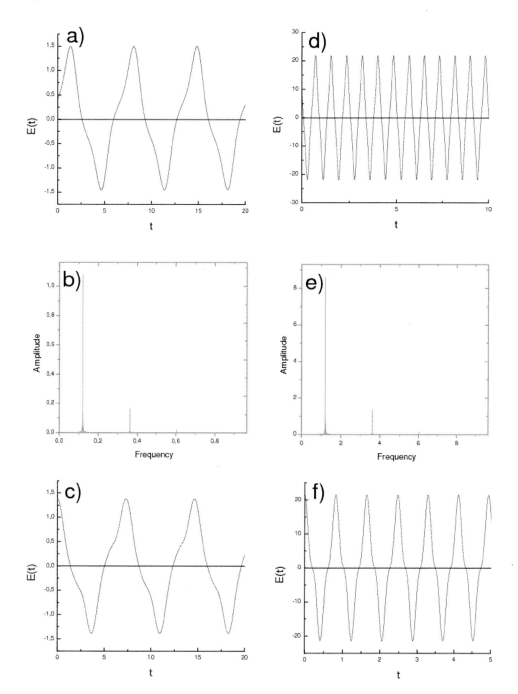

Figure 23. Numerical solutions and their analytical counterparts obtained with far apart control parameters. a) $\kappa = 100$, $\gamma = 1$, $E_s = 0.75$, b) corresponding frequency spectrum, c) analytical solution drawn with Eq. 38a d) $\kappa = 5$, $\gamma = 1$, and $E_s = 13$, e) frequency spectrum of d), and e) analytical solution obtained from Eq. 38b.

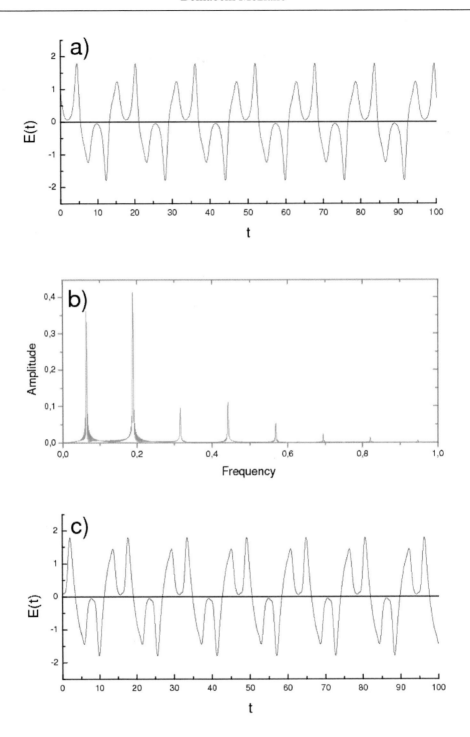

Figure 24. a) numerical solution with a double peak structure, obtained with $\kappa = 100$, $\gamma = 1$, and $E_s = 0.795$, b) corresponding frequency spectrum, and c) analytical solution following Eq. 39.

In the course of our investigation, numerous examples were simulated along the strong harmonic lines presented in this paper, and in each case, the corresponding analytical solution is found to perfectly match the numerical counterpart.

XIII. Instability Switch-Off

A question of concern, which intrigued the laser community, pertains to whether the unstable state persists for excitation levels well above the instability threshold. To the best of our knowledge, this question still remains open. In the case of the integro-differential equations (1), detailed data reveals that the second laser-threshold increases with decreasing κ and increasing $\wp$ values. The maximum allowed value for the cavity decay rate is the radiative limit $\wp=2$. This value together with $\kappa=2.5$ yields an instability threshold $E_{2th} \cong 2.43$, corresponding to $r \cong 3.3$. An excitation increase, above E_{2th}, results in a series of regular oscillations around steady state that evolve towards the regular long-term self-pulsing solutions of Fig. 3. The pulsing state persists up to $E_s \cong 14.18$ before undergoing an inverse hopf-bifurcation at $E_s \cong 14.19$, yielding, again, an oscillatory solution around steady-state.

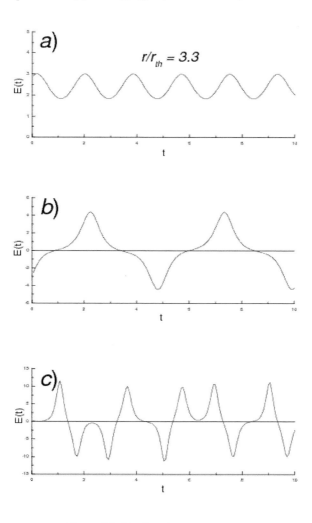

Figure 25. Continued on next page.

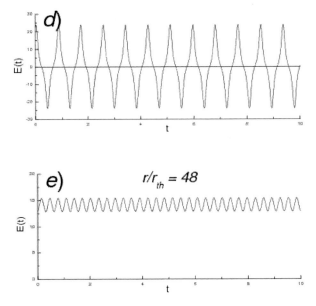

Figure 25. A hierarchy of solutions extending from instability onset to instability switch-off, obtained with $\wp=2$ and $\kappa=2.5$; a) almost pure harmonic oscillation around steady state obtained at r/r$_{th}$=3.3, b)-d) time evolution of pulsing oscillations around zero-mean values, obtained with increasing excitation levels, e) faster pure sinusoidal oscillations, obtained just before instability switch off at r/r$_{th}$=48. For r/r$_{th}$>48, the solution is a stable straight line.

The solution settles again in a stable with constant intensity output at $E_s \cong 14.20$, following the steady-state relation (Eq. 4). A few steps of the solution evolution from instability threshold to instability switch-off are represented in Fig. 25. Note that the excitation parameter evolves from $r \cong 3.3$ to $r \cong 48$! Indeed, the unstable zone extends over much higher excitation ranges for lower $\wp$'s and/or lower κ values. This may explain the lack of any report pertaining to instability switch-off.

The numerical example of Fig. 25 constitutes the first demonstration of instability switch-off in the SMIB dynamical system, and to the best of our knowledge, in other fundamental models.

We expect the same behavior to characterize the system for other control parameters. However, the much higher levels of excitation complicate and lengthen the numerical calculations, because the higher the excitation, the lower the discrete-time steps to be used in order to optimize the calculations, must be.

XIV. Conclusion

Beyond resuming well-established properties of the weak-side band approach which have already been largely reported in the SMIB laser instabilities issue, this chapter has put forward its main drawbacks and proposed much better readjustments to deal with the laser dynamics issue. From the presented analysis, it transpires that the weak-side band approach contains much of the information of the transient properties of SMIB laser dynamics, both in the stable and unstable regime of operation. In addition, it is well adapted for the description of the unstable solutions when these consist of small amplitude oscillation around non-zero mean values, which usually characterize the onset of unstable operation in small range of excitation levels beyond the instability threshold. However, as amply discussed and demonstrated in this chapter, when dealing with the permanent self pulsing solutions, the weak-side band approach is revealed to be quite inoperative. Starting from the recognition of a regular amplitude field switching around zero mean values, we have given the essential steps leading to a genuine extension of the side-band approach to deal with the permanent state of the unstable solutions that consist of regular oscillations around the time axis for the field and polarization amplitudes. Such a behavior is the rule for a large domain of the parameter space $(\kappa, \wp, E_s)$, and its description was shown to be contained in a strong harmonic expansion technique. The main features of the regular solutions (oscillation frequencies, details of the time traces, etc.) were derived, and analytical solutions obtained. Beyond the regular and symmetric period-one solutions studied in this paper, the method applies equally well to asymmetric period-one oscillations and higher order periodic solutions (period two, four, etc.). In order to avoid redundancy, an example of a period-doubling hierarchy will be analyzed in the next chapter. As it will transpire, the differential equations that describe the dynamics of a homogeneously broadened laser are characterized by the same analytical solutions as those of Eqs. (1). The much simpler set of equations will allow for the obtaining of much simpler closed form relations for the transient and long term-frequencies. In addition, the calculations in field-amplitude evaluations will be performed up to the fifth order.

Returning back to the case studied in this paper, it is well known that amplitude field-switching does not always characterize the solutions of Eqs. (1). The analytical solutions in this case can be obtained by leaving the constraint of zero mean-value. In these cases, the small harmonic expansion method is closer to the weak-side band approach, and the characteristics of the long-term solutions can be obtained provided the harmonic expansion does not constrain the system to an evolution around the stationary state.

This paper also demonstrates that the laser-matter equations that govern the dynamic properties of single-mode inhomogeneously broadened systems are characterized by a dissipation effect, stemming from anomalous dispersion. This dissipation phenomenon results in less average output intensities as compared to stable operation. This ensues in discontinuous output versus pump characteristics, at the instability threshold, while, at high excitation levels (well beyond threshold), the average output intensity exceeds that at steady state. This behavior results from an energy transfer from the lateral part of the gain profile that does not efficiently participate to the field-matter interaction during stable operation and when the frequency of the laser field coincides with the center of the spectral profile.

With respect to dynamic gain structuring, the study also delimited the applicability of the Casperson-Hendow-Sargent small sideband analyses. In connection to the fact that the small-signal analysis only applies at the onset of instability, in the small-amplitude oscillatory domain we have shown that the corresponding gain profiles also evolve following small oscillations of the components situated in the vicinity of the gain-curve center. In this small region of the instability domain, the small signal analysis does apply to describe dynamic gain profiling. However, when extended to the instability domain inside which the laser fields consist of pulse trains, the small-signal analysis does not apply anymore. Starting from the recognition of an amplitude-field switching around zero-mean values, we have adapted the side-band method to a strong harmonic expansion and derived analytical expressions that give a remarkably good account of the strong distortion of the dynamic gain profiles occurring during pulse build-up. Our study deliberately focused on periodic solutions, since these can be described with simple Fourier series, whose inclusion into the integro-differential equations give ample information on the physics involved in the non-linear nature of light-matter interactions inside a laser medium operating in its dynamical regime.

Appendix A.
Closed-Form Expression for the Transient Frequency

Inserting the small-amplitude expansions (Eqs. 10) into Eqs. (1) we obtain a series of relations. Equalizing the first order terms in each relation, we extract the following set of equations

$$-\Delta p_{r1} = -p_{r2} + wp_{i2} + E_0 d_2 \tag{A1a}$$

$$\Delta p_{r2} = -p_{r1} + wp_{i1} + (E_0 d_1 + E_1 d_0) \tag{A1b}$$

$$\Delta p_{i1} = p_{i2} + wp_{r2} \tag{A1c}$$

$$\Delta p_{i2} = -p_{i1} - wp_{r1} \tag{A1d}$$

$$\Delta d_1 = \wp(d_2 + E_0 p_{r2}) \tag{A1e}$$

$$\Delta d_2 = -\wp(d_1 + E_0 p_{r1} + E_1 p_{r0}) \tag{A1f}$$

From (A1e) and (A1f), we extract

$$d_1 = -\alpha(E_0 p_{r1} + E_1 p_{r0}) + \frac{\Delta}{\wp}\alpha E_0 p_{r2} \tag{A2a}$$

and,

$$d_2 = -\alpha E_0 p_{r2} - \frac{\Delta}{\wp}\alpha(E_0 p_{r1} + E_1 p_{r0}) \tag{A2b}$$

where:

$$\alpha = \frac{\wp^2}{\wp^2 + \Delta^2} \tag{A2c}$$

Inserting these relations into Eqs. (A1c) and (A1d) yields

$$p_{i1} = -\frac{w}{1 + \Delta^2} p_{r1} + \frac{w\Delta}{1 + \Delta^2} p_{r2} \qquad (A3a)$$

$$p_{i2} = -\frac{w\Delta}{1 + \Delta^2} p_{r1} - \frac{w}{1 + \Delta^2} p_{r2} \qquad (A3b)$$

Equations (A1a) and (A1b) transform into

$$-\Delta(1 + \Delta^2 - w^2)p_{r1} = -(1 + \Delta^2 + w^2)p_{r2} + (1 + \Delta^2)E_0 d_2 \qquad (A4a)$$

$$(1 + \Delta^2 + w^2)p_{r1} = -\Delta(1 + \Delta^2 - w^2)p_{r2} + (1 + \Delta^2)\left(E_0 d_1 + E_1 d_0\right) \qquad (A4b)$$

Inserting relations (A2a) and (A2b) into (A4a) and (A4b), we obtain the following pair of relations for the polarization components

$$\left\{-\Delta(1 + \Delta^2 - w^2) + \alpha(1 + \Delta^2)\frac{\Delta}{\wp}E_0^2\right\}p_{r1} + \left\{1 + \Delta^2 + w^2 + \alpha(1 + \Delta^2)E_0^2\right\}p_{r2} =$$

$$-\alpha(1 + \Delta^2)\frac{\Delta}{\wp}p_{r0}E_0 E_1$$

$$(A5a)$$

$$\left\{1 + \Delta^2 + w^2 + \alpha(1 + \Delta^2)E_0^2\right\}p_{r1} - \left\{-\Delta(1 + \Delta^2 - w^2) + \alpha(1 + \Delta^2)\frac{\Delta}{\wp}E_0^2\right\}p_{r2} =$$

$$(1 + \Delta^2)\left\{d_0 - \alpha\, p_{r0}E_0\right\}E_1$$

$$(A5b)$$

in which p_{r0} is related to the steady state field through

$$p_{r0} = \frac{d_0}{1 + w^2} E_0 \qquad (A6)$$

Injecting expressions (A5a) and (A5b) into Eq. (1a) yield analytical relations for the gain and dispersion effects experienced by the side-mode component E_1.

$$G_{sb}(\Delta) = 2C \int_{-\infty}^{+\infty} dw g(w) d_s(w)\frac{A_1 B_1 + A_2 B_2}{A_1^2 + A_2^2} = \int_{-\infty}^{+\infty} dw p_1(w, \Delta) = 1 \qquad (A7a)$$

and

$$D_{sb}(\Delta) = 2C \int_{-\infty}^{+\infty} dw g(w) d_s(w)\frac{A_1 B_2 - A_2 B_1}{A_1^2 + A_2^2} = \int_{-\infty}^{+\infty} dw p_2(w, \Delta) = -\frac{\Delta}{\kappa} \qquad (A7b)$$

where

$$A_1 = (1 + \Delta^2)\left(1 + \frac{\wp^2}{\Delta^2 + \wp^2} E_s^2\right) + w^2 \tag{A7c}$$

$$A_2 = \Delta\left\{(1 + \Delta^2)\left(1 - \frac{\wp}{\Delta^2 + \wp^2} E_s^2\right) - w^2\right\} \tag{A7d}$$

$$B_1 = (1 + \Delta^2)\left(1 - \frac{\wp^2}{\Delta^2 + \wp^2}\frac{1}{1 + w^2} E_s^2\right) \tag{A7e}$$

$$B_2 = \frac{\Delta}{\wp}\frac{1 + \Delta^2}{1 + w^2}\frac{\wp^2}{\Delta^2 + \wp^2} E_s^2 \tag{A7f}$$

$$d_s = -\frac{1 + w^2}{1 + w^2 + E_0^2} \tag{A7g}$$

The ratio of (A7a) and (A7b) yields an equation that relates the transient frequency Δ to the excitation parameter E_s, given the values of the decay rate and the population-inversion relaxation rate

$$\frac{\Delta}{\kappa} = \frac{\int_{-\infty}^{+\infty} dw\, g(w) F_{X_s, \wp}(w, \Delta)}{\int_{-\infty}^{+\infty} dw\, g(w) G_{X_s, \wp}(w, \Delta)} \tag{A8}$$

Identical to Eq.(12).

Appendix B.
Closed-Form Expressions for the First and Third-Order Electric Field Components and for the Long-Term Frequency

In order to derive the first and third order terms of the strong harmonic expansion Eq. (14a), we impose, as a first trial solution

$$E(t) = E_1 \cos(\Delta t) \tag{B1}$$

driven by the field, the polarization components follow

$$p_r(w, t) = p_{r1} \cos(\Delta t) + p_{r2} \sin(\Delta t) \tag{B2a}$$

$$p_i(w, t) = p_{i1} \cos(\Delta t) + p_{i2} \sin(\Delta t) \tag{B2b}$$

and the population inversion can be written

$$d(t) = d_{0d} + d_1(t) \tag{B3}$$

Inserted into the polarization equations (Eq. 1b of the text), these yield the following first order terms

$$-\Delta p_{r1} = -p_{r2} + wp_{i2} \tag{B4a}$$

$$\Delta p_{r2} = -p_{r1} + wp_{i1} + E_1(d_{0d} + d_1(t)) \tag{B4b}$$

and,

$$\Delta p_{i1} = p_{i2} + wp_{r2} \tag{B5a}$$

$$\Delta p_{i2} = -p_{i1} - wp_{r1} \tag{B5b}$$

Solving (B4a) and (B4b), with respect to p_{r1} and p_{r2}, we obtain

$$(1 + \Delta^2 + w^2)p_{r1} + \Delta(1 + \Delta^2 - w^2)p_{r2} = (1 + \Delta^2)d_{0d}E_1 \tag{B6a}$$

$$\Delta(1 + \Delta^2 - w^2)p_{r1} - (1 + \Delta^2 + w^2)p_{r2} = 0 \tag{B6b}$$

(As a first approximation, population variations are neglected with respect to the *dc* term) with the solutions

$$p_{r1} = M_1 d_{0d} E_1 \tag{B7a}$$

$$p_{r2} = M_2 d_{0d} E_1 \tag{B7b}$$

where

$$M_1 = \frac{(1 + \Delta^2)(1 + \Delta^2 + w^2)}{(1 + \Delta^2 + w^2)^2 + \Delta^2(1 + \Delta^2 - w^2)^2} \tag{B7c}$$

and

$$M_2 = \frac{(1 + \Delta^2)(1 + \Delta^2 - w^2)}{(1 + \Delta^2 + w^2)^2 + \Delta^2(1 + \Delta^2 - w^2)^2} \tag{B7d}$$

The population equation writes

$$\frac{d}{dt}d(t) = -\wp\left\{d_{0d} + d(t) + 1 + E_1\cos(\Delta t)\left[p_{r1}\cos(\Delta t) + p_{r2}\sin(\Delta t)\right]\right\} \tag{B8}$$

we see that it contains higher order terms in the modulation frequency. Thus, a reasonable expansion of d(t) is in the form

$$d(t) = d_{0d} + d_1 \cos(\Delta t) + d_2 \sin(\Delta t) + d_3 \cos(2\Delta t) + d_4 \sin(2\Delta t) \qquad \text{(B9)}$$

One easily finds (as expected)

$$d_1 = d_2 = 0 \qquad \text{(B10)}$$

and

$$-2\Delta d_3 \sin(2\Delta t) + 2\Delta d_4 \cos(2\Delta t) = -\wp \left\{ \begin{array}{l} d_{0d} + 1 + d_3 \cos(2\Delta t) + d_4 \sin(2\Delta t) + E_1 p_{r1} \cos^2(\Delta t) \\ \qquad\qquad + E_1 p_{r2} \sin(\Delta t) \cos(\Delta t) \end{array} \right\}$$

$$= -\wp \left\{ d_{0d} + 1 + d_3 \cos(2\Delta t) + d_4 \sin(2\Delta t) + \frac{E_1}{2} p_{r1} \big(\cos(2\Delta t) + 1 \big) + \frac{E_1}{2} p_{r2} \sin(2\Delta t) \right\}$$

$$\text{(B11)}$$

yielding

$$d_{0d} = -1 - \frac{E_1}{2} p_{r1} \qquad \text{(B12)}$$

With the above expression (Eq. B7a) for $pr1$, B12 takes the form

$$d_{0d} = -\frac{1}{1 + M_1(w, \Delta)\dfrac{E_1^2}{2}} \qquad \text{(B13)}$$

This new value of d_{0d} undergoes a saturating effect from the first-field harmonic E_1
The second order population components, extracted from (B11), satisfy

$$2\Delta d_3 = \wp\, d_4 + \wp\, \frac{E_1}{2} p_{r2} \qquad \text{(B14a)}$$

$$2\Delta d_4 = -\wp\, d_3 - \wp\, \frac{E_1}{2} p_{r1} \qquad \text{(B14b)}$$

transforming into

$$d_3 = \frac{\wp^2}{\wp^2 + 4\Delta^2} \left\{ -p_{r1} + \frac{2\Delta}{\wp} p_{r2} \right\} \frac{E_1}{2} = \Gamma_4 \left\{ -M_1 + \frac{2\Delta}{\wp} M_2 \right\} d_{0d} \frac{E_1^2}{2} \qquad \text{(B15a)}$$

$$d_4 = -\frac{\wp^2}{\wp^2 + 4\Delta^2} \left\{ p_{r2} + \frac{2\Delta}{\wp} p_{r1} \right\} \frac{E_1}{2} = -\Gamma_4 \left\{ \frac{2\Delta}{\wp} M_1 + M_2 \right\} d_{0d} \frac{E_1^2}{2} \qquad \text{(B15b)}$$

The calculations can now be conducted to higher orders. To third order, the polarization components expansion write

$$p_r(w,t) = p_{r1} \cos(\Delta t) + p_{r2} \sin(\Delta t) + p_{r3} \cos(3\Delta t) + p_{r4} \sin(3\Delta t) \quad \text{(B16a))}$$

$$p_i(w,t) = p_{i1} \cos(\Delta t) + p_{i2} \sin(\Delta t) + p_{i3} \cos(3\Delta t) + p_{i3} \sin(3\Delta t) \quad \text{(B16b)}$$

When these expansions and the population expression

$$d(t) = d_{0d} + d_3 \cos(2\Delta t) + d_4 \sin(2\Delta t) \quad \text{(B17)}$$

are inserted into the polarization equations (Eq. 1b of the text), we are led to

$$
\begin{aligned}
& -\Delta p_{r1} \sin(\Delta t) + \Delta p_{r2} \cos(\Delta t) - 3\Delta p_{r3} \sin(3\Delta t) + 3\Delta p_{r4} \cos(3\Delta t) \\
& = -p_{r1} \cos(\Delta t) - p_{r2} \sin(\Delta t) - p_{r3} \cos(3\Delta t) - p_{r4} \sin(3\Delta t) \\
& + w\left(p_{i1} \cos(\Delta t) + p_{i2} \sin(\Delta t) + p_{i3} \cos(3\Delta t) + p_{i4} \sin(3\Delta t) \right) \\
& + E_1 \cos(\Delta t)\left(d_{0d} + d_3 \cos(2\Delta t) + d_4 \sin(2\Delta t) \right)
\end{aligned}
\quad \text{(B18a)}
$$

and

$$
\begin{aligned}
& -\Delta p_{i1} \sin(\Delta t) + \Delta p_{i2} \cos(\Delta t) - 3\Delta p_{i3} \sin(3\Delta t) + 3\Delta p_{i4} \cos(3\Delta t) \\
& = -p_{i1} \cos(\Delta t) - p_{i2} \sin(\Delta t) - p_{i3} \cos(3\Delta t) - p_{i4} \sin(3\Delta t) \\
& - w\left(p_{r1} \cos(\Delta t) + p_{r2} \sin(\Delta t) + p_{r3} \cos(3\Delta t) + p_{r4} \sin(3\Delta t) \right)
\end{aligned}
\quad \text{(B18b)}
$$

yielding

$$p_{r1} = \frac{1}{1+\Delta^2}\left\{ w\left(p_{i1} - \Delta p_{i2} \right) + E_1 d_{0d} + \frac{E_1}{2}(d_3 - \Delta d_4) \right\} \quad \text{(B19a)}$$

which transforms into

$$p_{r1} = \delta_1 d_{0d} E_1 \left\{ 1 + \delta_1 w^2 \left[-(1-\Delta^2) M_1 + 2\Delta M_2 \right] + \Gamma_4 \left[\left(\frac{2\Delta^2}{\wp} - 1 \right) M_1 + \left(\frac{2\Delta}{\wp} + \Delta \right) M_2 \right] \frac{E_1^2}{4} \right\} \quad \text{(B19b)}$$

and

$$p_{r2} = \frac{1}{1+\Delta^2}\left\{ w\left(p_{i2} + \Delta p_{i1} \right) + \Delta E_1 d_{0d} + \frac{E_1}{2}(d_4 + \Delta d_3) \right\} \quad \text{(B20a)}$$

which takes the form

$$p_{r2} = \delta_1 d_{0d} E_1 \left\{ \Delta - \delta_1 w^2 \left[2\Delta M_1 + (1-\Delta^2) M_2 \right] - \Gamma_4 \left[\left(\frac{2\Delta}{\wp} + \Delta \right) M_1 - \left(\frac{2\Delta^2}{\wp} - 1 \right) M_2 \right] \frac{E_1^2}{4} \right\} \quad \text{(B20b)}$$

For simplifying purposes we have introduced the following parameters

$$\delta_1 = \frac{1}{1+\Delta^2} \qquad (B21a)$$

$$\delta_3 = \frac{1}{1+9\Delta^2} \qquad (B21b)$$

$$\Gamma_4 = \frac{\wp^2}{\wp^2 + 4\Delta^2} \qquad (B21c)$$

Now let's turn back to the field variable. The third-order field expansion writes

$$E(t) = E_1 \cos(\Delta t) + E_3 \cos(3\Delta t) \qquad (B22a)$$

That we insert into the field equation

$$\frac{d}{dt} E(t) = -\kappa \left\{ E(t) + 2C \int_{-\infty}^{+\infty} dw g(w) p_r(w,t) \right\} \qquad (B22b)$$

along with

$$p_r(w,t) = p_{r1} \cos(\Delta t) + p_{r2} \sin(\Delta t) + p_{r3} \cos(3\Delta t) + p_{r4} \sin(3\Delta t) \qquad (B22c)$$

to obtain

$$-\Delta E_1 \sin(\Delta t) - 3\Delta E_3 \sin(3\Delta t)$$
$$= -\kappa \left\{ E_1 \cos(\Delta t) + E_3 \cos(3\Delta t) + 2C \int_{-\infty}^{+\infty} dw g(w) \left[\begin{array}{l} p_{r1} \cos(\Delta t) + p_{r2} \sin(\Delta t) \\ + p_{r3} \cos(3\Delta t) + p_{r4} \sin(3\Delta t) \end{array} \right] \right\} \qquad (B23)$$

Equalizing the first order field and polarization relations yields

$$\frac{\Delta}{\kappa} E_1 = 2C \int_{-\infty}^{+\infty} dw g(w) p_{r2} \qquad (B24a)$$

$$- E_1 = 2C \int_{-\infty}^{+\infty} dw g(w) p_{r1} \qquad (B24b)$$

that must be satisfied simultaneously for the unknown Field component E_1 and oscillation frequency Δ

The ratio of these two relations yields a frequency determining equation related to the first-order field amplitude as

$$\frac{\Delta}{\kappa} = -\frac{\int_{-\infty}^{+\infty} dw g(w) F_\wp(w, \Delta, E_1)}{\int_{-\infty}^{+\infty} dw g(w) G_\wp(w, \Delta, E_1)} \tag{B25}$$

with

$$G_\wp(w, \Delta, E_1) = \delta_1 d_{0d} \left\{ 1 + \delta_1 w^2 \left[-(1-\Delta^2)M_1 + 2\Delta M_2 \right] + \Gamma_4 \left[\left(\frac{2\Delta^2}{\wp} - 1 \right) M_1 + \left(\frac{2\Delta}{\wp} + \Delta \right) M_2 \right] \frac{E_1^2}{4} \right\} \tag{B26a}$$

and

$$F_\wp(w, \Delta, E_1) = \delta_1 d_{0d} \left\{ \Delta - \delta_1 w^2 \left[2\Delta M_1 + (1-\Delta^2)M_2 \right] - \Gamma_4 \left[\left(\frac{2\Delta}{\wp} + \Delta \right) M_1 - \left(\frac{2\Delta^2}{\wp} - 1 \right) M_2 \right] \frac{E_1^2}{4} \right\} \tag{B26b}$$

In order to determine E_1 and Δ for a given excitation level E_s, we solve graphically the following relations

$$\int_{-\infty}^{+\infty} dw g(w) F_\wp(w, \Delta, E_1) = \frac{\Delta}{\kappa} \int_{-\infty}^{+\infty} dw g(w) \frac{1}{1 + w^2 + E_s^2} \tag{B27a}$$

and

$$\int_{-\infty}^{+\infty} dw g(w) G_\wp(w, \Delta, E_1) = -\int_{-\infty}^{+\infty} dw g(w) \frac{1}{1 + w^2 + E_s^2} \tag{B27b}$$

Now let us focus on the third-order relations. From Eqs (B18a) and (B18b), we extract the third-order polarization components

$$p_{r3} = \frac{1}{1 + 9\Delta^2} \left\{ w \left(p_{i3} - 3\Delta p_{i4} \right) + \frac{E_1}{2} (d_3 - 3\Delta d_4) \right\} \tag{B28a}$$

which transforms into

$$p_{r3} = \delta_3 \left\{ -\delta_3 w^2 (1 - 9\Delta^2) p_{r3} + \delta_3 6 w^2 \Delta p_{r4} + \Gamma_4 \left\{ \left(\frac{6\Delta^2}{\wp} - 1 \right) M_1 + \left(\frac{2\Delta}{\wp} + 3\Delta \right) M_2 \right\} d_{0d} \frac{E_1^3}{4} \right\} \tag{B28b}$$

and

$$p_{r4} = \frac{1}{1 + 9\Delta^2} \left\{ w \left(p_{i4} + 3\Delta p_{i3} \right) + \frac{E_1}{2} (d_4 + 3\Delta d_3) \right\} \tag{B29a}$$

which takes the form

$$p_{r4} = \delta_3 \left\{ - \delta_3 6w^2 \Delta p_{r3} - \delta_3 w^2 (1 - 9\Delta^2) p_{r4} - \Gamma_4 \left\{ \left(\frac{2\Delta}{\wp} + 3\Delta \right) M_1 - \left(\frac{6\Delta^2}{\wp} - 1 \right) M_2 \right\} d_{0d} \frac{E_1^3}{4} \right\}$$

$$\text{(B29b)}$$

For easier algebraic manipulations, we set

$$A_1 = \delta_3^2 w^2 \left(1 - 9\Delta^2 \right) \tag{B30a}$$

$$B_1 = \delta_3^2 6w^2 \Delta \tag{B30b}$$

$$C_1 = \delta_3 \Gamma_4 \left(\frac{6\Delta^2}{\wp} - 1 \right) \tag{B30c}$$

$$D_1 = \delta_3 \Gamma_4 \left(\frac{2\Delta}{\wp} + 3\Delta \right) \tag{B30d}$$

With $$\delta_3 = \frac{1}{1 + (3\Delta)^2} \tag{B30e}$$

and $$\Gamma_4 = \frac{\wp^2}{\wp^2 + 4\Delta^2} \tag{B30f}$$

So that Eqs. (B28b) and (B29b) transform into two linear relations with respect to p_{r3} and p_{r4}

$$(1 + A_1) p_{r3} - B_1 p_{r4} = \left[C_1 M_1 + D_1 M_2 \right] d_{0d} \frac{E_1^3}{4} \tag{B31a}$$

$$B_1 p_{r3} + (1 + A_1) p_{r4} = \left[- D_1 M_1 + C_1 M_2 \right] d_{0d} \frac{E_1^3}{4} \tag{B31b}$$

Whose solutions write

$$p_{r3} = \frac{(1 + A) \left[CM_1 + DM_2 \right] d_{0d} + B \left[- DM_1 + CM_2 \right] d_{0d}}{(1 + A)^2 + B^2} \frac{E_1^3}{4} \tag{B32a}$$

$$p_{r4} = - \frac{(1 + A_1) \left[- D_1 M_1 + C_1 M_2 \right] d_{0d} - B_1 \left[C_1 M_1 + D_1 M_2 \right] d_{0d}}{(1 + A_1)^2 + B_1^2} \frac{E_1^3}{4} \tag{B32b}$$

So that according to the field equation Eq. (B23), we obtain the following relations

$$\frac{3\Delta}{\kappa} E_3 = 2C \int_{-\infty}^{+\infty} dw g(w) p_{r4} \tag{B33a}$$

$$- E_3 = 2C \int_{-\infty}^{+\infty} dw g(w) p_{r3} \tag{B33b}$$

That must also be satisfied simultaneously

These relations clearly indicate, as expected from the non-linear nature of the field-matter equations, that the third order field component is related to the third power of the first order field-amplitude component.

The ratio of Eqs. (B32a) and (B32b)

$$\frac{p_{r4}}{p_{r3}} = - \frac{(1+A_1)\left[D_1 - C_1 \dfrac{M_2}{M_1}\right] + B_1\left[C_1 + D_1 \dfrac{M_2}{M_1}\right]}{(1+A_1)\left[C_1 + D_1 \dfrac{M_2}{M_1}\right] + B_1\left[-D_1 + C_1 \dfrac{M_2}{M_1}\right]} \tag{B34}$$

shows no dependence on electric field.

For simplifying purpose, we further set

$$\frac{M_2}{M_1} = \Delta \frac{(1+\Delta^2 - w^2)}{(1+\Delta^2 + w^2)} \quad \frac{M_2}{M_1} = \Delta Q(w,\Delta) \tag{B35}$$

So that the ratio of Eqs. (B33a) and (B33b), writes

$$\frac{3\Delta}{\kappa} = \frac{\int_{-\infty}^{+\infty} dw g(w) F(w,\Delta)}{\int_{-\infty}^{+\infty} dw g(w) G(w,\Delta)} \tag{B36a}$$

With

$$F(w,\Delta) = \delta_3 \Gamma_4 \Delta \left[\left(1 + \delta_3^2 w^2 (1 - 9\Delta^2)\right)\left(\frac{2}{\wp} + 3 - \left(\frac{6\Delta^2}{\wp} - 1\right)Q\right) + 6\delta_3^2 w^2 \left(\frac{6\Delta^2}{\wp} - 1 + \left(\frac{2\Delta}{\wp} + 3\Delta\right)\Delta Q\right) \right] \tag{B36b}$$

$$G(w,\Delta) = \delta_3 \Gamma_4 \left[\left(1 + \delta_3^2 w^2 (1 - 9\Delta^2)\right)\left(\frac{6\Delta^2}{\wp} - 1 + \left(\frac{2\Delta}{\wp} + 3\Delta\right)\Delta Q\right) + 6\delta_3^2 w^2 \Delta^2 \left(-\left(\frac{2}{\wp} + 3\right) + \left(\frac{6\Delta^2}{\wp} - 1\right)Q\right) \right] \tag{B36c}$$

Eliminating identical terms, that contain no w dependence, from the ratio $\dfrac{G(w,\Delta)}{G(w,\Delta)}$, we obtain:

$$F(w,\Delta) = \Delta\left[\left(1+\delta_3^2 w^2(1-9\Delta^2)\left(\frac{2}{\wp}+3-\left(\frac{6\Delta^2}{\wp}-1\right)Q\right)+6\delta_3^2 w^2\left(\frac{6\Delta^2}{\wp}-1+\left(\frac{2}{\wp}+3\right)\Delta^2 Q\right)\right]$$

(B36d)

$$G(w,\Delta) = \left[\left(1+\delta_3^2 w^2(1-9\Delta^2)\left(\frac{6\Delta^2}{\wp}-1+\left(\frac{2}{\wp}+3\right)\Delta^2 Q\right)+6\delta_3^2 w^2\Delta^2\left(-\left(\frac{2}{\wp}+3\right)+\left(\frac{6\Delta^2}{\wp}-1\right)Q\right)\right]$$

(B36e)

Let us give a brief outline and comment on the physical implications of the results of this appendix:

The third-order relation, Eq. (B36a) shows no dependence on excitation level. For a given inhomgeneous gain profile $g(w)$, it relates exclusively to the material and cavity relaxation rates $\wp$ and κ. The frequency value evaluated from Eq. (36a) is considered as giving the system's eigenpulsation or resonant frequency, typical of any oscillatory system.

In addition, the response to an external excitation mechanism, through the excitation parameter E_S, is rooted in the first order relation, Eq. (B25), which may be reorganised to include its dependence on pumping level, as

$$\frac{\Delta}{\kappa} = -\frac{\int_{-\infty}^{+\infty} dw g(w) F_\wp(w,\,\Delta,\,E_S)}{\int_{-\infty}^{+\infty} dw g(w) G_\wp(w,\,\Delta,\,E_S)}$$

(B37)

For a given set of control parameters $\wp$, κ and E_S, this equation can be solved graphically (as was done in Fig. 8a) to give the operating permanent frequency. For a typical excitation level, the value obtained with Eq. (B37) matches the third-order value extracted from Eq. (B36a). In that case, the system is put into resonance.

Furthermore, once the operating frequency and the first order field component are evaluated, the third-order field component straightforwardly follows either from Eq. (B33a) or (B33b).

These elements are more easily understood with the simpler case of the Lorenz-Haken system presented in the next chapter. We strongly advice the reader to first focus on the following chapter, before giving back some more attention to these crucial points. Indeed, the inhomogeneous and the homogenous case bear strong similarities. For example, as one may easily verify, the inhomogeneous eigenfrequency, Eq. (B36a) transforms into a closed form expression that gives the homogeneous aigenpulsation. Setting $w=0$, yields $Q=1$, and Eq. (B36a) straightforwardly transforms into

$$\frac{3\Delta}{\kappa} = -\frac{2\Delta(1+2\wp-3\Delta^2)}{\wp(1-3\Delta^2)-8\Delta^2}$$

(B38)

yielding the eigen-pulsation of the Lorenz-Haken model (Eq. (14b) of the next chapter). The above relation constitutes a compelling test of validity for Eq. (B36a).

Appendix C.
Delimiting the Control Parameter Space for Regular Field Oscillations around Zero-Mean Values

The upper bound limit is imposed by the relative evolution of the in-phase and out-of phase population components in the small side-band approach

$$d_1 = \alpha \left\{ \frac{\Delta}{\wp} E_0 p_{r2} - E_0 p_{r1} - E_1 p_{r0} \right\} \tag{C1a}$$

$$d_2 = -\alpha \left\{ \frac{\Delta}{\wp} E_0 p_{r1} + E_0 p_{r2} + \frac{\Delta}{\wp} E_1 p_{r0} \right\} \tag{C1b}$$

With

$$p_{r0} = d_0 E_0 \tag{C2a}$$

$$p_{r1} = \frac{A_1 B_1 + A_2 B_2}{A_1^2 + A_2^2} E_1 \tag{C2b}$$

and

$$p_{r2} = \frac{A_1 B_2 - A_2 B_1}{A_1^2 + A_2^2} E_1 \tag{C2c}$$

where

$$A_1 = (1 + \Delta^2)(1 + \alpha E_0^2) \tag{C3a}$$

$$A_2 = \Delta(1 + \Delta^2) \left(1 - \frac{\alpha}{\wp} E_0^2 \right) \tag{C3b}$$

$$B_1 = d_0 (1 + \Delta^2)(1 - \alpha E_0^2) \tag{C3c}$$

$$B_2 = d_0 \frac{\Delta}{\wp} (1 + \Delta^2) \alpha E_0^2 \tag{C3d}$$

So that

$$A_1 B_1 + A_2 B_2 = d_0 (1 + \Delta^2)^2 \left(1 - \alpha^2 E_0^4 + \frac{\Delta^2}{\wp} \alpha E_0^2 - \frac{\Delta^2}{\wp^2} \alpha^2 E_0^4 \right) \tag{C4a}$$

$$A_1 B_2 - A_2 B_1 = d_0 \Delta \left(1+\Delta^2\right)^2 \left(\frac{\alpha}{\wp} E_0^2 + \frac{\alpha^2}{\wp} E_0^4 - 1 + \alpha E_0^2 + \frac{\alpha}{\wp} E_0^2 - \frac{\alpha^2}{\wp} E_0^4 \right) = d_0 \Delta \left(1+\Delta^2\right)^2 \left(\frac{2\alpha}{\wp} E_0^2 - 1 + \alpha E_0^2 \right)$$

(C4b)

$$A_1^2 + A_2^2 = \left(1+\Delta^2\right)^2 \left(1 + 2\alpha E_0^2 + \alpha^2 E_0^4 + \Delta^2 - \frac{2\alpha}{\wp} \Delta^2 E_0^2 + \frac{\alpha^2}{\wp^2} \Delta^2 E_0^4 \right) \qquad \text{(C4c)}$$

With (C2a, (C2b, and (C2c), Eq (C1a) transforms into

$$d_1 = \alpha d_0 E_0 E_1 \left\{ \frac{\dfrac{\Delta^2}{\wp^2} 2\alpha E_0^2 - \dfrac{\Delta^2}{\wp} + \dfrac{\Delta^2}{\wp} \alpha E_0^2 - 1 + \alpha^2 E_0^4 - \dfrac{\Delta^2}{\wp} \alpha E_0^2 + \dfrac{\Delta^2}{\wp^2} \alpha^2 E_0^4}{1 + 2\alpha E_0^2 + \alpha^2 E_0^4 + \Delta^2 - \dfrac{2\alpha}{\wp} \Delta^2 E_0^2 + \dfrac{\alpha^2}{\wp^2} \Delta^2 E_0^4} - 1 \right\}$$

(C5a)

$$d_1 = \frac{\alpha d_0 E_0 E_1}{\left(1+\alpha E_0^2\right)^2 + \Delta^2 \left(1 - \dfrac{\alpha}{\wp} E_0^2\right)} \left\{ \frac{\Delta^2}{\wp^2} 2\alpha E_0^2 - \frac{\Delta^2}{\wp} - 1 - \left(1 + 2\alpha E_0^2 + \Delta^2 - \frac{2\alpha}{\wp} \Delta^2 E_0^2\right) \right\}$$

(C5b)

$$d_1 = \frac{\alpha d_0 E_0 E_1}{\left(1+\alpha E_0^2\right)^2 + \Delta^2 \left(1 - \dfrac{\alpha}{\wp} E_0^2\right)} \left\{ \left(\frac{\Delta^2}{\wp^2} 2\alpha + \frac{2\alpha}{\wp} \Delta^2 - 2\alpha \right) E_0^2 - \frac{\Delta^2}{\wp} - 2 - \Delta^2 \right\} \qquad \text{(C5c)}$$

and Eq. (C1b) becomes

$$d_2 = -\alpha d_0 E_0 E_1 \left\{ \frac{\dfrac{\Delta}{\wp} 2\alpha E_0^2 - \Delta + \Delta \alpha E_0^2 + \dfrac{\Delta}{\wp} - \dfrac{\Delta}{\wp} \alpha^2 E_0^4 + \dfrac{\Delta^3}{\wp^2} \alpha E_0^2 - \dfrac{\Delta^3}{\wp^3} \alpha^2 E_0^4}{1 + 2\alpha E_0^2 + \alpha^2 E_0^4 + \Delta^2 - \dfrac{2\alpha}{\wp} \Delta^2 E_0^2 + \dfrac{\alpha^2}{\wp^2} \Delta^2 E_0^4} + \frac{\Delta}{\wp} \right\}$$

(C6a)

$$d_2 = -\frac{\alpha d_0 E_0 E_1}{[\ ]} \left\{ \frac{\Delta}{\wp} 2\alpha E_0^2 - \Delta + \Delta \alpha E_0^2 + \frac{\Delta}{\wp} + \frac{\Delta^3}{\wp^2} \alpha E_0^2 + \frac{\Delta}{\wp} \left(1 + 2\alpha E_0^2 + \Delta^2 - \frac{2\alpha}{\wp} \Delta^2 E_0^2\right) \right\}$$

(C6b)

$$d_2 = -\frac{\alpha d_0 E_0 E_1 \Delta}{\left(1+\alpha E_0^2\right)^2 + \Delta^2\left(1-\frac{\alpha}{\wp}E_0^2\right)}\left\{\frac{4\alpha}{\wp}E_0^2 -1 + \alpha E_0^2 + \frac{2}{\wp} - \frac{\Delta^2}{\wp^2}\alpha E_0^2 + \frac{\Delta^2}{\wp}\right\} \quad \text{(C6c)}$$

$$d_2 = -\frac{\alpha d_0 E_0 E_1 \Delta}{\left(1+\alpha E_0^2\right)^2 + \Delta^2\left(1-\frac{\alpha}{\wp}E_0^2\right)}\left\{\left(\frac{4\alpha}{\wp} + \alpha - \frac{\Delta^2}{\wp^2}\alpha\right)E_0^2 -1 + \frac{2}{\wp} + \frac{\Delta^2}{\wp}\right\} \quad \text{(C6d)}$$

Threshold for Oscillations (Instability Threshold)

When we represent the in-phase population component with respect to Δ (for a fixed value of E_s), the obtained curve shows a maximum value. This maximum value corresponds to the onset of unstable operation. In order to obtain the position of this maximum, let us derive d_1 with respect to Δ

$$\frac{d}{d\Delta}d_1 \propto 2\wp^2\left(\frac{2\Delta}{\wp^2}+\frac{2\Delta}{\wp}\right)E_0^2 - 2\Delta\left(\frac{\Delta^2}{\wp}+2+\Delta^2\right)-\left(\wp^2+\Delta^2\right)\left(\frac{2\Delta}{\wp}+2\Delta\right) \quad \text{(C7)}$$

the solution of

$$\frac{d}{d\Delta}d_1 = 0 \quad \text{(C8)}$$

is obtained as

$$3(1+\wp)\Delta^2 - 4\wp(1+\wp)E_0^2 + 2\wp(1+\wp+\wp^2) = 0 \quad \text{(C9)}$$

showing that non-zero values of Δ (oscillations) exist if the following relation is satisfied

$$E_0^2 > \frac{1}{2}\left(1+\frac{\wp^2}{1+\wp}\right) \quad \text{(C10)}$$

This gives the onset of oscillations. First it shows no dependence on the cavity decay rate. This result demonstrates that, indeed, population decay rate plays the key role in the origin of unstable behavior in inhomogeneously broadened systems. Indeed, as the numerical solutions show, the cavity decay rate plays no important role in the features of the solutions, apart from a few details in pulse shaping during the rise and fall of the field variable.

References

[1]   W. E. Lamb Jr., *Phys. Rev. A* **134**, 1429 (1964).

[2]   U. S. Idiatulin and A. V. Uspenskii, *Radiotech. Electron.* **18**, 422 (1973).

[3]   L. W. Casperson and A. Yariv, *IEEE J. Quantum Electron.* **QE-8**, 69 (1972).

[4]   L. W. Casperson, *IEEE J. Quant. Electron.* **QE-14**, 756 (1978).

[5]   *Phys. Rev. A* **21**, 911 (1980).

[6]   L. W. Casperson, *Phys. Rev. A* 23, 248 (1981).

[7]   L. W. Casperson, *J. Opt. Soc. Am. B* **2**, 62 (1985).

[8]   L. W. Casperson, *J. Opt. Soc. Am. B* **2**, 73 (1985).

[9]   S. T. Hendow and M. Sargent III, *Optics Comm.* **43**, 59 (1982).

[10] S. T. Hendow and M. Sargent III, *J. Opt Soc. Am. B* **2**, 84 (1985).

[11] S. T. Hendow and M. Sargent III, *Optics Comm.*, **40** No 5, 385 (1985).

[12] M. F. Tarroja and al., *Phys. Rev. A* **34**, 3148 (1986).

[13] L. Luigiato, *Optics Comm.*, **46** No 2, 115 (1983)

[14] R. Graham and Y. Cho, *Optics Comm.* **47**, 52 (1983).

[15] N. B. Abraham and al., *J. Opt. Soc. Am. B* **2**, 35 (1985).

[16] N. B. Abraham and al. *J. Opt. Soc. Am. B* **2**, 56 (1985)

[17] M. L. Shih, P. W. Milonni and J. R. Ackerhalt, *J. Opt. Soc. Am. B* **2**, 130 (1985).

[18] Ji-yue Zhang, H. Haken and H. Ohno, *J. Opt. Soc. Am. B* **2**, 141 (1985).

[19] M. Sargent III, M. O. Scully, and W. E. Lamb, Jr., *Laser Physics* (Addison-Wesley, Reading, Mass., 1974) Chap. XX.

[20] H. Haken, *Light*, Vol. 2, *Laser Light Dynamics*, (North Holland, Amsterdam, 1985) Chap. 8.

[21] C. M. Bowden and G. P. Agrawal, *Optics Comm.* **100**, 147 (1993).

[22] Jun Yao, G. P. Agrawal, P. Gallion, and C. M. Bowden, *Optics Comm.* **119**, 246 (1995).

[23] C. M. Bowden and G. P. Agrawal, *Phys. Rev. A*, Vol. 51, No 5, 4132 (1995).

[24] B. Meziane, *Construction et Etude de Modèles Simples en Dynamique des Lasers: Réduction des Systèmes de Dimensions Infinies,* PhD dissertation, Université de Rennes I (1992).

[25] B. Meziane, *Optics Comm.* **75**, 287 (1990).

[26] B. Meziane and H. Ladjouze, *Phys. Rev. A*, Vol. 45, No 5, 3150 (1992).

[27] B. Meziane, *Phys. Rev. A,* vol.47, N° 5, 4430(1993).

[28] B. Meziane, *Phys. Rev. A*, Vol. 48, No 3, 2346 (1993).

[29] B. Meziane, *Optics Comm.* **128**, 377 (1996).

[30] B. Meziane, *Opt. and Quant. Electron.*, **30**, 99 (1998).

[31] S. Banerjee and al., *Physics Letters A* **291**, 103 (2001).

[32] M. Wolfrum and al., *Optics Comm.* **212**, 127 (2002).

[33] T . Sylvestre and al. *Optics Letters* Vol. 27, No. 7, 482 (2002).

[34] P. Chenkosol and L. W. Caspesron, JOSA B, Vol. 20, N°12, 2539-2547 (2003).

[35] G.J de Valcarcel, E. Roldan, and F. Prati, *Optics Comm.* **216**, 203 (2003).

[36] M.Schetzen and R. Yildirim, *Optics Comm.* **219**, 341 (2003).

[37] J. Jayabalan and al., *Optics and Laser Technology* **35**, 613 (2003).

[38] E. Cabrera and and al., *Phys. Rev. A* **70**, 063808 (2004).

[39] B. Meziane and S. Ayadi, *Opt. And Quant. Electron.* http://dx.doi.org/10.1007/s11082-007-9068-6 (2007).

[40] S. Ayadi and B. Meziane, *Opt. And Quant. Electron.* http://dx.doi.org/10.1007/s11082-007-9065-9 (2007).

[41] B. Meziane, *J. Phys. B: At. Mol. Opt. Phys.* **40**, pp3343-3355 (2007).

[42] B. Meziane, *IEEE J. Quant. Electron.* Vol. 43, n°11, pp1048-1054 (2007).

In: Atomic, Molecular and Optical Physics…
Editor: L.T. Chen, pp. 61-107

ISBN: 978-1-60456-907-0
© 2009 Nova Science Publishers, Inc.

Chapter 2

PULSE STRUCTURING IN LASER-LIGHT DYNAMICS: FROM WEAK TO STRONG SIDEBAND ANALYSES; II. THE LORENZ-HAKEN MODEL

Belkacem Meziane[*]

Université d'Artois, UCCS Artois, UMR CNRS 8181, Rue Jean Souvraz, SP18, 62307, Lens Cedex, France

Abstract

The present chapter extends the strong harmonic expansion method of the preceding report to the much simpler Lorenz-Haken model that describes self-pulsing in single-mode homogeneously broadened lasers. The study focuses on a typical pulse-structuring hierarchy of periodic solutions, which include a period-doubling cascade that builds up with increasing excitation levels. Despite the uncomplicated formalism they display, the Lorenz-Haken equations are shown to carry the same qualitative features as those of the much more complex integro-differential system. Remarkable analogies are found between a few typical solutions of both systems as direct one-to-one comparisons reveal. The obvious advantage of a simpler set of equations lies in the fact that it renders the analytical and numerical handling much easier to carry out. As a consequence, new and ampler information with respect to pulse structuring are more easily extracted. In particular, the iterative harmonic-expansion algorithm is carried out, up to the fifth-order in field amplitude, giving persuasive and explicit credit to the strong-sideband ability to provide an accurate description of pulse structuring in laser-light dynamics.

I. Introduction

The standard laser equations describe semi-classical atom-field interactions inside a unidirectional ring cavity. In the case of a single mode homogeneously broadened (SMHB) system, these transform, after adequate approximations, into a simple set of three non-linearly coupled differential equations, the so-called Lorenz-Haken equations. Despite their amazing

---

[*] E-mail address: belkacem.meziane@univ-artois.fr

simplicity, these deliver a remarkably broad range of solutions, ranging from regularly periodic to erratic time traces in their unstable regime of operation. The details of the solutions are usually accessible with the help of numerical integration, only. These details closely depend on the exact values of four control parameters: the decay rates that characterize the three interacting variables (electric field, population inversion and polarization of the medium) and a pumping term, which transforms an initially absorbing material into an amplifying medium. Usually, the medium and the geometry of the cavity fix all three decay rates. In that case, the only variable that may be freely adjusted is the excitation parameter. It constitutes the exclusive degree of freedom of the laser. The system becomes unstable whenever the (fixed) decay rates satisfy the bad cavity condition (i.e., that the field relaxation-rate exceeds the sum of the inversion and polarization decay rates) and the pump parameter is raised beyond a critical value (referred to as the second laser threshold), whereas the first laser threshold refers to the level of excitation required to initiating coherent laser action.

For the past thirty years, the dynamic properties of the Lorenz-Haken equations received considerable attention. References dealing with the subject are so numerous that it seems impossible to give a complete list without omitting quite a few good contributions. Instead of giving an exhaustive list, our reference section provides a list that spreads across the past four decades, including a few books that give the complete state of the art of laser dynamics at each stage of the growth of the subject. The first investigations were initiated with the recognition, by Haken, of the equivalence between the Lorenz model that describes fluid turbulence [1], already known to predict deterministic chaos, and the equations of a single-mode homogeneously broadened laser [2]. The richness of the solutions is so wide that their analysis fills extensive review papers [3, 4] and entire books [5-7]. A detailed description of the instability problem requires extensive numerical handling [8-17], while the analytical approach is often limited to a standard linear stability analysis, from which closed-form expressions that give the position of the instability threshold and the oscillation frequency at the onset of instability are derived. However, the obtained oscillation frequency formula only describes the development of the unstable solution when departing from the stationary-state, at, and only at, the instability threshold. In addition, its value bears no resemblance to the frequency of the long-term solutions, yet it can be used as a modulating parameter in order to provide an external feedback control to stabilize some chaotic behavior of the system [18-22]. These features that are typical to both the homogeneous and the inhomogeneous systems continue to be the focus of numerous contributions [23-32]. However, despite these frequent and current analyses, no method has ever been proposed to handle the properties of the long-term solutions, apart from our recent works [33, 36]. In this chapter, we will apply the iterative harmonic-expansion method of the preceding chapter to the Lorenz-Haken equations to obtain analytical solutions in the control parameter space inside which periodicity holds. Typical examples show an excellent match with their numerical counterparts, as in the more complex model of the preceding chapter.

While the complexity of the integro-differential "Maxwell-Bloch" equations limited our calculations to third-order in field amplitude, the much simpler SMHB equations will allow carrying the calculations, to obtain higher-order components for all three interacting variables. Despite the very lengthy and quite often cumbersome algebra involved in the calculations, we extend the results up to the fifth order terms. In addition, we will take advantage of the indications given by the corresponding frequency spectra, obtained with the application of a Fast-Fourier Transform algorithm to the temporal signals, to construct the complete analytical solutions. A direct quantification of the higher-order terms follows from a simple evaluation of each of the components as these appear in the frequency spectrum of each solution. With these elements and strategies in mind, we will focus on a typical hierarchy of periodic solutions and show how the method allows for a straightforward construction of typical hierarchies of periodic pulse structures. These include the single, the double, as well as the fourth periodic orbits. Such a periodic flow constitutes the first steps of the well-known period-doubling cascade and route to chaos.

The presentation is organized as follows. First, we give a brief outline of the Laser-Lorenz equations in section II. In section III, we recall the important steps of the standard linear stability analysis (LSA), which is known to contain the necessary features of the unstable state. Its application to any set of differential equations constitutes a powerful mean to extract both the position of the second laser threshold and the frequency of the transient signal developing from the stationary state, following the application of small perturbations to the steady-state variables. Since LSA only holds at the onset of instability, section IV applies a small harmonic expansion method to obtain a new analytical expression of the transient oscillation-frequencies whose validity extends to the entire control-parameter space. At the onset of instability, this new expression is shown to converge towards the LSA result. A typical transient solution in the vicinity of the instability threshold shows that the three interacting variables bear a sinusoidal evolution with a noticeable time-lag between the field, polarization and population inversion. This is a clear signature of an out-of-phase development, which is taken into account in the small harmonic expansion procedure but not in LSA.

However, while the transient frequency spectrum exhibits the same single and small component for all three variables, validating a small harmonic handling, the long-term solution shows quite distinct features: The field and polarization time traces always orbit around zero-mean values while their frequency spectra exhibit a good number of odd-order harmonics with decreasing amplitudes. On the other hand, the population inversion shows a behavior of its own, since it oscillates, around a mean-value distinct from its steady-state level, with a fundamental frequency at twice the fundamental frequency of the field and polarization variables. Indeed, these elements go along the structure of the non-linear equations, imposing the field spectrum to contain odd harmonics, while imposing the population-inversion spectrum to carry even harmonics only. The recognition of these fundamental properties constitutes the basis of the strong-harmonic expansion method. Section V focuses on a typical hierarchy of numerical pulsing-solutions obtained with increasing excitation levels. Starting from the instability threshold, the solution first consists of symmetric period-one oscillations. These persist over a large range of excitation levels before transforming into an asymmetric solution of period-one, period two, period four, etc, with a sequential appearance, in the corresponding field frequency-spectra, of even and sub-harmonic components. Section VI presents the essential steps of the strong harmonic expansion method used to obtain an expression of the long-term pulsing frequencies, whose validity extends to high excitation levels for periodic states, and to derive analytical expressions for the first and the third harmonic components of the field variable. Section VII is devoted to the analytical construction of the same hierarchy of solutions obtained numerically in section V. Some physical insights will be extracted from the details of these solutions providing complementary guidelines to reach the best possible accuracy between the numerical and the analytical solutions. In addition to the description of regular period-one signals, the strong harmonic expansion method will yield analytical expressions for the first few steps of the famous period-doubling cascade, known to ultimately yield a chaotic state. Section VIII focuses on two examples with extremely different values of the decay rates. These provide a further proof of the reliability of the strong-harmonic expansion method to construct analytical solutions at any repetition rate, inside the control parameter space of periodic solutions. Section IX is devoted to localizing the control-parameter space embedding periodic eigen-solutions. Finally, section X gives some conclusions and perspectives towards future research in the laser dynamics issue that extends beyond the SMHB and the SMIB systems presented in these chapters.

In addition to the bulk of the report, we devote three appendixes to outline the important steps of the lengthy calculations. A closed form expression that gives the transient frequencies as a function of the decay rates and excitation level is derived in Appendix A, while the expression of the long term frequencies is derived in Appendix B. Also in Appendix B, we derive the third-order terms of the harmonic expansions. Finally, Appendix C gives the essential steps that lead to the derivation of the fifth-order closed form expression along with its numerical evaluation for a typical periodic solution.

II. Quick Review of the Lorenz-Haken Equations

In the mean-field, slowly varying envelope approximation, the properties of a single-mode unidirectional ring laser containing a homogeneously broadened medium with two-level atoms are described with a set of three non-linearly coupled differential equations. Assuming exact tuning between the atomic and cavity frequency, these take the following normalized form [2, 12, 16, 33, 35, 36]:

$$\frac{d}{dt}E(t) = -\kappa\{E(t) + 2CP(t)\} \tag{1a}$$

$$\frac{d}{dt}P(t) = -P(t) + E(t)D(t) \tag{1b}$$

$$\frac{d}{dt}D(t) = -\wp\{D(t) + 1 + E(t)P(t)\} \tag{1c}$$

where $E(t)$, $P(t)$, $D(t)$ represent, respectively, the laser field-amplitude, the polarization, and the population inversion of the amplifying medium, κ and $\wp$ are, respectively, the cavity decay-rate and the population relaxation rate, both scaled to the polarization relaxation-rate, and $2C$ is an excitation parameter that quantifies the external pumping mechanism with respect to its level at lasing threshold. No attempt is made, here, to recall the full properties of these equations, since these were the subject of numerous full and consistent papers as well books and book chapters [3-5, 15, 22, 32]. We just mention that the transition point from stable to unstable operation is a Hopf bifurcation of the sub-critical type. This means that the departure from the stable state is characterized by a transient period of time during which the signal evolves with small oscillations whose amplitude gradually grows before reaching the final long-term solution. The transient and the long-term signals evolve with quite distinct features, as shown in the typical example represented in Fig. 1, obtained just beyond the instability threshold, with the following parameter values: 2C=10, κ=3, and $\wp$=0.1. In many examples, given in the following sections, we shall try to focus our analysis on these fixed values of κ and $\wp$, the excitation level 2C remaining the only control parameter that may be varied.

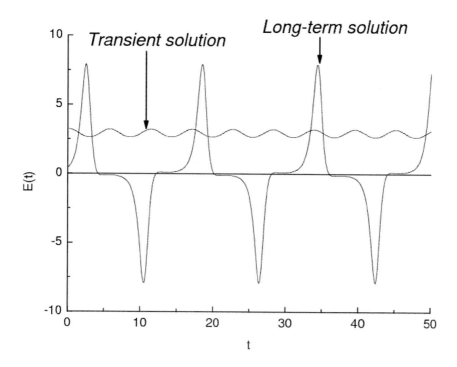

Figure 1. Comparison between the transient and the long term development of the electric field amplitude, solutions of Eqs (1), obtained at the boundary of the instability domain for κ = 3, γ = 0.1 and 2C = 10. Note that while the transient signal oscillates around steady-state according to a small-harmonic expansion in the form given in Eq. (7a), the long term time trace evolves around the temporal axis, with a zero-mean value, following the strong-harmonic expansion of Eq. (11a).

In the following section, we recall typical results obtained when the standard linear stability analysis is applied to Eqs (1) in order to characterize the transient oscillations that precede the long-term solution of Fig. 1, at the instability threshold.

III. Standard Linear Stability Analysis

The well-known linear stability analysis (LSA) is rooted in the application of small perturbations to Eqs (1) in order to delimit the boundary region inside which the system possesses unstable solutions with respect to its control parameters κ, $\wp$, and 2C. The method that can be found in any laser textbook that deals with laser theory and instabilities (see Refs. 15, 16 and 33, for example), consists in superimposing small deviations to the stationary solutions, predetermined when the time derivatives appearing in Eqs. 1 are set equal to zero. Each of the three interacting variables $E(t)$, $P(t)$, and $D(t)$ is replaced with

$$X(t) = X_o + \delta x(t) \tag{2}$$

in Eqs (1), to obtain new equations for the small perturbations $\delta x(t)$. The stability problem is then investigated by assuming solutions of the form

$$\delta x(t) = \delta x_o \exp(\lambda t) \tag{3}$$

and seeking the conditions under which a Hopf bifurcation occurs, that is when $\lambda = i\omega$.
    Such an analysis yields a first expression for the second laser threshold

$$2C_{2th} = 1 + \frac{(\kappa+1)(\kappa+1+\wp)}{(\kappa-1-\wp)} \tag{4}$$

and a second expression for the pulsation frequency of the perturbed signal, developing from its stationary state

$$\Delta_t = \sqrt{\frac{2\kappa\wp(\kappa+1)}{\kappa-1-\wp}} \tag{5}$$

    Inferring the values $\kappa = 3$ and $\wp = 0.1$, into Eq. (5) yields $\Delta_t = 1.1$. The value of $\Delta_t/2\pi = 0.175$ is close to the frequency of the transient oscillation of Fig. 1, which is obtained near the instability threshold. One may rapidly check from Eq.(4) that $2C_{2th} = 9.63$, while $2C = 10$, in Fig. 1.
    The bad cavity condition characterizing unstable operation follows from both relations, and reads

$$\kappa > 1 + \wp \tag{6}$$

    Relation (5) only holds at the instability threshold, in the vicinity of the corresponding steady-state solution. Indeed, the closed form expression shows no dependence on excitation level. As a consequence, linear stability analysis provides information that are connected exclusively to the initial oscillations of the transient solution as it leaves the stable steady-state. In order to obtain the transient frequency that characterizes the perturbed stationary solutions of Eqs. (1), at any level of excitation, we apply a weak-harmonic expansion method based on the small side-band approach, initiated by Casperson [7, 8], also readjusted and largely studied by Hendow and Sargent [9, 10]. However, we insist on the fact that neither the original method of Casperson nor the readjusted one of Hendow and Sargent have ever been extended to extract any information beyond the onset of instability. In this context, the weak side band approach has always been considered as a different approach to analyzing the onset of instability in both the SMHB and the SMIB systems, yielding the same analytical conditions as LSA with respect to the occurrence of unstable solutions. The approach presented in the following section is an extension of the Hendow-Sargent small side-band analysis, adapted to the competing nature of the interacting variables inside a laser medium, which yields new analytical information of the unstable state inside a much broader range of the control parameter space.

IV. Weak-Sideband Analysis: Generic Expression for the Transient Angular Frequency

Our starting point proceeds from the small-side band method of Hendow and Sargent [9, 10]. The main difference with their approach consists in the introduction of out-of phase components to the medium's variables, with respect to the small-amplitude harmonic oscillations of the field variable. Indeed, it will transpire from the following analysis that the transient signals of the interacting variables are not phase-locked but evolve with a permanent time-lag, giving clear evidence of the competing nature of light-matter interactions inside the laser medium when self-pulsing sets in. As a consequence, adapted to the transient part of the unstable solutions are the following weak-harmonic expansions

$$E(t) = E_0 + e\cos(\Delta t) \tag{7a}$$

$$P(t) = P_0 + p_1 \cos(\Delta t) + p_2 \sin(\Delta t) \tag{7b}$$

$$D(t) = D_0 + d_1 \cos(\Delta t) + d_2 \sin(\Delta t) \tag{7c}$$

where E_0, P_0, D_0 represent the stationary solution obtained when the derivatives in Eqs (1) are set equal to zero. From the derived expressions, we extract the dependence of each variable on excitation parameter according to

$$E_0 = \sqrt{2C-1}, \tag{8a}$$

$$P_0 = -\frac{\sqrt{2C-1}}{2C}, \tag{8b}$$

$$D_0 = -\frac{1}{2C} \tag{8c}$$

In Eqs (7) e is the small perturbation of the field-amplitude, p_1 and p_2 are, respectively, the in-phase and the out-of phase components of the polarization perturbation, d_1, and d_2, the in-phase and the out-of phase components of the population perturbation, with respect to the field temporal evolution, taken as a reference.

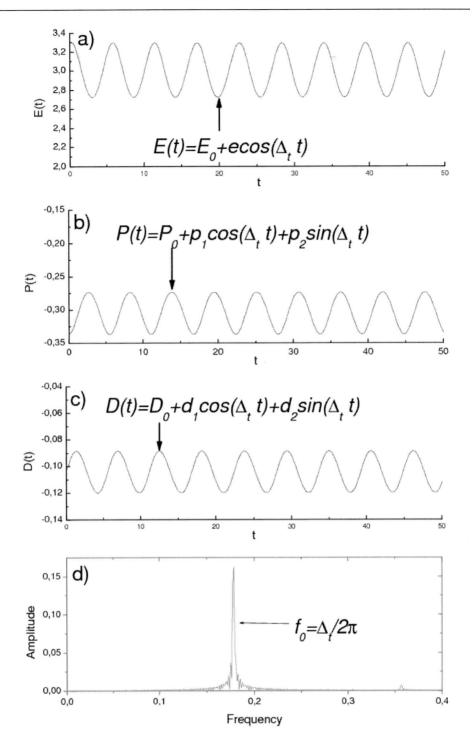

Figure 2. Typical time dependence of the interacting variables in the transient regime, a) electric field amplitude, b) polarization, c) population inversion (note the time lag between the three signals), and d) corresponding frequency spectrum obtained with a Fast Fourier Transform algorithm of the electric-field signal. A FFT applied to the traces of Figs b) and c) exactly matches the spectrum of Figure (d), an indication of the presence of a unique frequency during the transient stage of light-matter interactions.

When Eqs. (7) are inserted into Eqs. (1), we obtain (after lengthy, but straightforward calculations) an expression for the transient pulsation frequency as (see Appendix A for the essential steps of the derivation)

$$\Delta_t = \sqrt{\frac{(2C-1)\wp^2 + (2C-1)\kappa\wp(2+\wp) - (\kappa+1)\wp^2}{(2C-1)\wp + \kappa + 1}} \tag{9}$$

This relation gives the pulsation frequency of the transient signal that characterizes the temporal evolution of the perturbed stationary state towards its long-term solution (either stable or unstable). Under stable operating conditions, Eq. (9) represents the value of the well-known relaxation oscillations, exponentially decaying towards steady-state when the system undergoes a hard excitation mechanism. The transient solution consists of an almost pure sinusoidal function in the vicinity of the second laser-threshold were the variables experience critical slowing-down (that means that the transient part of the unstable solution occurs with a slow motion before reaching the final long-term state). An example of such a solution is represented in Fig. 2a, for the laser field amplitude. The corresponding fast Fourier transform demonstrates the presence of a single component in the frequency spectrum. As shown in Fig. 2b and 2c, the polarization and population-difference also evolve with the same frequency as the field-amplitude. This is also demonstrated in Fig 2d, representing the Fast-Fourier transform of all three variables. Note the time lag between the three signals which cannot be ignored for a correct description of light-matter interactions, since it represents a signature of the competing effects between the three interacting variables. Here and in other examples to follow, we focus on typical control parameter values that yield regular and irregular long-term pulsing solutions in the unstable domain of operation. For $\kappa = 3$ and $\gamma = 0.1$, the instability threshold is $2C_{2th} = 9.63$, while, in order to avoid unnecessary computer time consumption that occurs during critical slowing down, the transient solutions of Fig. 2 were obtained just above this second laser-threshold, with $2C = 10$. The oscillation frequency of the signals exactly matches the value given by Eq. (9) i.e. $f_0 = \Delta_t / 2\pi \approx 0.18$. The same close value is also inferred from Eq. (5), since we are close to the instability threshold. Let us note that the solutions represented in Fig 2 were simulated with Eqs (1). Needless to emphasize on the fact that each ot the variable evolves according to a pure sinusoïdal harmonic of the form given in Eqs (7).

Figure 3 (upper trace) represents the evolution of Δ_t with respect to the excitation level $2C$ as given by Eq. (9), with the same fixed values of the decay rates $\kappa = 3$ and $\gamma=0.1$. Along with the curve are a few scattered bold-points. These correspond to the pulsation values evaluated numerically with the simulated traces of Eqs. 1. Note the exact match for both low (stable zone) and high (unstable region) excitation levels, an indication (as already mentioned) of the validity of Eq. (9). Extensive simulations and comparisons reveals that the corespondance between the transient frequencies of numerically simulated signals with those of Eq. (9) extends to the whole control-parameter space ($\kappa, \wp, 2C$), both under stable and unstable conditions.

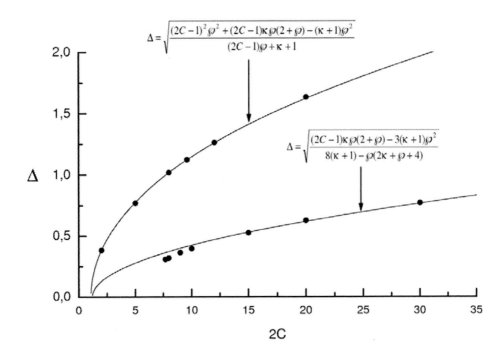

Figure 3. Comparison between the predicted transient (upper trace, solid line) and long-term pulsing frequencies (lower trace, solid line), as respectively given by Eqs. (9) and (10), with those of the calculated values (dots) from the numerical solutions of Eqs. (1). A perfect match between the analytical and the numerical counterparts is the rule at any excitation level for the transient curve, whereas some slowing down effect below the instability threshold deviates the numerical dots from the corresponding analytical curve.

On the other hand, just as in the case of linear stability analysis, Eq. (9) gives only the transient frequencies of Eqs. (1), and cannot be extended to describe the long-term solutions. This is easily understandable, in view of the fact that expansions (7) assume that the system orbits around steady-sate, while, as will be shown in the following, the long-term solutions show an orbiting around zero-mean values for the field and polarization variables, and around a *dc* component, distinct from the steady-state value, for the population inversion. In fact, just as in the case of the SMIB system, a kind of fluctuation-dissipation effect, characterizes the unstable solutions of Eqs. (1) forbidding any orbiting around steady-state.

Figure 4 represents a typical long-term solution, for the field (Fig. 4a), polarization (Fig. 4b) and population inversion (Fig 4c), obtained with 2C=10. The corresponding stationary-state obtained from Eqs. (8) reads: $E_0 = 3$, $P_0 = -0.3$, $D_0 = -0.1$, while the represented long-term signals satisfy

$$< E(t) >= 0 ,$$
(10a)

$$< P(t) >= 0 ,$$
(10b)

$$< D(t) >= d_{dc} \neq D_0$$
(10c)

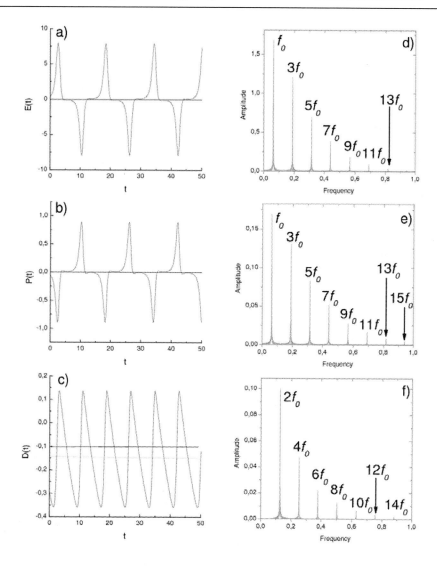

Figure 4. Long-term time-traces of the interacting variables along with their corresponding frequency spectra, a),d) electric-field amplitude, b), e) polarization, c), f) population inversion. Note that while both the field and polarization oscillate around zero mean-values, the population inversion oscillates around a *dc* value (dotted line in Figure c), different from its steady-state level (solid line). We also notice that both the field and polarization spectra exhibit odd frequency components while the population inversion contains even components only, in accordance to the non-linear nature of Eqs (1) and to the developments (11a)-(11c).

In addition, contrary to the frequency spectra of the transient solutions, the Fast-Fourier transform of the long-term signals exhibit strong harmonic components of odd-order for the field (Fig. 4d) and polarization (Fig. 4e), and of even order for the population inversion (Fig. 4f). Thus, a redistribution of the variable amplitudes between higher order terms occurs during unstable operation. Note that the values of the amplitude of each component may directly be determined from the frequency spectra. Indeed, the tedious algebra involved in the determination of high-order harmonics will require the use of the fast-Fourier spectrum to evaluate the corresponding amplitudes that appear beyond the fifth order.

The elements presented thus far are indicative of the weaknesses of the weak signal methods to handle the long-term part of the unstable solutions of the Lorenz-Haken equations. The challenge for another approach towards the investigation of the dynamic properties of Eqs (1), which must go beyond the developments (7), emerges undoubtedly. We devote the remaining part of this chapter to give the elements of a strong-harmonic expansion technique that allows for the determination of the long-term operating frequencies and further leads towards the construction of quite satisfactory analytical solutions in the self-pulsing regime of operation. The amplitudes of the first, third and fifth-order terms for the field variable will be determined analytically, while the amplitudes of the higher-order terms will be inferred directly from the corresponding frequency spectra. Before digging into the analytical method, we first focus on a typical example of numerically simulated period-doubling hierarchy and characterize the field intensity through its average value. It is shown that that the average output intensity is always lower than the steady-state intensity E_0^2. This allows for a clear representation of a hysteresis effect that goes along the output versus pump-input characteristic for adiabatic increasing and decreasing scans of the excitation parameter, in the transition region between stable and unstable operation.

V. Typical Hierarchy of Pulsing Solutions

Regular and irregular pulsing solutions are known to characterize Eqs (1) for low values of the population inversion decay rate. However, the boundary regions inside which such solutions likely occur still remains an open question which will be discussed in some detail, in Sec. IX.

The essential features of the pulsing solutions may again be rapidly grabbed with the help of the typical example of Fig. 4. The temporal traces of the field, polarization and population inversion are represented with their corresponding frequency spectra, obtained with a Fast-Fourier Transform algorithm. Note again that the field and polarization undergo a perpetual switching around a zero-mean value while the population inversion oscillates with a *dc* component (dotted line in Fig. 4d) distinct from its value at steady-state (straight line in Fig.4d). The corresponding frequency spectra exhibit sharp odd-order components at Δ, 3Δ, 5Δ etc for the field and polarization, while the population-inversion spectrum exhibits even components at 2Δ, 4Δ, 6Δ etc. This first example shows the essential features of the pulsing solutions that will allow for a construction of the corresponding analytical solutions with an extraordinarily good fit.

We have carried numerical simulations with $\kappa = 3$ and $\wp = 0.1$ and delimited the region of unstable behavior under hard excitation mechanism and adiabatic fine-scans of the pump parameter 2C. For each value of the excitation level, we have calculated the average output intensity $< E(t)^2 >$. The results are represented in Fig. 5. An increase of the excitation parameter from laser threshold towards the instability threshold yields an output versus pump input characteristic following $E_0^2 = 2C - 1$, as indicated in Fig 5 (upper trace). At the instability threshold, the average output intensity abruptly falls to a lower value. When an adiabatic scan of the excitation parameter from the second laser threshold towards lower values is performed, the pulsing solutions persist far below the instability threshold. The corresponding average output intensity, represented in Fig. 5 (lower trace) is a straight line parallel to the steady-state characteristic. Thus, a clear hysteresis effect characterizes the output versus pump characteristic when the excitation parameter is scanned in the vicinity of the instability threshold. This ensues in a double solution (stable and unstable) zone that extends from $2C = 6.11$ to. $2C = 2C_{2th}$. Inside this zone, the solution may switch from one branch to another depending on the external excitation mechanism, however an adiabatic scan of the excitation level exclusively follows the path indicated by the arrows along the curve.

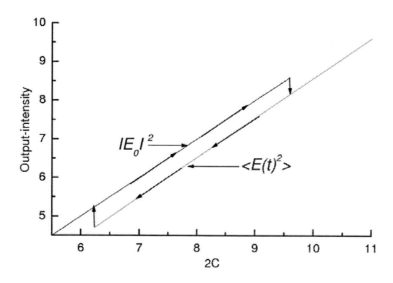

Figure 5. Output-intensity versus excitation characteristic, showing a hysteresis effect (with respect to average output intensity) in the excitation region where stable and unstable operation coexist. When the pump parameter is increased from below the instability threshold, the output intensity increases following the steady-state dependence of Eq. 8a. At the instability threshold, the average intensity undergoes an abrupt decrease. Adiabatic pump scans towards lower values maintain the pulsing solution (with lower average intensity) far below the instability threshold. This results in the depicted hysteresis phenomenon. Indeed, depending on the applied excitation mechanism, the system may be forced to jump from one branch to another, i.e. from stable to unstable operation following a hard excitation mechanism or from unstable to stable operation following an adiabatic switch-off and switch-on.

We will now focus on a period-doubling sequence obtained with increasing excitation level.

Figure 6 represents a typical hierarchy of a period-doubling sequence obtained with increasing excitation parameter. In each case the temporal evolution of the field variable is represented along with its corresponding frequency spectrum. Up to $2C=18.4$, the signal consists of a symmetric period one solution, while the frequency spectrum exclusively exhibits odd components at $\Delta/2\pi$, $3\Delta/2\pi$, $5\Delta/2\pi$ etc (Fig. 6a). Increasing the excitation parameter transforms the symmetric solution into an asymmetric signal with respect to the time axis. An example obtained with $2C=27.7$ is represented in Fig. 6b. In addition to the odd components of Fig (6a), the corresponding frequency spectrum displays even components at $2\Delta/2\pi$, $4\Delta/2\pi$, $6\Delta/2\pi$. This constitutes the first signature of a period-doubling cascade. With a further increase in excitation, at $2C =29.5$, the asymmetric signal shows period-two oscillations, while the frequency spectrum shows additional components at $1.5\Delta/2\pi$, $2.5\Delta/2\pi$, $3.5\Delta/2\pi$, etc.

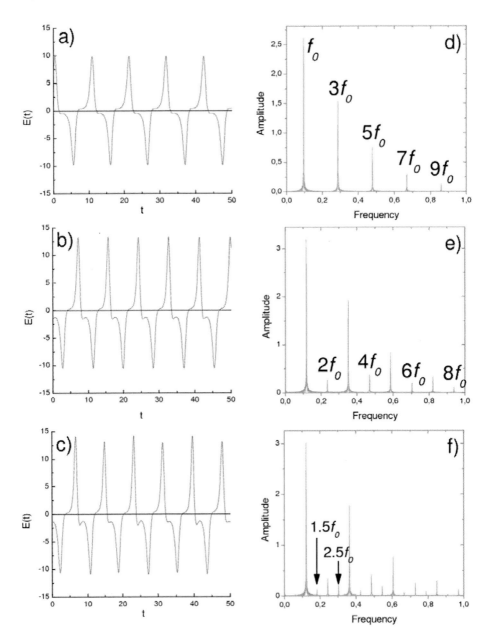

Figure 6. Typical hierarchy of the electric-field pulsing solutions, numerically simulated with increasing excitation parameter, along with their corresponding frequency spectra; a) symmetric period-one solution, obtained with 2C = 18.4, b) asymmetric period-one solution, obtained with 2C = 27.9, c) asymmetric period-two solution, obtained with 2C = 29.5, d) frequency spectrum of the period-one solution, e) frequency spectrum of the asymmetric solution, showing the emergence of small even harmonics, and f) frequency spectrum of the asymmetric period two solution, where, in addition to the even harmonics of e), additional peaks appear at intermediate values between the odd and even components (sub-harmonic components).

Note that, for all the solutions represented in Fig. 6, the field-amplitude undergoes perpetual switching around a zero-mean value. This behavior constitutes the basis of the strong-harmonic expansion method, presented in the following paragraph.

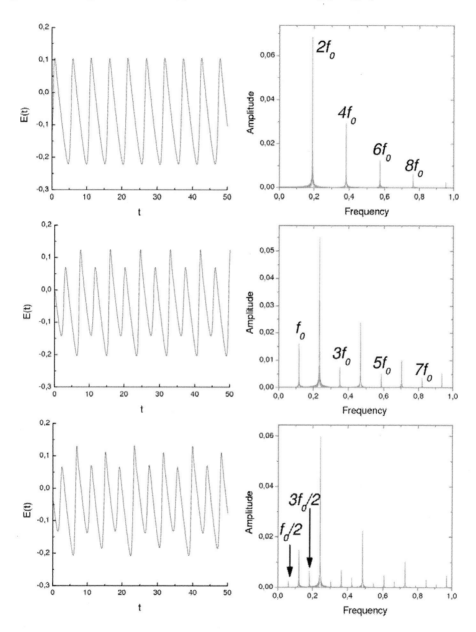

Figure 7. Hierarchy of the population inversion time-traces obtained along with the solutions of Figure 6, a) regular period-one solution at 2C = 18.4, b) period-two oscillations at 2C = 27.9, and c) period-four oscillations at 2C = 29.5. The corresponding frequency spectra respectively show d) regularly spaced even components, e) even and small odd components, and d) additional peaks at intermediate positions.

In order to completely characterize the period-doubling sequence, we have represented, in Fig. 7, the population inversion time traces, along with their corresponding frequency spectra, obtained with the same excitation parameters as those of Fig. 6. Up to 2C=18.4, the population inversion is a regular period one oscillation, with sharp even components in the frequency spectrum at $2\Delta/2\pi$, $4\Delta/2\pi$, $6\Delta/2\pi$, etc. (Fig. 7a). An increase of 2C transforms the period one solution into a period-two signal, an example obtained with 2C=27.9, is represented in Fig. 7b. The corresponding frequency spectrum shows additional odd components at $\Delta/2\pi$, $3\Delta/2\pi$, $5\Delta/2\pi$, etc. For 2C = 29.5, a period-four oscillations characterize the solution (Fig 7c) while the frequency spectrum consists of odd and even components with additional small peaks merging at $0.5\Delta/2\pi$, $1.5\Delta/2\pi$, $2.5\Delta/2\pi$, etc.

We now turn on to analytical considerations. First we will apply an adapted strong-harmonic expansion method to characterize the essential features of the pulsing solutions. We first obtain an analytical expression of the long-term frequencies in Sec. VI, and then construct the analytical counterparts of the above sequence of period-doubling periodic states, in Sec. VII. In fact, the main challenging aspect is the evaluation of the pulsing frequency. Once this is derived, it takes little more effort to find the first few harmonic-amplitudes of the signals.

VI. Strong-Sideband Analysis: Generic Expression for the Permanent Angular Frequency

Adapted to the long-term periodic solution of Fig.1 is the following strong-harmonic expansion

$$E(t) = \sum_n E_{2n+1} \cos\left[(2n+1)\Delta t\right] \tag{11a}$$

n=0,1, 2, etc.

$$P(t) = \sum_n \left\{ P_{ip(2n+1)} \cos\left[(2n+1)\Delta t\right] + P_{op(2n+1)} \sin\left[(2n+1)\Delta t\right] \right\} \tag{11b}$$

$$D(t) = d_{dc} + \sum_n \left\{ d_{ip(2n)} \cos\left[2n\Delta t\right] + d_{op(2n)} \sin\left[2n\Delta t\right] \right\} \tag{11c}$$

n=1,2,3, etc..

where the subscripts ip and op refer, respectively, to the in-phase and out-of phase components. When these are inserted into Eqs (1), one finds a hierarchical set of relations between the variable components. Here, we limit the calculations to third order for the field and polarization and to second order for the population inversion. These are shown to be the essential driving terms that allow for the determination of the operating long-term frequencies. From the obtained set of algebraic relations, the following third-order expressions of the polarization are straightforwardly derived (see Appendix B for the complete derivation)

$$P_{ip3} = \Gamma_d T_{2d} T_{3d} \left(\wp^2 (1 - 3\Delta^2) - 8\wp \Delta^2 \right) \frac{E_1^3}{4} \qquad (12a)$$

$$P_{op3} = \Gamma_d T_{2d} T_{3d} 2\wp \Delta \left(1 + 2\wp - 3\Delta^2 \right) \frac{E_1^3}{4} \qquad (12b)$$

where

$$\Gamma_d = \frac{1}{1 + \Delta^2 + \dfrac{E_1^2}{2}}, \qquad (12c)$$

$$T_{2d} = \frac{1}{\wp^2 + 4\Delta^2}, \qquad (12d)$$

$$T_{3d} = \frac{1}{1 + 9\Delta^2} \qquad (12e)$$

Equation (1a), with (11a) and (11b), yields the following relations

$$-\Delta E_1 = -\kappa 2 C P_2 \qquad (13a)$$

$$E_1 = -2 C P_1 \qquad (13b)$$

$$3\Delta E_3 = \kappa 2 C P_{op3} \qquad (13c)$$

$$E_3 = -2 C P_{ip3} \qquad (13d)$$

An expression for the operating long-term frequency follows from the ratio of the last two equations,

$$\frac{3\Delta}{\kappa} = -\frac{P_{op3}}{P_{ip3}} \qquad (14a)$$

with the use of Eqs. (12a) and (12b), the final expression takes the form

$$\Delta = \sqrt{\frac{2\kappa(1 + 2\wp) + 3\wp}{24 + 6\kappa + 9\wp}} \equiv \Delta_{lt0} \qquad (14b)$$

This formula shows no dependence on excitation level. It constitutes an expression of the eigen-frequency that characterizes a given set of κ and γ values that allow for periodic solutions. It is the equivalent of Eq. (5) for the transient oscillation near the instability threshold. Eq. (14b) may be considered as a frequency which is characteristic of the involved material and cavity decay rates whose values are fixed by construction, for a given laser system.

Indeed, as demonstrated in the numerical simulations, the frequency of the obtained signals increases with excitation level. The dependence on 2C is derived with the above first-order relations (Eqs. (13a) and (13b)) (see Appendix B for the main steps of its derivation) and takes the form:

$$\Delta = \sqrt{\frac{(2C-1)\kappa\wp(2+\wp)-3(\kappa+1)\wp^2}{8(\kappa+1)-\wp(2\kappa+\wp+4)}} \equiv \Delta_{lt} \tag{15}$$

Equalizing Eq. (14b) and Eq. (15), which must be consistent with each other, gives the value of the necessary excitation level, which is required to force the system into oscillating at the eigen-frequency $\Delta_{lt0}/2\pi$. For $\kappa = 3$ and $\gamma = 0.1$, we obtain 2C = 9.79, a value very close to the instability threshold.

Figure 3 (lower curve) gives a representation of relation (15) with respect to 2C. Along the curves are a few distributed bold points obtained with direct numerical simulations of Eqs (1). We see that beyond the instability threshold, the above relation exactly matches the numerical replications. However, the frequencies of the pulsing solutions that correspond to hard excitation (below the instability threshold) slightly deviate from relation (15). One may attribute these small deviations to the fact that these were obtained outside the self-pulsing domain, while Eq. (15) has been evaluated inside the unstable domain, beyond the instability threshold.

As demonstrated in Appendix B, the expressions of the first and third-order field-amplitudes are derived from Eqs. (13a)-(13b) and (13c)-(13d), respectively, to take the form

$$E_1 = 2\sqrt{\frac{(1+\kappa)\left(\wp^2+4\Delta^2\right)\left(1+\Delta^2\right)}{2\kappa\wp\left(1+\wp-\Delta^2\right)+\wp^2\left(1-\Delta^2\right)-4\wp\Delta^2}} \tag{16a}$$

$$E_3 = -\frac{T_{3d}\left(\wp^2(1-3\Delta^2)-8\wp\Delta^2\right)}{1-\Gamma_d T_{1d}\left(\wp^2(1-\Delta^2)-4\wp\Delta^2\right)\frac{E_1^2}{4}}\frac{E_1^3}{4} \tag{16b}$$

According to Eqs (12c)-(12e) and (15), these components also depend on the values of κ, $\wp$ and 2C. Note that (as expected from the non linear nature of Eqs (1)) E_3 scales to the third power in E_1, but that does not mean that the third-order amplitude is higher than the first one, since a saturating term, which scales as E_1^2, in the denominator prevents such an unphysical situation.

Leaning on the analytical features extracted above, we now show how a typical solution with increasing accuracy structures itself with the inclusion of increasing order terms before focusing on a sequence of analytical solutions that corresponds to the numerical counterparts of Fig. 4.

VII. Typical Hierarchy of Self-Pulsing Solutions

First, we focus on a typical example, obtained at the boundary of the instability domain. In order to make a one-to-one comparison with the numerical simulations, we construct a first solution in the vicinity of the instability threshold. The long term operating frequency is estimated from Eq. (15), while the first and third order field components are evaluated from Eqs. (16a) and (16b), respectively.

The values of these components, for $\kappa = 3$, $\wp = 0.1$ and $2C=10$, are $E_1 = 5.2$ and $E_3 = 1.73$ and the corresponding pulsation frequency is $\Delta_{lt} \approx 0.42$. Thus, to third order, the analytical field expansion writes

$$E(t) = 5.2\cos(0.42t) + 1.73\cos(3 \times 0.42t) \tag{17}$$

and is represented in Fig. (8a). One may see that a few details of the numerical counterpart of Fig. (1) are recovered. However there still remain a few differences between the analytical and the numerical solution. The long term time trace of Fig. (1) evolves with $\Delta_{lt} \approx 0.38$, and the pulses peak at $E_p = 7.96$, while $\Delta_{lt} \approx 0.42$ and $E_p = 6.95$ for the above, third order solution.

In order to obtain a better fit, we extend the calculations towards fifth order. As demonstrated in Appendix C, when the out-of-phase field contribution is not taken into account, the fifth order component is obtained in terms of the first and third order field amplitudes in the form

$$E_5 = -2C\Gamma_d\Gamma_4\Gamma_5\{f(\Delta)E_1^5 + g(\Delta)E_1^2E_3 + h(\Delta)E_1^4E_3 + q(\Delta)E_1E_3^2 + s(\Delta)E_1^3E_3^2\} \tag{18}$$

Its numerical value, evaluated in Appendix (C), is $E_5 = 0.8$, so that, to fifth order, the analytical field expansion writes:

$$E(t) = 5.2\cos(0.42t) + 1.73\cos(3 \times 0.42t) + 0.8\cos(5 \times 0.42t) \tag{19}$$

and is represented in Fig. (8b). As compared to the signal of Fig. (8a), the time trace of Fig. (8b) shows thinner and higher peak values approaching those of the numerical counterpart of Fig. 4a. However, the solution has undergone some distortion that rules out any improved resemblance. Higher order terms are still required to achieve much better accuracy. The very lengthy and time consuming calculations (with an increased rate of error-occurrence with increasing order term evaluation) required to obtain the fifth order terms is of no encouragement to attempt higher-order-terms determination any further. Instead, a much more convenient way to find the amplitudes of high-order terms consists in a direct evaluation of the components peak-heights appearing in the frequency spectra of the corresponding numerical solutions. Such a procedure straightforwardly yields the analytical solution up to the desired order. The highest order term is dictated by the importance of the corresponding frequency component merging from the frequency spectrum. The examples of Fig. 4, suggest an expansion up to the 11th term. Limiting the expansion to the ninth order still shows some differences with the exact numerical solution, as demonstrated in (Fig. 8c). The adapted expansion requires taking into account all the terms up to the eleventh-order.

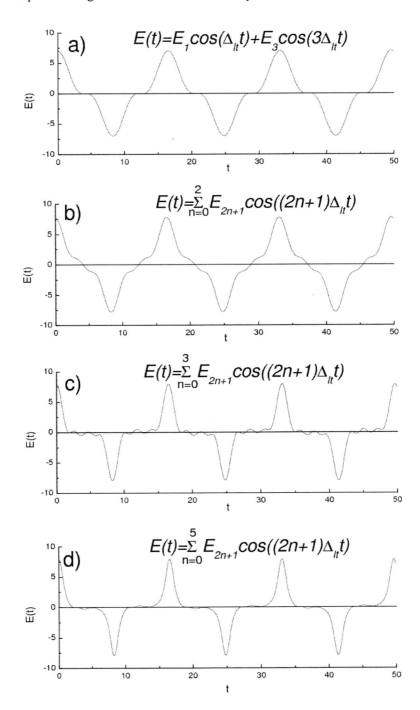

Figure 8. Typical hierarchy of analytical pulse structuring obtained with increasing order and accuracy, a) third-order, b) fifth order, c) seventh order, and d) eleventh order, in electric field amplitude.

In short, the final solution for the regular oscillations of Fig. 4a writes

$$E(t) = \sum_{n=0}^{n=5} E_n \cos((2n+1)\Delta t) \tag{20}$$

The numerical values of the different components, corresponding to an excitation parameter 2C=10, obtained from the frequency spectrum of Fig. (4a), are indicated in Table I (a).

Table I. Field-amplitudes of the spectral components of Figs. (6d)-(6f)

(a) *Odd-harmonic components*

2C	Δ	E_1	E_3	E_5	E_7	E_9	E_{11}
10	0.38	1.68	1.22	0.69	0.38	0.18	0.1
18.4	0.6	2.5	1.55	0.8	0.35	0.1	0.05
27.7	0.74	3.2	1.9	0.82	0.35	0.11	0.05
29.5	0.76	3.0	1.8	0.88	0.3	0.1	0.05

(b) *Even-harmonic components*

2C	E_2	E_4	E_6	E_8	E_{10}
27.7	0.3	0.4	0.2	0.1	0.05
29.5	0.32	0.4	0.25	0.1	0.05

(c) *Sub-harmonic components*

2C	$E_{1.5}$	$E_{2.5}$	$E_{3.5}$	$E_{4.5}$	$E_{5.5}$	$E_{6.5}$
29.5	0.11	0.25	0.1	0.2	0.06	0.01

The final analytical solution is represented in Fig. (8d). A one-to-one comparison with the solution of Fig. (4a) demonstrates an almost perfect match between the two signals.

The odd-order expansion (20) remains valid as long as the solution consists of symmetric period one oscillations, i.e. up to 2C = 18.4. Indeed, the field-amplitude components depend on the excitation parameter and must be evaluated separately. The values of these components obtained with 2C =18.4 are also given in Table I (a), when these values are inserted into Eq (20) one obtains the actual expansion of the corresponding analytical solution, represented in Fig. (9a) along with the corresponding FFT in Fig. (9d). Note the increase in field amplitude up to fifth-order component while higher order show a less contribution than those of the solution obtained with 2C = 10.

Now let us turn to the asymmetric and the period-two signals of Fig. 6.

The asymmetric solution of Fig. 6b stems from the emergence, in the corresponding frequency spectrum (Fig. 6e), of even components. Thus, obviously an asymmetric regular solution must contain even frequency components. A general asymmetric time trace should thus follow

$$E(t) = \sum_{n=1}^{n=11} E_n \cos(n\Delta t)$$ (21)

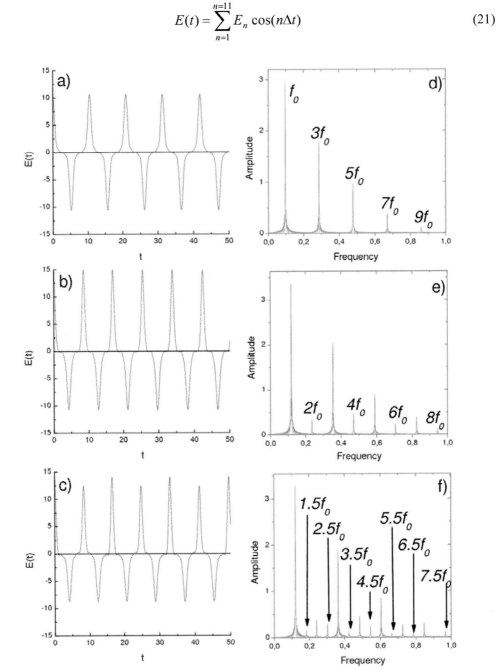

Figure 9. Analytical counterparts of the numerical solutions of Figure 6; a) symmetric period-one oscillations described by the field-expansion (20) b) asymmetric period-one oscillations described by the strong-harmonic field-expansion (21), and c) asymmetric period-two solution described with expansion (22). The corresponding frequency spectra represented respectively in d), e), f) show exactly the same features as those of Figure 6. The field amplitude-values are given in Table I. Note the symmetrical nature between the rise and fall of the pulses, stemming from phase-free components in expansions (20)-(22).

The amplitudes are evaluated from the Fast Fourier Transform of Fig. (6e) and are given in Table I (a) for the odd-harmonic components and Table I (b) for the even-harmonic components. The solution is represented if Fig (9b) with its corresponding FFT spectrum in Fig. (9e).

Finally, the period-two solution is constructed by inferring terms that contain frequency components of the form 1.5Δ, 2.5Δ, 3.5Δ etc. merging from the corresponding frequency-spectrum of Fig. (6f). The period-two solution of Fig. (6c) is described by the following field-expansion

$$E(t) = \sum_{n=1}^{n=11} E_n \cos(0.5 \times (n+1)\Delta t) \tag{22}$$

The amplitudes are obtained from the Fast Fourier Transform of Fig. (6f) and are supplied in Table I (a) for the odd-harmonics, in Table I (b) for the even components and in Table I (c) for the sub-harmonic components. The solution is represented in Fig. (9c) and its FFT in Fig. 9f.

A one-to-one comparison between the numerical solutions of Fig. (6) and the analytical hierarchy of Fig. (9) gives clear evidence of identical pulse frequencies and peak-heights. However, detailed inspections in the vicinity of the horizontal axis, where the field values are close to zero, show some discrepancy. One may see that while, the signals of Fig. (6) undergo some distortion appearing in the form of a symmetry-breaking evolution between the rise and fall of the pulses, the signals of Fig. (9) are perfectly symmetric. These differences stem from the fact that we have supposed and imposed an in-phase evolution between all the field harmonics. However, once the first harmonic is taken as a reference, there is no reason why the higher order harmonics should remain permanently locked to the first-order component. Thus, more accurate solutions should take into account some dephasing processes between high-order components. An immediate consequence is that once the first-order field is chosen in the form of a symmetric cosine with no phase term

$$E_1(t) = E_1 \cos(\Delta t) \tag{23}$$

The higher-order harmonics should evolve according to

$$E_n(t) = E_n \cos(n\Delta t + \varphi_n) \tag{24}$$

When small phase terms are inserted into expansions (20)-(22), we obtain solutions that carry much more obvious resemblance with the numerical counterparts. These are represented in the curves of Fig. (10). As a first glance verification reveals, even the small details that distort the signals in the vicinity of zero-field values, show up in the analytical solutions. Indeed, an accurate evaluation of the phase terms may be obtained with the method outlined in Sec VI. Here, we just scanned over a few values of φ_n until a best possible fit with the numerical solution is achieved.

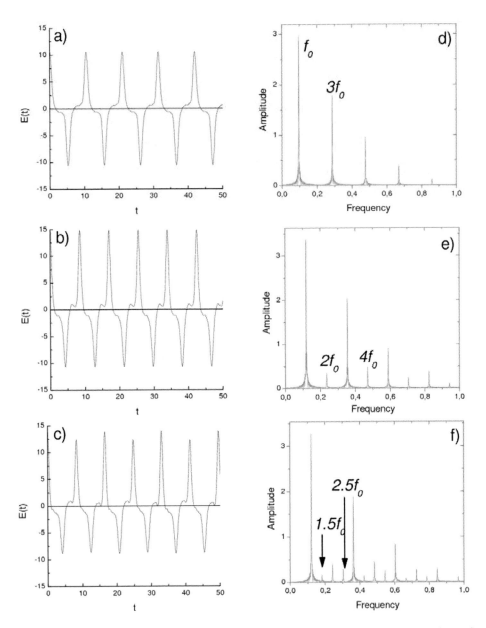

Figure 10. Improved analytical solutions, obtained with the same excitation levels as the series of Figure 9, that take into account phase factors between higher-order harmonics. The signal distortions that occur in the vicinity of zero-field values, as appearing in Figure 6, are fully retrieved in each case. This seems to indicate that dephasing processes do occur between high-order harmonics, and cannot be neglected.

As a further illustration of the accuracy of the strong-harmonic expansion method, we represent the field-intensity signals corresponding to the series of signals of both Fig. (6) and Fig. (10), for comparison. These are shown in Fig. (11). The first few period-doubling sequences show a perfect match between the numerical and the analytical solutions. One may thus conclude that the analytical period-doubling sequence peculiar to Eqs. (1) with the parameter values $\kappa = 3$, $\wp = 0.1$ has its analytical equivalence, as demonstrated in Fig. (11). Other examples of such a hierarchy are known to characterize Eqs. (1), these may also be described with the same strong harmonic method presented in this chapter.

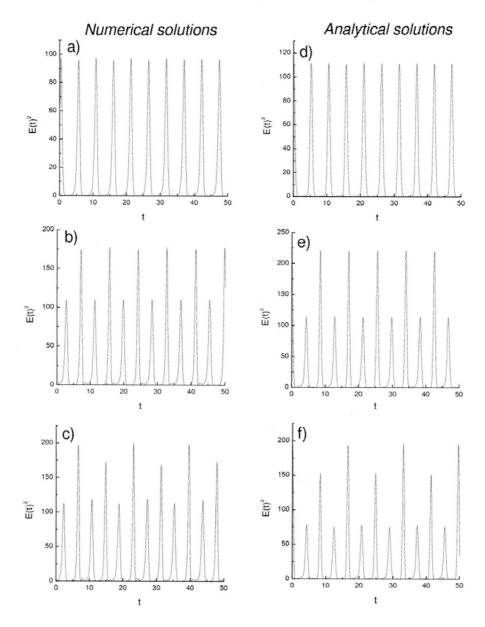

Figure 11. Comparison between two hierarchies of field-intensities, as obtained: a), b), c) numerically and d), e) f) analytically. The period–doubling sequence appears clearly in both series of traces. Again an obvious match between the numerical solutions and their analytical counterparts clearly transpires from the two series of traces.

As a final illustration, the following section gives additional analytical solutions for other values of the control parameters κ and $\wp$, deliberately chosen to lie far apart inside the control parameter space.

VIII. Further Examples of Solutions with Extreme Control-Parameter Values

As already mentioned, Eqs. (1) reveal periodic self-pulsing solutions in a large zone of the control parameter space (κ, $\wp$). We further test the accuracy of the presented approach against large scans of parameter values. We have found quite a large number of periodic solutions and in each case we derived the corresponding analytical field-expansions, always with the same precision as in the above examples. When the repetition rate of the pulsing signals remains close to $\Delta_t/3$, the field expansions were invariably found to be limited to the 11th order harmonic term. Higher order terms participate to less than 1% in field amplitude and can be neglected within this accuracy. However, for low repetition rates, many more terms must be evaluated as demonstrated in the following examples.

Table II. Field-amplitude-values of the spectral components of Fig. (12b)
(The solution pulsation is $\Delta = 0.16$)

E1	E3	E5	E7	E9	E11
2.76	2.39	1.77	1.17	0.72	0.42
E13	E15	E17	E19	E21	E23
0.23	0.12	0.06	0.04	0.02	0.01

The solution of Fig. (12a) was obtained with $\kappa = 1.2$, $\gamma = 0.01$ and $2C = 25$. The corresponding spectrum (Fig. 12b) exhibits harmonic-orders up to the 23rd component. Their amplitudes are given in Table II. The corresponding analytical solution writes:

$$E(t) = \sum_{n=0}^{n=11} E_{2n+1} \cos((2n+1)\Delta t) \tag{25}$$

and is represented in Fig. (12c).

The solution of Fig. (13a) was obtained with $\kappa = 1.1$, $\gamma = 0.001$ and $2C=50$.

The corresponding spectrum (Fig. 13b) exhibits persistent harmonic-components up to the 79th order! Their amplitudes are given in Table III. The analytical solution in this case, expands as

$$E(t) = \sum_{n=0}^{n=39} E_{2n+1}\cos((2n+1)\Delta t) \tag{26}$$

and is represented in Fig. (13c).

Apart from some fine structure, stemming from some little mismatch between the evaluated components and the exact ones, Fig. (13c) fits fairly well to the numerical solution of Fig. (13a).

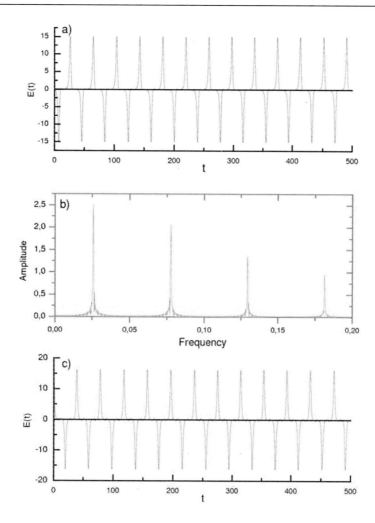

Figure 12. a) Symmetric period-one solution obtained with $\kappa = 1.2$, $\gamma = 0.01$, and 2C=25, b) corresponding frequency spectrum. The field amplitude values of the components are given in Table II, c) corresponding analytical solution described with Eq. (25).

Table III. Field-amplitude values of the spectral components of Fig. (13b)
(The solution pulsation is $\Delta = 0.038$)

E1	E3	E5	E7	E9	E11	E13	E15
2.26	1.98	2.03	2.13	1.98	1.61	1.74	1.7
E17	**E19**	**E21**	**E23**	**E25**	**E27**	**E29**	**E31**
1.48	1.23	1.29	1.19	0.96	0.86	0.85	0.75
E33	**E35**	**E37**	**E39**	**E41**	**E43**	**E45**	**E47**
0.58	0.55	0.52	0.43	0.32	0.31	0.30	0.24
E49	**E51**	**E53**	**E55**	**E57**	**E59**	**E61**	**E63**
0.20	0.19	0.17	0.13	0.12	0.11	0.09	0.07
E65	**E67**	**E69**	**E71**	**E73**	**E75**	**E77**	**E79**
0.07	0.06	0.05	0.04	0.04	0.036	0.030	0.020

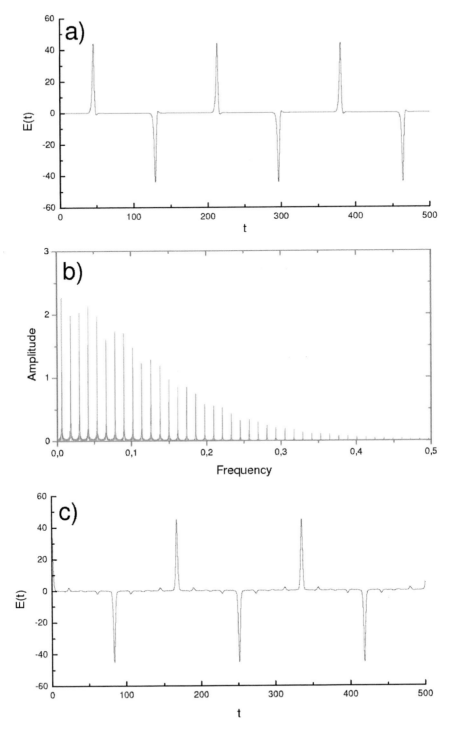

Figure 13. a) Symmetric period-one solution, obtained with $\kappa = 1.1$, $\gamma = 0.001$, and $2C = 50$, b) corresponding frequency spectrum. The field amplitude values of the components are given in Table III, c) corresponding analytical solution described with Eq. (26).

Table IV. Field-amplitude-values of the spectral components of Fig. (14b)
(The solution pulsation is $\Delta = 0.41$)

E1	E3	E5	E7	E9	E11
2.33	2.0	1.23	1.1	0.48	0.45
E13	E15	E17	E19	E21	E23
0.17	0.17	0.06	0.06	0.025	0.020

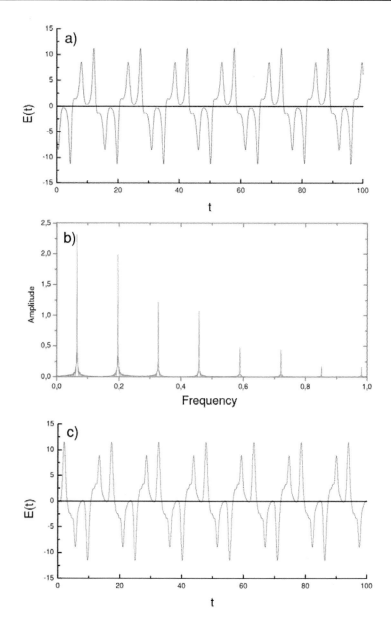

Figure 14. a) Peculiar solution showing a double peak structure obtained with $\kappa = 3$, $\gamma = 0.15$, and $2C = 22$; b) corresponding Fast Fourier Transform, and c) analytical solution described with Eq. (25), the corresponding field components are given in Table IV.

As a last and quite forceful example, we represent a solution with a double peak structure resembling Fig. (24) of the preceding chapter. The signal represented in Fig. (14a) was obtained with $\kappa = 3$, $\gamma = 0.15$ and 2C=22. The corresponding spectrum, represented in Fig. (14b), shows almost equal amplitudes between the first and the third-order components, the fifth and seventh-order terms as well as the ninth and the eleventh-order terms, etc. The analytical solution, represented in Fig. (14c), follows Eq. (25) with amplitude values given in Table IV. Note however, the quite distinct values between the control parameters that correspond to Fig. (24) of the preceding chapter (for which $\kappa = 100$, $\gamma = 1$, and 2C=1.3) with those of Fig. (14) (where $\kappa = 3$, $\gamma = 0.15$, and 2C=22).

IX. Localisation of the Parameter Space Involving Periodic Solutions

It is well known that Eqs. (1) include a wealth of regular and irregular solutions. The parameter zone, in terms of κ and $\wp$ values inside which periodic solutions are likely to occur has been tempted in a number reports. However the delimited zones always concerned a narrow region of these control parameters. The findings converge towards the fact that periodic solutions likely occur for small values of the population decay-rate with respect to the polarization relaxation rate, i.e. $\wp \ll 1$. The strong harmonic expansion is of quite some help in delimiting the parameter space for regular solutions in a much broader parameter zone. The expressions of the pulsing eigen-frequencies, derived in Sec. VI, allow for the localization of the $(\kappa, \wp)$ parameter space that contain regular pulsing solutions, as demonstrated in the following.

Combining Eq. (14b) with Eq. (15) yields a formula that gives the excitation level corresponding to periodic solutions in terms of κ an d $\wp$, in the form

$$(2C-1)=\frac{[2\kappa(1+2\wp)+3\wp][8(\kappa+1)-\wp(2\kappa+\wp+4]}{\kappa\wp(2+\wp)[24+6\kappa+9\wp]}+\frac{3(\kappa+1)\wp^2}{\kappa\wp(2+\wp)} \qquad (27)$$

The excitation level required to obtain periodic solutions will be denoted $2C_{pth}$.

Periodic solutions occur in the unstable domain with respect to the control parameters set $(\kappa, \wp)$, where the following relation

$$2C_{pth} > 2C_{2th} \qquad (28)$$

must be satisfied. The inequality imposes pulsing solutions above the instability threshold. A condition that seems quite obvious, yet pulsing solution also exist below the instability threshold but these can only be excited with a hard excitation mechanism, as already and amply discussed.

The use of relations (4) and (27) in (28) delimits the parameter space $(\kappa, \wp)$ where periodic solutions, above the instability threshold, most likely occur.

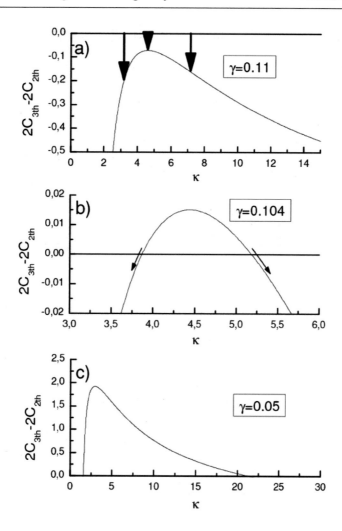

Figure 15. Boundaries of the pulsing solutions for different values of γ. Pulsing solutions develop for parameter values that satisfy $2C_{3th}-2C_{2th} \geq 0$, a) for $\gamma = 0.11$, the boundary of pulsing solutions lies below the instability threshold. A hard excitation mechanism (indicated by the downward arrows) may excite these solutions, this most likely explains the persistence of pulsing solutions below the instability threshold as found in Figure 5 , b) for $\gamma = 0.104$, the crossing points of $2C_{3th}-2C_{2th}$ with the horizontal axis delimits the region of κ values allowing for pulsing solution for fixed γ's, adiabatic scans along the curve from above to below the instability threshold allows the pulsing solution to persist below the instability threshold, following the curve, as indicated by the arrows. c) for $\gamma = 0.05$, the range of κ values satisfying Eq. (31) extends from 1.9 to 21.

Figures (15) represents a series of curves $2C_{pth}(\kappa) - 2C_{2th}(\kappa)$ parameterized with $\wp$, indicating regions of κ values that allow for pulsing solutions. For $\wp = 0.11$, the curve, Fig. (15a), lies below the instability threshold. Pulsing solutions are forbidden in that case, unless a hard excitation mechanism is applied, as indicated with the downward arrows. For $\wp = 0.104$, the curve merges beyond the horizontal axis for a small range of κ values extending from $\kappa = 3.7$ to $\kappa = 5.5$, pulsing solutions are restrited to this zone, while for $\wp = 0.05$, Fig.

(15c) pulsing solutions extend from κ = 1.9 to κ = 21. Decreasing $\wp$ extends the κ values to larger ranges for pulsing solutions.

One may see that in the whole range of K values, $\wp$ never exceeds by far the value 0.1. Thus, relation (27) may be simplified into

$$2C_{pth} - 1 = \frac{4(1+2\wp)}{3\wp(\kappa+4)} \qquad (29)$$

So that in terms of K and $\wp$ relation (28) may be transformed into

$$f(\kappa,\wp) = \frac{4(1+2\wp)}{3\wp(\kappa+4)} - \frac{\kappa+1+\wp}{\kappa-1-\wp} \geq 0 \qquad (30)$$

A Three-dimensional representation of $f(\kappa,\wp)$ delimits the parameter space inside which pulsing solutions occur. These correspond to the merging points from the plane $f(\kappa,\wp) = 0$, shown in Fig. (16). One may see that the self-pulsing zone constitutes a narrow space with decreasing $\wp$ as K increases.

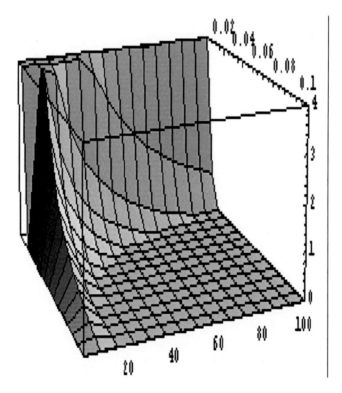

Figure 16. Three dimensional representation of $f(\kappa,\wp)$, delimiting the parameter space (κ, γ) inside which pulsing solutions likely occur. This parameter zone represents a "hill" clearly merging from the plane $f(\kappa,\wp) = 0$. One may see that the pulsing zone is concentrated in the regions (low κ's, high γ's) and (high κ's, low γ's).

Close re-inspection of Fig. (15) also inspires the following comments with respect to the persistent pulsing solutions below the second-laser threshold under hard excitation mechanisms or with an adiabatic scan of the excitation level from beyond to below the instability threshold. It seems that unstable solutions persist below the instability domain because a decrease in excitation level drives the solution along the path of $2C_{pth}$ towards levels where

$$2C_{pth} < 2C_{2th} \tag{31}$$

Pulsing-solutions do exist below the instability threshold. They just cannot be excited with an adiabatic increase of excitation, but these may be generated with a hard excitation mechanism as demonstrated in the output versus excitation level curve of Fig. 5.

X. Conclusion

This second chapter was devoted to presenting an in-depth analysis of a strong harmonic expansion technique applied to the Laser-Lorenz equations that describe the self-pulsing regime of single-mode homogeneously broadened lasers operating in a bad-cavity configuration. The method presented here is fairly general to be applicable to other differential sets of equations that qualitatively possess the same pulsing solutions. In particular, the well-known integro-differential "Maxwell-Bloch" equations adapted to describe the pulsing behavior of single-mode inhomogeneously broadened lasers of the first chapter has been, as well, handled with the same approach. However, the involved algebra was much more complicated, yet the analytical calculations were carried out up to third order in field amplitude.

The simpler model of this second chapter has delimited the extent of validity of the usual small side-band method, which was shown to be restricted to the transient part of the unstable solution. Yet, a readjustment of the method that takes into account some unavoidable dephasing mechanism between the light and matter variables, which takes place during the non-linear interactions, was shown to be necessary in order to find closed-form expression of the transient frequencies that match those of the numerical signals inside the whole control parameter space both below and above the instability threshold.

Analytical information of the long-term part of the solution was shown to be contained in a strong-sideband analysis that carries the essential properties of pulse structuring during pulse build-up. The recognition of these properties, namely a perpetual field and polarization switching around zero-mean values, has allowed for the extraction of an analytical expression of the long-term oscillation frequency of the pulsing signals. In addition, the first few components of the electric-field variable have been evaluated, likewise analytically, while the complete solutions called for the application of a Fast Fourier Transform algorithm to the temporal traces of the pulsing signals.

However, even in the simpler case presented in this chapter, the very lengthy and quite tedious algebra involved in the determination of high-order terms has not allowed us to evaluate higher than the fifth-order terms. Despite these limitations, leaning on the analytical information that we extracted from the long-term regular pulse structures, we have proposed a straightforward scheme that leads to the construction of the complete solutions by inferring the field amplitudes directly from the field-amplitude spectra obtained with a Fast-Fourier transform of the temporal signals. The method has been applied to describe, analytically, the first few solutions of a period-doubling sequence, peculiar to the Lorenz equations and which builds subsequently to an increase of the excitation parameter. Even though most of the analysis focused on fixed control parameters (cavity-decay rate and population inversion relaxation rate), we have shown that the method carries a strong enough validity to be extended to the entire parameter-space that exhibits regular pulsing solutions. Indeed, the examples chosen in section VIII show prevailing evidence of the efficacy of the strong harmonic expansion technique, which calls for no further proofs. In addition to the single-mode homogeneously and inhomogeneously broadened lasers, we expect the method to be applicable to non-autonomous systems, such as the modulated rate-equations [37], describing good-cavity lasers, in which the polarization adiabatically follows the electromagnetic field, or the Lang and Kobayashi delayed rate-equations which describe the dynamical properties of the semiconductor laser with optical feedback [38]. These examples should be the focus of forthcoming investigations along the strong-versus-weak-harmonic-expansion lines of thought presented in these two chapters.

Appendix A.
Closed-Form Expression for the Transient Frequency

Inserting the small-amplitude expansions (7) into Eqs (1) yield the following relations:

$$-\Delta e \sin(\Delta t) = -\kappa\left\{E_0 + e\cos(\Delta t) + 2CP_0 + 2Cp_1\cos(\Delta t) + 2Cp_2\sin(\Delta t)\right\} \quad \text{(A1a)}$$

$$\begin{aligned}-\Delta p_1 \sin(\Delta t) + \Delta p_2 \cos(\Delta t) &= -P_0 - p_1\cos(\Delta t) - p_2\sin(\Delta t) \\ &+ \left(E_0 + e\cos(\Delta t)\right)\left(D_0 + d_1\cos(\Delta t) + d_2\sin(\Delta t)\right)\end{aligned} \quad \text{(A1b)}$$

$$\begin{aligned}-\Delta d_1 \sin(\Delta t) + \Delta d_2 \cos(\Delta t) &= -\wp\left\{D_0 + 1 + d_1\cos(\Delta t) + d_2\sin(\Delta t)\right\} \\ &- \wp\left(E_0 + e\cos(\Delta t)\right)\left(P_0 + p_1\cos(\Delta t) + p_2\sin(\Delta t)\right)\end{aligned} \quad \text{(A1c)}$$

Equalizing terms of the same-order in each relation yields

$$\Delta e = \kappa 2Cp_2 \quad \text{(A2a)}$$

$$e = -2Cp_1 \quad \text{(A2b)}$$

$$-\Delta p_1 = -p_2 + E_0 d_2 \quad \text{(A2c)}$$

$$\Delta p_2 = -p_1 + E_0 d_1 + D_0 e \quad \text{(A2d)}$$

$$-\Delta d_1 = -\wp d_2 - \wp E_0 p_2 \quad \text{(A2e)}$$

$$\Delta d_2 = - \wp \, d_1 - \wp \, E_0 p_1 - \wp \, P_0 e \tag{A2f}$$

Solving (A2e) and (A2f) with respect to d_1 and d_2, we obtain

$$d_1 = -\frac{\wp^2}{\wp^2 + \Delta^2}\left\{ E_0 p_1 + P_0 e - \frac{\Delta}{\wp} E_0 p_2 \right\} \tag{A3a}$$

$$d_2 = -\frac{\wp^2}{\wp^2 + \Delta^2}\left\{ E_0 p_2 + \frac{\Delta}{\wp}\left(E_0 p_1 + P_0 e \right) \right\} \tag{A3b}$$

So that (A2c) and (A2d) transform into

$$\left\{ -\Delta + \frac{\Delta}{\wp}\Gamma E_0^2 \right\} p_1 + \left\{ 1 + \Gamma E_0^2 \right\} p_2 = -\frac{\Delta}{\wp}\Gamma E_0 P_0 e \tag{A4a}$$

$$\left\{ 1 + \Gamma E_0^2 \right\} p_1 - \left\{ -\Delta + \frac{\Delta}{\wp}\Gamma E_0^2 \right\} p_2 = \left(-\Gamma E_0 P_0 + D_0 \right) e \tag{A4b}$$

where, for simplifying purposes, we have introduced the following parameter

$$\Gamma = \frac{\wp^2}{\wp^2 + \Delta^2} \tag{A5}$$

The ratio of (A2a) and (A2b) yields

$$\frac{\Delta}{\kappa} = \frac{\left(-\Delta + \dfrac{\Delta}{\wp}\Gamma E_0^2 \right)\left(-\Gamma E_0 P_0 + D_0 \right)e + \left(1 + \Gamma E_0^2 \right)\dfrac{\Delta}{\wp}\Gamma E_0 P_0 e}{\left(1 + \Gamma E_0^2 \right)\left(-\Gamma E_0 P_0 + D_0 \right)e - \left(-\Delta + \dfrac{\Delta}{\wp}\Gamma E_0^2 \right)\dfrac{\Delta}{\wp}\Gamma E_0 P_0 e} \tag{A6}$$

After adequate simplifications, the stationary relations given in Eqs (8) of the text transform (A6) into

$$\frac{1}{\kappa} = \frac{\Delta^2 + \wp^2 - (2C - 1)(2 + \wp)\wp}{\left((2C - 1)^2 - 1 \right)\wp^2 - \left(1 + (2C - 1)\wp \right)\Delta^2} \tag{A7}$$

from which we derive the following final relation

$$\Delta^2 = \frac{(2C-1)^2 \wp^2 + (2C-1)\kappa\wp(2+\wp) - (\kappa+1)\wp^2}{(2C-1)\wp+\kappa+1} \tag{A8}$$

Let us evaluate (A8), at the instability threshold, given in Eq. (4) of the text as

$$2C_{2th} = 1 + \frac{(\kappa+1)(\kappa+1+\wp)}{(\kappa-1-\wp)} \tag{A9}$$

that we insert into (A8) to obtain

$$\Delta^2 = \frac{\dfrac{(\kappa+1)(\kappa+1+\wp)^2}{(\kappa-1-\wp)^2}\wp^2 + \dfrac{(\kappa+1+\wp)}{(\kappa-1-\wp)}\kappa\wp(2+\wp) - \wp^2}{\dfrac{(\kappa+1+\wp)}{(\kappa-1-\wp)}\wp + 1} \tag{A10}$$

yielding, after a few simple transformations

$$\Delta^2 = \frac{2\kappa\wp(\kappa+1)}{\kappa-1-\wp} \tag{A11}$$

equivalent to Eq. (5).

This constitutes a test for the validity of Eq (A8), in addition to the excellent agreement with the numerical evaluations demonstrated in Fig. 3.

Appendix B.
Derivation of the Pulsing Angular Frequency and Other Fundamental Closed-Form Expressions for the Permanent State

The derivation of closed-form expressions for the long-term frequency along with the first and third order components of the electric-field expansion is based on an iterative procedure that calls for time-consuming but straightforward algebraic manipulations. Starting with the first order component of the electric-field expansion Eq. (11a)

$$E(t) = E_1 \cos(\Delta t) \tag{B1}$$

That we insert into Eq. 1a to obtain

$$P(t) = P_1 \cos(\Delta t) + P_2 \sin(\Delta t) \tag{B2}$$

Inserting both relations into Eq. 1b, and limiting, as a first step in the iterative procedure, the population inversion to its *dc* part in the driving term *E(t)D(t)*, yields

$$- \Delta P_1 \sin(\Delta t) + \Delta P_2 \cos(\Delta t) = -P_1 \cos(\Delta t) - P_2 \sin(\Delta t) + E_1 \cos(\Delta t) d_{dc} \quad \text{(B3)}$$

Through simple identifications, we obtain

$$\Delta P_1 = P_2 \quad \text{(B4a)}$$

and

$$\Delta P_2 = -P_1 + E_1 d_{dc} \quad \text{(B4b)}$$

Transforming into

$$P_1 = \frac{d_{dc}}{1 + \Delta^2} E_1 \quad \text{(B4c)}$$

and

$$P_2 = \frac{\Delta d_{dc}}{1 + \Delta^2} E_1 \quad \text{(B4d)}$$

With Eqs. (B1) and (B2), the driving term in Eq. (1c) writes

$$E(t)P(t) = P_1 E_1 \cos^2(\Delta t) + P_2 E_1 \sin(\Delta t)\cos(\Delta t)$$
$$= \frac{P_1 E_1}{2} + \frac{P_1 E_1}{2}\cos(2\Delta t) + \frac{P_2 E_1}{2}\sin(2\Delta t) \quad \text{(B5)}$$

Forcing the population inversion to evolve according to

$$D(t) = d_{dc} + d_1 \cos(2\Delta t) + d_2 \sin(2\Delta t) \quad \text{(B6)}$$

With Eqs. (B5) and (B6), Eq. (1c) transforms into

$$-\frac{2\Delta}{\wp} d_1 \sin(2\Delta t) + \frac{2\Delta}{\wp} d_2 \cos(2\Delta t) = -d_{dc} - d_1 \cos(2\Delta t) - d_2 \sin(2\Delta t) - 1$$
$$- \frac{P_1 E_1}{2} - \frac{P_1 E_1}{2}\cos(2\Delta t) - \frac{P_2 E_1}{2}\sin(2\Delta t) \quad \text{(B7)}$$

Yielding, through simple identifications

$$d_{dc} = -1 - \frac{P_1 E_1}{2} = -1 - \frac{d_{dc}}{1+\Delta^2} \frac{E_1^2}{2} \tag{B8a}$$

In which relation (B4c) has been taken into account, and from which we readily extract

$$d_{dc} = -\frac{\left(1+\Delta^2\right)}{1+\Delta^2 + \dfrac{E_1^2}{2}} = -\left(1+\Delta^2\right)\Gamma_d \tag{B8b}$$

The parameter

$$\Gamma_d = -\left(1+\Delta^2 + \frac{E_1^2}{2}\right)^{-1} \tag{B8c}$$

is introduced for simplifying purposes only.
Additional identifications in Eq. (B7) give

$$\frac{2\Delta}{\wp} d_1 = d_2 + \frac{P_2 E_1}{2} \tag{B9a}$$

$$\frac{2\Delta}{\wp} d_2 = -\left(d_1 + \frac{P_1 E_1}{2}\right) \tag{B9b}$$

That transform, with Eqs. (B4c) and (B4d), into

$$d_1 = \frac{\left(\dfrac{2\Delta^2}{\wp} - 1\right)}{\left(1+\dfrac{4\Delta^2}{\wp^2}\right)} \frac{d_{dc}}{1+\Delta^2} \frac{E_1^2}{2} \tag{B10a}$$

and

$$d_2 = -\frac{\Delta\left(1+\dfrac{2}{\wp}\right)}{\left(1+\dfrac{4\Delta^2}{\wp^2}\right)} \frac{d_{dc}}{1+\Delta^2} \frac{E_1^2}{2} \tag{B10b}$$

In a second step, the driving term $E(t)D(t)$ in Eq. (1b) must take into account the second-order expression (B6) for the population inversion. So that, with Eq. (B1), we are led to

$$E(t)D(t) = d_{dc}E_1 \cos(\Delta t) + \frac{d_1 E_1}{2}\cos(\Delta t) + \frac{d_1 E_1}{2}\cos(3\Delta t) + \frac{d_2 E_1}{2}\sin(\Delta t) + \frac{d_2 E_1}{2}\sin(3\Delta t)$$

(B11)

Implying that $P(t)$ in Eq. (1b) should contain third-harmonic components and evolve according to

$$P(t) = P_1 \cos(\Delta t) + P_2 \sin(\Delta t) + P_3 \cos(3\Delta t) + P_4 \sin(3\Delta t)$$

(B12)

Inserting Eqs (B11) and (B12) into Eq. (1b) yields the following relation

$$-\Delta P_1 \sin(\Delta t) + \Delta P_2 \cos(\Delta t) - 3\Delta P_3 \sin(3\Delta t) + 3\Delta P_4 \cos(3\Delta t)$$
$$= -P_1 \cos(\Delta t) - P_2 \sin(\Delta t) - P_3 \cos(3\Delta t) - P_4 \sin(3\Delta t) +$$
$$d_{dc}E_1 \cos(\Delta t) + \frac{d_1 E_1}{2}\cos(\Delta t) + \frac{d_1 E_1}{2}\cos(3\Delta t) + \frac{d_2 E_1}{2}\sin(\Delta t) + \frac{d_2 E_1}{2}\sin(3\Delta t)$$

(B13)

From which, the following equalities between first and-third-order polarisation components are extracted

$$-\Delta P_1 = -P_2 + \frac{d_2 E_1}{2}$$

(B14a)

$$\Delta P_2 = -P_1 + d_{dc}E_1 + \frac{d_1 E_1}{2}$$

(B14b)

$$-3\Delta P_3 = -P_4 + \frac{d_1 E_1}{2}$$

(B14c)

$$3\Delta P_4 = -P_3 + \frac{d_1 E_1}{2}$$

(B14d)

The first-order polarisation relations transform into

$$\left(1 + \Delta^2\right)P_1 = d_{dc}E_1 + \left(d_1 - \Delta d_2\right)\frac{E_1}{2}$$

(B15a)

and

$$\left(1 + \Delta^2\right)P_2 = \Delta d_{dc}E_1 + \left(d_2 + \Delta d_1\right)\frac{E_1}{2}$$

(B15b)

While the third-order polarisation relations take the form

$$\left(1+9\Delta^2\right)P_3 = \left(d_1 - 3\Delta d_2\right)\frac{E_1}{2} \tag{B16a}$$

and

$$\left(1+9\Delta^2\right)P_4 = \left(d_2 + 3\Delta d_1\right)\frac{E_1}{2} \tag{B16b}$$

With the second-order population inversion expressions, evaluated in Eqs (B10a) and (B10b), we are led to

$$\left(1+\Delta^2\right)P_1 = d_{dc}E_1 + \frac{\left(\dfrac{4\Delta^2}{\wp}+\Delta^2-1\right)}{\left(1+\dfrac{4\Delta^2}{\wp^2}\right)}\frac{d_{dc}}{1+\Delta^2}\frac{E_1^3}{4} \tag{B17a}$$

$$\left(1+\Delta^2\right)P_2 = \Delta d_{dc}E_1 - 2\Delta\frac{\left(1+\dfrac{1}{\wp}-\dfrac{\Delta^2}{\wp}\right)}{\left(1+\dfrac{4\Delta^2}{\wp^2}\right)}\frac{d_{dc}}{1+\Delta^2}\frac{E_1^3}{4} \tag{B17b}$$

For the first-order polarization components, and

$$\left(1+9\Delta^2\right)P_3 = \left\{\left(\frac{2\Delta^2}{\wp}-1\right)+3\Delta^2\left(1+\frac{2}{\wp}\right)\right\}\frac{1}{\left(1+\dfrac{4\Delta^2}{\wp^2}\right)}\frac{d_{dc}}{1+\Delta^2}\frac{E_1^3}{4} \tag{B18a}$$

$$\left(1+9\Delta^2\right)P_4 = \left\{-\Delta\left(1+\frac{2}{\wp}\right)+3\Delta\left(\frac{2\Delta^2}{\wp}-1\right)\right\}\frac{1}{\left(1+\dfrac{4\Delta^2}{\wp^2}\right)}\frac{d_{dc}}{1+\Delta^2}\frac{E_1^3}{4} \tag{B18b}$$

For the third-order components
The ratio of (B18a) and (B18b) yields

$$\frac{P_3}{P_4} = \left[\left(\frac{2\Delta^2}{\wp}-1\right)+3\Delta^2\left(1+\frac{2}{\wp}\right)\right]\Bigg/\left[3\Delta\left(\frac{2\Delta^2}{\wp}-1\right)-\Delta\left(1+\frac{2}{\wp}\right)\right] \tag{B19}$$

Now, in order to derive a closed form expression for the pulsing frequency, let us turn back to Eq. (1a), into which we insert the first order expansion Eq. (B1), for the electric field and Eq. (B2) for the polarisation. We obtain the following first-order relations

$$\frac{\Delta}{\kappa} E_1 = 2CP_2 \tag{B20a}$$

and

$$E_1 = -2CP_1 \tag{B20b}$$

Their ratio, transforms, with Eqs. (B17a), (B17b) and (B8b) into

$$\frac{\Delta}{\kappa} = -\frac{P_2}{P_1} = -\Delta \frac{\left(\wp^2 + 4\Delta^2\right)\left(1 + \Delta^2\right) - \wp\left(1 + \wp - \Delta^2\right)\dfrac{E_1^2}{4}}{\left(\wp^2 + 4\Delta^2\right)\left(1 + \Delta^2\right) - \left[\wp^2(1 - \Delta^2) - 4\wp\Delta^2\right]\dfrac{E_1^2}{4}} \tag{B21}$$

From which we extract the first-order field amplitude in the form

$$\frac{E_1^2}{4} = \frac{(1 + \kappa)\left(\wp^2 + 4\Delta^2\right)\left(1 + \Delta^2\right)}{2\kappa\wp\left(1 + \wp - \Delta^2\right) + \wp^2\left(1 - \Delta^2\right) - 4\wp\Delta^2} \tag{B22}$$

i.e., Eq. (16a) of the text.

From Eq. (B20b) and (B17a), the excitation parameter relates to the first-order field amplitude through

$$E_1 = -2CP_1 = 2C\Gamma_d\left\{1 - T_{1d}\left(\wp^2(1 - \Delta^2) - 4\wp\Delta^2\right)\frac{E_1^2}{4}\right\}E_1 \tag{B23a}$$

With Γ_d, given in (B8c) and

$$T_{1d} = \frac{1}{\left(1 + \Delta^2\right)\left(\wp^2 + 4\Delta^2\right)} \tag{B23b}$$

also introduced to simplify the algebraic manipulations.

Equation (B23a) transforms into

$$2C = \frac{1}{\Gamma_d\left\{1 - T_{1d}\left(\wp^2(1 - \Delta^2) - 4\wp\Delta^2\right)\dfrac{E_1^2}{4}\right\}} \tag{B24}$$

As a last step, we extract the evolution of the long term- frequency with respect to the excitation parameter by replacing the different variables, including the first order field as given by Eq. (A22), with their respective expressions. We arrive at

$$\Delta^2 = \frac{(2C-1)\kappa\wp(2+\wp) - 3(\kappa+1)\wp^2}{8(\kappa+1) - \wp(2\kappa+\wp+4)} \tag{B25}$$

i.e. Eq. (15) of the text.

For small values of $\wp$ ($\wp \ll 1$), the expression is identical to Eq. (3.3b) of Narducci and al. [12], i.e.

$$\Delta^2 = \frac{(2C-1)\kappa\wp}{4(\kappa+1)} \tag{B26}$$

In order to evaluate the third-order electric field component and the system's eigen frequency, let us turn back to the third-order relations (B18a) and B(18b), which are readily transformed into

$$P_3 = \Gamma_d T_{3d}\left(\wp^2(1-3\Delta^2) - 8\wp\Delta^2\right)\frac{E_1^3}{4} \tag{B27a}$$

$$P_4 = \Gamma_d T_{3d}\, 2\wp\Delta\left(1 + 2\wp - 3\Delta^2\right)\frac{E_1^3}{4} \tag{B27b}$$

Driven with this third-order polarization, the electric field variable evolves according to

$$E(t) = E_1\cos(\Delta t) + E_3\cos(3\Delta t) \tag{B28}$$

Which, when inserted into Eq. (1) yields

$$E_3 = -2CP_3 \tag{B29a}$$

$$-3\Delta E_3 = -\kappa 2CP_4 \tag{B29b}$$

The ratio of these last to equations, along with Eqs. (B27a) and (B27b) yields

$$3\Delta = -\kappa\frac{P_4}{P_3} = -\kappa\frac{2\wp\Delta\left(1 + 2\wp - 3\Delta^2\right)}{\wp^2\left(1 - 3\Delta^2\right) - 8\wp\Delta^2} \tag{B30}$$

and the angular frequency of the locked solution to third order

$$\Delta^2 = \frac{2\kappa(1 + 2\wp) + 3\wp}{24 + 6\kappa + 9\wp} \tag{B31}$$

i.e. Eq (14b) in the text.

As for the inhomogeneously broadened case, the third-order relation, Eq. (B31) shows no dependence on excitation level. It relates exclusively to the material and cavity relaxation rates $\wp$ and κ. The frequency value evaluated from Eq. (B31) is considered as giving the system's eigenpulsation or resonant frequency, typical of any oscillatory system.

The third-order field component readily follows from Eq. (B29a) and (B27a), to write

$$E_3 = -\frac{T_{3d}\left(\wp^2(1 - 3\Delta^2) - 8\wp\Delta^2\right)}{1 - \Gamma_d T_{1d}\left(\wp^2(1 - \Delta^2) - 4\wp\Delta^2\right)\frac{E_1^2}{4}} \frac{E_1^3}{4} \tag{B32}$$

i.e. Eq. (16b) of the text, in which we introduced the following additional parameter

$$T_{3d} = \frac{1}{(1 + 9\Delta^2)(\wp^2 + 4\Delta^2)} \tag{B33}$$

Let us note, according to Eq. (B32) that the third-order field-component scales, as expected, to the third power in the first-order component, but, indeed, an intensity-dependent saturating term prevents any divergence of the solution.

Appendix C.
Determination of the Fifth-Order Field Component

Here, we shall not develop the calculations, since these would take quite a few pages. We only give the final results that lead to the calculation of the fifth-order field. Inserting Eqs. (11) with n = 0, 1, 2 for the field and polarization developments and n=2, 4, 6 for the population inversion components, we obtain a series of relations as in Appendix B, from which we find, after lengthy but straightforward calculations

$$E_5 = -2C\Gamma_d\Gamma_4\Gamma_5\left\{f(\Delta)E_1^5 + g(\Delta)E_1^2E_3 + h(\Delta)E_1^4E_3 + q(\Delta)E_1E_3^2 + s(\Delta)E_1^3E_3^2\right\} \tag{C1}$$

where

$$f(\Delta) = \frac{1}{16}\left[\left(5\Delta - \frac{\wp}{4\Delta}\right)\alpha_3 + \left(\frac{5\wp}{4} + 1\right)2\wp\Delta\alpha_4\right]T_{3d} \tag{C2}$$

$$g(\Delta) = -\frac{1}{4}\left[12\Delta - \frac{3\wp}{4\Delta} + \frac{15\Delta\wp}{4}\right] \tag{C3}$$

$$h(\Delta) = \frac{1}{16}\left[\left(10\Delta - \frac{3\wp}{4\Delta}\right)T_{1d}\alpha_1 + \left(\frac{15\wp}{4} + 2\right)T_{1d}\wp\Delta\alpha_2 + \left(5\Delta - \frac{\wp}{2\Delta}\right)T_{3d}\alpha_3 + \left(\frac{5\wp}{2} + 1\right)T_{3d}\,2\wp\Delta\alpha_4\right]$$

(C4)

$$q(\Delta) = \frac{1}{4}\left[-(5\Delta - 1) + \Delta\left(\frac{5\wp}{2} + 1\right)\right]$$

(C5)

$$s(\Delta) = \frac{1}{16}\left[(5\Delta - 1)T_{1d}\alpha_1 - \left(\frac{5\wp}{2} + 1\right)T_{1d}\wp\Delta\alpha_2\right]$$

(C6)

Table V. Closed-form expressions of the various parameters appearing in Appendixes B and C

Γ_d	Γ_2	Γ_4	Γ_5	T_{1d}
$\dfrac{1}{1+\Delta^2+\dfrac{E_1^2}{2}}$	$\dfrac{2\wp\Delta}{\wp^2+4\Delta^2}$	$\dfrac{4\wp\Delta}{\wp^2+16\Delta^2}$	$\dfrac{1}{1+25\Delta^2}$	$\dfrac{1}{(1+\Delta^2)(\wp^2+4\Delta^2)}$
α_1	α_2	α_3	α_4	T_{3d}
$\wp^2(1-\Delta^2)-4\wp\Delta^2$	$1+\wp-\Delta^2$	$\wp^2(1-3\Delta^2)-8\wp\Delta^2$	$1+2\wp-3\Delta^2$	$\dfrac{1}{(1+9\Delta^2)(\wp^2+4\Delta^2)}$

Table VI. Numerical values of the parameters appearing in Table V (2C=10, $\kappa = 3$, $\gamma = 0.1$)

Γ_d	Γ_2	Γ_4	Γ_5	T_{1d}
0.068	0.123	0.062	0.2	1.326
α_1	α_2	α_3	α_4	T_{3d}
-0.0556	0.94	-0.1228	0.72	0.63
$f(\Delta)$	$g(\Delta)$	$h(\Delta)$	$q(\Delta)$	$s(\Delta)$
-0.0068	-1.190625	-0.0164	-0.125	-0.0085

The expressions of the parameters appearing in the above relations are given in table V. These are related to the population decay rate γ and to the excitation parameter $2C$ through Δ.

The numerical values corresponding to the following parameters $\kappa = 3$, $\gamma = 0.1$ and $2C = 10$, are given in Table VI, and the corresponding fifth-order field component, evaluated from (C1), is

$$E_5 = 1.818 \qquad\qquad (C7)$$

References

[1]   E. N. Lorenz, "Deterministic nonperiodic flow", *J. Atmos. Science* **20**, 130 (1963).

[2]   H. Haken, "Analogy between higher instabilities in fluids and lasers", *Phys. Lett. A* **53**, 77 (1975).

[3]   K.A. Robbins; "Periodic solutions and bifurcation structure at high R in the Lorenz Model." *SIAM J. Appl. Math.*, vol. 36, N°3, 457-472 (1979).

[4]   A.C. Fowler, J. D. Gibon, a,d M. J. McGuiness, "The complex Lorenz equations", *Phys. D* **4**, 139-163 (1982).

[5]   C. Sparrow, *"The Lorenz equations: bifurcations, chaos, and strange attractors"*, Applied Mathematical Sciences, vol. 41, Springer-Verlag, Heidelberg (1982).

[6]   H. Haken, *"Evolution of Order and Chaos in Physics, Chemistry, and Biology"*, *Springer series in Synergetics*, vol. 17, (1982).

[7]   L. W. Casperson, "Spontaneous pulsations in lasers", in *Laser Physics*, J.D. Harvey and D.F. Walls, eds. (Springer Verlag, Heidelberg, 1983, pp. 107-131.

[8]   L. W. Casperson, "Spontaneaous coherent pulsations in ring-laser oscillators: simplified models", *Opt. Soc. Am. B*, vol. 2, N°1, pp.73-80 (1985)

[9]   S. Hendow and M. Sargent III, "The role of population pulsations in single-mode laser instabilities", *Opt. Comm.* **40**, pp385-390 (1982).

[10] S. Hendow and M. Sargent III, "Theory of single-mode laser instabilities", *J. Opt. Soc. Am. B,* vol. 2, N°1, pp.84-101 (1985).

[11] J. W. Swift and K. Wiesenfeld, "Suppression of period doubling in symmetric systems", *Phys. Rev. Lett.* **52**, 705-708 (1984).

[12] L. M. Narducci, H. Sadiky, L. A. Lugiato, and N. B. Abraham, "Experimentally accessible periodic pulsations of a single-mode homogeneously broadened laser (The Lorenz model), *Optics Comm.*, vol. 55, N° 5, 370-376 (1985).

[13] C. O. Weiss and J. Brock, "Evidence for Lorenz-Type Chaos in a laser", *Phys. Rev. Lett. Vol.* 57, N° 22, pp2804-2806 (1986).

[14] M. A. Dupertuis, R.R.E Salomaa, and M.R. Siegrust, "The conditions for Lorenz chaos in an optically-pumped far-infrared laser", *Optics Comm.*, vol. 57, N° 6, pp410-414 (1986).

[15] L. M. Narducci and N. B. Abraham; " *Laser physics and laser instabilities"*, World Scientific Publishing Co. Singapore, USA (1988)

[16] B. Meziane, *"Construction et etude de modèles simples en Dynamique des lasers: Réduction des systèmes de dimensions infinies"* Phd dissertation, ENSSAT, Université de Rennes I (1992).

[17] CP. Smith and R. Dykstra, "Lorenz like chaos in a Gaussian mode laser with a radially dependent gain", *Optics. Comm.* **117**, pp 107-110 (1995).

Belkacem Meziane

[18] L. Shil'nikov, " Homoclinic Phenomena in Laser Models", *Computers Math. Applica.* Vol. 34, N° 2-4, pp. 245-251 (1997).

[19] G. H. M. van Tartwijk and G. P. Agrawal, "Nonlinear dynamics in the generalized Lorenz-Haken model"; *Optics Comm.* **133**, pp 565-577 (1997).

[20] Chi-Chuan Hwang and al, "A nonlinear feedback control of the Lorenz equations", *Int. Journ. of Engin. Science* **37**, pp 1893-1900 (1999).

[21] M. Clerc, P. Collet, and E. Tirapegui, "The Maxwell-Bloch description of 1/1 resonances", *Optics Comm.* **167**, pp 159-164 (1999).

[22] Kenju Otsuka *"Nonlinear dynamics in optical complex systems"*, KTK Scientific publisher systems, Tokyo; Kluver Academic Publishers, Dordrecht, London, Boston, (1999) Chapter I

[23] F. Patti, E. M. Pessina, G. J. de Valcarel, and E. Roldan; " Lorenz-Haken instability in a laser with arbitrary mirrors reflectivity" *Optics Comm.* **185**, 153-157 (2000).

[24] S. Banerjee, P. Saha, and A. R. Chowdhury, "Chaotic aspects of lasers with host-induced nonlinearity and its control*", Physics Letters A* **291**, pp. 103-114 (2001)

[25] G. Kociuba and N.R. Heckenberg, "Controlling the complex Lorenz equations by modulation Source", *Phys. Rev. E*, vol.66, n°2 (2002)

[26] D. Viswanath, "Symbolic dynamics and periodic orbits of the Lorenz attractors"; Nonlinearity: (Bristol), vol.16, N°3, pp 1035-1056 (2003)

[27] Grigorenko and E. Grigorenko, "Chaotic dynamics of the fractional Lorenz system", *Phys. Rev. Lett.*, vol.91, n°3 (2003)

[28] El-Gohary and F. Bukhari, "Optimal control of Lorenz system during different time intervals"; *Applied Math. and Comp.* **144**, pp 337-351 (2003).

[29] Yu. Pogromsky, G. Santoboni, and H. Nijmeijer; "An ultimate bound on the trajectories of the Lorenz system and its applications" *Nonlinearity*: (Bristol.Print), vol. 16, N° 5, pp. 1597-1605 (2003).

[30] Yusuf Oysal, *"An Intelligent Control of Chaos in Lorenz System with a Dynamic Wavelet Network"*, in *Lecture notes in Computer Science*, Springer Berlin /Heidelberg (2004).

[31] J. Yang[1] and M. Zhang[2], "The instability of chaotic synchronization in coupled Lorenz systems: from the Hopf to the Co-dimension two bifurcation", *Eur. Phys. J. B* **47**, 251-254 (2005), doi: 10.1140/epjb/e2005-00315-0.

[32] Ya I. Khanin, *"Fundamental laser dynamics"* Cambridge Int. Science Publ. (2006), Chapter 3.

[33] S.Ayadi and B. Meziane, "Weak versus strong harmonic-expansion analyses of self-pulsing lasers: I-The Laser-Lorenz model", *Opt. Quant. Electron.* (2007) doi:10.1007/s11082-007-9065-9.

[34] Meziane and S. Ayadi, "Weak versus strong harmonic-expansion analyses of self-pulsing lasers: I-The inhomogeneously broadened system", *Opt. Quant. Electron.* (2007) doi:10.1007/s11082-007-9065-6.

[35] B. Meziane and S. Ayadi, "Third-order laser-field expansion analysis of the Lorenz-Haken equations", Optics Commun., Vol. 281, 4061-4067 (2008).

[36] B. Meziane, "Mechanic-like resonance in the Maswell-Bloch equations", *Eur. J. Phys.* **29**, 781-797 (2008)

[37] M. Ikezouhene, B. Meziane, and G. Stephan, *Optical and Quantum Electronics* **28**, 1029-1038 (1996)

[38] B. Meziane, P. Besnard, and G. Stephan, *IEEE J. Quantum Electronic*, Vol. 31, No 4, 617 (1995).

In: Atomic, Molecular and Optical Physics...   ISBN: 978-1-60456-907-0
Editor: L.T. Chen, pp. 109-146   © 2009 Nova Science Publishers, Inc.

Chapter 3

Rydberg Atoms, Molecules, Ionization, Radiation Trapping and Surface Physics in Laser Excited Dense Vapours of In and Ga Confined in Quartz Cells

P. Bicchi and S. Barsanti

Department of Physics, University of Siena, Via Roma 56, 53100 Siena, Italy

Abstract

This chapter gives an update of the study of laser assisted collisions in dense resonantly laser-excited In or Ga homo/heteronuclear vapours. In this way, Rydberg levels, as well as autoionizing levels and collisional ionization, have been evidenced for each element and the relative cross sections have been measured via laser induced fluorescence spectroscopy (LIF). The presence of a large density of excited atoms favours the collision of an atom in the fundamental state and another one in the first excited state to give a molecule directly in an excited electronic state from which it radiates. At the same time, the presence of a rather large density of ions can also favour the onset of molecular ions via collisions, which can radiate after electron/ion recombination and/or eventually dissociate. All of these processes produce molecular signals in the fluorescence spectrum following the resonant atomic excitation. In this way, some unresolved molecular bands and excimer emission have been detected. Special emphasis is dedicated to the description of the most recent LIF experiments performed in a heteronuclear In-Ga vapour by the resonant excitation of either element. Besides the evidence of homonuclear molecules, the presence of the GaIn molecules is inferred from the time-resolved analysis of some fluorescence features. The experiments are carried out in the 900–1100°C temperature range with the vapors confined in fused silica cells. The permanent modifications induced in the silica optical properties by the migration of the In/Ga atoms inside the silica matrix are also described. In this chapter, a section is dedicated to the analysis of how the latter effect, added to all the processes mentioned previously, affects the basic physical properties of laser-excited dense vapors as the radiation trapping, which turns out to be greatly reduced, if not quenched.

1. Introduction

1.a. General Remarks

Laser radiation gave a great stimulus to atomic and molecular spectroscopy and a huge amount of work, both theoretical and experimental, was produced on the subject. In spite of this there are still some elements that have been studied much less than others. Among these are gallium and indium. There is more than one reason for this occurrence: their resonance excitation wavelengths are not in immediately accessible spectral ranges, and their very limited vapour pressure even at very high temperatures [1] are accompanied by an extremely high chemical reactivity, which makes it very difficult to find an appropriate container for their vapours [2]. In contrast, these two elements exhibit several features of primary interest both for fundamental and applicative physics. A rich database of their atomic as well as molecular parameters is useful in spectroscopy, astrophysics etc., and their atomic structure makes them quite attractive to study the collisions between excited atoms [3] as well as their interaction with other atomic species [4,5] or with surfaces [6]. In addition, their vapour composition has a remarkable importance for the fabrication of various semiconductor devices. Up to now there has been very limited experimental knowledge of the composition of Ga and In vapours or of their mixtures, as the information on the electronic state structure of In_2, Ga_2 and GaIn as well as of other cluster species of these elements is quite scarce. The molecular constants are necessary to evaluate the right thermodynamic functions which allow the determination of the molecules' composition temperature dependence. Also, the Ga or In atomic parameters, which have been studied and classified in the past decades to some extent with absorption techniques [7,8] or two-photon laser spectroscopy [9-13], remain in part obscure and continue to present some irregularities to explain, and about which only hypotheses could be advanced. The theoretical investigation of some of these fundamental atomic parameters has been going on up to the present day [14]. A great step forward in the study of these elements was achieved with the application of the resonant laser assisted collisions technique.

1.b. Resonant Laser Assisted Collisions

The terminology "resonant laser assisted collisions (RLAC)" indicates a vast class of phenomena whose common trait is that they take place only when there is the simultaneous presence of the resonant laser field and the atomic collision. Quite frequently RLAC is referred to collisions taking place between two atoms transferred to their first excited state by resonant laser photons crossing an atomic dense vapour. This situation is referred to in the literature as "energy pooling reactions (EPR)" [3]. The results of EPR are several and to describe them the following symbolism is used throughout the chapter: Z denotes an atom in the ground state, Z* indicates an atom in the first excited state, Z** stands for an atom in a Rydberg level and Z^+ is an atomic ion. The superscripts *, ** and + have the same meaning when applied to molecules.

The outcomes of such collisions are both atomic and molecular. In the first case the common starting point is always the transfer of the internal energy of one of the colliding atoms to the other, followed by its relaxation to the ground state. The atom that collects the total amount of available energy can undergo several modifications. If it jumps to a Rydberg level, the process is called "energy pooling collision (EPC)" [15]. In case it ionizes with the production of an electron, the process is known as "energy pooling ionization (EPI)"[16] while if it transfers to an autoionizing level the process is indicated as "energy pooling autoionization (EPA)" [3]. The formation of a molecular ion during the collision is also possible in which case the process is referred to as "associative ionization (AI)" [17]. All these events are summarized in the following reactions:

EPC $\qquad$ $Z^* + W^* \rightarrow Z + W^{**} \pm \Delta E_{EPC}$ $\qquad$ (1)

AI $\qquad$ $Z^* + W^* \rightarrow ZW^+ + e^- \pm \Delta E_{AI}$ $\qquad$ (2)

EPI $\qquad$ $Z^* + Z^* \rightarrow Z + Z^+ + e^- \pm \Delta E_{EPI}$ $\qquad$ (3)

EPA $\qquad$ $Z^* + Z^* \rightarrow Z + Z^{**} \text{ (a.l.)} \pm \Delta E_{EPA}$ $\qquad$ (4)

Where Z^{**} (a.l.) stands for an atom in an autoionizing level and ΔE is, in any case, the absolute value of the energy balance. The latter has to be exchanged with the atomic system as kinetic energy and strongly influences the cross sections of the reactions which quickly fall off anytime ΔE gets bigger than a few kT, T being the experiment temperature.

Processes (1) and (2) have been known for decades and studied in a large variety of both homonuclear and heteronuclear vapours [3, 15, 18, 19] while processes (3) and (4) have been discovered more recently [20] and can take place only in gallium, indium and thallium as will be explained in section 1. c. The laser power density, P_L, and the atomic vapour density, N, are the essential parameters for EPR to be doubtlessly identified among other processes that are simultaneously present in the laser/vapour interaction such as multiphoton excitation and ionization. Typical values are 10^{12} cm^{-3} $\leq$ N $\leq 10^{14}$ cm^{-3} and, for pulsed laser excitation, 10^2 W/cm^2 $\leq P_L \leq 10^5$ W/cm^2.

EPR provide a powerful tool to extend the database of atomic and molecular parameters as well as the knowledge of both atomic and molecular interactions as they allow the population of levels not directly reachable by optical excitation and the measurements of their cross sections supply some news about the interatomic potentials in the middle - to long-distance range. This importance is enhanced when the semiconductor elements Ga and In are considered. In fact for these elements EPC directly and EPI followed by recombination produce population in very high Rydberg levels regardless of the selection rules or the exciting wavelengths [19, 21, 22]. In addition EPA allows the analysis of interconfiguration transitions [19, 23] while EPI ensures a way to produce ions from atoms in very deeply bound states [24, 25], provides a new approach to determine the electron - ion recombination coefficient [24] and plays an important role in the laser ablation of group III-IV compounds aiming to the realization of semiconductor devices [26]. The study of collisions in a resonantly excited dense Ga or In vapour has been the only way, up to now, to get some experimental information on the In$_2$ [27, 28] and Ga$_2$ [29] dimers.

At the end of this section it is worthy to mention also the "sensitized fluorescence (SF)" process [30] as belonging to RLAC considering it is an energy transfer collision process under any respect. In fact it consists in the energy transfer of the internal energy of one excited atom to an unexcited one, of the same or different species, during a collision, according to the reaction:

SF $$Z^* + W \rightarrow Z + W^* \pm \Delta E_{SF} \tag{5}$$

where ΔE_{SF} is again the energy defect or excess to be exchanged with the vapour, whose value can not exceed a few kT. SF is explicitly recalled in this context as it is important to explain some results obtained in the heteronuclear Ga+In vapours as detailed in the dedicated section.

1.c. Gallium and Indium

Gallium and indium are, together with thallium, the heavy elements of the third group. They share a notable concurrence of favourable items, summarized in fig. 1, which make them unique to be studied with the EPR technique. They have a ns^2np fundamental configuration resulting in a $n^2P_{1/2,3/2}$ ground state and a $(n+1)^2S_{1/2}$ first excited level. They possess the unique characteristic of having the first excited level at an energy equal to about half the ionization energy. In addition they exhibit an autoionizing level, arising from the $nsnp^2$ configuration, whose energy is close to the ionization one. All these features together with the presence of a fine structure in the ground state make possible the realization of the EPC, EPI and EPA processes schematized in fig. 2 and expressed in the following formulas, where H is any of the mentioned atoms:

$$H^* + H^* \rightarrow H_{1/2} + H^+ + e^- + \Delta E_{EPI} \tag{6}$$

$$H^* + H^* \rightarrow H_{3/2} + H^{**} \pm \Delta E_{EPC} \tag{7}$$

$$H^* + H^* \rightarrow H_{1/2} + H^{**} \text{ (a.l.)} \pm \Delta E_{EPA.} \tag{8}$$

Ga and In physical properties

	Ga	In
Atomic number	31	49
Atomic weight	69.723	114.818
Isotopes	^{69}Ga (60%), ^{71}Ga (40%)	^{115}In (95.7%), ^{113}In (4.3%)
Nuclear spin	3/2	9/2
Electronic configuration	[Ar] $3d^{10}$ $4s^2$ $4p$ $4^2P_{1/2}$	[Kr] $4d^{10}$ $5s^2$ $5p$ $5^2P_{1/2}$
Fine structure of the ground state	826.2 cm^{-1}	2212.6 cm^{-1}
Hyperfine structure of the ground state	89.328×10^{-3} cm^{-1}	0.38 cm^{-1}
First excited state	$5^2S_{1/2}$	$6^2S_{1/2}$
First excited state lifetime	6.2 ns	7.4 ns
First excited state energy	24788.6 cm^{-1} (3.1 eV)	24372.9 cm^{-1} (3.0 eV)
Ionization energy	48387.6 cm^{-1} (6.0 eV)	46670.1 cm^{-1} (5.8 eV)
Autoionizing state	[Ar] $3d^{10}$ $4s$ $4p^2$ $4^2D_{3/2, 5/2}$	[Kr] $4d^{10}$ $5s$ $5p^2$ $5^2D_{3/2, 5/2}$
Autoionizing state energy	53796.9cm^{-1} (6.7 eV)	49276.8 cm^{-1} (6.1 eV)
Melting point	29.76 °C	156.61 °C
Boiling point	2204 °C	2080 °C

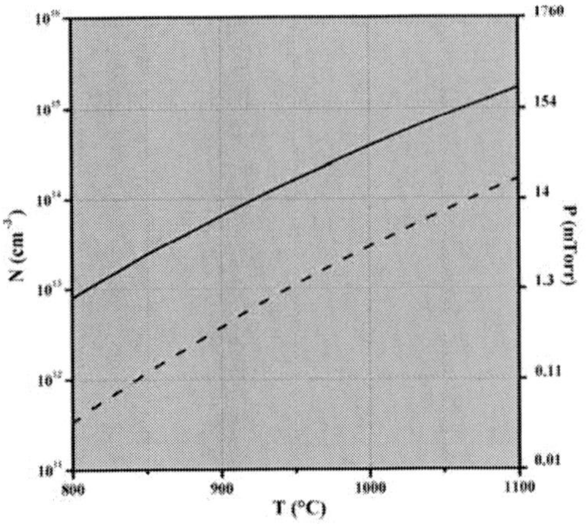

Figure 1. Ga and In physical properties and plot of their saturated vapour pressure and atomic density versus temperature. In (solid line); Ga (dashed line).

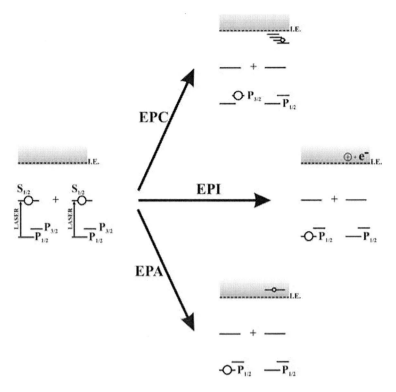

Figure 2. Graphic representation of EPR in Ga and In. The P state is the ground state; the S state is the first excited state and I.E. indicates the ionisation energy.

Process (6) is made evident by detecting directly the electrons produced during the collision. Processes (7) and (8) are revealed by recording the fluorescences arising from the Rydberg or the autoionizing levels after the resonant excitation of the vapour. No matter what the detection is, the EPR produced signal shows always a typical mark. It has a square dependence both on the laser power and on the atomic density. This means that it is detectable only under the simultaneous presence of the resonant photons and of a large atomic density, larger than the one needed to see signals arising from processes involving only one excited atom, such as the fundamental fluorescence.

It is worthy to spend few other words for the EPI process as, in principle, it can take place in any collision between atoms whose global internal energy is larger than the ionization energy. So it can be observed in any element provided that a population of its Rydberg levels is produced in whatever way. In the latter case, anyway, EPI would be masked by other processes, even more relevant, such as the interaction with the blackbody radiation and the Rydberg population redistribution due to collisions with the other atoms or to radiative decay. Instead when the EPI collision happens between two atoms in the first excited state, no such secondary process is present and the EPI identification is unique and straightforward. As a consequence of the latter consideration, the essential requisite for EPI to be studied is that twice the energy of the first excited state has to be about equal to the ionization energy. This requirement is fulfilled by gallium and indium with the only addition of thallium.

We focused our attention on In and Ga due to the fact that, besides the interesting spectroscopic features, they manifest some semiconductor characteristics of large applicative importance.

In the following we report about the experiments performed in our laboratory in the homonuclear Ga and In vapours kept in quartz cells and resonantly excited by pulsed laser radiation. Rydberg levels population and production of ions with the relative cross sections determinations are reviewed together with the analysis of the interconfiguration transitions and the molecules formation. Dedicated sections deals with the strong interaction arising between the Ga or In atoms and fused silica at the vapour/cell boundary surface and with the resulting anomalous radiation trapping. A section is devoted to the very last achievements accomplished in one of the very first experiments performed with a heteronuclear Ga+In vapour.

When needed, the results found in Ga or In vapours kept in a stainless steel heat pipe oven (HPO) [31] are also recalled.

2. Experimental Apparatus

The experimental set up comprises three main blocks: excitation, detection and acquisition and is described in detail in [19]. Here we recall the main features and describe the modifications or improvements introduced in recent years. For the sake of clarity, the apparatus is sketched in fig. 3.

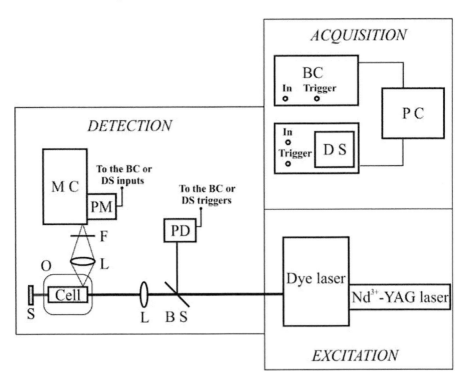

Figure 3. Scheme of the experimental apparatus. *DETECTION*: BS = beam splitter, L = lens, F = filter, PD = photodiode, MC = monochromator, PM = photomultiplier, O = oven, S = stop. *ACQUISITION*: PC = personal computer, DS = digital oscilloscope, BC = boxcar averager. *EXCITATION*: Dye laser pumped by a Nd^{3+}:YAG laser.

2.a. Excitation

The excitation block is formed by a pulsed dye laser + Nd^{3+}-YAG pump laser system. The dye laser active medium is a solution of the DCM dye in methanol which is pumped by the second harmonic at 532 nm of the Nd^{3+}-YAG laser. The resonant excitations of the fundamental transitions $5^2P_{1/2} \rightarrow 6^2S_{1/2}$ at 410.3 nm for In (see fig.4) and $4^2P_{1/2} \rightarrow 5^2S_{1/2}$ at 403.4 nm for Ga (see fig.11) are obtained by mixing the dye laser output at, respectively, 667 nm and 650 nm with the residual of the Nd^{3+}-YAG laser output at 1064 nm. The older results have been obtained with laser pulses of 10 Hz repetition rate, 10 ns duration and laser bandwidth of 24 GHz, essentially due to the bandwidth (21 GHz) of the Nd^{3+}-YAG laser. The substitution of the latter made possible the excitation with laser pulses with the same characteristics with the exceptions of the duration reduced to 8 ns and the bandwidth reduced to 4.8 GHz. The laser beam is partially focused to a spot of 4 mm diameter and the peak power is maintained in the range 1 kW – 300 kW, with the lower values used for electronic detection and the higher ones for fluorescence detection.

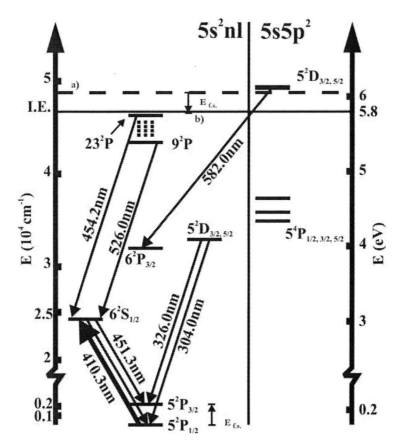

Figure 4. Simplified In energy level scheme with the main spectroscopical features relevant in this article. E_{fs} indicates the fine structure of the ground level; a) indicates twice the value of the $6^2S_{1/2}$ level energy and b) the a) value minus E_{fs}.

2.b. Detection

The exciting laser beam crosses, along its axis, a cylindrical quartz cell of 5 cm length and 3 cm diameter with a side arm in which metallic gallium and/or indium of 99.999% purity have been distilled under vacuum. A buffer gas can be added inside the cell when required. The cell is placed inside an oven that can be heated up to 1100 °C with ± 1°C stabilization. Such high temperatures are needed to reach the proper densities for these experiments due to the very low saturated vapour densities of these elements [1]. The fluorescence spectrum following the resonant excitation is collected perpendicularly to the cell axis and is dispersed in the range 250–700 nm by a 1000 mm monochromator with 18 cm^{-1} resolution connected to a photomultiplier. The output of the latter feeds either a box car integrator, when frequency resolved spectra are required, or a fast digital oscilloscope, when time resolved signals are requested.

The fluorescence detection is adequate to study EPC, but it revealed suitable also for EPI identification [32] even if it does not allow the EPI cross section determination. The direct detection of the EPI produced electrons is imperative for this purpose and this achievement requires the vapour to be kept in a stainless steel cross shaped HPO. As this chapter deals with Ga and In vapours kept in quartz cells, we skip the description of the electron detection and refer the interested reader to [3, 19, 24, 25].

2.c. Acquisition

The fluorescence signals acquisition can be done, as mentioned before, both in the time domain and in the frequency domain by using a fast digital oscilloscope or a box car integrator respectively. A small fraction of the exciting laser beam triggers both of them so that the acquisition is synchronous with the laser pulses. When a time resolved fluorescence signal is requested, the photomultiplier load resistor is 50 Ω and its output is sent to a 500 MHz bandwidth digital scope that averages the signal over a definite number of trigger pulses and whose output feeds a personal computer for data processing and analysis. For a frequency resolved fluorescence spectrum, instead, the output of the photomultiplier, loaded with 1 MΩ resistor, is sent to the boxcar whose gate and delay have been appropriately chosen. The signal is averaged over several trigger pulses and then is acquired by the personal computer which drives the wavelength scan of the monochromator synchronously with the data acquisition.

3. Homonuclear Vapours

3.a. Indium

3.a.1. EPC, EPI and EPA

Indium has the electronic configuration [Kr] $4d^{10}5s^25p$ resulting in the $5^2P_{1/2,3/2}$ ground state with a fine structure $E_{f.s.} = 2212.6$ cm^{-1}. The first excited state is the $6^2S_{1/2}$ at an energy of 24372.9 cm^{-1} very close to half the ionization one (46670.1 cm^{-1}). In addition the [Kr]

$4d^{10}5s5p^2$ configuration gives rise to the autoionizing levels $5s5p^2$ $5^2D_{3/2,5/2}$ at an energy, averaged over the J multiplets, of 49276.8 cm^{-1}. These features allow all the three EPC, EPI and EPA processes which in indium become:

$$\text{In } (6^2S_{1/2}) + \text{In } (6^2S_{1/2}) \rightarrow \text{In}^{**} (nL) + \text{In } (5^2P_{3/2}) \pm \Delta E \qquad (9)$$

$$\text{In } (6^2S_{1/2}) + \text{In } (6^2S_{1/2}) \rightarrow \text{In}^+ + e^- + \text{In } (5^2P_{1/2}) + 2075.8 \text{ cm}^{-1} \qquad (10)$$

$$\text{In } (6^2S_{1/2}) + \text{In } (6^2S_{1/2}) \rightarrow \text{In}^{**} (5s5p^2\ 5^2D) + \text{In } (5^2P_{1/2}) - 554 \text{ cm}^{-1} \qquad (11)$$

where In** (nL) indicates any state above the $6^2S_{1/2}$ one.

An In energy level diagram, comprising the main features relevant for this review, is sketched in fig. 4.

Table 1. List of the fluorescences from the Rydberg levels detected in In vapour kept in a quartz cell at T = 950°C (kT = 850 cm^{-1}) and excited by resonant laser pulses at 410.3 nm. λ = transition wavelength, E = energy of the Rydberg level, $\Delta E = [2E(6^2S_{1/2})-E_{f.s.}-E]$.

Transition	λ (nm)	Upper level	E$^{(a)}$ (cm^{-1})	ΔE (cm^{-1})	$\Delta E/kT$
9P $\rightarrow$ 6S$_{1/2}$	526.0	9P	43384.3	3148.9	3.70
10P $\rightarrow$ 6S$_{1/2}$	502.2	10P	44284.7	2248.5	2.65
11P $\rightarrow$ 6S$_{1/2}$	488.1	11P	44859.5	1673.7	1.97
12P $\rightarrow$ 6S$_{1/2}$	479.0	12P	45249.3	1283.9	1.51
13P $\rightarrow$ 6S$_{1/2}$	472.8	13P	45525.0	1008.2	1.19
14P $\rightarrow$ 6S$_{1/2}$	468.0	14P	45742.1	791.1	0.93
15P $\rightarrow$ 6S$_{1/2}$	465.0	15P	45880.5	652.7	0.77
16P $\rightarrow$ 6S$_{1/2}$	462.4	16P	45999.0	534.2	0.63
17P $\rightarrow$ 6S$_{1/2}$	460.4	17P	46092.9	440.3	0.52
18P $\rightarrow$ 6S$_{1/2}$	458.8	18P	46168.2	365.0	0.43
19P $\rightarrow$ 6S$_{1/2}$	457.5	19P	46229.9	303.3	0.36
20P $\rightarrow$ 6S$_{1/2}$	456.4	20P	46280.8	252.4	0.30
21P $\rightarrow$ 6S$_{1/2}$	455.6	21P	46323.3	209.9	0.25
22P $\rightarrow$ 6S$_{1/2}$	454.8	22P	46359.3	173.9	0.20
23P $\rightarrow$ 6S$_{1/2}$	454.2	23P	46389.9	143.3	0.17
24P $\rightarrow$ 6S$_{1/2}$	453.6	24P	46416.2	117.0	0.14
25P $\rightarrow$ 6S$_{1/2}$	453.2	25P	46438.9	94.3	0.11
26P $\rightarrow$ 6S$_{1/2}$	452.8	26P	46458.6	74.6	0.09
27P $\rightarrow$ 6S$_{1/2}$	452.4	27P	46476.4	56.8	0.07
5D$_{3/2}$ $\rightarrow$ 5P$_{1/2}$	304.0	5D$_{3/2}$	32892.1	13641.1	16.05
5D$_{3/2}$ $\rightarrow$ 5P$_{3/2}$	326.0	5D$_{3/2}$	32892.1	13641.1	16.05
5D$_{5/2}$ $\rightarrow$ 5P$_{3/2}$	325.7	5D$_{5/2}$	32915.4	13617.8	16.02

(a) from references [11, 34]

E for the nP levels is averaged over the J multiplets.

Equation (9) is energetically resonant with n $\approx$ 30 and allows the population of Rydberg levels [33] via the unusual way of an elementary collision. As a matter of fact, when the temperature of the cell containing the In vapour is raised to 950°C (atomic densities $\approx 10^{14}$

cm^{-3}), the atomic fluorescences corresponding to the $n^2P \to 6^2S_{1/2}$ series with $9 \le n \le 23$ are detected after irradiation with resonant laser pulses at 410.3 nm together with the fluorescences corresponding to the $5^2D_{3/2,5/2} \to 5^2P_{1/2,3/2}$ transitions at 304.0 and 326.0 nm. Table 1 summarizes all the observed transitions, their wavelengths, the upper level involved, its energy E [11, 34], the corresponding value $\Delta E = [2E(6^2S_{1/2}) - E_{f.s.} - E]$ and the ratio ΔE /kT at T = 950°C (kT = 850 cm^{-1}). The portions of the fluorescence spectrum with the observed lines, taken under different temperature and excitation conditions, are shown in fig. 5. Specifically:

5(a) the laser beam is resonant at 410.3 nm and the temperature is T = 700°C
5(b) the laser beam is out of resonance and the temperature is T = 950°C
5(c) the laser beam is resonant at 410.3 nm and the temperature is T = 950°C.

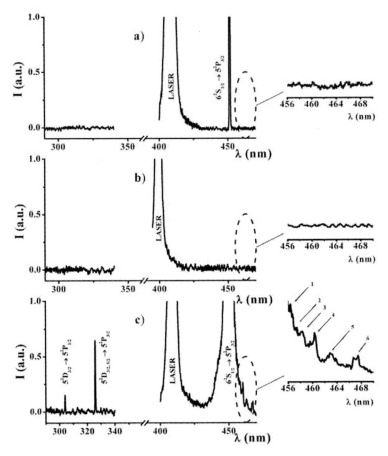

Figure 5. Portions of the fluorescence spectrum recorded in a quartz cell with In under different excitation conditions. a) laser resonant at 410.3 nm and T = 700°C; b) laser out of resonance and T = 950°C; c) laser resonant at 410.3 nm and T=950°C. The inserts reproduce, enlarged by a factor 3, the dotted sections. The laser power density is always $P_L = 100$ kW/cm^2. Peak 1 corresponds to the transition $20^2P \to 6^2S_{1/2}$, Peak 2 corresponds to the transition $19^2P \to 6^2S_{1/2}$, Peak 3 corresponds to the transition $18^2P \to 6^2S_{1/2}$, Peak 4 corresponds to the transition $17^2P \to 6^2S_{1/2}$, Peak 5 corresponds to the transition $16^2P \to 6^2S_{1/2}$, Peak 6 corresponds to the transition $15^2P \to 6^2S_{1/2}$.

It is evident that all the lines show the typical mark of an energy pooling generated signal. In fact when there are the resonant photons but the temperature is still so low to give a density of atoms ($\approx 10^{11}$ cm^{-3}) only larger than the threshold necessary to observe the fundamental fluorescence, only the latter is present, fig. 5a. When the temperature is high enough to give an atomic density higher than the threshold needed to observe non linear effect ($10^{13} \div 10^{14}$ cm^{-3}) but the resonant photons are absent, no signal is observed, fig. 5b. But when there is the simultaneous presence of both the resonant photons and the very high atomic density, all the energy pooling produced fluorescences appear, fig. 5c.

A further proof of process (9) being the origin of the observed lines is given by the investigation of the fluorescence intensities versus the atomic density of the $6^2S_{1/2}$ level, $N(6^2S_{1/2})$, and the laser power density, P_L. In fact two $6^2S_{1/2}$ atoms and two resonant photons are required in the collision (9). As a consequence, any signal originated by this collision must show a quadratic dependence on both $N(6^2S_{1/2})$ and P_L. The results, for the intensity of the transition $17P \rightarrow 6^2S_{1/2}$, I_{17P}, are reported in fig.6. In fig. 6a, I_{17P} is plotted versus the square of the intensity, $I_{451.3}$, of the fundamental transition $6^2S_{1/2} \rightarrow 5^2P_{3/2}$, $\lambda = 451.3$ nm, which depends linearly on $N(6^2S_{1/2})$, for several temperature values in the range $820 - 870°C$. Actually the linearity between $I_{451.3}$ and $N(6^2S_{1/2})$ is based on the assumption that the radiative emission prevails on any other depopulation mechanism. The processes (9), (10) and (11) contribute, as a matter of fact, to the depopulation of the $6^2S_{1/2}$ level, but in the mentioned temperature range, even present, they are still negligible if compared with the radiative one [24, 32]. In this situation almost the total population of the level contributes to the fluorescence intensity and the linearity between $I_{451.3}$ and $N(6^2S_{1/2})$ holds. The plot of the square root of I_{17P} versus P_L at T = 950° is reproduced in fig. 6b and the linearity manifests evidently.

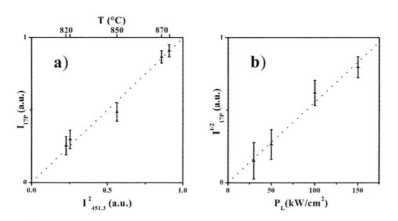

Figure 6. (a) Intensity of the fluorescence of the transition $17^2P \rightarrow 6^2S_{1/2}$, I_{17P}, versus the square of the intensity of the fundamental fluorescence at 451.3 nm, $I_{451.3}$. The exciting laser is resonant at 410.3 nm and its power density is 100 kW/cm^2. (b) Square root of the fluorescence intensity of the transition $17^2P \rightarrow 6^2S_{1/2}$ versus the laser power density recorded at T = 950°C. The laser is resonant at 410.3 nm.

From all these checks it is confirmed that the origin of most of the florescences listed in table 1 come from the EPC (9). The fluorescences $5^2D_{3/2,5/2} \rightarrow 5^2P_{1/2,3/2}$ can not be generated by EPC as ΔE in equation (9) is > 16 kT even at 950°C, a value that excludes the direct population of the levels $5^2D_{3/2,5/2}$ via EPC. The latter are populated via EPI followed by electron/ion recombination, which produce population in the sets of Rydberg levels immediately underneath the ionization limit, and subsequent cascade transitions. This procedure is confirmed by a temporal analysis of the signal as, while the EPC produced fluorescences have to be synchronous with the laser excitation [22], that is with the fundamental fluorescence, the signals corresponding to the $5^2D_{3/2,5/2} \rightarrow 5^2P_{1/2,3/2}$ transitions must be delayed due to their population dynamics [32]. This is just what happens as shown in fig. 7 where the intensities of the transitions $5^2D_{3/2,5/2} \rightarrow 5^2P_{3/2}$, I_{5D}, $13P \rightarrow 6^2S_{1/2}$, I_{13P}, and $I_{451.3}$ are plotted versus time. Fluorescence so, is also an indirect method to detect EPI, whose cross section, σ_{EPI}, comes out $\sigma_{EPI} = (1.2 \pm 0.6) \times 10^{-16}$ cm^2 when calculated from the direct detection of the electrons produced in reaction (10) inside a HPO [24].

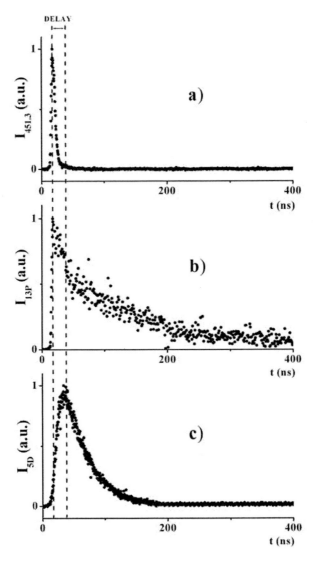

Figure 7. Intensity of the In fluorescences at 451.3 nm, $6^2S_{1/2} \rightarrow 5P_{3/2}$, (a) at 472.8 nm, I_{13P}, $13^2P \rightarrow 6^2S_{1/2}$, (b) and at 326.0 nm, I_{5D}, $5^2D_{3/2,\ 5/2} \rightarrow 5^2P_{3/2}$, (c) plotted versus time. The signals are recorded in the quartz cell with In at T = 950°C following resonant excitation at 410.3 nm.

Few words are to be spent to clarify why we detect a limited number of lines in spite of ΔE in equation (9) being less than 0.1 kT up to n ≈ 50. For large n values, the radiative emission is strongly influenced by other level depopulation mechanisms such as the collisions with other atoms and the interaction with the blackbody radiation which may prevail. Fluorescences from these levels are detected only when the radiative rate, Ψ_{rad}, is larger than the collisional rate, Ψ_{coll}, and the blackbody radiation one, Ψ_{blk}. The latter is given by the formula $\dfrac{4\alpha^3 kT}{3n^{*2}}$ [35], where α is the fine structure constant and n* is the principal effective quantum number [9]. Ψ_{blk} comes out two orders of magnitude lower than Ψ_{rad} for n ≥ 6 [22] and plays no role. As to the collisions, Ψ_{coll} is given by $\Psi_{coll} = Nv_m\sigma$, where N is the atomic density inside the cell, $v_m = 1.13\sqrt{\dfrac{2kT}{M}}$ is the mean interatomic velocity and σ is the kinetic cross section. The inverse of the level lifetime, that scales as n^{*3}, equals Ψ_{rad}. Collisions with the other atoms become particularly relevant when a buffer gas is added inside the cell. The collisional rate in this case has been named Ψ_{buf} and the values calculated when 10^3 Pa of Ar (atomic density = 3.2×10^{17} cm^{-3}) are added inside the cell, are plotted in fig. 8 together with Ψ_{coll} and Ψ_{rad} versus n at T = 950°C.

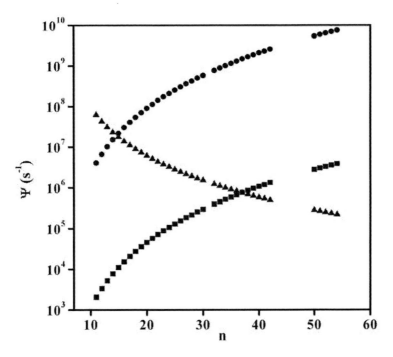

Figure 8. Order of magnitude at T = 950°C of the depopulation radiative rate Ψ_{rad} (7), of the depopulation collisional rate Ψ_{buf} in presence of 10^3 Pa of buffer gas (,) and of the depopulation collisional rate Ψ_{coll} in absence of buffer gas (!), as function of the principal quantum number n in a cell with In.

In absence of a buffer gas, Ψ_{rad} prevails for the levels with n < 36, while those with n > 36 are depopulated mainly by collisions. The fluorescences relative to transitions from the n^2P

levels with $24 \leq n \leq 36$ can not be recorded as they are overlapped by the In fundamental transition at 451.3 nm due to their wavelengths falling within the fundamental fluorescence linewidth. In presence of 10^3 Pa of buffer gas inside the cell, Ψ_{buf} prevails on Ψ_{radf} already for $n \geq 14$ and most of the lines disappear as testified in fig. 9 which shows the same portion of the spectrum reported in fig.5c, recorded both in a cell with 10^3 Pa of Ar, fig. 9a, and in a cell without buffer gas, fig. 9b, at T = 950°C under the same excitation conditions. The quenching of the fluorescences from the Rydberg levels with $n \geq 15$ is evident.

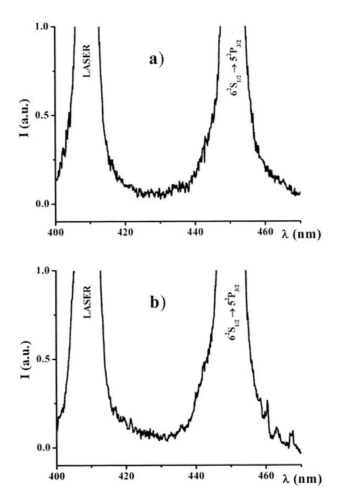

Figure 9. Portion of the fluorescence spectrum recorded at T = 950°C in a quartz cell with In + 10^3 Pa of Ar (a) and in a quartz cell with In (b) following resonant excitation at 410.3 nm. The laser power density is 100 kW/cm^2 in both cases.

The EPC cross section can be derived by the intensities of the detected fluorescences in a way detailed in [22, 32] which is here summarized. The starting point is the rate equation representing the time evolution of any Rydberg level population:

$$\frac{dN(nL)}{dt} = N^2(6^2S_{1/2})\xi_{nL} - \sum_{n'L'} N(nL)A(nL \rightarrow n'L') + \qquad (12)$$

$$+ \sum_{n''L''} A(n''L'' \rightarrow nL)N(n''L'') + N^+\xi' - N(nL)\xi''$$

where N is the atomic density of the considered level, N^+ is the density of the ions produced by the EPI process (10), $\xi_{nL} = \sigma_{nL}v_m$ is the EPC rate for the nL level, σ_{nL} is the EPC cross section for the nL level, $A(\varphi \rightarrow \psi)$ is the spontaneous transition probability for $|\varphi\rangle \rightarrow |\psi\rangle$, n'L' is any state lying below the nL level, n''L'' is any excited state with energy higher than that of the nL level, ξ' is the recombination rate and ξ'' is the rate for depopulation of the Rydberg levels but for radiative transitions.

Equation (12) can be solved following the same procedure detailed in [32], to give σ_{nP} for the n^2P levels under consideration. To reach this goal some simplifications are possible in equation (12). The last term in eq. (12) expresses the effect of the collisional depopulation and the interaction with the black body radiation which can be neglected in respect to the spontaneous decay (see fig.8). The effects of the recombination which follows EPI, fourth term in eq. (12), and the contribution from the radiative transitions from the n''L'' excited states, third term in eq. (12), are both negligible because these processes would introduce a delay between the fluorescences from the n^2P and the $6^2S_{1/2}$ levels which is not observed (see fig.7a and 7b).

With these simplifications the solution of eq. (12) gives

$$\sigma_{nP} = 2i_{nP}\delta_{nP} \frac{I(n^2P \rightarrow 6^2S_{1/2})v_{6^2S_{1/2} \rightarrow 5^2P_{3/2}}}{I(6^2S_{1/2} \rightarrow 5^2P_{3/2})v_{n^2P \rightarrow 6^2S_{1/2}}N(6^2S_{1/2})v_m\tau} \qquad (13)$$

where the statistical factor 2 is due to the homonuclear colliding partners [22], τ is the effective lifetime of the $6^2S_{1/2}$ level, which has to be measured in any specific experimental condition, i_{nP} is an instrumental factor that takes into account the different instrumental response to the different wavelengths, δ_{nP} is the branching ratio of the analysed transition whose intensity I is related to the transition probability between the two involved states by the relation:

$$I(\varphi \rightarrow \psi) = i_{\varphi\psi}A(\varphi \rightarrow \psi)N_\varphi v_{\varphi \rightarrow \psi} \qquad (14)$$

$v_{\varphi \rightarrow \psi}$ being the frequency of the transition.

The results of this calculations give values of σ_{nP} that converge to the EPC cross section $\sigma_{EPC} = (1.4 \pm 0.5) \times 10^{-15}$ cm^2 [22].

All the lines so far analysed correspond to transitions between levels of the same configuration. In addition the fluorescence spectrum reveals also a very weak, broad line at 582 nm that is due to the interconfiguration transition $5s5p^2$ $^2D \rightarrow 5s^26p$ $6^2P_{3/2}$. The level $5s5p^2$ ^2D is the autoionizing level populated via the EPA process represented by eq. (11). The mentioned interconfiguration transition manifests all the typical mark of a pooling produced signal, fig. 10, and its intensity is about 30 times smaller than the intensity of the line starting from the 11P level. From the intensity ratio of the two lines, the lower limit of the EPA cross section can be estimated and we obtain $\sigma_{EPA} \sim 10^{-17}$ cm^2 [36]. The actual cross section is anyhow larger as the direct ionization, which is the main source of depopulation for an autoionizing level, has been totally neglected in the σ_{EPA} evaluation.

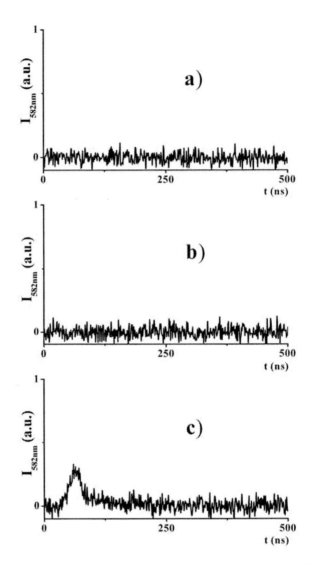

Figure 10. Intensity of the line at 582 nm, corresponding to the transition $5s5p^2 \; 5^2D \rightarrow 5s^26p \; 6^2P_{3/2}$, versus time recorded in a cell with In in different experimental condition: a) laser resonant at 410.3 nm, $T = 700°C$; b) laser detuned from resonance, $T = 950°C$; c) laser resonant at 410.3nm, $T = 950°C$.

3.a.2. Molecules

Experiments with a resonantly excited dense In vapour can also cast some light on the still rather unknown In_2 molecules. The above described experiments are characterized by a large production of In atoms in the first excited state which makes possible the formation of a molecule directly in an excited state in the collision of an excited atom with a ground state one. The consequent molecular emission appears in the fluorescence spectrum following the resonant excitation of the atomic vapour and can be analyzed. It was in this way that the first LIF investigation of the indium dimer has been done and some unresolved bands and excimer emission have been discovered [27, 28]. Here we show in fig. 11 the latest results obtained by illuminating a cell with In vapour at 1000°C with 200 kW/cm^2 pulsed laser radiation at 410.3 nm. Around the two fundamental fluorescences at 410.3 and 451.3 nm there are two molecular bands appearing essentially as blue wing shoulders of the atomic transitions. Some peaks can be resolved and the bands extend from 395 to 422 nm (25316 ÷ 23697 cm^{-1}) and from 435 to 458 nm (22988 ÷ 21834 cm^{-1}), respectively. On the basis of the very limited theoretical calculations available for the In$_2$ electronic state structure [37] an identification can be advanced. The first band is likely due to the transition $^3\Pi_g$ (II) $\rightarrow$ $^3\Pi_{2u}$, for which the quoted calculation yields T = 25234 cm^{-1}, and the second is likely due the transition $^1\Delta_u \rightarrow$ $^1\Delta_{2g}$, for which the quoted calculation yields T = 21991 cm^{-1}. The state $^3\Pi_g$ (II) correlates with the atomic states In ($6^2S_{1/2}$) + In ($5^2P_{1/2,3/2}$) and is so energetically accessible. The state $^1\Delta_u$ correlates with the atomic states In ($6^2P_{1/2,3/2}$) + In ($5^2P_{1/2,3/2}$) but it can be accessed via the crossing with the state $^3\Pi_g$ (II) [37].

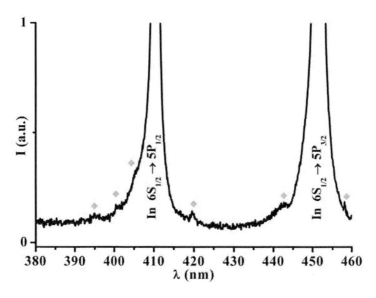

Figure 11. Portion of the fluorescence spectrum, following excitation resonant with the In fundamental transition at 410.3 nm, detected at T = 1000°C in a quartz cell containing In. ♦ indicates molecular peaks.

3.b. Gallium

3.b.1. EPI and EPA

Gallium has the electronic configuration [Ar] $3d^{10}4s^24p$ resulting in the $4^2P_{1/2,3/2}$ ground state with a fine structure of 826.2 cm^{-1}. The first excited state is the $5^2S_{1/2}$ at an energy of 24788.6

cm^{-1} which is very close to half the ionization one (48387.6 cm^{-1}). In addition the [Ar] $3d^{10}4s4p^2$ configuration gives rise to the autoionizing levels $4s4p^2$ $4^2D_{3/2,5/2}$ at an energy, averaged over the J multiplets, of 53796.9 cm^{-1}. The Ga atomic level structure, including the autoionizing level at an energy close enough to the ionization one, is quite similar to that of In. Two are the differences that introduce some peculiarities in the Ga behaviour. The first is the lower saturated vapour pressure, see fig. 1, which requires a temperature of $\approx 1100°C$ to reach Ga atomic densities $\approx 10^{14}$ cm^{-3}. The second is related to the fine structure of the ground state which is so small that only EPI and EPA can take place in the resonantly excited dense Ga vapour according to these reactions:

$$Ga\,(5^2S_{1/2}) + Ga\,(5^2S_{1/2}) \rightarrow Ga^+ + e^- + Ga\,(4^2P_{1/2}) + 1189.5 \text{ cm}^{-1} \tag{15}$$

$$Ga\,(5^2S_{1/2}) + Ga\,(5^2S_{1/2}) \rightarrow Ga^+ + e^- + Ga\,(4^2P_{3/2}) + 363.3 \text{ cm}^{-1}. \tag{16}$$

$$Ga\,(5^2S_{1/2}) + Ga\,(5^2S_{1/2}) \rightarrow Ga^{**}\,(4s4p^2\ 4^2D) + Ga\,(4^2P_{1/2}) - 4219.7 \text{ cm}^{-1} \tag{17}$$

A simplified Ga energy level scheme comprising the main features relevant in this review, is drawn in fig. 12.

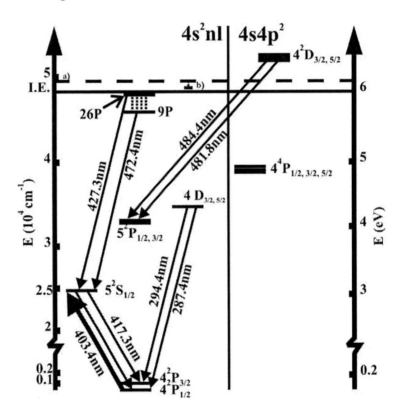

Figure 12. Simplified Ga energy level scheme with the main spectroscopical features relevant in this chapter. a) indicates twice the value of the $5^2S_{1/2}$ level energy and b) the a) value minus the fine structure of the ground state.

The fact that both reactions (15) and (16) give rise to EPI is the reason for the latter to be more efficient than in In. The EPI cross section in Ga, again determined by collecting directly the electrons produced in a HPO, comes out $\sigma_{EPI} = (1.1 \pm 0.5) \times 10^{-15}$ cm^2 [25], that is one order of magnitude larger than in In.

Table 2. List of the fluorescences from the Rydberg levels detected in Ga vapour kept in a quartz cell at T = 1050°C (kT = 920 cm^{-1}) and excited by resonant laser pulses at 403.4 nm. λ = transition wavelength, E = energy of the Rydberg level, ΔE = ionisation energy minus that of the upper level.

Transition	λ (nm)	Upper level	E$^{(a)}$ (cm^{-1})	ΔE (cm^{-1})	ΔE/kT
9P $\rightarrow$ 5S$_{1/2}$	472.4	9P	45958.8	2428.8	2.64
10P $\rightarrow$ 5S$_{1/2}$	459.5	10P	46551.0	1836.6	2.00
11P $\rightarrow$ 5S$_{1/2}$	451.3	11P	46948.2	1439.4	1.56
12P $\rightarrow$ 5S$_{1/2}$	445.6	12P	47229.0	1158.6	1.26
13P $\rightarrow$ 5S$_{1/2}$	441.6	13P	47434.9	952.7	1.04
14P $\rightarrow$ 5S$_{1/2}$	438.6	14P	47589.9	797.7	0.87
15P $\rightarrow$ 5S$_{1/2}$	436.3	15P	47710.7	676.9	0.74
16P $\rightarrow$ 5S$_{1/2}$	434.5	16P	47805.8	581.8	0.63
17P $\rightarrow$ 5S$_{1/2}$	433.0	17P	47882.4	505.2	0.55
18P $\rightarrow$ 5S$_{1/2}$	431.9	18P	47944.4	443.2	0.48
19P $\rightarrow$ 5S$_{1/2}$	430.9	19P	47995.8	391.8	0.43
20P $\rightarrow$ 5S$_{1/2}$	430.1	20P	48038.5	349.1	0.38
21P $\rightarrow$ 5S$_{1/2}$	429.4	21P	48074.9	312.7	0.34
22P $\rightarrow$ 5S$_{1/2}$	428.9	22P	48105.7	281.9	0.31
23P $\rightarrow$ 5S$_{1/2}$	428.4	23P	48132.5	255.1	0.28
24P $\rightarrow$ 5S$_{1/2}$	428.0	24P	48155.2	232.4	0.25
25P $\rightarrow$ 5S$_{1/2}$	427.6	25P	48175.4	212.2	0.23
26P $\rightarrow$ 5S$_{1/2}$	427.3	26P	48193.7	193.9	0.21
4D$_{3/2}$ $\rightarrow$ 4P$_{3/2}$	294.5	4D$_{3/2}$	34781.7	13605.9	14.79
4D$_{5/2}$ $\rightarrow$ 4P$_{3/2}$	294.4	4D$_{5/2}$	34787.8	13599.8	14.78
4D$_{3/2}$ $\rightarrow$ 4P$_{1/2}$	287.4	4D$_{3/2}$	34781.7	13605.9	14.79

(a) from reference [13].

E for the nP levels is averaged over the J multiplets.

When the Ga vapour is confined in a quartz cell, in the low-density plasma produced by EPI, electron/ion recombination is favoured and a high degree of Rydberg states population is reached, both directly in the recombination or after cascade transitions following the former, also for this element as testified by the atomic fluorescences present in the spectrum recorded after resonant excitation of the vapour at 403.4 nm when the temperature is raised to 1050°C. In this way the lines corresponding to the n^{2}P series with $9 \leq n \leq 26$ and to the 4D$_{3/2,5/2}$ $\rightarrow$ 4^{2}P$_{1/2,3/2}$ are detected. Their characteristics are listed in table 2 and a portion of the spectrum is shown in fig. 13 for the three essential experimental conditions that reveal these fluorescences to manifest all the signatures of an energy pooling produced effect. The details are indicated in the figure caption, but once more the signals are present only when there is the simultaneous presence of the resonant photons and of an atomic density larger than the threshold value necessary to produce collisional non linear effects. All the lines underwent the same checks described for indium including the time resolved analysis. In fact the electron/ion recombination provides an immediate population only in those Rydberg levels whose energy is within few kT to the ionization energy. For the levels $4D_{3/2,5/2}$, due to their lying more than 14 kT from the ionization, see table 2, cascade transitions must be accounted for with the introduction of a delay in their population, as it was the case for the $5D_{3/2,5/2}$ levels in In. This behaviour is testified in fig. 14 where the intensities of the transitions $4^2D_{3/2,5/2} \rightarrow 5^2P_{3/2}$, I_{4D}, and $26P \rightarrow 5^2S_{1/2}$, I_{26P}, are plotted versus time together with the intensity $I_{417.3}$ of the fundamental transition $5^2S_{1/2} \rightarrow 4^2P_{3/2}$. The similarity with In is completed by the presence in the fluorescence spectrum of two very week lines at 481.8 and 484.4 nm corresponding to the interconfiguration transitions $4s4p^2\ 4^2D_{3/2} \rightarrow 4s^25p\ 5^2P_{1/2}$, and $4s4p^2\ 4^2D_{3/2} \rightarrow 4s^25p\ 5^2P_{3/2}$, respectively, which show all the typical mark of a pooling produced signal [23].

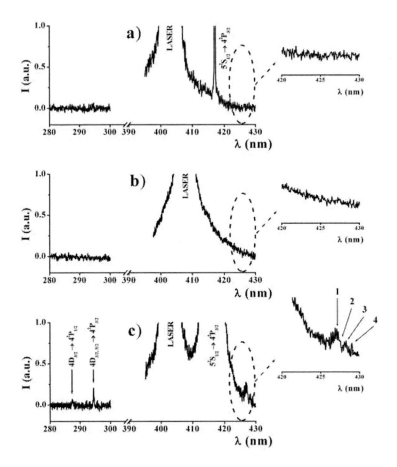

Figure 13. Portions of the fluorescence spectrum recorded in a quartz cell with Ga under different excitation conditions. a) laser resonant at 403.4 nm and T = 800°C; b) laser out of resonance and T = 1050°C; c) laser resonant at 403.4 nm and T = 1050°C. The inserts reproduce, enlarged by a factor 3, the dotted sections. The laser power density is always 110 kW cm^{-2}. Peak 1 corresponds to the transition $26^2P \rightarrow 5^2S_{1/2}$, peak 2 to the transition $25^2P \rightarrow 5^2S_{1/2}$, peak 3 to the transition $24^2P \rightarrow 5^2S_{1/2}$ and peak 4 to the transition $23^2P \rightarrow 5^2S_{1/2}$.

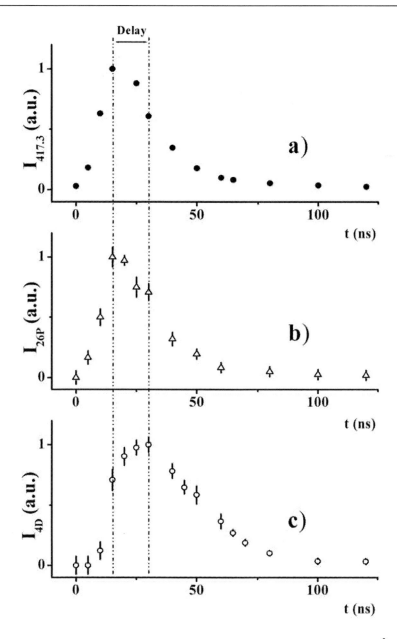

Figure 14. Intensity of the Ga fluorescences at 417.3 nm, $I_{417.3}$, (a), at 427.3 nm, I_{26P}, $26^2P \to 5^2S_{1/2}$, (b) and at 294.4 nm, I_{4D}, $4^2D_{3/2, 5/2} \to 4^2P_{3/2}$, (c) plotted versus time. The signals are taken in a quartz cell with Ga at $T = 1050°C$ following resonant excitation at 403.4 nm.

3.b.2. Molecules

Experiments with a resonantly excited dense Ga vapour may provide also some insight in the molecular characteristics of the gallium dimer in the same way and for the same reasons detailed in section *3.a.2* for In. In fact fig. 15 reproduces a portion of the fluorescence spectrum recorded, in the latest performed experiments, from a quartz cell with Ga at 1100°C after irradiation with 200 kW/cm^2 pulsed laser radiation at 403.4 nm. In addition to the fundamental atomic lines, a band extended from 389 to 414 nm (24155 – 25640 cm^{-1}) appears evident. From the comparison with the available theoretical data of some electronic states of Ga$_2$ [38, 39], this band is likely due to the transition from the excited molecular state $^1\Delta_u$ to the ground molecular state $^1\Delta_g$ for which the calculation yields T = 24498 cm^{-1} [38]. The population in the excited molecular state presumably results from the collisional combination between excited and ground state Ga atoms. The $^1\Delta_u$ state as a matter of fact correlates with Ga (4P$_{1/2, 3/2}$) + Ga (5P$_{1/2, 3/2}$) separated atoms, but it can be populated via the crossing with the state $^3\Pi_g$(II) which correlates with Ga (5^{2}S$_{1/2}$) + Ga (4^{2}P$_{1/2, 3/2}$) [38] and is so energetically accessible.

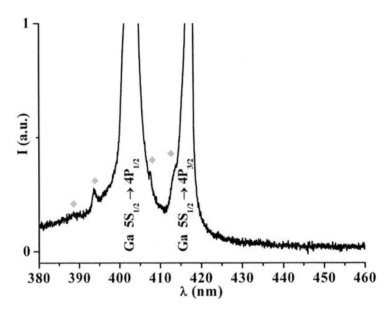

Figure 15. Portion of the fluorescence spectrum, following excitation resonant with the Ga fundamental transition at 403.4 nm, detected at T = 1100°C in a quartz cell containing Ga. ♦ indicates molecular peaks.

4. In(Ga)/SiO$_2$ Interaction

Among the many peculiar characteristics of In and Ga, a special significancy is laid, in this contest, upon their high chemical reactivity at the high temperatures. This is the reason why it is difficult to find a container for the high dense vapours of these elements. Stainless steel HPO become porous after some hours of operation [2] but also the fused silica cells get permanently "contaminated" by the In/Ga atoms. What happens is that beyond a threshold temperature (T$_{In}$ = 850°C for In and T$_{Ga}$ = 950°C for Ga) the atoms are adsorbed at the silica surface and migrate inside the silica matrix to give permanent inclusions which modify persistently the silica optical properties. This has a direct manifestation in the emission spectrum recorded after resonant excitation of the vapour. In fact emission lines in wavelength ranges where there are no atomic emissions, appear after the cells have been operated at T ≥ T$_{In}$/T$_{Ga}$, respectively. These lines are not present when the high temperature is reached for the first time and are not present in the brand new cells. Instead, once they appear, they are permanently present in the emission spectrum even at room temperature. Examples are given in fig. 16 for In and in fig. 17 for Ga.

This effect is quite important for itself as, considering that it shows up at the same temperatures at which the EPR processes become relevant, it has important implications in the In and Ga basic physical properties such as the radiation trapping, see next section, due to the induced subtraction of atoms from the vapour which results in an effective atomic density, N_{eff}, lower than the saturated one N_{sat}. Values of N_{eff} about one order of magnitude less than N_{sat} have been measured both for In at T = 950°C [6] and for Ga at T = 1050°C [40]. Possible interesting applications of this "contaminated" silica are foreseen in the manufacturing of optoelectronic devices due to the peculiar induced optical properties. All these reasons have triggered a dedicated study of the inclusions of In/Ga inside the silica matrix in our as well as in other laboratories.

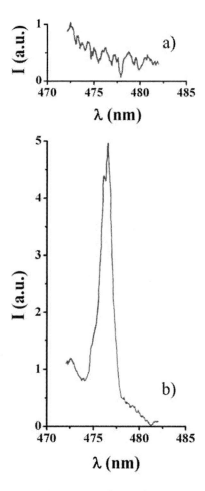

Figure 16. Portion of the fluorescence spectrum detected at T = 900°C in a quartz cell with In illuminated by a resonant laser beam tuned to 410.3 nm. a) the cell temperature is reached for the first time; b) the cell has been operated at this temperature for one day. The sensitivity in a) is five times larger than in b).

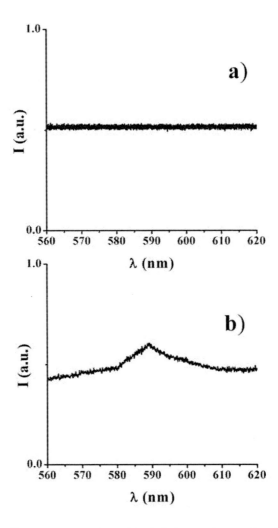

Figure 17. Portion of the fluorescence spectrum detected at room temperature in a quartz cell with Ga illuminated by a resonant laser beam tuned to 403.4 nm. a) brand new cell; b) the cell has been operated for days at temperatures above 950°C.

We have recently found that the absorption spectrum at room temperature of a sample of Herasil 3 (H3) fused silica taken from a cell containing In vapour and operated for at least one day at T = 900°C reveals a strong absorption band peaked around 200 nm and extending to the red up to 240 nm together with a much weaker one extending from 320 to 450 nm, which are extremely weak or absent in a brand new sample of the same H3 fused silica, fig. 18 a,b. Besides we have analysed a sample of commercial H3 thermally treated in a controlled manner in an oven in presence of In vapour. The sample was heated at 1000°C for 8 hours and than it was taken back to room temperature. After excitation with laser photons at 410.3 nm, the emission spectrum already evidenced the peak at 475 nm which appeared in the emission spectrum recorded from the cell with In vapour, see fig. 16.

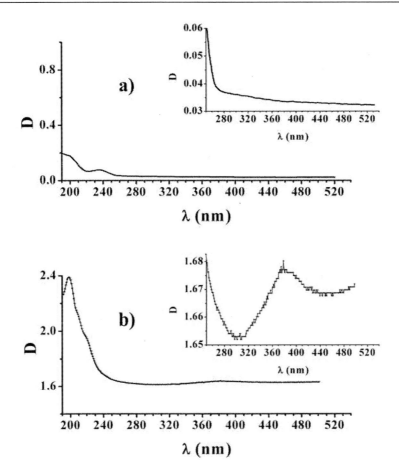

Figure 18. Absorption spectra of a H3 brand new fused silica sample (a) and of a H3 fused silica sample taken from a cell with In operated for one day at T = 900 °C (b). The quantity reported in the vertical axis is the optical density D. The inserts reproduce in an amplified scale the region centered around 380 nm.

The Ga/SiO₂ interaction has been studied more extensively both in H3 samples taken from the cells used for the EPR experiments and in commercial H3 samples thermally treated in a controlled way. The thermal treatment consists in heating the samples in an oven at 1050 °C for 8 hours. One of the samples is heated in presence of Ga vapours (H3/Ga), while another receives the same thermal treatment in presence of an inert He atmosphere (H3/He) and is used for comparison. All the checks are then performed at room temperature. When the H3/Ga samples are irradiated with laser light tuned to 403.4 nm, their emission spectrum always presents the peak at 588 nm which appeared in the emission spectrum recorded from the cell with Ga vapour, see fig. 17, and which is absent in the fluorescence spectra taken from the H3/He samples as well as in the never treated ones. In addition to this the H3/Ga samples and those taken from the operated cells manifest a weak absorption increase above 400 nm and a strong absorption below 235 nm which is related to a strong emission band centered at 377 nm (26500 cm⁻¹) shown in fig. 19 together with its maximum excitation profile. All these features are absent in the brand new H3 samples and in the H3/He ones [41]. Further analyses of the emission spectra from the H3/Ga samples illuminated by the 403.4 nm photons, evidenced also the presence of peaks exhibiting a typical Raman scattering behaviour [42]. Furthermore some indication of an electron paramagnetic resonance signal present only in the H3/Ga samples [41] is a further hint of the Ga doping whose role would be that of inducing defects in the silica matrix responsible of unpaired electrons. These experiments are still under way but there are already hints of similar effects in Suprasil 1 silica samples thermally treated at 1050°C in presence of Ga vapours [42].

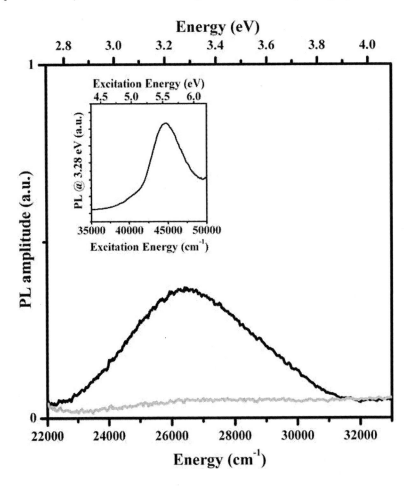

Figure 19. Comparison between the emission spectra measured under excitation at 225 nm (44400 cm^{-1}) in H3/Ga (black line) and H3/He (grey line) samples. The insert shows the excitation profile corresponding to the PL band at 377 nm (26500 cm^{-1}) detected in the H3/Ga sample. From ref. [41].

All the spectral features reported above do not match any emission from known defects in silica and so they are attributed to the In/Ga doping. As a last comment it is noticeable to underline that aluminium (an element of the same group of Ga and In) defects in silica have been known and studied for some time [43] and emission bands arising from the creation of Al_2O_3 in the interaction of Al atoms and fused silica are reported in the literature since the sixties [44].

5. Radiation Trapping

In a resonantly excited atomic vapour kept inside a fused silica cell, the radiation emitted by an atom can be reabsorbed by one of its neighbours, just to be reemitted a little later until the photon escapes from the cell. This is the basic idea of the radiation trapping [45] which can modify by many orders of magnitude the natural lifetime of the excited level. Such diffusion–like process of the radiation, in its classical formulation, is described by the Holstein equation [46, 47] from whose solution the effective lifetime of the excited level can be derived.

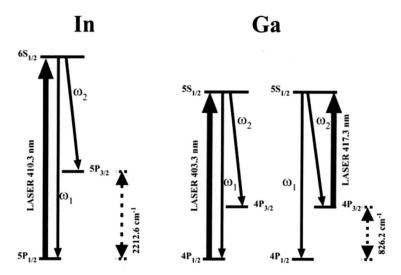

Figure 20. Λ level structure formed by the ground and the first excited states in In and Ga with the possible excitation schemes.

The Holstein equation can be solved, for particular geometries, when two levels are involved in the trapping process. A more complicated situation arises when the radiation is trapped in a 3 – level Λ scheme configuration [48] involving the ground states and the first excited state. This is the case of In and Ga that have a $n^2P_{1/2, 3/2}$ ground state and a $(n+1)$ $^2S_{1/2}$ first excited state which form the 3 – level Λ scheme sketched in fig. 20. The fundamental resonance radiation trapping in dense vapours of both elements resonantly excited by laser radiation, show unexpected results when they are kept in fused silica sealed cells. These results are discussed in the framework of the outstanding influence of the EPR and of the already mentioned strong In(Ga)/SiO$_2$ interaction.

5.a. In Homonuclear Vapours

The natural lifetime of the first excited state $6^2S_{1/2}$ is 7.4 ns [49]. The analysis of the resonant radiation trapping is done by measuring the effective lifetime, τ_{eff}, of the $6^2S_{1/2}$ state by fitting the decay of the time resolved fundamental fluorescence signal at 451.3 nm following the resonant excitation with laser pulses tuned to 410.3 nm. The temperature range explored is $700 \div 950°C$ corresponding to an atomic density in the range $10^{11} \div 10^{14}$ cm^{-3}. Such growth in the atomic density gives rise in a two level system to several orders of magnitude increase in the level lifetime. This is not the case in In where τ_{eff} increases at most a factor 6 as evident in fig. 21 where the time resolved fluorescence signal at 451.3 nm are plotted versus time for different temperatures, fig. 21a, and the corresponding measured τ_{eff} values are plotted versus temperature in fig. 21b.

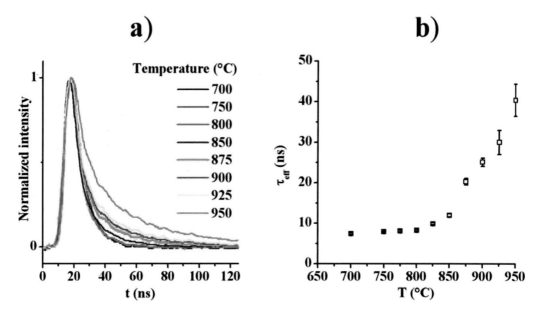

Figure 21. Fluorescence signals at 451.3 nm following resonant excitation at 410.3 nm recorded versus time at different temperatures in a quartz cell with In (a); corresponding τ_{eff} values obtained from the fit of the signals decay, plotted versus temperature (b).

Due to the fine structure of the ground state, the population density ratio between the $5^2P_{1/2}$ and the $5^2P_{3/2}$ ground sublevels is close to 10:1 even for T = 950°C. The presence of the much less populated sublevel offers an "escape channel" with a much weaker absorption coefficient, for the photons at 410.3 and 451.3 nm emitted from the $6^2S_{1/2}$ level. This results in an escaping competition between the two emitted photons with the consequence of an effective fluorescence decay time strongly different from the one expected for two-level atoms. The theoretical evaluation of the effective lifetime of the first excited state is not straightforward because of the three-level Λ scheme of indium. Anyhow, an analytical solution of the Holstein equation is found for a cylindrical geometry by taking into account the $6^2S_{1/2} \rightarrow 5^2P_{1/2}$ and the $6^2S_{1/2} \rightarrow 5^2P_{3/2}$ transitions separately [6] following the Holt method [50]. This calculation yields a τ_{eff} value still more than one order of magnitude larger than the measured one at the highest temperature. The further reduction is to be attributed to an effective atomic density lower than the saturated one due to the strong In//SiO$_2$ interaction which becomes efficient at temperatures higher than 850 °C as explained in section 4.

5.b. Ga Homonuclear Vapours

The natural lifetime of the first excited state $5^2S_{1/2}$ is 6.2 ns [49]. The temperature range explored is 800 ÷ 1100 °C corresponding to an atomic density in the range $10^{11} \div 10^{14}$ cm^{-3}.

Due to the low fine structure of the ground state, both its J sublevels are essentially equally populated in this temperature interval. This means that both the resonant excitations at 403.4 nm ($4^2P_{1/2} \rightarrow 5^2S_{1/2}$ transition) and at 417.3 nm ($4^2P_{3/2} \rightarrow 5^2S_{1/2}$ transition) can be used to populate the $5^2S_{1/2}$ state. The analysis of the resonant radiation trapping is done by measuring the effective lifetime, τ_{eff}, of the $5^2S_{1/2}$ state by fitting the decay of the time resolved fundamental fluorescence signals. When the excitation is at 403.4 nm, the fluorescence at 417.3 nm is investigated and vice versa.

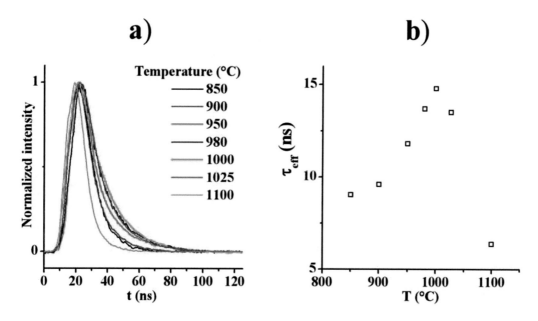

Figure 22. Fluorescence signals at 417.3 nm following resonant excitation at 403.4 nm recorded versus time at different temperatures in a quartz cell with Ga (a); corresponding τ_{eff} values obtained from the fit of the signals decay, plotted versus temperature (b). The errors are within the squares dimensions.

As both the J sublevels of the 4P ground state are equally populated, a reduction of the radiation trapping, similar to what detected in In, is not expected as no one of the ground sublevels provides an "escape channel" with a much weaker absorption coefficient. A situation similar to that of a two level atom should arise. In spite of this some unexpected results manifest as shown in fig. 22 where the time resolved fluorescence signal at 417.3 nm are reported for different temperatures, fig.22a, and the corresponding measured τ_{eff} values are plotted versus temperature, fig. 22b. In the first place the radiation trapping results unexpectedly small if compared with what happens for a two level atom for the same atomic density variation [51, 52]. The τ_{eff} value reaches a maximum of only about three times the natural lifetime when T is slightly smaller than 1000°C, atomic density $\sim 10^{13}$ cm^{-3}. In addition, when T exceeds 1000°C, the effective lifetime start decreasing to go back to the natural one, so testifying of a quenching of the radiation trapping. This result is independent of the presence, kind and pressure of a buffer gas inside the cell, of the exciting transition being saturated or not, and of the excitation wavelength [40]. These results differ from the similar ones obtained when the Ga vapour is confined in a stainless steel HPO. In the latter case too the effective lifetime increase is limited but it continues over all the range of temperatures explored as illustrated in fig. 23 where the time resolved fluorescence signals are reported for different temperatures, fig. 23a, and the correspondent τ_{eff} measured values, fig. 23b, are plotted versus T.

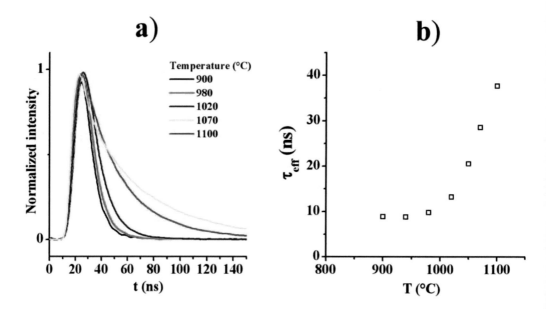

Figure 23. Ga fluorescence signals at 417.3 nm following resonant excitation at 403.4 nm recorded versus time at different temperatures in a HPO (a); corresponding τ_{eff} values obtained from the fit of the signals decay, plotted versus temperature (b). The errors are within the squares dimensions.

The limited increase of the radiation trapping can be explained with the high efficiency of the EPR processes in Ga, particularly the EPI one [25] which reduces itself the effective lifetime of the $5^2S_{1/2}$ level as it takes away from this level the atoms with a longer lifetime. When these processes take place in a quartz cell, the effect of the prominent Ga/SiO$_2$ interaction which takes away atoms from the vapour for temperatures $\geq 950°C$ adds to the former. The cooperative effects of an atomic density lower than the saturated one and of EPI produce the quenching of the radiation trapping when T becomes $\geq 1000°C$. This analysis is confirmed by a Monte Carlo simulation of the EPR in dense Ga vapours kept in quartz cells and their consequences which is detailed in [53] and summarized in the following.

The Monte Carlo simulation follows the path of a fixed number of resonant photons until they leave the vapour confined in the quartz cell taking into account the absorption/reemission processes, the influence of EPI and the interaction of the atoms with the cell boundaries under the following assumptions:

- the atomic system can be considered as a three-levels system,
- the photons time of flight between one emission and the following reabsorption is negligible compared to the Ga $5^2S_{1/2}$ state natural lifetime,
- the diffusion of the atoms inside the volume is negligible during the timescale of the effective lifetime of the $5^2S_{1/2}$ level,

- the frequency of the emitted photons is calculated according to the complete frequency redistribution (CFR) approximation [46],
- the cell boundaries are completely transparent.

These assumptions fit the conditions in which the experiment has been performed. As in any other Monte Carlo simulation, the physical quantities are taken at random according to the suitable probability distribution functions. As the density of atoms in the ground state is inhomogeneous along the path of the photons, the cylindrical volume is divided into sub-cells in which the distribution of absorbers can be considered constant. This division is sketched in fig. 24 together with the coordinate system used in the simulation to calculate the spatial and angular quantities. The laser beam direction and versus are also indicated.

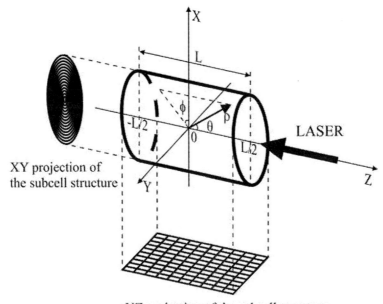

Figure 24. Scheme of the cylindrical cell and of its sub cells division considered by the Monte Carlo simulation. The coordinate system is also plotted.

The algorithm, after the input of the main experimental parameters such as the laser power density, the temperature of the vapour, both the laser beam and the cylindrical cell dimensions and the density of the Ga atoms corrected by the effects of the Ga/SiO$_2$ interaction, keeps track of the absorption/reemission processes of a photon bundle along its path to the detector. The number of electrons produced via EPI is calculated along this path and the weight of the photon bundle is changed accordingly. As the opacity "seen" by the photon bundle changes along its path from sub-cell to sub-cell, the photon bundle path length is rescaled anytime the photons cross the sub-cells borders. The simulation keeps track of the changing in the distribution of the excited state in each sub-cell where the absorption process takes place, and uses the updated distribution for the next photon bundle run. In each run the algorithm calculates the probability distribution p_i that a photon bundle has left the vapour after exactly i absorption/reemission processes. The simulation stops when a sufficient accuracy in the probability distribution p_i is achieved. The effective lifetime of the $5^2S_{1/2}$ state can be derived by the p_i distribution according to [45]:

$$\tau_{eff} = \tau \sum_i i p_i$$

In this way it is possible to extract from the simulation information on the resonant radiation trapping in the Ga Λ-level system under the influence of non-linear processes, such as the EPI process and the Ga/SiO$_2$ interaction, in different experimental conditions as is shown in fig. 25 where the simulated and measured values of the effective lifetime of the Ga first excited state, when the vapour is confined in a quartz cell, are plotted versus temperature. The comparison is quite satisfactory as the reproduction of the experimental behaviour is evident.

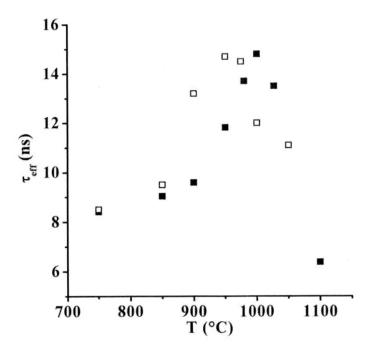

Figure 25. Effective lifetime of the $5^2S_{1/2}$ level plotted versus temperature when the Ga vapour is kept in a quartz cell and the exciting laser is resonant with the Ga fundamental transition at 403.4 nm. (■) experimental results, (□) Monte Carlo simulation. The errors are within the squares dimensions.

6. Heteronuclear Ga+In Vapours

As mentioned in the introduction, no information is available for the GaIn dimer or in general for the Ga$_x$In$_y$ complex. In the literature there is, to our knowledge, only one mention of a possible Ga$_x$In$_y$ complex emission without any analysis or attribution [54]. Very recently we found that the presence of the GaIn molecule could be inferred from the time resolved analysis of some atomic fluorescences appearing in the emission spectrum recorded from a cell with In+Ga following resonant excitation of either element in the temperature range 900 ÷ 1100°C.

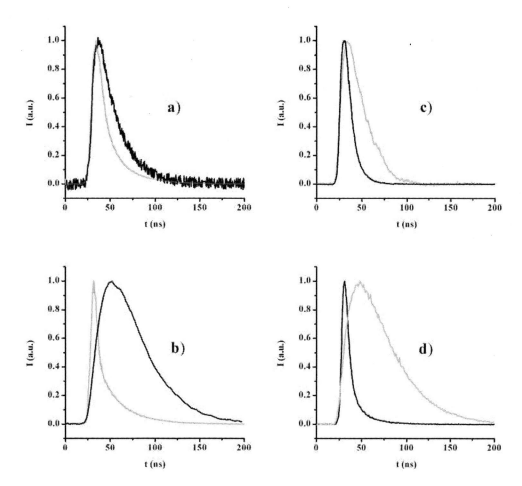

Figure 26. Intensity of the fluorescence signals at 417.3 nm (black line) corresponding to the Ga $5^2S_{1/2} \to 4^2P_{3/2}$ transition and at 451.3 nm (grey line) corresponding to the In $6^2S_{1/2} \to 5^2P_{3/2}$ transition recorded versus time in a quartz cell containing Ga + In when: the exciting laser is resonant with the In fundamental transition at 410.3 nm and T = 900°C (a), the exciting laser is resonant with the In fundamental transition at 410.3 nm and T = 1100°C (b), the exciting laser is resonant with the Ga fundamental transition at 403.4 nm and T = 900°C (c) and the exciting laser is resonant with the Ga fundamental transition at 403.4 nm and T = 1100°C (d).

Fig. 26a shows the time resolved signals of the In fundamental transition at 451.3 nm together with the Ga fundamental fluorescence at 417.3 nm recorded after excitation of the In fundamental transition $5^2P_{1/2} \to 6^2S_{1/2}$ at 410.3 nm, at T = 900°C. The laser power density is ~ 300 kW/cm². Fig. 26b displays the same situation for T = 1100°C. Fig. 26c exhibits the time resolved signals of the Ga fundamental transition at 417.3 nm together with the In fundamental fluorescence at 451.3 nm recorded after excitation of the Ga fundamental transition $4^2P_{1/2} \to 5^2S_{1/2}$ at 403.4 nm, at T = 900°C. The laser power density is ~ 300 kW/cm². Fig. 26d illustrates the same situation for T = 1100°C.

It is evident that at 900°C the fundamental fluorescences of the laser unexcited species appear and are synchronous with the resonantly excited species ones, figs. 26a and 26c. This behaviour is caused by the sensitized fluorescence effect (5):

$$\text{In } (6^2S_{1/2}) + \text{Ga } (4^2P_{1/2}) \rightarrow \text{In } (5^2P_{1/2}) + \text{Ga } (5^2S_{1/2}) - 417 \text{ cm}^{-1} \qquad (18)$$

$$\text{Ga } (5^2S_{1/2}) + \text{In } (5^2P_{1/2}) \rightarrow \text{Ga } (4^2P_{1/2}) + \text{In } (6^2S_{1/2}) + 417 \text{ cm}^{-1} \qquad (19)$$

as the kT value at this temperature (815 cm^{-1}) is almost twice ΔE_{SF} in equations (18) and (19). SF gives rise to emissions from the unexcited species synchronous with the excitation as the energy transfer is produced by collisions which are instantaneous in the nanosecond time scale.

When the temperature is raised to 1100°C, the fluorescence from the unexcited species is delayed in respect to those originated from the excited ones, figs. 26b and 26d. SF is even more effective at 1100°C as the kT value increases to 954 cm^{-1}, but it can not be the main source of the population of the first excited level of the unexcited element for the temporal considerations reported above. The principal contribution of the Ga ($5^2S_{1/2}$) population, in case of In excitation, and of the In ($6^2S_{1/2}$) population, in case of Ga excitation, must proceed via a different route that requires some time to develop after the laser pulses. Let us discuss the case of Ga excitation. The resonant excitation, at 1100°C, produces a large amount of Ga ions as well as a considerable population in the Rydberg levels as detailed in sections 3.b.1. This means that there is a large probability of collisions with the ground In ($5^2P_{1/2,\ 3/2}$) atoms to produce very excited GaIn molecular states (GaIn**) either directly or via the formation of molecular ions with subsequent recombination. In both cases the In ($6^2S_{1/2}$) level can be populated either via cascade transitions to lower lying molecular levels correlating with In ($6^2S_{1/2}$), or via molecular dissociation to high lying In atomic state followed by cascade transitions to In ($6^2S_{1/2}$). Both these processes would introduce the observed delay. Exactly the same explanation holds for the case of In excitation.

A further hint of the presence and influence of the GaIn molecule comes from the temporal analysis of other atomic fluorescences recorded at 1100 °C under resonant Ga excitation at 403.4 nm. The time resolved spectrum shows, beside the fundamental Ga line at 417.3 nm, also the fundamental In fluorescence at 451.3 nm, a signal at 294.4 nm originating from the Ga ($4^2D_{3/2,5/2}$) levels and a weak signal at 326.0 nm starting from the In ($5^2D_{3/2,5/2}$) levels. They are all plotted versus time in fig. 27. The In resonance fluorescence is delayed for the reasons postulated above, the signal at 294.4 nm is delayed due to the Ga ($4^2D_{3/2,5/2}$) dynamics formation, see section 3.b.1 and fig. 14, but the presence and temporal behaviour of the line at 326.0 nm can not be immediately explained. The latter can not be ascribed to the EPI In collision (10) followed by cascade transitions as in the case of homonuclear In vapour resonantly excited, see section 3.a.1, as the In ($6^2S_{1/2}$) density is now too low to allow this process and in addition the signal would be delayed in respect to the excitation, fig. 7. In this case, instead, the In $5^2D_{3/2,5/2} \rightarrow 5^2P_{3/2}$ transition is synchronous with the Ga excitation. This occurrence can be explained by postulating that the In ($5^2D_{3/2,5/2}$) levels are populated via the same route postulated before to explain the delay between the Ga fundamental fluorescence and the In one at 451.3 nm. The GaIn** levels involved in this case would be the unbound ones that directly correlate with the In ($5^2D_{3/2,5/2}$) separated atoms. This hypothesis finds some support in a similar result obtained in an experiment done in a dense In+Hg vapour to study the HgIn molecule. In that case too a signal at 326.0 nm, synchronous with the resonant In excitation, was observed and this behaviour was attributed to the molecular presence [4]. In the present experiment the role played in that circumstance by Hg could be played by Ga. Similar results, if not identical, to the ones here summarized have been obtained in a vapour mixture of Ga and In kept in a HPO [29].

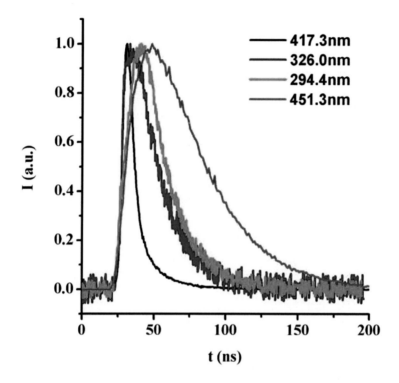

Figure 27. Intensity of the fluorescence signals at 417.3 nm corresponding to the Ga $5^2S_{1/2} \rightarrow 4^2P_{3/2}$ transition, at 451.3 nm corresponding to the In $6^2S_{1/2} \rightarrow 5^2P_{3/2}$ transition, at 294.4 nm corresponding to the Ga $4^2D_{3/2, 5/2} \rightarrow 4^2P_{3/2}$ transition and at 326.0 nm corresponding to the In $5^2D_{3/2, 5/2} \rightarrow 5^2P_{3/2}$ transition, recorded versus time in a quartz cell containing Ga + In when the exciting laser is resonant with the Ga fundamental transition at 403.4 nm and T = 1100°C.

The total lack of any calculation or experimental data on the GaIn molecule prevents any definitive identification, but this approach seems very promising and further experiments are in progress to attempt to enrich the available data.

7. Conclusions

This chapter gives a summarized review of the processes taking place when dense vapours of Ga and In are confined in a quartz cell and are resonantly excited. The peculiar spectroscopic characteristics of these elements, their high chemical reactivity at high temperatures, and their very low saturated vapour pressure give rise to several unique phenomena whose study has been going on for the last twenty years. The most recent results are presented in detail, while a complete bibliography covering all of the aspects of this research is listed in the reference section.

The RLAC applied to In and Ga kept in quartz cells offer an unusual way to approach different physics research fields. The main ones are recalled in the following list:

- The presence of EPC and/or EPI inside the vapour produce ions, atoms in the Rydberg levels and electrons with an energy less than 1 eV in a confined volume. This provides, as a consequence, the possibility of performing some typical Rydberg physics such as Rydberg-Rydberg collisions [55] and low-energy electrons–Rydberg atoms impact [56].
- The time resolved analysis of several processes taking place in dense vapours, which is usually prevented by the strong radiation trapping, can be carried out in In and Ga, as in their vapours the radiation trapping is strongly reduced if not totally quenched.
- The frequency and time resolved study of the LIF spectra emitted by the dense vapours offers a new route to throw some light into the scarcely known molecular aspect of these elements. It is worth recalling that since the beginning of the last century the papers dealing with this subject total a one-digit number!
- Last, but not least, the opportunity ensured by the dense vapours of these elements to produce, in a rather easy way, defects inside fused silica that originate in the latter can lead to new and very interesting optical properties for the use of such doped silica as a possible active medium or for optoelectronic device manufacturing.

The findings listed above are evidence of the importance of the study of resonantly excited Ga and In vapours, which still presents new branches to be explored—not to mention the many results, some probably unexpected, that could arise from the investigation of cold collisions of such elements.

References

[1]   Nesmeyanov, A. N. In *Vapour Pressure of the Chemical Elements*; Elsevier: Amsterdam, 1963; p 447.
[2]   Niemax, K.; Weber, K. H. *Appl. Phys. B* 1985, 36, 177-180.
[3]   Bicchi, P. *Riv. Nuovo Cimento* 1997, 20 (7), 1-74.
[4]   Bicchi, P.; Marinelli, C.; Bernheim, R. A. *Phys. Rev. A* 1997, 56, 2025-2031.
[5]   Gallagher, A.; In *Excimer Lasers;* Rhodes C. K.; Ed.; Springer – Verlag: Berlin, 1984, p 139.
[6]   Bicchi, P.; Marinelli, C.; Mariotti, E.; Meucci, M.; Moi, L. *Opt. Commun.* 1994, 106, 197-201.
[7]   Garton, W. R. S. *Proc. Phys. Soc.* 1952, 65, 268-276.
[8]   Garton, W. R. S.; Codling, K. *Proc. Phys. Soc.* 1961, 78, 600-606.
[9]   Mirza, M. Y.; Duley, W. W. *Proc. R. Soc. A* 1978, 364, 255-263.
[10] Neijzen, J. H. M.; Donszelmann, A. *Physica* 1981, 111C, 127-135.
[11] Neijzen, J. H. M.; Dönszelmann, A. *Physica* 1982, 114C, 241-250.
[12] Kasimov, A.; Tursunov, A. T.; Tukhlibaev, O. *Opt. Spectrosc.* 1996, 81, 19-21.
[13] Young, W. A.; Mirza, M. Y.; Duley, W. W. *Opt. Commun.* 1980, 34, 353-356.

[14] Safronova, U. I.; Safronova, M. S.; Kozlov, M. G. *Phys Rev. A* 2007, 76, 022501.

[15] Kopystynska, A.; Moi, L. *Phys. Rep.* 1982, 92, 135-181.

[16] Bicchi, P.; Meucci, M.; Moi, L. In *LASERS '87*; Duarte, F.J.; Ed.; STS Press: Mc Lean, VA, USA, 1988 pg.959-965.

[17] Bezuglov, N. N.; Klucharev, A. N.; Sheverev, V. A. *J. Phys. B: At. Mol: Opt. Phys* 1987, 20, 2497-2513.

[18] Gabbanini, C.; Biagini, M.; Gozzini, S.; Lucchesini, A.; Kopystynska, A.; Moi,L. *J. Quant. Spectrosc. Radiat. Transfer* 1992, 47, 103-112.

[19] Bicchi, P.; Barsanti, S.; Favilla, E. In *Recent Research Developments in Physics Vol. 5 Part III*; Masashi, K.; Nasser, A.; Randriamampianina, A.; Ed.; Transworld Research Network publisher: Trivandrum, India, 2004; pg.839-857.

[20] Bicchi, P.; Kopystynska, A.; Meucci, M.; Moi, L. *Phys. Rev. A - Rapid Commun.* 1990, 41, 5257-5260.

[21] Barsanti, S.; Bicchi, P. *J. Phys. B: At. Mol. Opt. Phys.* 2002, 35, 4553 – 4563.

[22] Favilla, E.; Barsanti, S.; Bicchi, P. *J. Phys. B: At. Mol. Opt. Phys.* 2005, 38, S37–S50.

[23] Bicchi, P.; Barsanti, S.; Favilla, E. *Laser Physics* 2004, 14, n.2, 144–149.

[24] Bicchi, P.; Marinelli, C.; Mariotti, E.; Meucci, M.; Moi, L. *J. Phys. B: At. Mol. Opt. Phys.* 1997, 30, 473–482.

[25] Barsanti, S.; Bicchi, P. *J. Phys. B: At. Mol. Opt. Phys.* 2001, 34, 5031 – 5040.

[26] Wang, L.; Ledingham, K. W. D.; McLean, C. J.; Singhal, R. P. *Appl. Phys. B* 1992, 54, 71-75.

[27] Bicchi, P.; Marinelli, C.; Bernheim, R.A. *J. Chem. Phys.* 1992, 97, 8809-8810.

[28] Bicchi, P.; Marinelli, C.; Bernheim, R.A. *Chem. Phys.* 1994, 187, 107-110.

[29] Favilla, E.; Barsanti, S.; Bicchi, P. *Rad. Phys. Chem* 2007, 76, 440–444.

[30] Mitchell, A. C. G.; Zemansky, M. W. *Resonance Radiation and excited atoms*; Cambridge University Press: London, 1971, pp 59-65.

[31] Finlay, I. C.; Green, D. B. *J. Phys. E* 1976, 51, 1026-1035.

[32] Bicchi, P.; Marinelli, C.; Mariotti, E.; Meucci, M.; Moi, L. *J. Phys. B: At. Mol. Opt. Phys.* 1993, 26, 2335-2344.

[33] Connerade, J. P. *Highly Excited Atoms*; Cambridge University Press: Cambridge, 1998.

[34] Moore, C.E. *Atomic Energy Levels. Natl. Bur. Stand. Circ. No 467 vol. III*; US GPO: Washington, DC, 1994.

[35] Stebbings, R. F.; Dunning, F. B. *Rydberg States of Atoms and Molecules*; Cambridge University Press: Cambridge, 1983, p 170.

[36] Bicchi, P.; Marinelli, C.; Mariotti, E.; Meucci, M.; Moi, L. In *Plasma Collective Effects in Atomic Physics*; Giammanco, F.; Spinelli, N.; Ed.; ETS: Pisa, 1996, pp 188-220.

[37] Balasubramanian, K.; Li, J. *J. Chem. Phys.* 1988, 88, 4979-4986.

[38] Balasubramanian, K. *J. Chem. Phys.* 1990, 94, 7764-7768.

[39] Das, K. K. *J. Phys. B: At. Mol. Opt. Phys.* 1997, 30, 803-809.

[40] Bicchi, P.; Barsanti, S. *Rad. Phys. Chem.* 2003, 68, 91–95.

[41] Barsanti, S.; Cannas, M.; Bicchi, P. *J. Non- Cryst. Solids* 2007, 353, 679-683.

[42] Barsanti, S.; Cannas, M.; Favilla, E.; Bicchi, P. *Rad. Phys. Chem* 2007, 76, 508-511.

[43] Devine, R. A. B.; Duraud, J. P.; Dooryhé, E. *Structure and Imperfections in Amorphous and Crystalline Silicon Dioxide*; Wiley: Chichester, 2000.

[44] Schulman, J. H.; Compton, W. D. *Color center in Solids*; Pergamon Press: London, 1962, pp 291-324.

[45] Molisch, A. F.; Oehry, B. P. *Radiation trapping in atomic vapours*, Clarendon Press: Oxford, 1998.

[46] Holstein, T. *Phys. Rev.* 1947, 72, 1212-1233.

[47] Holstein, T. *Phys. Rev.* 1951, 83, 1159-1168.

[48] Molish, A. F.; Shupita, W.; Oehry, B. P.; Sumetsberger, B.; Magerl, G. *Phys. Rev. A* 1995, 51, 3576-3583.

[49] Radzig, A. A.; Smirnov, B. M. *Reference Data on Atoms, Molecules and Ions*, Springer-Verlag: Berlin, 1985, pp 239-242.

[50] Holt, H. K. *Phys. Rev. A*, 1976, 13, 1442-1447.

[51] Garver, W. P.; Pierce, M. R.; Leventhal, J. J. *J. Chem. Phys.* 1982, 77, 1201-1205.

[52] Molish, A. F.; Shupita, W.; Oehry, B. P.; Magerl, G. *J. Phys. B: At. Mol. Opt. Phys.* 1997, 30, 1879-1891.

[53] Barsanti, S.; Favilla, E.; Bicchi, *Laser Physics* 2005, 15, 233 - 237.

[54] Ginter, D. S.; Ginter, M. L.; Innes, K. K. *J. Phys. Chem.* 1965, 69, 2480-2483.

[55] Vitrant, G.; Raimond, J. M.; Gross, M.; Haroche, S. *J. Phys. B: At. Mol. Opt. Phys.* 1982, 15, L49-L55.

[56] Nagesha, K.; Mac Adam, K. B. *Phys. Rev. Lett.* 2003, 91, 113202.

In: Atomic, Molecular and Optical Physics... ISBN 978-1-60456-907-0
Editor: L.T. Chen, pp. 147-203 © 2009 Nova Science Publishers, Inc.

Chapter 4

EXCITATION OF ATOMS AND IONS BY ELECTRON IMPACT

S.S. Tayal

Dept. of Physics, Clark Atlanta Univ., Atlanta, GA

Abstract

Electron-ion collision processes play an important role in the understanding of energy balance in various types of astrophysical plasmas and in controlled thermonuclear fusion plasmas. Electron collisional excitation rates and transition probabilities are important for computing electron temperatures and densities, ionization equilibria and for deriving elemental abundances from emission lines formed in the collisional and photoionized plasmas. Over the last several years our research work has been focused on accurate calculations of electron-ion and atom excitation processes using the close-coupling methods for application to astrophysical and laboratory plasmas. Our effort has been to benchmark theory against experiment for the electron impact excitation of selected atomic systems for transitions of diagnostic importance. Accurate representation of target wave functions that properly account for the important correlation and relaxation effects and inclusion of coupling effects including coupling to the continuum are essential components of a reliable collision calculation. The resonant excitation processes are important in the low energy region. This chapter presents a discussion of the state of our knowledge of the electron impact excitation processes together with some recent results for neutral atoms and singly and multiply charged ions.

1. Introduction

Electron collisional excitation and ionization of atoms and atomic ions play an important role in the determination of energy loss in controlled thermonuclear fusion plasmas. The atomic impurities in fusion plasma may include B, C, O, Mg, Al, Si, Fe and Ni. All ionization stages of these atomic species may be observed and used for the diagnostic purposes in fusion plasmas. Collisions of electrons with atoms and ions are the major excitation mechanism in a wide range of astrophysical objects such as planetary nebulae, H II regions, stellar and planetary atmospheres, active galactic nuclei, novae and supernovae. The plasma diagnostic techniques based on spectroscopic line intensities, profiles and wavelengths have been used to determine temperatures, densities, emission measures, ionization equilibria, mass motion and elemental abundances. In many cases the accuracy of the astrophysical analysis is limited by the inadequacy of atomic data as highlighted by the high quality of spectroscopic data from the recent flight instruments on space missions. Spectral synthesis codes have been developed to generate synthetic spectrum by converting a database of atomic parameters into a model spectrum that can be determined generating isothermal models for each ion. The oscillator strengths and electron impact excitation cross sections and rates of ions are important for the determination of column densities and velocity structure of the interstellar and intergalactic matter. Studies of temperatures, densities and turbulence and flow velocities of the outer atmospheres of individual stars and their variation with global stellar properties such as effective temperature, surface gravity and dust content contribute to the understanding of evolution of stars. A large amount of atomic data are needed to compute the energy transport through the outer layers of a star. The computation of true opacity of stellar matter requires both continuum and line opacities. Hot plasmas can be diagnosed using both emission and absorption lines.

Cross sections and rate parameters of inelastic processes for electron-ion and atom collisions can be calculated by the use of close-coupling methods. The physical effects that are known to be important in the low-energy electron scattering to obtain cross sections should be included in a reliable calculation. These effects include sophisticated target description and channel coupling between many channels to attain convergence. The target wave functions can be described either by orthogonal orbitals or by non-orthogonal orbitals technique. Accurate representation of target wave functions in open-shell atomic systems is complicated by the large term dependence of valence orbitals and by large correlation corrections. The core correlation, core-valence correlation, interaction between different Rydberg series and interaction between Rydberg series and perturber states should be considered. An important test of the quality of target wave functions is provided by the calculated excitation energies and by the oscillator strengths between the target levels. The calculated results may be compared with measured values and other reliable calculations. In many cases target description can be improved by using the flexible non-orthogonal atomic orbitals to account for the term dependence of valence orbitals and correlation and relaxation effects [1]. The near-threshold results for electron impact cross sections may be sensitive to the treatment of inner-core short-range correlation effect in the target description as well as on a proper account of the target polarizability. Consequently, accurate representation of target wave functions may require extensive configuration expansions. The spectroscopic bound states, autoionizing states and pseudostates with significant coupling are included in the close-coupling expansion. The pseudostates are chosen to approximate the loss of flux into the infinite number of bound and continuum states that are coupled with low-lying spectroscopic bound states.

2. Electron Impact Excitation Processes

Electron impact excitation of an ion may occur in a direct excitation (DE) process when the incident electron gives a part of its energy to the target ion A in the charge state q to excite it from an initial state i to a final state f,

$$e^- + A_i^{q+} \rightarrow A_f^{q+} + e^-, \tag{1}$$

or in a two-step resonant excitation (RE) process. In the first step the incident electron is captured by the target ion into a short-lived intermediate compound ion,

$$e^- + A_i^{q+} \rightarrow [A_j^{(q-1)+}], \tag{2}$$

which decays either by autoionization or by radiative transition as a second step. The autoionization of the doubly excited intermediate states (indicated by brackets) is given by

$$[A_j^{(q-1)+}] \rightarrow A_f^{q+} + e^-. \tag{3}$$

The RE process is more probable than the DE process in the near-threshold region and gives rise to Rydberg series of resonances converging to various excitation thresholds. In many cases the resonance structures may enhance the direct excitation cross sections substantially by over an magnitude. At the second step involving doubly excited autoionizing states there is a competition between the autoionization and radiative decay which depends up on the relative probabilities for autoionization and radiative decay. An autoionizing state can decay by a radiative transition to a pure bound state in the so called dielectronic recombination process. In cases where autoionizing states may also undergo radiative decay, the autoionizing resonance structures will be reduced depending on the branching between autoionization and radiative decay. The radiative stabilization to a pure bound states may be important for some transitions in highly ionized ions and should be taken into account in an accurate electron scattering calculation.

3.   Close-Coupling Approach

The close-coupling methods can include both DE and RE processes and are believed to be the most accurate for the calculation of electron impact excitation cross sections at low incident electron energies. The distorted-wave methods can also include both processes but most of the available distorted-wave calculation in the literature only include the DE process and thus are not as accurate as the close-coupling calculations. In many cases the contribution of the RE process can be substantial in the near-threshold region. Among the close-coupling methods the R-matrix method [2] has been very successful in producing large amounts of electron impact cross section data for many ions of diagnostic importance to astrophysical and fusion plasmas over the last three decades. Many of the earlier R-matrix calculations in the literature were performed either in LS-coupling or in LS-coupling plus an algebraic transformation to intermediate coupling. The relativistic effects are not significant for ions with low nuclear charge. For ions where relativistic effects may be significant, these were included through the use of term coupling coefficients [3]. This approach generally has two main limitations. The fine-structure splitting of the target terms is neglected and in the resonance region, where there are both open and closed channels, only those components of the term coupling coefficients are used for which channels are open. Both approximations may lead to serious errors in the position of resonances and may ignore many of the resonance Rydberg series. However, this approach is computationally very fast. The intermediate-coupling frame transformation (ICFT) method [4] to obtain electron impact collision strengths between fine-structure levels by transforming the LS-coupled K-matrices to K-matrices in an intermediate coupling proved to be more reliable and accurate. This method employs multi-channel quantum defect theory to generate unphysical K-matrices in LS coupling and these are transformed to intermediate coupling. The physical K-matrices are then obtained from the intermediate-coupled unphysical K-matrices.

The Breit-Pauli R-matrix method and the related computer packages [2] represent state-of-the-art numerical and computational methods to produce a variety of atomic data including electron excitation collision strengths. The Breit-Pauli R-matrix method takes into account the one-body mass correction, Darwin and spin-orbit relativistic terms directly in the scattering equations. The non-convergence of the close-coupling expansion may cause large uncertainties in the calculated cross sections for some transitions and atomic systems. Sufficient number of target terms should be included in the close-coupling expansion to account for all channel couplings as well as to provide rates for all relevant cascade transitions. The inclusion of higher excited states in the expansion brings the lower resonances in the threshold energy regions to correct positions, and also allows the inclusion of higher resonance series converging to these states. A set of well-chosen pseudostates can be included in the close-coupling expansion, in addition to physical states, to allow for loss of flux in the continuum for atomic systems where coupling to the continuum is important. The pseudostates are chosen to simulate the continuum target states.

The total wave function representing the collision of electrons with an ion for each total angular momentum J and parity π is expanded in R-matrix basis as

$$
\Psi_k(J\pi) = A \sum_{ij} c_{ijk} \Phi_i^J(x_1, x_2, \ldots, x_N; x_{N+1}, \sigma_{N+1}) u_{ij}(x_{N+1})
$$
$$
+ \sum_j d_{jk} \phi_j^J(x_1, x_2, \ldots, x_{N+1}), \qquad (4)
$$

where A is the antisymmetrization operator, Φ_i^J are channel functions representing the spectroscopic and pseudo fine-structure target levels coupled with angular and spin functions of the scattered electron to form channel functions of J and π, the ϕ_j^J are (N+1)-electron configurations formed from the atomic orbitals and are included to ensure completeness of the total wave function expansions and to allow for short-range correlations and the u_{ij} are the orthogonal set of continuum basis. The coefficients c_{ijk} and d_{jk} are obtained by diagonalization of the (N+1)-electron Breit-Pauli Hamiltonian in the inner region. A reliable scattering calculation requires a balance between the N- and (N+1)-electron short-range correlation terms.

The B-spline R-matrix method offers certain advantages to investigate the electron impact of the neutral atoms and low ionization atomic systems. In the B-spline R-matrix method, a B-spline basis is used to represent the continuum functions in the internal region and no orthogonality constraint is imposed between the continuum functions and the valence spectroscopic and correlated atomic orbitals [1, 5, 6, 7]. The use of non-orthogonal orbitals considerably simplifies the structure of the bound part of the close-coupling expansion to avoid pseudo-resonances. The total wave function in the internal region for the total

$(N+1)$-electron function is expanded in terms of energy-independent basis functions

$$\Psi_k = A \sum_{ij} a_{ijk} \overline{\Phi_i} u_j(r), \tag{5}$$

where $\overline{\Phi_i}$ are channel functions formed from the multi-configurational functions of the target levels and u_j are the radial basis functions describing the motion of the scattering electron. The operator A antisymmetrizes the wave function and expansion coefficients a_{ijk} are determined by diagonalizing the (N+1)-electron Hamiltonian. The radial functions u_j are expanded in the B-spline basis as

$$u_j(r) = \sum_i \overline{a_{ij}} B_i(r), \tag{6}$$

and the coefficients $\overline{a_{ij}}$ (which now replace a_{ijk}) are determined by diagonalizing the (N+1)-electron Hamiltonian inside the R-matrix box that contained all bound atomic orbitals used for the description of the target levels. Here the B_i are B-splines used to describe the scattering electron. There are no $(N+1)$-electron bound configurations in expansion (5). Such terms are usually included in Breit-Pauli R-matrix calculations [2] to compensate for the orthogonality constraints imposed on the continuum orbitals. The (N+1)-electron bound configurations which allow for the short-range correlation effect can still be included. These additional bound configurations help to bring the low-lying resonances to correct positions. Care should be taken to avoid linear dependence problem. A very limited orthogonality constraint is imposed, which do not affect the completeness of the total trial wave function. The scattering orbitals are normally constructed orthogonal to the core orbitals. Using the B-spline basis leads to a generalized eigenvalue problem of the form

$$\mathbf{H}c = E\mathbf{S}c \tag{7}$$

where $\mathbf{S}$ is the overlap matrix which in the case of the usual orthogonal conditions on scattering orbitals reduces to the banded matrix, consisting of overlaps between individual B-splines, but in the more general case of non-orthogonal orbitals has a more complicated structure [5].

In order to make the interaction matrix Hermitian in the internal region, the Bloch operator is added to the Hamiltonian. The amplitude of the wave functions at the boundary, which are needed to construct the R-matrix, are simply given by the coefficient of the last spline, the only nonzero spline at the boundary. The choice of B-splines as basis functions has certain advantages. They have been widely used due to their excellent numerical approximation properties. The B-splines are bell-shaped piecewise polynomial functions of order k_s and degree k_s - 1, defined by a given set of points in some finite radial interval. The characteristic feature of B-splines is that the Schrodinger equation is solved within a box, which is very similar to the inner region in the R-matrix method, and it can be expected that B-splines will be very effective for forming the R-matrix basis. The important property of B-spline basis is that they form an effectively complete basis on the interval spanned by the knot sequence. The completeness of the B-spline basis ensures that no Buttle correction to the R-matrix elements is required.

An appropriate number of continuum basis functions for each angular momentum is chosen. This number is defined by the number of B-splines, which in turn is defined by the choice of grid. In the continuum calculations the wavelength of the scattering particle cannot be smaller than the grid step, otherwise the B-spline basis hardly describes the oscillating behavior of the wave function. It approximately limits the maximum grid step by the value of 1/k, where k is the maximum linear momentum of the incident electron. On the other hand, the R-matrix box should contain all atomic orbitals used for the construction of target states. The above requirement somewhat limits the use of the B-spline basis to low energy scattering and to excitation of states which are too spatially spread to keep the calculations manageable within the available computational resources. The calculation of interaction and overlap matrices in secular equation (7) on the basis of non-orthogonal orbitals can be carried with a general computer code [5]. The non-orthogonal orbital technique leads to more time consuming calculations in comparison to orthogonal orbitals. However, the data banks of matrix elements may considerably reduce the computational efforts [5].

The R-matrix box of appropriate radius is chosen so that the box well contained all atomic orbitals that are used for the construction of target levels. The relativistic effects in the scattering calculations are incorporated in the Breit-Pauli Hamiltonian through the use of the Darwin, mass correction and spin-orbit operators. The scattering parameters are then found by matching the inner solution at $r = a$ to the asymptotic solutions in the outer region. The calculations are carried out for all scattering symmetries of J that are estimated to give converged collision strengths for forbidden transitions. The higher partial waves contributions for the dipole-allowed transitions can be calculated using the Coulomb-Bethe approximation [8]. The energy dependent collision strength $\Omega(i \rightarrow j)$ and cross section $Q(i \rightarrow j)$ are related as follows:

$$Q(i \rightarrow j) = \frac{\pi \Omega(i \rightarrow j)}{(2J_i + 1)k_i^2}, \tag{8}$$

where k_i is the wave number of the colliding electron incident on the target ion in level i and $(2J_i + 1)$ is the statistical weight of level i.

In many astrophysical applications it is convenient to use excitation rate coefficients or thermally averaged collision strengths as a function of electron temperature. The excitation rates are obtained by averaging collision strengths over a Maxwellian distribution of electron energies. The excitation rate coefficient for a transition from state i to state f at electron temperature T_e is given by

$$C_{if} = \frac{8.629 \times 10^{-6}}{g_i T_e^{1/2}} \gamma_{if}(T_e) exp(\frac{-\Delta E_{if}}{kT_e}) \ cm^3 s^{-1}, \tag{9}$$

where g_i is the statistical weight of the lower level i, $\Delta E_{if} = E_f - E_i$ is the excitation energy, and γ_{if} is a dimensionless quantity called effective collision strength given by

$$\gamma_{if}(T_e) = \int_0^\infty \Omega_{if} exp(\frac{-E_f}{kT_e}) d(\frac{E_f}{kT_e}), \tag{10}$$

where E_f is the energy of incident electron with respect to the upper level f. If the collision strength is assumed to be independent of the incident electron energy, we have γ_{if}

$= \Omega_{if}$. The effective collision strengths are calculated by integrating collision strengths for fine-structure levels over a Maxwellian distribution of electron energies. The integration in equation (10) should be carried out using energy dependent collision strengths from threshold to infinity. The calculations of effective collision strength needs to be carried out including collision strengths in the high energy limits as described in details by Burgess and Tully [9]. The collision strengths at higher energies are particularly important for the allowed transitions. The energy dependence of collision strengths for allowed transitions are properly accounted for by using extrapolation technique. In the asymptotic region, the collision strengths follow a high energy limiting behavior for the dipole-allowed transitions

$$\Omega_{if}(E) \sim_{E \to \infty} d \, ln(E), \qquad (11)$$

where the parameter d is proportional to the oscillator strength. The collision strengths vary smoothly in high energy region and exhibit an increasing trend for dipole-allowed transitions. The collision strength increases more rapidly for the stronger dipole-allowed transitions than the weaker transitions.

In the case of the multipole optically forbidden transitions such as electric quadruple or magnetic dipole transitions

$$\Omega_{if}(E) \sim_{E \to \infty} constant, \qquad (12)$$

where the value of *constant* can be obtained from the Bethe approximation. The inter-combination transitions have the asymptotic form

$$\Omega_{if}(E) \sim_{E \to \infty} \frac{1}{E^2}. \qquad (13)$$

The rapid decrease of collision strengths with incident electron energies in the case of spin-forbidden transitions is due to the fact that these transitions can only occur through electron exchange.

4.   Cross Sections for Atoms and Ions

In the past decade, several theoretical and experimental studies have been carried out of the electron impact elastic and inelastic cross sections for neutral atoms and ions in various charge states. The available experimental measurements on the electron impact excitation of atoms and ions are still a very few. On the other hand, theoretical methods for electron impact excitation studies are very well developed, and have been used to produce a wealth of integral and differential cross sections for the dipole-allowed, spin-forbidden and forbidden transitions in neutral atoms, singly-charged and multiply-charged ions. Most of the calculations reported in the literature have been based on the close-coupling, distorted-wave and Coulomb-Born approximations. The reliability of cross sections produced by the methods primarily depends on the electron impact energy, type of transition and charge state of the ion. We use the Breit-Pauli R-matrix and B-spline R-matrix methods, both close-coupling methods, in our investigations of electron-ion and atom collisions.

Accurate description of target wave functions is an essential part of a reliable scattering calculation. The quality of target wave functions is assessed by carrying out self-consistency tests and by comparing computed excitation energies and oscillator strengths with experiments and other reliable calculations. Checks are also performed on the convergence of close-coupling expansions for the main transitions of interest in an ion or atom. Whenever possible, it is also important to use experiments to provide a benchmark of theory for selected atomic systems and transitions. Most of the measured absolute excitation cross sections in the near-threshold energy region are based on the crossed-beams and merged-beams electron energy-loss techniques. The quoted total systematic uncertainties of these measurements are typically in the range 20-40%. The measurements of absolute excitation cross sections of multiply-charged ions using the crossed-beams technique is difficult because of low detection efficiency and low signal rates. However, these difficulties of the crossed-beams technique are not present in the merged-beams experiments. The energy resolution in the merged-beams experiments is typically in the range 125-250 meV. In recent years this technique has been successfully used to measure absolute excitation cross sections of multiply charged ions in the near-threshold energy region. Over the last several years we have carried out collaborative research with the experimental groups at the Jet Propulsion Laboratory for ions of astrophysical importance to benchmark theory against experiment. There is a very good agreement between the measurements and close-coupling calculations for several ions. The agreement between theory and experiment provides a support to the accuracy of the theoretical approach and scattering models.

4.1.   Electron Scattering from Neutral Atoms

Accurate transition probabilities and electron excitation collision strengths of neutral carbon, nitrogen, oxygen, sulfur and of many other atoms are needed for the interpretation and analysis of spectra from the planetary and stellar atmospheres and from a variety of other astrophysical objects obtained in the ground and space based observations. Many resonance transitions in these atomic systems occur in the ultraviolet and far ultraviolet wavelength domains of the spectrum. Strong atomic oxygen and sulfur emission features have been observed in the spectra from Jupiter's satellite Io and the Io torus and in stellar atmospheres in several space based observations. The electron impact excitation of the $2p^4\,^1D$ and ^1S states of atomic oxygen gives rise to emission features at 6300 and 2959 $\mathring{A}$ and the resonance lines at 1027, 989, 878 and 792 $\mathring{A}$ arise due to the transitions $2p^4\,^3P$ - $2p^33d\,^3D^o$, $2p^4\,^3P$ - $2p^33s\,^3D^o$, $2p^4\,^3P$ - $2p^33s\,^3P^o$ and $2p^4\,^3P$ - $2s2p^5\,^3P^o$ appear in the spectra of Earth, Mars and Venus. Several nitrogen prominent lines have been observed in the atmospheres of Earth and Titan as well as in solar coronal spectrum. The sulfur lines have been observed in cometary comae as well as in interstellar gas. The atomic oxygen emission line ratios can be used to infer the electron distributions in the daytime Earth's thermosphere and aurora. The oxygen and sulfur lines offer valuable diagnostic tool to study the structure of cool and low-gravity stars.

In this section we review the current status of electron collisions with atomic oxygen, nitrogen and sulfur, and present some examples of cross sections for selected transitions of diagnostic importance. A detailed review of experimental and theoretical studies of collisions of electrons with atomic oxygen was published in 2005 [10]. There are a number of experimental and theoretical investigations of electron collisions with these atomic systems published in the literature. Atomic oxygen is one of most studied atomic systems because of its importance in the modeling of various astrophysical plasmas. Discrepancies exist not only between theory and experiment for several transitions, but also among various experiments as well as among various theories. The uncertainties in the measurements are up to 50%, and the cross sections are measured at a limited number of energy points. The measured cross sections are needed at a large number of energies to reproduce the full behavior of excitation cross sections as a function of energy. Accurate measured cross sections with error bars within 10% for elastic scattering of atomic oxygen in the low-energy region have been reported [11]. On the theoretical side, the studies of low-energy electron collisions with these open-shell atomic systems present a challenging task. First, the target wave functions show strong term dependence of the one-electron orbitals and large correlations. Second, the convergence of close-coupling expansion is slow because of a large number of strongly coupled bound, autoionizing, and continuum states. The main part of the polarizability of low-lying states arises from transitions to the continuum. This implies that the close-coupling expansion should also take into account the possible loss of flux into the continuum. The contribution of the ionization continuum is significant for resonance transitions.

4.1.1. Atomic Oxygen

Integral Cross sections for excitation of the forbidden $2s^2 2p^4\ ^3P$ - ^1D, $2s^2 2p^4\ ^3P$ - ^1S and $2s^2 2p^4\ ^1D$ - ^1S transitions in the low electron energy region for atomic oxygen have been presented in the literature [1, 12, 13, 14, 15]. The R-matrix with long-range polarization pseudostates approach was used by Tayal [12], Thomas et al. [13] and Plummer et al. [15] while Zatsarinny and Tayal [1] used the B-spline R-matrix with pseudostates method. The six spectroscopic plus log-range polarization pseudostates were used by Plummer et al. [15] in the close-coupling expansion. Tayal and Zatsarinny [1] first carried out calculations in 3-, 16- and 26-state close-coupling approximations together with Born approximation to check the convergence of close-coupling expansion, and then they added 17 pseudostates to the 26 spectroscopic bound and autoionizing states to check the effect of coupling to the continuum on the excitation cross sections. They selected pseudostates to represent the polarization of the $2p^4$ states due to excitation to the low-lying continuum. The 26-state close-coupling approximation contained the lowest 21 spectroscopic states: $2s^2 2p^4\ ^3P$, ^1D, ^1S, $2s^2 2p^3(^4S^o)ns$ (n = 3-5) $^5S^o$, $^3S^o$, $2s^2 2p^3(^4S^o)np$ (n = 3-4) ^5P, ^3P, $2s^2 2p^3(^4S^o)nd$ (n = 3-4) $^5D^o$, $^3D^o$, $2s^2 2p^3(^4S^o)4f\ ^5F^o$, $^3F^o$, $2s^2 2p^3(^2D^o)3s\ ^3D^o$, $^1D^o$ and five autoionizing states: $2s^2 2p^3(^2P^o)3s\ ^3P^o$, $^1P^o$, $2s^2 2p^3(^2P^o)3d\ ^3P^o$, $2s 2p^5\ ^3P^o$, $^1P^o$ which have strong dipole connection with the $2s^2 2p^4$ terms. The three pseudostates to represent the polarizability of the ground state together with three spectroscopic target states in the close-coupling expansion were used by Tayal [12]. Zatsarinny and Tayal [1] emphasized on the accuracy of target wave functions by using non-orthogonal radial wave functions and did not impose any orthogonality constraint between continuum functions and the spectroscopic or correlated orbitals to avoid any inconsistency between the continuum and bound

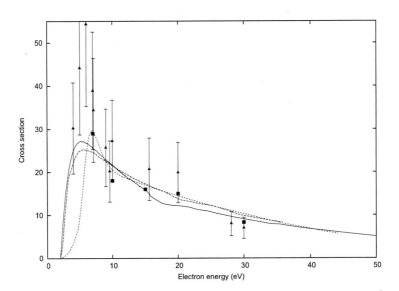

Figure 1. The excitation cross section ($10^{-18} cm^2$) for the $2s^2 2p^4\ ^3P$ - ^1D transition in O I as a function of electron energy. Solid curve, 26-state B-spline R-matrix [1]; long-dashed curve, Berrington and Burke [14]; short-dashed curve, Thomas et al. [13]. Experimental results: solid rectangles, Shyn et al. [17]; solid triangles, Doering [16].

parts of the close-coupling expansions. The results from Zatsarinny and Tayal [1] supersede the calculation of Tayal [12]. Thomas et al. [13] included 19 spectroscopic states and 12 pseudostates in the close-coupling expansion. The pseudostates were chosen to represent the polarizability of all terms of the ground $2s^2 2p^4$ configuration and thus to account for the possible loss of flux into the continuum. The main feature of the calculation of Thomas et al. [13] is the large reduction in the excitation cross sections in the near-threshold region which they attributed to large polarization effects. The cross sections for the ^3P - ^1D transition are not very sensitive to the scattering model and to the long-range polarization effects due to the insignificance of coupling to higher excited bound and continuum states [1].

The excitation cross sections for the ^3P - ^1D transition from the calculations of Thomas et al. [13], Zatsarinny and Tayal [1] and Berrington and Burke [14] together with the experimental results [16, 17] have been compared in Fig. 1. The 3-, 16- and 26-state B-spline R-matrix results in the work of Zatsarinny and Tayal [1] differed only slightly from each other indicating insignificance of coupling to higher excited states for this transition. They noted a small smooth dip in the cross section in energy region around 20 eV which is caused by the inclusion of the $2s2p^5\ ^1P^o$ autoionizing state in the close-coupling expansion. The total uncertainties in the experimental values are up to 50%. The three calculations and experimental results agree with each other at higher energies above about 10 eV, but show significant differences at low electron impact energies in the near-threshold energy region where normally a balance between the correlation configurations used in the N-electron tar-

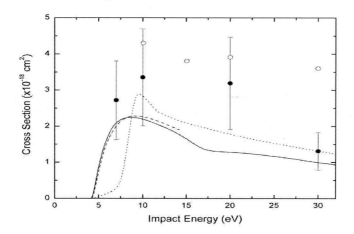

Figure 2. The excitation cross section ($10^{-18}cm^2$) for the $2s^2 2p^4$ ^3P - ^1S transition in O I as a function of electron energy. Solid curve, 26-state B-spline R-matrix [1]; long-dashed curve, Plummer et al. [15]; short-dashed curve, Thomas et al. [13]. Experimental results: solid circles, Shyn and Sharp [18]; open circles, Doering and Gulcicek [19].

get and (N+1)-electron scattering system is important. The calculations of Berrington and Burke [14], Zatsarinny and Tayal [1] and Plummer et al. [15] (not shown) agree very well with each other. However, serious discrepancies are noted with the calculation of Thomas et al. [13] which shows a much slower rate of increase and a narrower peak in the cross section opposed to a steeper rise in the cross section and a broader peak in the results of other calculations. It appears that the distortion in the cross section in the near-threshold energy region in the calculation of Thomas et al. [13] is caused by an imbalance in the N- and (N+1)-electron system correlations. The calculations of Zatsarinny and Tayal [1] and Plummer et al. [15] do not support the claim made by Thomas et al. [13] that the near-threshold energy behavior is caused by the influence of polarization. The $^2P^o$ partial wave contributes about 85% to the integrated cross section for the ^3P - ^1D transition. Zatsarinny and Tayal [1] argued that this corresponds to an interaction defined by the exchange integrals $R_0(2p, kp; kp, 2p)$ and $R_2(2p, kp; kp, 2p)$ in the perturbation theory. It means that the polarization can only influence through the changes in the scattering orbital kp, and there is no direct influence of the dipole polarization on the quadrupole interaction. The existing experimental results have large error bars and, therefore, cannot resolve discrepancies in the theoretical results. However, the cross sections seem to be fairly established at higher energies above 10 eV.

The various theoretical and experimental cross sections for the forbidden ^3P - ^1S transition have been displayed in Fig. 2. It is clear from the figure that the cross sections for this transition are much smaller in magnitude but similar in shape to that of the ^3P - ^1D transition. The cross sections for the two transitions differ by about an order of magnitude. The comparison between 3-, 16- and 26-state B-spline calculations [1] indicates that the coupling effects to the autoionizing states $2s^2 2p^3(^2P^o)3s$ $^1P^o$, $2s^2 2p^3(^2D^o)3d$ $^1P^o$ and $2s2p^5$ $^1P^o$ is important and their inclusion causes a reduction in cross section. Again there is a good agreement between the calculations of Zatsarinny and Tayal [1] and Plummer et al. [15], but significant discrepancies exist with the calculation of Thomas et al. [13] at lower energies below 12 eV where their cross section exhibits a much slower rise close to the threshold and a sharper peak around 10 eV compared to other two calculations. This is again judged to be caused by an imbalance between the number of N-electron and (N+1)-electron configurations. The experimental data of Shyn and Sharp [18] contain 50% uncertainties, and these results are larger than all the theoretical predictions. At some energies the experimental values are more than a factor two larger than the theory. The experimental cross sections of Doering and Gulcicek [19] have uncertainties of 35% and are available at 5, 10, 20 and 30 eV. The theoretical cross sections are within their uncertainties at electron energies of 5, 10 and 30 eV and lie outside at 20 eV. The experimental results with smaller uncertainties and large number of incident electron energies are needed to resolve discrepancies in theoretical calculations.

Significant discrepancies in magnitude between various calculations for the forbidden ^1D - ^1S transition were noted by Zatsarinny and Tayal [1] which they attributed to the differences in target description. They also found a slow convergence of the partial wave expansion for this transition and emphasized the need of higher partial wave contributions. There is a strong resonance feature in the cross section for the ^1D - ^1S transition predicted by various calculations which was identified as the $2s2p^6$ ^2S resonance state by Zatsarinny and Tayal [1] at the position of 14.816 eV and a width of 820 meV. The calculations of Thomas et al. [13] and Plummer et al. [15] predicted the resonance at the energy around 15.1 eV and 14.3 eV. There are no experimental results available for this transition to confirm the theoretical prediction of this resonance. This resonance feature should be measurable in a photodetachment measurement or in a electron scattering experiment which will require a sufficiently large ^1D target population. Bell et al. [20] reported collision strengths for transitions between the fine-structure levels of the ground ^3P state which were calculated in an R-matrix matrix approach by including three spectroscopic ^3P, ^1D, ^1S states plus three pseudostates in the close-coupling expansion.

The resonance transitions due to the excitation of the $2p^33s$ $^3S^o$, $2p^33d$ $^3D^o$, $2p^33s$ $^3D^o$, $2p^33s$ $^3P^o$ and $2s2p^5$ $^3P^o$ states in the oxygen atom give rise to the prominent emission features. Measurements of the cross sections for the resonance transitions have been carried out at the Johns Hopkins University by Doering and co-workers and at the Jet Propulsion Laboratory by Kanik and co-workers in several efforts as improvements in technique and normalization standards became available. Most experimental attention was paid to the ultraviolet emission line at 1304 Å due to the resonance $2p^4$ ^3P - $2p^33s$ $^3S^o$ transition. Doering and Yang [21] collected all the measured cross sections by their group for the $2p^4$ ^3P - $2p^33s$ $^3S^o$ transition and provided cross section as a function of electron energy in the region from 13.4 to 100 eV. The various data sets accumulated by their group showed some scatter in the threshold to peak energy region. A "best guess" curve was provided based on the fit of a Bethe line to the data. The latest measurement from this group by Doering and Yang [21] is quoted to contain an overall uncertainty of $\pm 20\%$ in the cross section.

Differential and integral direct excitation cross sections for the resonance $2p^4$ ^3P - $2p^33s$ $^3S^o$, $2p^4$ ^3P - $2p^33d$ $^3D^o$, $2p^4$ ^3P - $2p^33s$ $^3D^o$ and $2p^4$ ^3P - $2p^33s$ $^3P^o$ transitions have been measured by Kanik et al. [22] at 30, 50 and 100 eV using electron energy loss method. Johnson et al. [23] then extended these measurements to lower energies below 30 eV for the $2p^4$ ^3P - $2p^33s$ $^3S^o$ transition at 1304 Å. The measured integrated cross sections are quoted to be uncertain in the 25-29% range and are obtained by integrating the measured differential cross sections (DCS) in the 0° - 25° angular range and extrapolated DCS for unmeasured angles. The scaled 22-state R-matrix theoretical DCS of Kanik et al. [22] were used to guide extrapolation procedure at large angles. They indicated that the extrapolated DCS at unmeasured angles contribute, on an average 16% to integrated cross sections at 30, 50 and 100 eV. The absolute excitation cross sections of the $2p^4$ ^3P - $2p^33s$ $^3S^o$ transition were used to normalize the cross sections of other transitions. In addition to direct excitation cross sections, emission cross sections for atomic oxygen lines at 1304, 1027, 989 and 878 Å have also been measured. The recent emission measurements include emission cross sections of the $2p^4$ ^3P - $2p^33s$ $^3S^o$ transition at 1304 Å by Noren et al. [24] and electron impact-induced emission cross sections of the 1304, 1027, 989 and 878 Å atomic oxygen lines by Johnson et al. [25].

The slow convergence of the close-coupling expansion because of a large number of coupled bound states and particularly strong coupling to the continuum poses a serious theoretical challenge. The effect of coupling to the continuum target states can be included in a pseudostates approach where the pseudostates are chosen to simulate the continuum target states. Recently Tayal [26, 27] investigated the importance of coupling of bound states to the continuum states for the resonance transitions using the R-matrix with pseudostates (RMPS) approach. The contribution of the ionization continuum to the cross section was estimated by comparing the calculations with and without pseudostates. The effect of coupling to the continuum target states was found to be far greater on resonance transitions than on the forbidden transitions. Tayal [26] included 22 spectroscopic bound and autoionizing states together with 19 pseudostates in the close-coupling expansion in his 41-state RMPS calculation, and found the ionization continuum to contribute 5-15% for the excitation of the $2p^4$ ^3P - $2p^33s$ $^3S^o$ transition and 5-27% for the excitation of the $2p^4$ ^3P - $2p^33d$ $^3D^o$ transition. Good agreement with the latest experiment of Kanik et al. [22] was achieved for the $2p^4$ ^3P - $2p^33s$ $^3S^o$ transition, but discrepancies for the $2p^4$ ^3P - $2p^33d$ $^3D^o$ transition remained. Tayal [27] further attempted to improve the cross sections for the resonance $2p^4$ ^3P - $2p^33s$ $^3S^o$, $2p^4$ ^3P - $2p^33d$ $^3D^o$, $2p^4$ ^3P - $2p^33s$ $^3D^o$ and $2p^4$ ^3P - $2p^33s$ $^3P^o$ transitions by adding an extensive set of pseudostates to a set of physical bound and autoionizing states. He included 18 bound states, 4 autoionizing states, and 30 pseudostates in the close-coupling expansion in his 52-state RMPS calculation. He obtained an accurate description of target wave functions on the basis of configuration-interaction wave functions that were constructed with spectroscopic-type orbitals and pseudo-orbitals to adequately account for the main correlation corrections and the interaction between different Rydberg series $2s^22p^3nl$ as well as between Rydberg series and the $2s2p^5$ perturbers.

The quality of target wave functions in the work of Tayal [27] was assessed by carrying out calculations of excitation energies and oscillator strengths using the spectroscopic and pseudo-orbitals that were used in the construction of target wave functions. The results from his calculation of ab initio excitation energies of the spectroscopic states are compared with the experimental values from the National Institute of Standards and Technology [http://nist.phys.gov] compilation and the calculations of Zatsarinny and Tayal [1] and Hibbert et al. [29] in Table 1. The pseudostates have energies in the range 17.24 - 74.27 eV.

Table 1. Excitation energies relative to the ground state

State	Energy (eV)			
	Observed[a]	Tayal[b]	ZT^c	Hibbert[d]
$2s^22p^4\,^3P$	0.00	0.00	0.00	0.00
$2s^22p^3(^4S^o)3s\,^5S^o$	9.14	9.25	9.20	8.87
$2s^22p^3(^4S^o)3s\,^3S^o$	9.51	9.52	9.57	9.25
$2s^22p^3(^4S^o)3p\,^5P$	10.73	10.47	10.78	10.43
$2s^22p^3(^4S^o)3p\,^3P$	10.98	11.16	11.05	10.66
$2s^22p^3(^4S^o)4s\,^5S^o$	11.83	11.90	11.88	11.49
$2s^22p^3(^4S^o)4s\,^3S^o$	11.92	11.93	11.97	11.59
$2s^22p^3(^4S^o)3d\,^5D^o$	12.07	12.22	12.11	11.73
$2s^22p^3(^4S^o)3d\,^3D^o$	12.08	12.26	12.13	11.74
$2s^22p^3(^4S^o)4p\,^5P$	12.28	12.45	12.32	11.93
$2s^22p^3(^4S^o)4p\,^3P$	12.35	12.55	12.40	12.01
$2s^22p^3(^2D^o)3s\,^3D^o$	12.53	12.84	12.57	12.98
$2s^22p^3(^4S^o)5s\,^5S^o$	12.65	12.90	12.69	
$2s^22p^3(^4S^o)5s\,^3S^o$	12.69	12.94	12.73	
$2s^22p^3(^4S^o)4d\,^5D^o$	12.72	12.97	12.78	12.39
$2s^22p^3(^4S^o)4d\,^3D^o$	12.75	13.04	12.79	12.40
$2s^22p^3(^4S^o)4f\,^5F$	12.76	13.06	12.79	
$2s^22p^3(^4S^o)4f\,^3F$	12.76	13.19	12.79	
$2s^22p^3(^2P^o)3s\,^3P^o$	14.12	14.15	14.12	
$2s2p^5\,^3P^o$	15.65	15.49	15.70	
$2s^22p^3(^2D^o)3d\,^3P^o$		15.89		
$2s^22p^3(^2D^o)4d\,^3P^o$		17.08		

[a]National Institute of Standard and Technology database
($http://nist.phys.gov$)
[b]Tayal [27]
[c]Zatsarinny and Tayal [1]
[d]Hibbert et al. [29]

The results of his calculation show excellent agreement with the experimental values and high-quality theoretical results of Zatsarinny and Tayal [1] indicating a very good quality of target wave functions. The calculated values of Hibbert et al. [29] are normally lower than their results and observed values. It may be caused by the over correlated representation of the excited states relative to the ground state representation.

The length values of oscillator strengths for resonance transitions from the calculations of Tayal [27], Bell and Hibbert [34], Hibbert et al. [29] and Zatsarinny and Tayal [1] are listed in Table 2 where comparison is made with the experimental values of Doering et al. [30], Jenkins [31], Goldbach and Nollez [32] and Brooks et al. [33]. There is a very good agreement between the experimental values and the calculations of Tayal [27], Bell and Hibbert [34], Hibbert et al. [29] and Zatsarinny and Tayal [1]. Bell and Hibbert [34] used the R-matrix method and Hibbert et al. [29], Tayal [27] and Zatsarinny and Tayal [1] used configuration-interaction method with orthogonal and non-orthogonal orbitals, respectively. Doering et al. [30] reported electron-impact energy-loss measurements of oscillator strengths, while Jenkins [31] and Brooks et al. [33] reported absorption measurements. Goldbach and Nollez [32] used emission from a wall-stabilized arc in their relative measurements. The various calculations yield similar values of length oscillator strengths for the $2p^4\ ^3P$ - $2p^33s\ ^3S^o$, $2p^4\ ^3P$ - $2p^34s\ ^3S^o$ and $2p^4\ ^3P$ - $2p^33d\ ^3D^o$ transitions. The calculated results for the $2p^4\ ^3P$ - $2p^33s\ ^3S^o$ transition are in excellent agreement with the measured value of Jenkins [31], while for the $2p^4\ ^3P$ - $2p^33d\ ^3D^o$, $2p^4\ ^3P$ - $2p^33s\ ^3D^o$, $2p^4\ ^3P$ - $2p^33s\ ^3P^o$ and $2p^4\ ^3P$ - $2s2p^5\ ^3P^o$ transitions the calculated results show excellent agreement with the measurement of Doering et al. [30]. The close agreement between various results provides an additional test of the reliability of target wave functions used in the calculation of Tayal [27].

Table 2. Oscillator strengths of resonance transitions in atomic oxygen

Transition	$\lambda(\mathring{A})$	Tayal[a]	ZT[b]	Hibbert[c]	Experiment
$2s^22p^4\ ^3P$ - $2s^22p^33s\ ^3S^o$	1304	0.053	0.053	0.053	0.048^d; 0.053^e
$2s^22p^4\ ^3P$ - $2s^22p^34s\ ^3S^o$	1040	0.009	0.009	0.009	0.010^d; 0.0096^f
$2s^22p^4\ ^3P$ - $2s^22p^33d\ ^3D^o$	1027	0.020	0.022	0.021	0.019^d; 0.029^g
$2s^22p^4\ ^3P$ - $2s^22p^33s\ ^3D^o$	989	0.063	0.051	0.058	0.061^d; 0.052^f
$2s^22p^4\ ^3P$ - $2s^22p^34d\ ^3D^o$	972	0.010	0.011	0.012	0.016^d
$2s^22p^4\ ^3P$ - $2s^22p^33s\ ^3P^o$	878	0.080	0.085		0.086^d
$2s^22p^4\ ^3P$ - $2s2p^5\ ^3P^o$	792	0.061	0.092		0.070^d
$2s^22p^4\ ^3P$ - $2s^22p^33d\ ^3P^o$	780	0.080	0.013		
$2s^22p^4\ ^3P$ - $2s^22p^34d\ ^3P^o$	726	0.013	0.021		

[a]Tayal [27]
[b]Zatsarinny and Tayal [1]
[c]Hibbert et al. [29]
[d]Doering et al. [30]
[e]Jenkins [31]
[f]Goldbach and Nollez [32]
[g]Brooks et al. [33]

We have listed integrated cross sections in units of $10^{-18}cm^2$ for excitation of the resonance $2p^4\ ^3P$ - $2p^33s\ ^3S^o$ (1304 Å), $2p^4\ ^3P$ - $2p^33d\ ^3D^o$ (1027 Å), $2p^4\ ^3P$ - $2p^33s\ ^3D^o$ (989 Å), $2p^4\ ^3P$ - $2p^33s\ ^3P^o$ (878 Å) and $2p^4\ ^3P$ - $2s2p^5\ ^3P^o$ (787 Å) transitions from the 52-state RMPS calculation of Tayal [27] at incident electron energies from 15 eV to 100 eV. The cross sections for these resonance transitions exhibit similar energy behavior. The cross sections are all zero at threshold and then rise sharply in the near-threshold energy region. After reaching a maximum, the cross sections exhibit a slow fall-off at higher energies. The uncertainty in theoretical cross sections may be caused by any inaccuracy in target wave functions and convergence of close-coupling expansion. The excellent agreement between the calculated and measured transition energies and between the calculated and experimental values of oscillator strengths suggest an uncertainty of 5-10% in the cross sections due to target wave functions. Tayal [27] indicated that the convergence of close-coupling expansion in the 52-state RMPS calculation may have been achieved to about 10%, and total uncertainties in the cross sections for the resonance transitions may be in the range of 15-20%. The overall uncertainty in the cross sections for the $2p^4\ ^3P$ - $2s2p^5\ ^3P^o$ transition was estimated to be 20-25%. The 52-state RMPS results supersede previous 41-state RMPS calculation [26], 26-state B-spline R-matrix calculation [28] and 22-state R-matrix calculation [22] for the resonance transitions.

Table 3. Excitation cross sections ($10^{-18}cm^2$) for strong transitions in atomic oxygen

Energy (eV)	$^3P - 3s\ ^3S^o$ 1304 Å	$^3P - 3d\ ^3D^o$ 1027 Å	$^3P - 3s\ ^3D^o$ 989 Å	$^3P - 3s\ ^3P^o$ 878 Å	$^3P - 2s2p^5\ ^3P^o$ 787 Å
15.0	5.82	1.49			
16.5	6.75	1.79			
17.5	7.39	2.05	3.67		
20.0	8.28	2.22	5.02	3.90	4.75
22.5	8.40	2.28	5.55	4.78	5.59
25.0	8.49	2.33	5.78	5.26	6.39
30.0	8.40	2.35	5.88	5.91	8.16
50.0	8.15	2.18	5.78	6.63	9.30
100.0	7.22	1.84	4.82	5.47	6.26

A comparison between the theoretical cross sections for excitation of the resonance $2p^4\ ^3P$ - $2p^3 3s\ ^3S^o$ transition from the 52-state and 41-state RMPS calculations together with the 22-state R-matrix without pseudostates calculation is made in Fig. 3 as a function of incident electron energy from the threshold at 9.51 eV to 100 eV. The cross sections rise sharply at energies close to threshold and then exhibit a slow fall-off at higher energies after a broad maximum around 25 eV. The three theoretical results approach each other at higher energies, and are within 8% of each other at 100 eV. The cross sections in Born approximation agree in shape with other calculations, but are larger in magnitude. The effect of coupling to higher lying bound plus continuum target states through 19 pseudostates in the 41-state RMPS calculation is to reduce cross sections from the 22-state calculation by 5-15%. The inclusion of additional 11 pseudostates in the 52-state RMPS calculation leads to a further reduction in the cross sections in the range 3-10%. The considerable reduction in cross sections in the near-threshold energy region indicates the importance of polarization effect in the low-energy region.

In Fig. 4 we compare the 52-state RMPS excitation cross sections for the $2p^4\ ^3P$ - $2p^3 3s\ ^3S^o$ transition with the absolute direct excitation measured cross sections of Johnson et al. [23] and Kanik et al. [22]. The best guess fit to the measured cross sections of Doering and co-workers is also included in Fig. 4. The 52-state theoretical results are in excellent agreement with the measurements of Johnson et al. [23], Kanik et al. [22] and

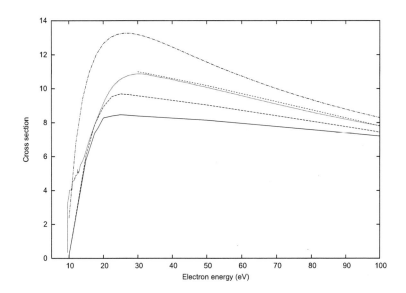

Figure 3. The excitation cross section $(10^{-18} cm^2)$ for the $2s^2 2p^4 \, ^3P$ - $2s^2 2p^3 3s \, ^3S^o$ transition in O I as a function of electron energy. Solid curve, 52-state RMPS [27]; long-dashed curve, 41-state RMPS [26]; dotted curve, 22-state R-matrix [22]; short-dashed curve, 26-state B-spline R-matrix [1]; dash-dotted curve, Born approximation.

Vaughan and Doering [35] over the entire energy range, but are somewhat lower than the measurements of Doering and Yang [21] and Gulcicek and Doering [36]. Johnson et al. [10] recommended integral cross sections for the $2p^4 \, ^3P$ - $2p^3 3s \, ^3S^o$ excitation by using the available experimental and theoretical cross sections. In arriving at the recommended data they used averages of the Vaughan and Doering [35] and Kanik et al. [22] experimental values at 30, 50 and 100 eV and took these as the recommended values at these energies. These experimental data sets from two independent measurements show an excellent agreement with each other. The theoretical cross sections of Tayal [27] were used as a guide at energies below 30 eV to select cross sections that fell smoothly through the measured cross sections. The recommended cross sections at 30, 50 and 100 eV were extrapolated to obtain recommended cross sections out to 200 eV. A B-spline interpolation was then used to produce smooth cross sections in the entire energy range. The recommended cross sections of Johnson et al. [10] are also displayed in Fig. 4. The theoretical cross sections appear to be converged and may contain at the most 10% uncertainty due to coupling effects for this transition. Excellent agreement between theory and experiment for the $2p^4 \, ^3P$ - $2p^3 3s \, ^3S^o$ transition indicates the cross sections for this transition are now very well established.

The position of maximum excitation cross section is around 25 eV in the theoretical 52-state RMPS calculation, and it agrees well with the measurement of Johnson et al. [23]. The theoretical peak value is $8.5 \times 10^{-18} \, cm^2$ which is in excellent agreement with the measured value of $(8.7 \pm 2.2) \times 10^{-18} cm^2$. The best guess fit excitation function of Doering and Yang [21] predicted a maximum in cross sections at 20 eV. Though the measured cross sections

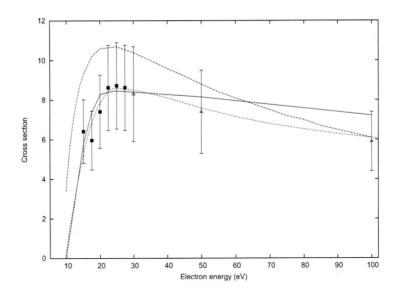

Figure 4. The excitation cross section ($10^{-18}cm^2$) for the $2s^2 2p^4\ {}^3P$ - $2s^2 2p^3 3s\ {}^3S^o$ transition in O I as a function of electron energy. Solid curve, 52-state RMPS [27]; long-dashed curve, best fit to the measured data [21]; short-dashed curve, recommended data [10]; solid rectangles, measured results [23]; open triangles, measured data [22].

from the Doering group are consistently higher than the experiments of Johnson et al. [23] and Kanik et al. [22] in the entire energy range, the cross sections from both groups are within combined error bars at all energies.

Theoretical cross sections for the resonance $2p^4\ {}^3P$ - $2p^3 3d\ {}^3D^o$ transition in the 52-state, 41-state, 22-state R-matrix and 26-state B-spline R-matrix approximations are compared in Fig. 5 as a function of incident electron energy from threshold at 12.07 eV to 100 eV. The measured cross sections of Kanik et al. [22] and Vaughan and Doering [37] are also displayed in this figure. The results from the various calculations differ, implying the sensitivity of cross sections to coupling and correlation effects. Zatsarinny and Tayal [38] emphasized that there is strong interaction between the $2p^3 nd\ {}^3D^o$ and $2p^3 ns\ {}^3D^o$ Rydberg series, and the cross sections for this transition is sensitive to the description of these interactions. The inclusion of coupling to the continuum in the 41-state RMPS approximation reduces the cross sections by up to about 27%. A further addition of 11 pseudostates in the close-coupling expansion to take a better account of coupling to the continuum and use of the pseudo-orbital $\overline{6d}$ to better represent the $2p^3 nd\ {}^3D^o$ series and thus the interaction between the $2p^3 nd\ {}^3D^o$ and $2p^3 ns\ {}^3D^o$ Rydberg series resulted in a further reduction in cross sections by up to about 10%. The peak of cross sections in the 52-state calculation is quite broad and appears to be around 30 eV.

The experimental results from both groups are available at 30, 50 and 100 eV. The 52-state theory differed with the experiment of Kanik et al. [22] at all three energies, lower at 30 and 50 eV and higher at 100 eV. These differences are translated into substantial discrepancy

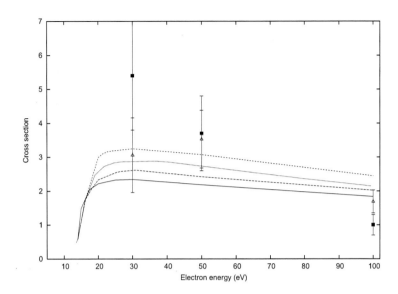

Figure 5. The excitation cross section ($10^{-18} cm^2$) for the $2s^2 2p^4\ ^3P$ - $2s^2 2p^3 3d\ ^3D^o$ transition in O I as a function of electron energy. Solid curve, 52-state RMPS [27]; long-dashed curve, 41-state RMPS [26]; short-dashed curve, 22-state R-matrix [22]; dotted curve, 26-state B-spline R-matrix [38]; open triangles, measured data [22]; solid rectangles, measured data [37].

in the trend of cross sections as a function of energy. After a steep rise of cross sections in the near-threshold energy region, the theoretical cross sections exhibit a very slow fall-off at higher energies; in sharp contrast to a rapid fall-off of measured cross sections of Kanik et al. [22]. The 52-state theory is in excellent agreement with the experiment of Vaughan and Doering [37] at 30 and 100 eV, but are lower at 50 eV. The measured cross sections of Kanik et al. [22] are uncertain by 25-29% and the cross sections of Vaughan and Doering [37] have total uncertainty of 20-36%. The two sets of measurements are within the combined error bars at 30 and 50 eV and slightly outside the combined error bar at 100 eV. However, the two measurements appear to lead to different trends in the shape of cross sections as a function of energy. Experimental data at larger number of energies are needed to reproduce the full energy behavior of excitation cross sections.

The excitation cross sections for the $2p^4\ ^3P$ - $2p^3 3s\ ^3D^o$ transition in various theoretical models together with the measured direct excitation cross sections of Kanik et al. [22] and Doering group [35, 37] are shown in Fig. 6 as a function of incident electron energy from threshold at 12.53 to 100 eV . The cross sections in the near-threshold region differ significantly in various approximations indicating the importance of polarization effect in the low-energy region. The 52-state results are up to 8% lower than the 41-state results. Tayal [27] noticed that the lower partial waves results (L = 0-5) are reduced in 52-state and 41-state RMPS calculations relative to the 22-state calculation but the intermediate partial waves (L = 6-14) are increased and thus making the total cross sections in the 52-state and

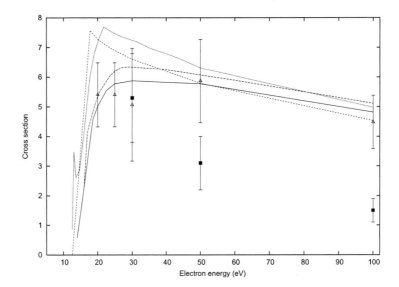

Figure 6. The excitation cross section ($10^{-18} cm^2$) for the $2s^2 2p^4\ {}^3P$ - $2s^2 2p^3 3s\ {}^3D^o$ transition in O I as a function of electron energy. Solid curve, 52-state RMPS [27]; long-dashed curve, 41-state RMPS [26]; short-dashed curve, 22-state R-matrix [22]; dotted curve, 26-state B-spline R-matrix [58]; open triangles, measured data [22]; solid rectangles, measured data [36].

41-state RMPS calculations larger than the 22-state calculation at E $\geq$ 50 eV. The 52-state model results show excellent agreement with the measurements of Doering group [35, 37], but normally disagrees with the experiment of Kanik et al. [22].

The calculation of the excitation cross sections for the $2p^3 3s\ {}^3P^o$ and $2s2p^5\ {}^3P^o$ states which lie above the ionization continuum at 13.61 eV is a challenging task. The excitation of the autoionizing $2p^3 3s\ {}^3P^o$ and $2s2p^5\ {}^3P^o$ states from the ground state is even more effective than the excitation of lower bound states. These states make substantial contribution to the total polarizability of the ground state. It is important to include these autoionizing states in the close-coupling expansion to account for the loss of flux for the resonance transitions. The excitation functions of these transitions exhibit trend similar to other resonance transitions discussed above. Various calculations differ by up to almost a factor of 2. Most of these differences in the various models are perhaps caused by the difference in the treatment of interaction between the ${}^3P^o$ Rydberg series and the $2s2p^5\ {}^3P^o$ perturber state. The excitation cross sections for the $2p^4\ {}^3P$ - $2p^3 3s\ {}^3P^o$ transition from the 52-state RMPS calculation show excellent agreement with the experiment of Kanik et al. [22] and are lower than the measured values of Vaughan and Doering [37] (see Tayal [27]). Tayal [27] also pointed out that the 22-state R-matrix results with orthogonal orbitals and 26-state B-spline R-matrix results with non-orthogonal orbitals are mostly caused by differences in the target wave functions as also reflected in calculated oscillator strengths from the two calculations.

Significant discrepancies in shape and magnitude of cross sections for several prominent observed atomic oxygen transitions exist. The cause of discrepancies can be attributed to the difficulties in dealing with atomic oxygen both theoretically and experimentally. Every reliable scattering calculation requires an accurate description of the target wave functions and inclusion of a large number of coupled bound and autoionizing states and strong coupling to the continuum. More accurate measurements of excitation cross sections are needed to resolve several outstanding issues.

Recent theoretical studies of low-energy elastic electron scattering by atomic oxygen have been mostly motivated by the need for an accurate interpretation of the reliable experimental data [11]. The recent elaborate calculations of the elastic cross sections include 16 or 26 physical target states plus pseudostates B-spline R-matrix calculations of Zatsarinny and Tayal [1], 39-state R-matrix calculation of Wu and Yuan [39], 6-state R-matrix calculations with several target models performed by Plummer et al. [15] and finally the B-spline R-matrix calculations of Zatsarinny et al. [40] who carried out a detailed investigation regarding the dependence of the predicted partial and total cross sections on the scattering model and the accuracy of the target description. The calculations of Zatsarinny and Tayal [1] and Wu and Yuan [39] showed good agreement with experiment for higher energies, but predicted a large resonance just above the elastic threshold which does not exist in the experiment. Plummer et al. [15] performed several numerical experiments with different target and scattering models to investigate the discrepancies between the various theoretical predictions. Plummer et al. [15] did not arrive at any final conclusion regarding the existence of the near-threshold resonance, because some models produced it while others did not. However, they concluded that the existence of the near-threshold resonance does not appear to be related to the accuracy of the representation of the long-range diploe polarization in the calculation. Zatsarinny et al. [40] performed most extensive and detailed investigations of the elastic scattering problem to date by using term-dependent non-orthogonal orbitals in the description of the target states and by exploring the influence of the target polarization through the use of an extensive set of pseudostates to account for channel coupling to the discrete and continuum spectra of atomic oxygen. They used four scattering models in their 3-state B-spline R-matrix (BSR-3), 16-state B-spline R-matrix (BSR-16), 55-state B-spline R-matrix (55-BSR) and 67-state B-spline R-matrix (BSR-67) calculations. The BSR-3 is the simplest model which does not account for any polarization of the target due to the scattering electron. On the other hand, BSR-67 is the most extensive model accounting for the major long-range and short-range correlations.

The elastic cross sections at low electron energies can be extremely sensitive to the representation of the target states and the scattering functions [40]. It has long been recognized that there should be a balance between the correlation configurations used in the N-electron target states and the (N+1)-electron scattering system. As discussed by Zatsarinny et al. [40] and others, the lack of a minimum principle for eigenphase sum due to the use of approximate target states in many electron systems has an important consequence. Generally the theoretical cross section at a particular energy is too large, then passes through the correct value and then becomes too small as more correlation terms are included. The spurious near-threshold resonances in the cross section can occur as the correlation in the scattering functions is increased due to the lowest eigenvalues of the partial-wave R-matrix Hamiltonian approaching zero relative to the ground state energy.

The elastic cross sections in the work of Zatsarinny et al. [40] in four scattering models using most correlated target description have been compared as a function of electron energy. The theoretical cross sections have also been compared with the experimental data [11]. The cross sections from the simplest scattering model BSR-3 are larger than all other models and experiment which means that the channel coupling and polarization effects are very important for low energy elastic electron scattering with atomic oxygen. The BSR-16 model contains target states that are strongly coupled to the ground state and that represent about 25% polarizability of the ground state. The BSR-55 model contains about 93% of the ground state polarizability. The cross sections in BSR-16 and BSR-55 models are reduced compared to the BSR-3 model and show a resonance feature around 8.8 eV which is identified as a Feshbach $2s^2 2p^3 (^4S) 3s^2 \, ^4S^o$ resonance [40]. The reduction in the BSR-55 model is larger than in the BSR-16 model, bringing BSR-55 model cross sections closer to the experiment. The decrease in cross sections is clearly due to the proper account of polarization effects. The BSR-67 model accounts for the same amount of ground-state polarizability as the BSR-55 model, but further improves on the short-range correlation effects. The BSR-67 model further improves the agreement with experiment. The experimental data are available for the total cross sections as well as elastic cross sections generated from a partial wave analysis of their angle-differential scattering data [11]. The total cross section includes excitation of the $2s^2 2p^4 \, ^1D$ and ^1S states. The contributions of these states are relatively small, but their inclusion improves the agreement between theory and experiment. There is good agreement between theory and experiment (see Zatsarinny et al. [40]). They concluded that the predicted near-threshold resonance feature in the earlier calculations is judged to be an artifact of an imbalance in correlation in the N-electron target description and the (N+1)-electron scattering system.

4.1.2. Atomic Nitrogen

Electron impact excitation cross sections of atomic nitrogen are needed for the interpretation and analysis of observed spectra from the solar and planetary atmospheres. The atomic nitrogen 5200, 1200 and 1135 Å lines due to the forbidden $2s^2 2p^3 \, ^4S^o$ - $^2D^o$ and resonance $2s^2 2p^3 \, ^4S^o$ - $2s^2 2p^2 3s \, ^4P$ and $2s 2p^4 \, ^4P$ transitions have been observed in the spectra of Earth's upper atmosphere. The forbidden line due to the $2s 2p^3 \, ^4S^o$ - $^2D^o$ transition appears in the optical part of the spectrum and the resonance lines due to the $2s 2p^3 \, ^4S^o$ - $2s 2p^2 3s \, ^4P$, $2s 2p^3 \, ^4S^o$ - $2s 2p^4 \, ^4P$, $2s 2p^3 \, ^4S^o$ - $2s 2p^2 3s \, ^4P$ and $2s 2p^3 \, ^4S^o$ - $2s 2p^2 3d \, ^4P$ transitions occur in the ultraviolet and far ultraviolet wavelength range.

Several experimental and theoretical studies of oscillator strengths and electron excitation cross sections for atomic nitrogen have been reported in the literature. The configuration-interaction calculations of excitation energies and oscillator strengths have been performed by Robinson and Hibbert [41], Hibbert et al. [42], Tayal and Beatty [43] and Tong et al. [44]. A brief description of the earlier theoretical and experimental studies of oscillator strengths and electron excitation cross sections can be found in Tayal and Beatty [43]. Absolute measured direct electron impact excitation cross sections for the forbidden $2s^2 2p^3 \, ^4S^o$ - $^2D^o$ transition and for the resonance $2s^2 2p^3 \, ^4S^o$ - $2s^2 2p^2 3s \, ^4P$ and $2s 2p^4 \, ^4P$ transitions have been reported in the literature [50, 51, 52]. The recent theoretical calculations for electron impact excitation have been reported by Ramsbottom and Bell [45], Tayal and Beatty [43], Tayal [49], Tayal and Zatsarinny [48] and Tayal [53]. Tayal and Beatty [43] reported differential and integral cross sections for the forbidden $2s^2 2p^3\ {}^4S^o$ - ${}^2D^o$ transition from a 11-state R-matrix approach. Tayal [49] extended the 11-state R-matrix calculation by including 7 additional target states in his 18-state R-matrix calculation to obtain thermally averaged effective collision strengths.

More recently, Tayal and Zatsarinny [48] reported cross sections for the scattering of electrons from atomic nitrogen using the B-spline R-matrix with pseudostates approach. Cross sections for the forbidden $2s^2 2p^3\ {}^4S^o$ - ${}^2D^o$, ${}^2P^o$ and ${}^2D^o$ - ${}^2P^o$ and resonance $2s^2 2p^3\ {}^4S^o$ - $2s^2 2p^2 3s\ {}^4P$, $2s2p^4\ {}^4P$, $2s^2 2p^2 4s\ {}^4P$ and $2s^2 2p^2 3d\ {}^4P$ transitions were reported for incident electron energies from threshold to 120 eV. The 24 spectroscopic bound and autoioinizing states together with 15 pseudostates were included in the close-coupling expansion. The pseudostates were chosen to approximate the loss of flux into the infinite number of bound and continuum states that are dipole coupled with ground configuration terms. The 18-state R-matrix calculation of Tayal [49] has been noted to suffer a distortion in cross sections at low incident electron energies close to threshold due to inconsistencies in the two sums of the standard close-coupling expansion. The (N+1)-electron bound configurations in the second summation of close-coupling expansion were generated by omitting configurations with small expansion coefficients in the calculation of Tayal [49]. The calculation of Tayal and Zatsarinny [48] represents significant improvement over the previously available calculations. Tayal [53] calculated electron impact collision strengths and rates for fine-structure transitions over the temperature range that is suitable for astrophysical modeling calculations.

Tayal and Zatsarinny [48] improved the theoretical aspect of electron impact excitation of atomic nitrogen by improving the description of target wave functions and by including spectroscopic bound and autoionizing states and pseudostates in the close-coupling expansion to account for all significant couplings. The transitions to the continuum states make significant contributions to the polarizabilities of the low-lying states. In addition to pseudostates, the autoionizing $2s2p^4$ and $2s2p^3 3p$ configurations have strong dipole connection with the ground terms. The autoionizing states to some extent may simulate a part of the coupling to the continuum. The contribution of the ionization continuum can be estimated by comparing calculations with and without pseudostates. Tayal and Zatsarinny [48] chose pseudostates to account for the static dipole polarizabilities of the ground $2s^2 2p^3\ {}^4S^o$ state and the metastable ${}^2D^o$ and ${}^2P^o$ states. They used a B-spline basis for the description of continuum functions and did not impose any orthogonality constraint between continuum functions and the valence spectroscopic and correlated radial functions. The atomic nitrogen wave functions exhibit strong term dependence of the valence orbitals. It is necessary to take account of the term dependence of one-electron orbitals as well as correlation corrections and interactions between target terms.

The oscillator strengths are an important indicator of the quality of target wave functions and thus of the reliability of predicted excitation cross sections. The trend of the predicted cross sections for the dipole-allowed transitions particularly at large incident electron energies depends on oscillator strengths and any error in the oscillator strengths will be reflected in the excitation cross sections. Tayal and Zatsarinny [48] assessed the accuracy of their oscillator strengths by comparing with other fairly extensive oscillator strength calculations of Hibbert et al. [29], Robinson and Hibbert [41], Tong et al. [44] and Tayal and Beatty [43].

Table 4. Oscillator strengths of resonance transitions in atomic nitrogen

Transition	Tayal[a] f_L/f_V	Tong[b] f_L/f_V	Hibbert[c] f_L/f_V	Robinson[d] f_L/f_V	Experiment
$2p^3\ {}^4S^o$ - $2p^23s\ {}^4P$	0.252/0.273	0.284/0.296	0.251/0.272	0.241/0.199	0.271^e; 0.266^f
$2p^3\ {}^4S^o$ - $2s2p^4\ {}^4P$	0.100/0.118	0.066/0.069	0.081/0.095	0.101/0.088	0.080^g; 0.085^h
$2p^3\ {}^4S^o$ - $2p^24s\ {}^4P$	0.024/0.023	0.025/0.026	0.032/0.030	0.034/0.028	0.030^g; 0.027^h
$2p^3\ {}^4S^o$ - $2p^23d\ {}^4P$	0.071/0.064	0.076/0.076	0.066/0.060	0.065/0.063	0.067^g; 0.075^h

[a]Tayal and Zatsarinny [48]
[b]Tong et al. [44]
[c]Hibbert et al. [42]
[d]Robinson and Hibbert [41]
[e]Goldbach et al. [54]
[f]Smith et al. [55]
[g]Lugger et al. [56]
[h]Goldbach et al. [57]

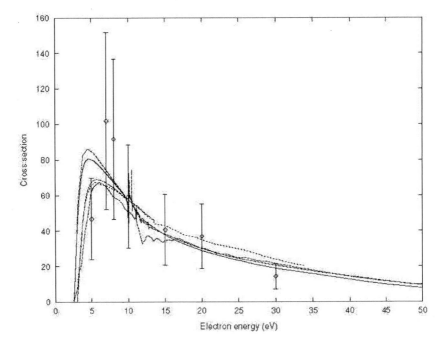

Figure 7. The excitation cross section ($10^{-18}cm^2$) for the $2s^22p^3\ {}^4S^o$ - ${}^2D^o$ transition in N I as a function of electron energy. Solid curve, 39-state B-spline RMPS [48]; long-dashed curve, 21-state B-spline R-matrix [48]; short-dashed curve, 8-state R-matrix [46]; dash-dotted curve, 7-state R-matrix [45]; dotted curve, 11-state R-matrix [43]. Experimental results: diamonds, Yang and Doering [50].

The comparison of the various length (f_L) and velocity (f_V) theoretical and experimental results from the paper of Tayal and Zatsarinny [48] has been displayed in Table 4. There is an overall good agreement between the length and velocity values of their theoretical oscillator strengths, and also a good agreement between the theoretical results and available measured values. This implies that the oscillator strengths for these transitions given in Table 4 have been established very well within a few per cent.

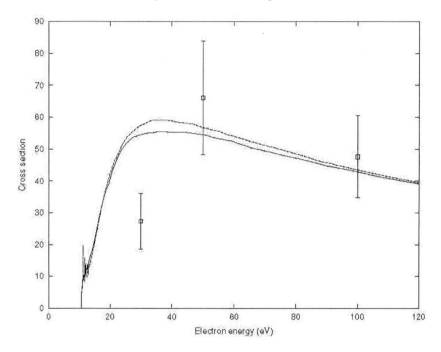

Figure 8. The excitation cross section ($10^{-18} cm^2$) for the $2s^2 2p^3 \, {}^4S^o$ - $2s^2 2p^2 3s \, {}^4P$ transition in N I as a function of electron energy. Solid curve, 39-state B-spline RMPS [48]; long-dashed curve, 21-state B-spline R-matrix [48]. Experimental results: rectangles, Doering and Goembel [51].

Tayal and Zatsarinny [48] compared excitation cross sections for the forbidden and resonance transitions from their 39-state B-spline R-matrix with pseudostates (RMPS) and 21-state B-spline R-matrix without pseudostates calculations to estimate the contribution of ionization continuum. They found that the effect of coupling to the continuum is small for the forbidden transition and is rather large for the resonance transitions. They also noticed that the effect of coupling to the continuum increases with increasing excitation energy of the upper excited state of the resonance transitions. The reduction in cross sections from the 21-state calculation was found to be up to 6, 16, 24 and 33% for the $2s^2 2p^3 \, {}^4S^o$ - $2s^2 2p^2 3s \, {}^4P$, $2s^2 2p^3 \, {}^4S^o$ - $2s2p^4 \, {}^4P$, $2s^2 2p^3 \, {}^4S^o$ - $2s^2 2p^2 4s \, {}^4P$ and $2s^2 2p^3 \, {}^4S^o$ - $2s^2 2p^2 3d \, {}^4P$ transitions respectively around the peak cross sections. However, at higher energies the two calculations approach each other. A part of discrepancies between the two calculations is also expected from the differences in wave functions.

The predicted integral excitation cross sections for the forbidden ${}^4S^o$ - ${}^2D^o$ transition in various theoretical models together with measured cross sections [50] have been displayed in Fig. 8 as a function of incident electron energy from threshold to 50 eV. The excitation cross sections rise sharply close to threshold and then decreases in the higher energy region after a broad peak in the low energy region around 5 eV and show a resonance structure in the 10.0 - 13.0 eV energy region. There is a satisfactory agreement in shape and magnitude of the predicted cross sections. The measured cross sections have $\pm49\%$ uncertainties which are rather large to resolve any discrepancies in the predicted results. The predicted peak in the 39-state RMPS calculation [48] is 80.6×10^{-18} cm^{-2} which is within the experimental uncertainties of the measured value $96\pm32\times10^{-18}$ cm^{-2}.

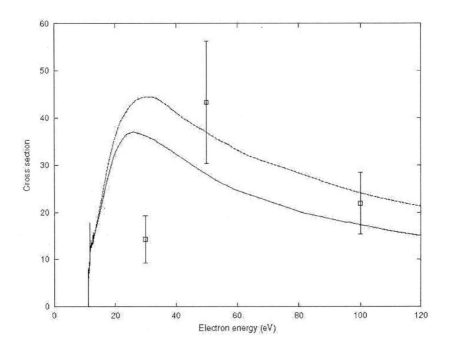

Figure 9. The excitation cross section $(10^{-18}cm^2)$ for the $2s^22p^3$ $^4S^o$ - $2s2p^4$ ^4P transition in N I as a function of electron energy. Solid curve, 39-state B-spline RMPS [48]; long-dashed curve, 21-state B-spline R-matrix [48]. Experimental results: rectangles, Doering and Goembel [52].

The theoretical results from the 39-state RMPS and 21-state B-spline R-matrix calculations together with measured values for the excitation of the resonance $2s^22p^3$ $^4S^o$ - $2s^22p^23s$ ^4P and $2s^22p^3$ $^4S^o$ - $2s2p^4$ ^4P transitions [51, 52] have been compared in Figs. 9 and 10 in the energy region from threshold to 120 eV. There is a sharp rise in the predicted cross sections in near-threshold energy region and after a broad maximum exhibit a slow fall-off at higher energies. The inclusion of polarization effect leads up to 6% and 16% reduction in cross sections for the $2s^22p^3$ $^4S^o$ - $2s^22p^23s$ ^4P and $2s^22p^3$ $^4S^o$ - $2s2p^4$ ^4P transitions respectively. There are significant discrepancies between the two calculations for the $2s^22p^3$ $^4S^o$ - $2s2p^4$ ^4P transition which are partly caused by a better representation of the interaction of the $2s2p^4$ ^4P state with other states of the ^4P symmetry. The measured cross sections have been reported at 30, 50 and 100 eV and contain uncertainties in the range 30-35%. The experimental data at a larger number of energies with smaller uncertainties are needed to verify the full energy behavior of excitation cross sections.

4.1.3. Atomic Sulfur

Prominent atomic sulfur emissions due to the $3p^4\,^3P$ - $3p^3(^4S^o)4s\,^3S^o$, $3p^4\,^3P$ - $3p^3(^2D^o)4s\,^3D^o$ and $3p^4\,^3P$ - $3p^3(^4S^o)3d\,^3D^o$ transitions have been observed in the vicinity of Io and in Io plasma torus. Electron collision excitation cross sections of atomic sulfur have not yet been measured and the first large scale scattering calculations were reported by Zatsarinny and Tayal [28, 58] using the B-spline R-matrix method with non-orthogonal orbitals. They presented integral cross sections for the elastic scattering and for excitation of the $3s^23p^4\,^1D$, ^1S, $3s^23p^34s\,^{3,5}S^o$, $^{1,3}P^o$, $^{1,3}D^o$, $3s^23p^33d\,^3D^o$ and $3s3p^5\,^3P^o$ states in the energy region from threshold to 50 eV. The calculations were performed in the LS coupling scheme and the close-coupling expansion included 27 spectroscopic bound and autoionizing states. They also performed several test calculations with different numbers of target physical states to check the convergence of the close-coupling expansion. The low-lying states in atomic sulfur belong to the $3s^23p^4$, $3s^23p^3(^4S^o,^2D^o,^3P^o)nl$ and $3s3p^5$ configurations. The states show different correlation patterns, and the s, p and d valence orbitals exhibit large term dependence which is due to both the total and intermediate terms. For example, the 3p orbital in the ground state and in the $3s^23p^3nl$ excited states is different with average radius equal to 2.06 and 1.88 au, respectively. The wave functions also exhibit large correlation corrections due to the core correlation, core-valence correlation and interaction between different Rydberg series. Zatsarinny and Tayal [28] indicated that the largest correlation effects arise from replacements $3s - d$, $3p^2 - d^2, p^2$. These correlation effects reflect large intershell 3s-3p and intrashell $3p^3$ correlation and small correlations in the $3s^2$ shell. The $3s^23p^3\,^2P^o$ states exhibit strong mixing with the $3p^5\,^2P^o$ state because they both belong to the same complex.

A large discrepancy existed between the prediction and measured values of oscillator strength for the $3p^4\,^3P$ - $3p^3(^2D^o)4s\,^3D^o$ transition for a long time. As discussed by Tayal [59], the calculated values were smaller by factors of 5-10 than the measured values. These discrepancies were first resolved by Tayal [59] and the results were later confirmed by other large scale calculations [28]. The accurate representation of the strong interactions of the Rydberg series $3s^23p^3(^2D^o)nd$ with the perturber $3s3p^5$ state is also a challenging problem in atomic sulfur. The $3s3p^5\,^3P^o$ state is found to be a highly mixed state with the composition of the $3s3p^5$ component being only 56% [28]. This component accounts for much of the oscillator strengths for transitions to the Rydberg series from the ground state and leads to significant reduction in the oscillator strengths.

Based on a comparison of 3-state, 15-state and 27-state scattering models, Zatsarinny and Tayal [28] concluded that the channel coupling effects are small for the forbidden ^3P - ^1D and ^3P - ^1S transitions. The simple 3-state calculation agreed with the elaborate 27-state calculation within 10% for the ^3P - ^1D transition and within 20% for the weaker ^3P - ^1S transition. The energy behavior of the cross sections for the forbidden transitions was found to be complicated and exhibits humplike structure. The energy behavior of the cross sections for the excitation of quintet states $3s^23p^34s\,^5S^o$, $4p\,^5P$ and $3d\,^5D^o$ shows a specific form of the exchange transitions. They exhibit a sharp maximum in near-threshold energy region and then a rapid decrease as energy increases. The exchange interaction was also found to dominate the allowed $3s^23p^4\,^3P$ - $3s3p^5\,^3P^o$ transition. On the other hand, Zatsarinny and Tayal [28] noted a significant reduction in cross sections in a large range of electron energies for the dipole-allowed transitions. The cross sections for the dipole-allowed transitions from the ground state to the $3s^2 3p^3 4s\,^3S^o$, $^3P^o$, $^3D^o$ and $3s^2 3p^3 3d\,^3D^o$ states have been found to be comparable in magnitude. Zatsarinny and Tayal [58] presented collision strengths in terms of simple parametric functions of scaled energy and Tayal [60] reported effective collision strengths for fine-structure transitions in atomic sulfur using the B-spline R-matrix approach.

4.2. Electron Scattering from Atomic Ions

In this section, we review the current status of electron impact excitation of the selected carbon, oxygen, sulfur and iron ions which are of considerable interest in the modeling of the astrophysical and fusion plasmas. The theoretical predictions from the close-coupling calculations have been compared with the electron energy-loss merged-beams experiments in the near-threshold region for these ions whenever possible. Generally a good agreement is obtained between the theoretical and experimental results, implying that the close-coupling methods together with accurate physical models can produce large amount of reliable electron excitation results for numerous transitions in many ions.

4.2.1. Carbon Ions

The intercombination lines due to the $2s^2 2p\,^2P^o_J$ - $2s2p^2\,^4P_{J'}$ transitions in C II can provide excellent density diagnostics in the range $10^9 \leq N_e \geq 10^{12}\ cm^{-3}$. The C II lines due to the $2s^2 2p\,^2P^o_J$ - $2s2p^2\,^2D_{J'}$ and $2s2p^2\,^4P_J$ - $2p^3\,^4S^o_{J'}$ transitions are also electron density sensitive. The C II intercombination lines $2s^2 2p\,^2P^o_J$ - $2s2p^2\,^4P_{J'}$ are useful to investigate turbulence and electron density in non-coronal stars. These lines are also important for the abundance determinations in both galactic and extragalactic stellar atmospheres and diffuse interstellar gas. Interstellar densities can be derived for H II regions by using C II 1037 $\mathring{A}$ line due to the $2s^2 2p\,^2P^o_{3/2}$ - $2s2p^2\,^2S_{1/2}$ transition.

The scattering of electrons from C II has been studied by several theoretical and experimental groups. On the theoretical side, the close-coupling calculations of Lennon et al. [61], Hayes and Nussbaumer [62] and R-matrix calculations of Keenan et al. [63], Luo and Pradhan [65], Blum and Pradhan [66, 67], Wilson and Bell [68] and Wilson et al. [69] have been reported in the literature. Blum and Pradhan [67] and Wilson et al. [69] included 10 and 16 LS target states respectively in the close-coupling expansion in their R-matrix calculations in LS coupling and then used an algebraic transformation to obtain collision strengths for fine-structure transitions. Measured and calculated excitation cross sections for the $2s^2 2p\,^2P^o$ - $2s2p^2\,^4P$, ^2D and ^2S transitions have been reported by Smith et al. (1996). The measurements of absolute cross sections were carried out using the electron energy-loss merged-beams method. The LS coupling R-matrix calculations were used to compare with the experiment as the fine-structure components were not resolved. A good agreement between theory and experiment was noted. Measured cross sections for the electron impact excitation of the $2s^2 2p\,^2P^o$ - $2s2p^2\,^2D$ transition in C II were reported by Lafyatis and Kohl [70] and Williams et al. [71].

The theoretical aspect of the electron impact excitation of C II can be improved by using an accurate description of the target wave functions and by properly including resonances in collision strengths. Higher degree of sophistication is required to accurately calculate the position of resonances, and a fine energy mesh is needed to resolve resonances. Tayal [72] used non-orthogonal orbitals that were optimized on separate states to account for the term dependence of target wave functions and used the B-spline basis for the description of continuum functions. No orthogonality constraint was imposed between continuum functions and the valence spectroscopic and correlated atomic orbitals to avoid potential inconsistencies between the continuum and bound parts of the close-coupling expansion. Tayal [72] considered 42 fine-structure levels arising from the lowest 23 LS target terms of the $2s^2 2p$, $2s2p^2$, $2s^2 3s$, $2s^2 3p$, $2p^3$, $2s^2 3d$, $2s^2 4s$, $2s^2 4p$, 2s2p3s, $2s^2 4d$, $2s^2 4f$, $2s^2 5s$, $2s^2 5p$, $2s^2 5d$, $2s^2 5f$ and $2s^2 6s$ configurations in the close-coupling expansion. The inclusion of these levels in the expansion ensures channel coupling effects reasonably well up to the excited $2s^2 5d\, ^2D_{3/2,5/2}$ levels around 1.7 Ryd.

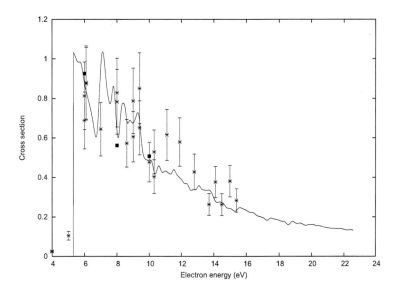

Figure 10. The excitation cross section ($10^{-16} cm^2$) for the $2s^2 2p\, ^2P^o$ - $2s2p^2\, ^4P$ transition in C II as a function of electron energy. Solid curve, 42-level B-spline BPRM [72]; asterisks, measured data [73]; solid rectangles, 8-state R-matrix [73].

Accurate description of target wave functions is important for a reliable scattering calculation. The quality of target wave functions in the calculation of Tayal [72] is very good as has been assessed by comparing computed excitation energies and oscillator strengths with experiments and other reliable calculations. He used the spectroscopic and correlation functions to construct configuration-interaction expansions for different target states by allowing one-electron and two-electron excitations from all the basic configurations $2s^2 2p$, $2s2p^2$, 2s2p3s, $2s^2 nl$ (n=3-5; l=0-2) and 2s2p3p. He found that the resonance structures are complex and enhance the collision strengths significantly. The non-resonant background collision strength for the allowed transitions was found to be larger than for the forbidden transitions. The resonance enhancement in collision strengths for the forbidden transitions is normally larger compared to allowed transitions. The relativistic effects appear to be small in C II ion.

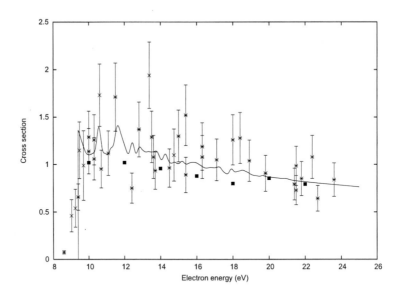

Figure 11. The excitation cross section $(10^{-16} cm^2)$ for the $2s^2 2p\ ^2 P^o$ - $2s2p^2\ ^2 D$ transition in C II as a function of electron energy. Solid curve, 42-level B-spline BPRM [72]; asterisks, measured data [73]; solid rectangles, 8-state R-matrix [73].

A good agreement with the measured absolute direct excitation cross sections for the intercombination $2s^2 2p\ ^2 P^o$ - $2s2p^2\ ^4 P$ and resonance $2s^2 2p\ ^2 P^o$ - $2s2p^2\ ^2 D, ^2 S$ transitions [73] has been obtained [72]. The fine-structure components of these multiplets were not resolved in the experiment because for very small separations between fine-structure levels. The uncertainty in experimental cross sections is about 20% and the energy resolution is 0.250 eV. The theoretical cross sections were convoluted to the experimental energy spread. The convoluted theoretical cross sections have been compared with the measured cross sections in Figs. 11-13. The convoluted theoretical cross sections for the intercombination transition in Fig. 11 are within experimental error bars for most incident electron energies except a few energies around 12 eV and 15 eV where experimental results are somewhat larger than the theory. The excitation cross sections for the resonance $2s^2 2p\ ^2 P^o$ - $2s2p^2\ ^2 D$ and $^2 S$ transitions have been compared with the measured cross sections in Figs. 12 and 13. The measured cross sections of Smith et al. [73] and Lafyatis and Kohl [70] are also shown in Fig. 12 for the $^2 P^o$ - $2s2p^2\ ^2 D$ transition. There is a good agreement between the predicted and measured cross sections at most incident electron energies. However, discrepancies exist at a few energies. Similar agreement between theory and experiment exists for the $^2 P^o$ - $2s2p^2\ ^2 S$ transition shown in Fig. 13. The theoretical cross sections are within experimental uncertainty for all energies except two incident electron energies around 14 eV. The 8-state R-matrix theoretical cross sections from the work of Smith et al. [73] have also been displayed in Figs. 11-13 by open squares. The two calculations normally agree to

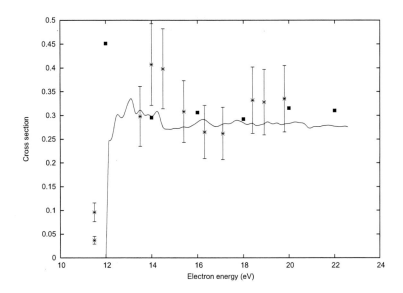

Figure 12. The excitation cross section ($10^{-16} cm^2$) for the $2s^2 2p\,^2P^o$ - $2s2p^2\,^2S$ transition in C II as a function of electron energy. Solid curve, 42-level B-spline BPRM [72]; asterisks, measured data [73]; solid rectangles, 8-state R-matrix [73].

about 20%. In the B-spline R-matrix calculation the use of non-orthogonal orbitals considerably simplified the structure of the bound part of the close-coupling expansion and thus allowed a substantial reduction in undesirable pseudoresonances.

The electron impact collision strengths and rates for fine-structure transitions among the $2s^2 2p\,^2P^o_{1/2,3/2}$, $2s2p^2\,^4P_{5/2,3/2,1/2}$, $2s2p^2\,^2D_{5/2,3/2}$, $2s2p^2\,^2P_{1/2,3/2}$, $2s2p^2\,^2S_{1/2}$, $2s^2 3l$ (l=0-2) $^2S_{1/2}$, $^2P^o_{1/2,3/2}$, $^2D_{3/2,5/2}$, $2p^3\,^4S^o_{3/2}$, $^2D^o_{3/2,5/2}$, $2s^2 4l$ (l=0-1) $^2S_{1/2}$ and $^2P^o_{1/2,3/2}$ levels and from these levels to other higher excitation levels up to the $2s^2 5d\,^2D_{3/2,5/2}$ levels have been calculated over a temperature range that is suitable for astrophysical plasma modeling calculations [72].

4.2.2. Oxygen Ions

Several prominent O II emission lines have been observed in the spectra of Earth's atmosphere, Io plasma torus, planetary nebulae and Seyfert galaxy. Accurate oscillator strengths and electron impact excitation cross sections of fine-structure transitions in O II are needed for the interpretation of the observed spectra to understand the physical processes and conditions. Several O II line intensity ratios are density and/or temperature sensitive.

Recently Tayal [74] and Montenegro et al. [75] investigated the importance of relativistic effects in electron impact excitation of forbidden transitions among the fine-structure levels of the ground $2s^2 2p^3$ configuration and concluded that the relativistic effects are negligible. This finding is in clear contrast with an earlier similar calculation reported by McLaughlin and Bell [76]. There are significant differences between the results of the two sets of calculations. The calculations of Montenegro et al. [75] and McLaughlin and Bell [76] have used the same Breit-Pauli R-matrix (BPRM) method and accompanying computer codes [2]. Montenegro et al. [75] included lowest 16 fine-structure levels arising from the O II $2s^2 2p^3$, $2s 2p^4$ and $2s^2 2p^2 3s$ configurations and McLaughlin and Bell [76] included all the 21 fine-structure levels belonging to these three configurations. Thus 17 levels of odd parity belonging to the $2s^2 2p^2 3p$ configuration below the highest $2s^2 2p^2 3s\ ^2S_{1/2}$ excited level of their list of 21 levels were ignored. The 11 LS $2s^2 2p^3\ ^4S^o$, $^2D^o$, $^2P^o$, $2s 2p^4\ ^4P$, ^2D, ^2S, ^2P, $2s^2 2p^2 3s\ ^4P$, ^2P, ^2D and ^2S target terms were represented by configuration-interaction wave functions. These were constructed using the spectroscopic 1s, 2s, 2p, 3s and pseudo $\overline{3p}$ and $\overline{3d}$ orbitals. A brief description of the earlier calculations can be found in the paper by Tayal [74]. Measured excitation cross sections of the forbidden $2s^2 2p^3$ $^4S^o$ - $^2D^o$ and resonance $2s^2 2p^3\ ^4S^o$ - $2s 2p^4\ ^4P$ transitions have been obtained using the electron-energy-loss and merged-beams approach [77].

Tayal [74] used the B-spline R-matrix method to investigate the electron impact excitation of forbidden and allowed transitions in singly ionized oxygen. The relativistic effects have been incorporated in the Breit-Pauli Hamiltonian. Flexible non-orthogonal sets of radial functions are used to obtain accurate target description and to represent the scattering functions. Tayal [74] attempted to further resolve the discrepancies and to provide definitive cross sections for electron impact excitation of forbidden and allowed transitions in O II. He carried out more extensive and detailed investigations of the dependence of cross sections on the scattering model and the accuracy of the target wave functions. He has attempted to improve the target wave functions by using flexible non-orthogonal orbitals to describe the term-dependence of valence orbitals as well as correlation and relaxation effects. A set of orthogonal orbitals was optimized for each atomic state separately. However, the different sets of orthogonal orbitals thus obtained for various states were not orthogonal to each other. Tayal [74] included 47 fine-structure levels arising from the lowest 22 LS target terms of the $2s^2 2p^3$, $2s 2p^4$, $2s^2 2p^2 3s$, $2s^2 2p^2 3p$ and $2s^2 2p^2 3d$ configurations in the close-coupling expansion. The inclusion of these levels in the expansion ensures channel coupling effects reasonably well perhaps up to the excited $2s^2 2p^2 (^1S) 3s\ ^2S_{1/2}$ level at 2.1 Ryd. The higher excited levels belonging to the $2s^2 2p^2 (^3P) 3d\ ^4F$, $2s^2 2p^2 (^3P) 3d\ ^4P$ and $2s^2 2p^2 (^1D) 3p\ ^2P^o$ terms may have significant coupling with higher excited levels that lie above these levels.

An independent check on cross sections for the forbidden and allowed transitions in O II was provided by performing a BPRM [2] calculation with orthogonal orbitals. In this calculation the scattering model contained 62 fine-structure levels arising from the 28 LS terms of the $2s^2 2p^3$, $2s 2p^4$, $2s^2 2p^2 3s$, $2s^2 2p^2 3p$, $2s^2 2p^2 3d$ and $2s^2 2p^2 4s$ configurations. The target wave functions were constructed with ten orthogonal one-electron orbitals 1s, 2s, 2p, 3s, 3p, 3d, 4s, 4p, 4d and 4f. The orthogonal orbitals are expressed in analytic form as a sum of Slater-type-orbitals.

The agreement between the length and velocity forms of oscillator strength may to some extent indicate the accuracy of wave functions and the convergence of the configuration-interaction expansions. However, it is not necessarily a sufficient condition for the reliability of the results. There is normally an excellent agreement between the length and velocity forms in the calculations of Bell et al. [78] and Tayal [74] and the two sets of results are normally within 10%. Bell et al. [78] used 12 one-electron orbitals and included 575 and

669 configurations in the configuration-interaction expansions of levels of odd parity and even parity, respectively. The relativistic effects were included through the spin-orbit, spin-other-orbit, spin-spin, mass and Darwin Breit-Pauli operators. Tayal [74] used the same one-body and two-body Breit-Pauli operators in the calculation of oscillator strengths and included 713 configurations for odd parity levels and 904 configurations for even parity levels which were constructed with 39 one-electron non-orthogonal orbitals. The close agreement between the two sets of calculations which have been carried out with two independent computer codes may suggest that the oscillator strengths for these fine-structure transitions are now well established. The agreement also reflects on the accuracy of the wave functions that have been used in the scattering calculations of Tayal [74].

The scattering calculations of Tayal [74] were carried out with a new general B-spline R-matrix code [7], in which non-orthogonal orbitals are used to describe both the target levels and the R-matrix continuum basis functions. The calculations include all scattering symmetries with $J \leq 35$ and were estimated to be converged for forbidden transitions. The higher partial waves contributions for the dipole-allowed transitions are calculated using the Coulomb-Bethe approximation [8]. The number of continuum basis functions for each angular momentum was 51 and the R-matrix box of radius a = 21.3 au was chosen. The R-matrix box well contained all atomic orbitals that were used for the construction of target levels. A total of 39 non-orthogonal one-electron orbitals were used for the description of target levels which were represented by 430 configurations. The relativistic effects in the scattering calculations have been incorporated in the Breit-Pauli Hamiltonian through the use of Darwin, mass correction and spin-orbit operators.

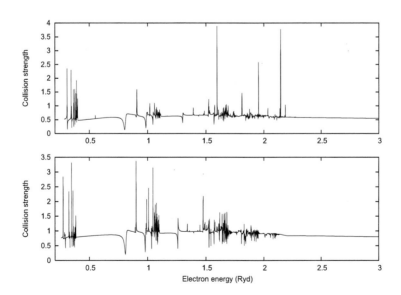

Figure 13. The collision strength for the $2s^2 2p^3 \, ^4S^o_{3/2}$ - $2s^2 2p^3 \, ^2D^o_{3/2}$ transition in O II as a function of electron energy. Upper panel, 62-level BPRM [74]; lower panel, 47-level B-spline BPRM [74].

In order to resolve the resonance structure Tayal [74] chose a fine-energy mesh of 0.0001 Ry in the $2s^2 2p^3\ {}^2D^o$ and ${}^2P^o$ thresholds region and 0.0002 Ry above 0.404 Ry in the thresholds energy region of other excited levels to calculate collision strengths. The collision strengths for the optically forbidden ${}^4S^o_{3/2}$ - ${}^2D^o_{5/2}$ transition from the 47 level (lower panel) and 62 level (upper panel) calculations have been compared in Fig. 14 as a function of incident electron energy from threshold to 2.5 Ryd. The resonance structures in the present two calculations are approximately similar except for a shift due to less accurate excitation thresholds in the 62 level BPRM calculation. The resonance enhancement in collision strengths for the forbidden transitions is about 10%. There are a few additional narrow and sharp resonances in the 62 level calculation in the common higher excitation thresholds region which are not present in the 47 level calculation. Resonances converging to the additional 15 target levels in 62 level calculation can also be seen in the figure. There are slight differences in the non-resonant background collision strengths in the two calculations. The discrepancies have been caused by the differences in the target wave functions utilized in the two calculations. The relativistic effects appear to have insignificant influence on collision strengths for forbidden transitions among the levels of the ground $2s^2 2p^3$ configuration. The electron correlation and relaxation effects in the target description influence the background collision strengths. The resonance structure is weak for the dipole-allowed transitions and the collision strengths exhibit an increasing trend with increasing energy as expected for a strong dipole-allowed transition. The electron excitation rates for fine-structure transitions in O II have been reported by Tayal [79].

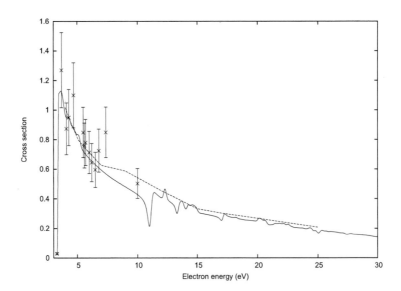

Figure 14. The excitation cross section ($10^{-16} cm^2$) for the $2s^2 2p^3\ {}^4S^o$ - $2s^2 2p^3\ {}^2D^o$ transition in O II as a function of electron energy. Solid curve, 47-level B-spline BPRM [74]; dashed curve, 11-state R-matrix [77]; crosses, measured data [77].

Measured cross sections for the forbidden ${}^4S^o$ - ${}^2D^o$ and resonance ${}^4S^o$ - $2s2p^4\ {}^4P$

transitions have been reported in the literature [77]. The quoted total uncertainty in experimental cross sections was about 20% and the energy resolution was 0.250 eV. The convoluted theoretical cross sections from the 47-level B-spline R-matrix calculation [74] have been compared with the measured values in Figs. 15 and 16 for these two transitions. The 11-state R-matrix theory from the work of Zuo et al. [77] has also been included in these figures for comparison. There is a reasonable agreement between theory and experiment at all energies except 7.4 eV for the forbidden transition. There is also a good agreement between the two sets of calculations. The convoluted theoretical cross sections for the resonance transition are also within the experimental error bars for most incident electron energies below 30 eV. The theoretical cross sections are systematically larger than the measured cross sections at energies E $\geq$ 30 eV. The cause of discrepancy at higher energies is not clear.

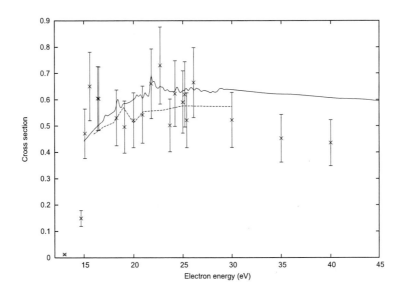

Figure 15. The excitation cross section ($10^{-16}cm^2$) for the $2s^22p^3$ $^4S^o$ - $2s2p^4$ ^4P transition in O II as a function of electron energy. Solid curve, 47-level B-spline BPRM [74]; crosses, measured data [77]; dashed curve, 11-state R-matrix [77].

The fine-structure transitions between levels of the ground $2s^22p$ and excited $2s^22p$, $2s2p^2$, $2p^3$, $2s^23l$, $2s^24l$ and 2s2p3l (l=s, p, d) configurations in OIV gives rise to emission lines in the infrared, ultraviolet and extreme ultraviolet wavelength regions. Several line intensity ratios involving these lines are density and/or temperature sensitive and can be used for diagnosing the physical conditions of planetary nebulae, Seyfert galaxies, symbiotic stars and solar transition region and corona. The line intensity ratios for the intercombination $2s^22p$ $^2P^o$ - $2s2p^4$ ^4P lines of O IV offer useful electron density indicator for the solar and stellar transition regions, novae and quasars. The lines due to the $2s^22p$ $^2P^o$ - $2s2p^2$ ^2D and $2s2p^2$ ^4P - $2p^3$ $^4S^o$ transitions are also electron density sensitive. The O IV $\lambda\lambda1338.61$, $\lambda\lambda1342.99$ and $\lambda\lambda1342.51$ lines due to the $2s2p^2$ ^2P - $2p^3$ $^2D^o$ transitions are useful to determine effective electron temperature.

The theoretical calculations for electron impact excitation collision strengths include close-coupling calculations of Hayes and Nussbaumer [80] and Hayes [81] and R-matrix calculations of Luo and Pradhan [65], Blum and Pradhan [67] and Tayal [82]. The experimental and theoretical electron impact excitation cross sections for the $2s^2 2p \; ^2 P^o$ - $2s 2p^4 \; ^4 P$ and $^2 D$ transitions were reported by Smith et al. [83]. The measurements of absolute direct cross sections were carried out using the electron energy-loss merged-beams method and the calculations were performed in a 25-state R-matrix approximation [83]. The 25-state R-matrix calculation [83] was performed in the LS coupling and a good agreement was obtained between the calculated and measured absolute excitation cross sections for the intercombination and resonance transitions. A brief description of the earlier calculations for energy levels, oscillator strengths and electron excitation cross sections is given by Tayal [82] who reported electron collision excitation strengths and rates for infrared and ultraviolet lines arising from transitions between the fine-structure levels of the $2s^2 2p$, $2s 2p^2$, $2p^3$, $2s^2 3s$, $2s^2 3p$, $2s^2 3d$, $2s2p3s$, $2s2p3p$, $2s2p3d$, $2s^2 4s$ and $2s^2 4p$ configurations of O IV have been calculated using the Breit-Pauli R-matrix approach. These configurations in O IV give rise to 54 fine-structure levels that were represented by configuration-interaction wave functions in the calculation of Tayal [82]. The important relativistic corrections in the Breit-Pauli Hamiltonian were included as a perturbation to the non-relativistic Hamiltonian and the scattering equations were solved with the mass, Darwin and spin-orbit relativistic corrections [2]. Rydberg series of resonances converging to all excited fine-structure levels were included. The collision strengths in the threshold energy regions were found to be dominated by resonance structures. The Breit-Pauli R-matrix method was used to calculate partial collision strengths from J = 0 to J = 25. The top-up contributions for the non-dipole transitions were estimated by assuming that the collision strengths form a geometric progression in J. The higher partial waves contributions for the dipole-allowed transitions were calculated using the Coulomb-Bethe approximation. The other R-matrix calculations of Luo and Pradhan [65] and Blum and Pradhan [67] included resonance structures converging to the lower 15 fine-structure levels. Tayal [82] improved upon the earlier calculations by using a better representation of target levels and by including a larger number of target levels in the close-coupling expansion to check on the importance of additional resonance structures converging to the higher excitation levels.

Correge and Hibbert [84] carried out two sets of atomic structure calculations for O IV with and without core correlation. They found that the inclusion of core correlation improves slightly the excitation energies and the agreement between the length and velocity values of oscillator strengths, but the calculations become computationally more involved. In the scattering calculations of Tayal [82] the $1s^2$ core shell was kept closed to keep collision calculation manageable and without loosing significant accuracy. The calculated values of excitation energies from the work of Tayal [82] were in good agreement with the measured values and other structure calculations.

A fine energy mesh for collision strength calculation in the thresholds energy region was used in the calculation of Tayal [82] to delineate the resonance structures. The collision problem in the external region was solved using an energy mesh of 0.0004 Ryd in the closed-channels energy regions and an energy mesh of 0.5 Ryd in the energy region of all open channels where collision strengths showed smooth variation. Tayal [82] found a

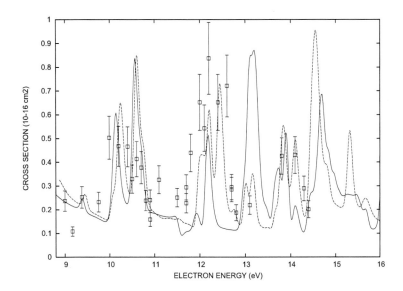

Figure 16. The excitation cross section for the $2s^22p\ ^2P^o$ - $2s2p^2\ ^4P$ transition in O IV as a function of electron energy. Solid curve, 54-level BPRM [82]; dashed curve, 25-state R-matrix [83]; open rectangles, measured data [83].

significant contribution made by resonance series converging to the additional excited levels which were not included in the previous R-matrix calculations. Resonance structures are quite dense in the energy region up to the $2s2p^2\ ^2P_{3/2}$ threshold around 1.6 Ryd.

The convoluted excitation cross sections for the spin-forbidden $2s^22p\ ^2P^o$ - $2s2p^2\ ^4P$ and resonance $2s^22p\ ^2P^o$ - $2s2p^2\ ^2D$ transitions from the calculation of Tayal [82] have been compared with the measured absolute direct excitation cross sections and with the other 25-state theory [83]. There is an overall good agreement between various theoretical cross sections for these two transitions [83, 74]. There are small discrepancies in the position of resonances and the background cross sections away from resonances among the different theoretical calculations. These discrepancies are generally caused by the differences in target wave functions. The total error in the measured cross sections have been estimated to be 18%. The theoretical results shown in Figs. 17 and 18 were convoluted using energy dependent electron energy of the experiment. The convolution resulted in a broadening of the theoretical sharp and densely packed resonances. The resonance enhancement in cross sections is clearly evident for these transitions. The excitation cross sections for the spin-forbidden transition are shown in Fig. 17 as a function of incident electron energy. The inclusion of spin-orbit interaction in the 54-level Breit-Pauli calculation [82] introduces additional Rydberg series of resonances converging at the increased number of thresholds than the non-relativistic 25-state LS calculation [83] which may cause discrepancies in the positions, widths and peaks of resonances as seen in Figs. 17 and 18. The comparison between the predicted and measured cross sections for the resonance $2s^22p\ ^2P^o$ - $2s2p^2\ ^2D$ transition has been displayed in Fig. 18 between the threshold and 24.0 eV. There is an

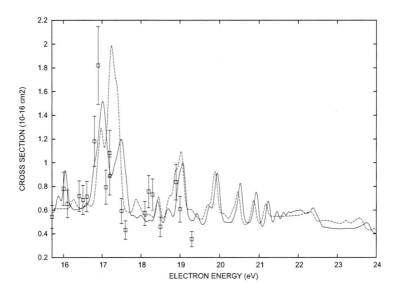

Figure 17. The excitation cross section for the $2s^2 2p\ ^2P^o$ - $2s2p^2\ ^2D$ transition in O IV as a function of electron energy. Solid curve, 54-level BPRM [82]; dashed curve, 25-state R-matrix [83]; open rectangles, measured data [83].

overall agreement except for the resonance feature around 17 eV, where the flux appears to be redistributed. Some differences can be noted between theory and experiment at a few incident electron energies.

The electron collision excitation of highly charged ions is responsible for visible, ultraviolet and x-ray emissions in astrophysical and magnetically confined fusion plasmas. The O VI lines are observed in the ultraviolet emission spectra of planetary nebulae and H II regions. The O VI doublet ($\lambda\lambda$ 1032, 1038) is the strongest emission feature in the far ultraviolet spectra of RR Tel and several other symbiotic stars obtained from the *Hopkin's Ultraviolet Telescope* (HUT) during Astro-2 mission. The strong stellar O VI emission lines near $\lambda\lambda$ 3811, 3834 are seen in the O VI sequence planetary nebula IC 2003. The O VI is one of the highest observed stages of ionization in stellar spectra and thus it can provide reliable diagnostic of high temperature plasma. The O VI 2s - 2p and 2s - 3p line pairs are density and/or temperature sensitive. The O VI gas diagnostic can provide important information on the ionization state and metallicity of the diffuse intergalactic gas. This information is important for understanding galaxy formation and evolution.

The effects of coupling to the highly excited bound and continuum target states in O VI are significant and these can be simulated by using a set of well-chosen pseudostates in the close-coupling expansion. Tayal [85] included 24 lowest physical bound fine-structure levels of the 2s-5s, 2p-5p, 3d-5d, 4f-5f and 5g configurations as well as 19 levels belonging to the $\overline{6s} - \overline{8s}\ ^2S$, $\overline{6p} - \overline{8p}\ ^2P^o$, $\overline{6d} - \overline{8d}\ ^2D$ and $\overline{6f}, \overline{7f}\ ^2F^o$ pseudo-states in the close-coupling expansion in his calculations for electron impact collision strengths and rates of O VI. These levels were represented by configuration-interaction wave functions. The one-body mass correction, Darwin and spin-orbit relativistic terms were considered in the Breit-Pauli approximation in the scattering equations [2]. Rydberg series of resonances converging to excited fine-structure levels were found to make significant contribution to collision strengths for many transitions. Other recent calculations of excitation cross sections for O VI include R-matrix with pseudostates (RMPS) calculations of Griffin et al. [86] and Lozano et al. [87]. These calculations were carried out in LS-coupling scheme and reported multiplet collision strengths. The RMPS calculation of Lozano et al. [87] considered nine physical states and 18 pseudostates, while the 41-state RMPS calculation of Griffin et al. [86] included nine physical and 32 pseudo-states in the close-coupling expansion. The excitation cross sections for the resonance 2s - 2p transition in a limited energy range close to threshold have been measured by Bell et al. [88] and Lozano et al. [87] using the electron-energy-loss technique. The calculated cross sections are in excellent agreement with each other and also with the recent experiments for the resonance 2s - 2p transition. The effects of coupling of the bound states with the target continuum states depend on the type of transition in the energy region above the ionization threshold and have been investigated in detail by Griffin et al. [86]. They noted that the coupling to the continuum has insignificant effect on the cross sections for the 2s - 2p transition. The cross sections for the resonance transitions are greatly influenced by contributions from higher partial waves which are insensitive to the coupling to higher excited and continuum target states. The effects of coupling to the target continuum have been shown to increase for transitions from the n = 2 states to the n = 3 and n = 4 states. These effects are estimated to be less than 10% for the n = 2 to n = 3 transitions and about 25% for the n = 2 to n = 4 transitions. The Rydberg series of resonances converging to the n = 4 and 5 excited states may also make significant contributions, in addition to the dominant resonances converging to the n = 3 excited states.

In the Breit-Pauli RMPS calculation of Tayal [85] 60 continuum orbitals for each angular momentum were included to obtain convergence in the energy range up to 25.0 Ry. The coupled equations were solved in the asymptotic region to yield K-matrices and then the collision strengths. The Breit-Pauli R-matrix method is used to calculate partial collision strengths from J = 0 to J = 57. These partial waves were judged to be sufficient to give converged collision strengths for the forbidden transitions in the energy range of interest. The higher partial waves contributions (J $\geq$ 58) for the dipole-allowed transitions are calculated using the Bethe approximation [8]. Tayal [85] found a complicated resonance structure due to the Rydberg series converging to the n = 3 excited level thresholds. The structure is particularly complicated because of the interference and overlapping of several Rydberg series of resonances. The resonance structure in the n = 4 and n = 5 thresholds is weak for the allowed transitions, but is more pronounced for the forbidden transitions. No theoretical or experimental data are currently available for electron impact excitation of fine-structure transitions for comparison.

The predicted and experimental cross sections from the work of Lozano et al. [87] together with the experimental data of Bell et al. [88] for electron excitation of the 2s ^2S - 2p $^2P^o$ transition in O VI are shown in Fig. 19. The electron energy resolution was about 125 meV. An excellent agreement between theory and experiment can be seen in Fig. 19. The theoretical results of Lozano et al. [87] and Griffin et al. [86] show excellent agreement with each other.

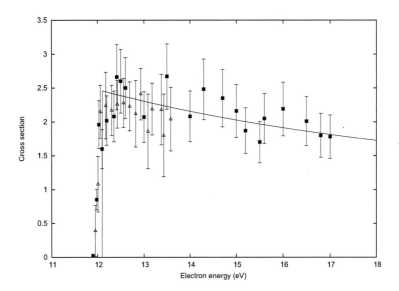

Figure 18. The excitation cross section ($10^{-16}cm^2$) for the $2s\ ^2S$ - $2p\ ^2P^o$ transition in O VI as a function of electron energy. Solid curve, 43-level BPRM [85]; solid rectangles, measured data [87]; open triangles, measured data [88].

4.2.3. Sulfur Ions

Electron excitation collision strengths for several ionization stages of sulfur are important to study physical processes and conditions such as temperature, density and elemental abundances in astrophysical plasmas. Emission lines of collisionally excited S II, S III and S IV ions have been detected in the spectra of Io torus and in the ultraviolet spectra of the Sun and stellar transition regions. The prominent features of the S III multiplets $3p^2\ ^3P$ - $3p4d\ ^3P^o$ and $^3D^o$ occur in the 480-489 $\mathring{A}$ wavelength region. The spin-forbidden emission lines of S IV from the multiplet $3s^23p\ ^2P^o$ - $3s3p^2\ ^4P$ are observed in the spectra of the Sun and of the transition region of Capella. The line intensity ratios within the multiplet $3s^23p\ ^2P^o$ - $3s3p^2\ ^4P$ can provide excellent density diagnostics in the electron density range 10^9 - 10^{12} cm^{-3}. Emission lines of S X in the ultraviolet, extreme ultraviolet and x-ray wavelength regions arise due to transitions between levels of the ground $2s^22p^3$ configuration and from the $2s^22p^3$ levels to the levels of the $2s2p^4$, $2p^5$ and $2s^22p^23l$ (l = s, p, d) configurations. Several line intensity ratios involving these lines are density and/or temperature sensitive. The importance of S X lines in the determination of density and temperature in the solar active region and flares have been discussed in details in the literature (see for example, references [89] and [90]). Keenan et al. [90] indicated that the intensity ratios involving $2s^22p^3$ - $2s2p^4$ transitions in the extreme ultrviolet wavelength range provide important density diagnostics for the solar flares.

Several calculations for electron impact excitation of sulfur ions have been reported in the literature. Fairly extensive calculations of the electron excitation cross sections have been reported for S II, S III, S IV and S X ions by Tayal [91], Ramsbottom et al. [92], Tayal and Gupta [93] and Tayal [94, 95] using the R-matrix method [2]. These publications also provide a summary of all the previous calculations for these ions. The calculations for S II [59, 92] and S III [93] were performed in LS R-matrix plus algebraic transformation to intermediate coupling approach. The same approach was applied in many other previous calculations for sulfur ions. The approximations used in this approach may lead to some inaccuracies in the results. The full Breit-Pauli R-matrix approach was used in the calculations of the excitation cross sections for S IV [94] and S X [95].

In the calculation for S III, 27 LS states of the $3s^2 3p^2$, $3s 3p^3$, $3s^2 3p 3d$, $3s^2 3p 4s$, $3s^2 3p 4p$ and $3s^2 3p 4d$ configurations were considered. These 27 LS states give rise to 49 fine-structure levels, and were constructed by fairly extensive configuration-interaction wave functions. These wave functions yield excitation energies that are in close-agreement with the measurement [96]. Rydberg series of resonances converging to excited state thresholds were found to make substantial enhancements in cross sections at low energies. The work of Tayal and Gupta [93] improved upon the accuracy of the calculation of Tayal [91] and supersedes the latter. The effective collision strengths were obtained from the total resonant and non-resonant collision strengths by integrating over a Maxwellian velocity distribution.

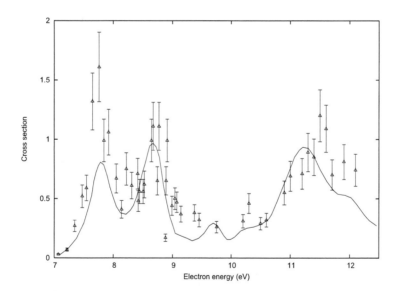

Figure 19. The excitation cross section ($10^{-16} cm^2$) for the $3s^2 3p^2 \ ^3P$ - $3s 3p^3 \ ^5 S^o$ transition in S III as a function of electron energy. Solid curve, 27-state R-matrix [59]; open triangles, measured data [104].

Experimental electron impact excitation cross sections for the $3s^2 3p^3 \ ^4 S^o$ - $3s^2 3p^3 \ ^2 D^o$, $^2 P^o$ and $3s 3p^4 \ ^4 P$ transitions in S II [103] and $3s^2 3p^2 \ ^3 P$ - $3s 3p^3 \ ^5 S^o$ transition in S III [104] in the low energy region have been reported in the literature. The experiments were performed using the electron energy-loss merged-beams technique. In these experiments the metastable states were produced in the ion beam sources which were assessed and minimized. The measured cross sections for transitions in S II were compared with the predicted cross sections from the 19-state R-matrix calculation of Tayal [59] and 18-state R-matrix calculation of Ramsbottom and Bell [92]. The theoretical results were convoluted with a 250 meV full-width-at-half-maximum electron energy width. The two calculations agree to within 20% for these transitions, and also exhibit reasonable agreement with the measured cross sections.

The experimental results and the convoluted 27-state R-matrix results have been compared in Fig. 20. The theoretical results were convoluted with a 100 meV electron energy distribution function. The experimental results confirm two strong predicted resonances in the near-threshold energy region of 7.5 - 8.5 eV and a third broader resonance near 11.5 eV. The theoretical cross sections also show some additional small resonances which are not present in the experiment. The experimental peak cross sections are larger than theory for the resonances around 7.8 and 11.5 eV. There are also some differences between theory and experiment in the positions of the three major resonances.

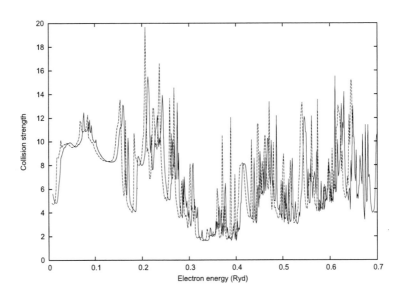

Figure 20. The collision strength for the $3s^23p\ ^2P^o_{1/2}$ - $3s23p\ ^2P^o_{3/2}$ transition as a function of electron energy. Solid curve, 52-level BPRM [94]; dashed curve, 24-state R-matrix [94].

Tayal [94] calculated electron excitation collision strengths of inelastic transitions among the 52 fine-structure levels of the ground $3s^23p$ and excited $3s3p^2$, $3s^23d$, $3s^24s$, $3p^3$, $3s^24p$, 3s3p3d, $3s^24d$, $3s^24f$ and 3s3p4s configurations in S IV using the full Breit-Pauli R-matrix method by including the mass, Darwin and spin-orbit one-body relativistic terms in the scattering equations. Saraph and Storey [97] performed a 21-state LS R-matrix plus term-coupling coefficients calculation to represent spin-orbit interaction as a perturbation. They demonstrated that the position of the resonances depend on the coupling to the higher states which translates to increase in effective collision strengths at lower temperatures. The collision strength for the $3s^2 3p\ ^2P^o_{1/2}$ - $3s^2 3p\ ^2P^o_{3/2}$ transition in S IV from the full Breit-Pauli R-matrix and LS R-matrix plus transformation to intermediate coupling have been compared in Fig. 21 as a function of electron energy from the threshold to 0.7 Ryd. The solid curve shows the full Breit-Pauli R-matrix calculation and the dashed curve shows the LS R-matrix plus transformation results. Some significant differences in the resonance structures from the two calculations may be seen. Most of the discrepancies have been caused by the well known limitations of the LS R-matrix plus transformation approach which causes inaccuracies in the results [94, 4].

Electron excitation collision strengths and rates for ultraviolet and x-ray lines arising from transitions between the fine-structure levels of the $2s^2 2p^3\ ^4S^o$, $^2D^o$, $^2P^o$ terms and from these levels to the levels of the $2s2p^4$, $2p^5$, $2s^2 2p^2 3s$, $2s^2 2p^2 3p$ and $2s^2 2p^2 3d$ configurations were calculated using the Breit-Pauli R-matrix method by Tayal [95]. The 72 fine-structure levels arising from the 34 LS terms ($2s^2 2p^3$, $^4S^o$, $^2D^o$, $^2P^o$, $2s2p^4\ ^4P$, ^2D, ^2P, ^2S, $2p^5\ ^2P^o$, $2s^2 2p^2(^3P)3l\ ^4P$, ^4D, ^4F, ^2P, ^2D, ^2F, $^4S^o$, $^4P^o$, $^4D^o$, $^2S^o$, $^2P^o$, $^2D^o$, $2s^2 2p^2(^1D)3l\ ^2S$, ^2P, ^2D, ^2F, ^2G, $^2P^o$, $^2D^o$, $^2F^o$, $2s^2 2p^2(^1S)3l\ ^2S$, $^2P^o$, ^2D, where l = s, p, d) of 15 symmetries were included in the close-coupling expansion. Configuration-interaction wave functions were used for an accurate representation of target levels. These wave functions give excitation energies and oscillator strengths which were found to be in close agreement with experiment and other accurate calculations. The Rydberg series of resonances converging to the n=2 and n=3 excited levels were explicitly included in the scattering calculation. Resonance structures make significant contribution to collision strengths. The other theoretical calculations for electron impact excitation collision strengths include distorted-wave calculations of Bhatia and Mason [98], Bhatia and Landi [99] and Zhang and Sampson [100] and R-matrix calculations of Bell and Ramsbottom [101, 102]. The calculation of Bell and Ramsbottom [101, 102] included 22 fine-structure levels belonging to the 11 LS target states of the $2s^2 2p^3$, $2s2p^4$, $2p^5$ and $2s^2 2p^2 3s$ configurations while Bhatia and Landi [99] included 72 fine-structure levels arising from the n=2 ($2s^2 2p^3$, $2s2p^4$ and $2p^5$) and n=3 ($2s^2 2p^2 3s$, $2s^2 2p^2 3p$ and $2s^2 2p^2 3d$) configurations. The R-matrix calculation of Bell and Ramsbottom [101, 102] included resonances converging to the n=2 and $2s^2 2p^2 3s$ excited levels in their calculation, while other calculations ignored resonances.

Tayal [95] included 24 continuum orbitals for each angular momentum to obtain convergence in the energy range up to 150.0 Ryd. The Breit-Pauli R-matrix method is used to calculate partial collision strengths from J = 0 to J = 18. These partial waves were sufficient to give converged collision strengths for the forbidden transitions in the energy range of interest. The higher partial waves contributions (J $\geq$ 18) for the dipole-allowed transitions were calculated using the Coulomb-Bethe approximation. The collision strengths above the resonance region for the forbidden $2s^2 2p^3\ ^4S^o_{3/2}$ - $^2D^o_{3/2}$, $2s^2 2p^3\ ^4S^o_{3/2}$ - $^2P^o_{1/2}$ and $2s^2 2p^3$ $^2D^o_{3/2}$ - $^2D^o_{5/2}$ transitions from the works of Tayal [95] (solid curve), Bhatia and Landi [99] (crosses), and Bell and Ramsbottom [101, 102] are displayed in Fig. 22 as a function of electron energy. The collision strengths vary smoothly in this energy region and exhibit decreasing trend as expected for forbidden transitions. The three calculations agree very well each other in this energy region.

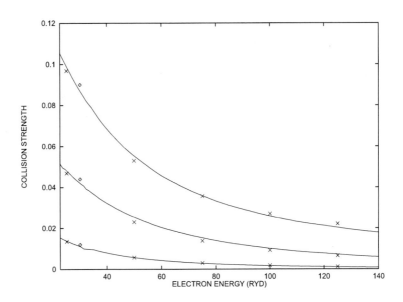

Figure 21. Electron impact excitation collision strengths for the forbidden $2s^2 2p^3\ ^4S^o_{3/2}$ - $^2P^o_{1/2}$ (lower curve), $2s^2 2p^3\ ^4S^o_{3/2}$ - $^2D^o_{3/2}$ (middle curve) and $2s^2 2p^3\ ^2D^o_{3/2}$ - $^2D^o_{5/2}$ (upper curve) transitions in S X as a function of electron energy. Solid curve, 72-level BPRM [95]; crosses, distorted-wave calculation [99]; diamonds, 11-state R-matrix [101, 102].

4.2.4. Iron Ions

Spectral lines of several stages of ionization of iron ions have been observed in the ultraviolet and extreme ultraviolet spectra solar spectra obtained by the (*Solar EUV Rocket Telescope and Spectrograph*) (SERTS) [105, 106] and by the Naval Research Laboratory's *skylab*. Emission lines of iron ions have also been observed in the spectra of late-type stars [108, 108] obtained by the *Extreme Ultraviolet Explorer* in the spectral range 170-400 Å. Electron excitation cross sections of Fe X - Fe XIV lines can provide plasma diagnostics for the solar flare and active region as well as for late-type stars such as Procyon and α Cen. Emission lines of Fe XIV in the optical and extreme ultraviolet wavelength regions arise due to transitions between levels of the ground $3s^2 3p$ and excited $3s3p^2$, $3s^2 3d$, $3p^3$, 3s3p3d and $3p^2 3d$ configurations. The coronal green line at 5303 Å due to the forbidden $3s^2 3p\ ^2P^o_{1/2}$ - $3s^2 3p\ ^2P^o_{3/2}$ transition offers a useful diagnostic of the solar corona. Emission lines of Fe XIV due to the $3s^2 3p\ ^2P^o$ - $3s^2 3d\ ^2D$ transitions at 211.32, 219.12 and 220.07 Å and due to the $3s^2 3p$ - $3s3p^2$ transitions at 252.20, 257.39, 264.78, 270.52, 274.21, 289.17, 353.83, 356.65, 429.54, 444.24 and 447.34 Å are observed from the SERTS in the solar extreme ultraviolet active region spectra [106]. Brickhouse et al. [109] reported predicted emissivities by solving a full set of statistical equilibrium equations and including all the available collision and radiative transition rates of Fe IX - Fe XXIV in the extreme ultraviolet spectral region. They compared the predicted spectra with observations from the SERTS [105] and noted large spread in the observed to predicted ratios for several iron ions. Several predicted line ratios of Fe XIV are factors of 2-3 different from the observations. The observed and predicted Fe XIV $\lambda 270.51/\lambda 257.38$ and $\lambda 270.51/\lambda 274.19$ ratios differ by a factor of 2. Young et al. [110] compared the atomic data from CHIANTI database [111] for various iron ions with the observed solar active region spectrum from the SERTS [105]. They noted serious discrepancies between the theoretical predictions and observations for Fe XIV. The Extreme ultraviolet Imaging Spectrograph on board Hinode satellite has observed the Fe XIV allowed lines near 200 $\mathring{A}$. The Fe XIV extreme ultraviolet lines observed by the Coronal Diagnostic Spectrometer on board SOHO can be used for electron density determinations.

Electron collision excitation strengths for transitions between fine-structure levels of the $3s^2 3p^5$, $3s3p^6$, $3s^2 3p^4 3d$, $3s3p^5 3d$ and $3s^2 3p^4 4s$ configurations in Fe X were calculated by Tayal [112, 113] using the Breit-Pauli R-matrix approach which takes into account one-body relativistic terms of the Breit-Pauli Hamiltonian. The configuration-interaction wave functions of the target states were constructed with the 1s, 2s, 2p, 3s, 3p, 3d, 4s, 4p, 4d and 4f radial functions. Mohan et al. [114] and Pelan and Berrington [115] used the R-matrix method, and included 14 lowest LS terms of the $3s^2 3p^5$, $3s3p^6$ and $3s^2 3p^4 3d$ configurations in the close-coupling expansion. The target states were represented by configuration-interaction wave functions formed with the 1s, 2s, 2p, 3s, 3p and 3d radial functions. They performed a transformation of LS coupled K-matrices to intermediate coupling to obtain collision strengths among fine-structure transitions. More recently Aggarwal and Keenan [116] performed Dirac Atomic R-matrix code calculation for Fe X. Measurement of absolute cross sections for excitation of the $3s^2 3p^5\ ^2 P_{3/2}$ - $3s^2 3p^5\ ^2 P_{1/2}$ fine-structure transition in Fe X was reported by Niimura et al. [117].

The convoluted 49-state Breit-Pauli R-matrix cross sections [113] for the excitation of the $3s^2 3p^5\ ^2 P_{3/2}$ - $3s^2 3p^5\ ^2 P_{1/2}$ fine-structure transition in Fe X have been compared with the measured absolute cross sections [117] in Fig. 23. The electron-energy resolution of the measured cross sections was 100 meV. The electron-energy loss merged-beams experiment has detected major resonance structures at 3.5 and 4.6 eV in the cross sections which were also predicted by the theory [113]. The theoretical cross sections show many other relatively narrow and small resonances in cross sections. In the convolution of theoretical cross sections with experimental energy spread the sharp and narrow resonances are smoothed out. The calculated position of the broad resonance is 4.35 eV which agrees well with the measured position of 4.6±0.1 eV. There is a slight shift in the measured resonance feature to higher energy than theory. The measured resonance at 3.5 eV is relatively weak than the resonance at 4.35 eV.

The calculated oscillator strengths of allowed transitions from the levels of the ground $3s^2 3p^5$ configuration to several excited levels of the $3s3p^6$, $3s^2 3p^4 3d$ and $3s^2 3p^4 4s$ configurations in Fe X [112] are listed in Table 5. There is a good agreement between the length and velocity values of oscillator strengths in the calculations of Tayal [112], Deb et al. [118] and the compilation of Fuhr et al. [119]. The good agreement indicates that the target wave functions used in the collision calculation [112] are of good quality to yield reliable collision strengths. A boundary radius of 6.4 au was chosen and twenty-two continuum orbitals for each angular momentum were included to obtain convergence in the energy range up to 90 Ryd. The coupled equations were solved in the asymptotic region using a perturbation method to yield K-matrices and then the collision strengths. A fine-energy mesh was

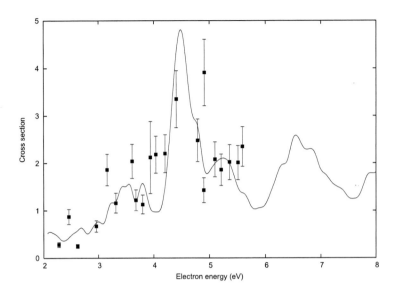

Figure 22. The excitation cross section $(10^{-16} cm^2)$ for the $3s^2 3p^5 \, {}^2P^o_{1/2}$ - $3s^2 3p^5 \, {}^2P^o_{3/2}$ transition in Fe X as a function of electron energy. Solid curve, 49-level BPRM [113]; solid rectangles, measured data [117].

chosen to delineate complicated resonance structures in collision strengths.

Fairly extensive Breit-Pauli R-matrix calculations for electron excitation of Fe XI were carried out by Gupta and Tayal [120, 121] who included 38 fine-structure levels arising from the 20 LS states of the $3s^2 3p^4$, $3s3p^5$, $3s^2 3p^3({}^4S^o)3d$, $3s^2 3p^3({}^2P^o)3d$ and $3s^2 3p^3({}^2D^o)3d$ configurations. The relativistic effects were considered in the Breit-Pauli approximation by including one-body mass correction, Darwin term and spin-orbit terms in the scattering equations. They reported total collision strengths and effective collision strengths for transitions between the levels of the ground $3s^2 3p^4$ configuration and from these levels to the excited levels of the $3s33p^5$ and $3s^2 3p^3 3d$ configurations. They also presented a summary of earlier works on Fe XI. Their results agreed within 20% with the previous calculation of Bhatia and Doscheck [124] for many transitions and showed large differences as well for a few transitions. The discrepancies were judged to be caused by the differences in target wave functions and in the scattering model. More recently effective collision strengths for transitions between the 48 fine-structure levels of the $3s^2 3p^4$, $3s3p^5$, $3s^2 3p^3 3d$ and $3p^6$ configurations in Fe XI were reported by Aggarwal and Keenan [125]. They employed Dirac Atomic R-matrix Code in their fairly extensive calculations which were performed in jj coupling scheme.

The collision strengths for electron impact excitation of fine-structure levels in Fe XIII were reported by Gupta and Tayal [126] and Aggarwal and Keenan [127]. Gupta and Tayal [126] used full Breit-Pauli R-matrix method in the excitation of the fine-structure levels of the ground $3s^2 3p^2$ configuration and from the levels of the ground $3s^2 3p^2$ configuration to the levels of the excited $3s3p^3$ and $3s^2 3p3d$ configurations. They considered 14 LS $3s^2 3p^2$

Table 5. Oscillator strengths of allowed transitions in Fe X

Transition	Tayal[a]		DGM[b]	Fuhr[c]
	f_L	f_V		
$3s^23p^5\,{}^2P^o_{1/2}$ - $3s3p^6\,{}^2S_{1/2}$	0.028	0.026		
$3s^23p^5\,{}^2P^o_{1/2}$ - $3s^23p^4({}^1D)3d\,{}^2S_{1/2}$	0.235	0.219	0.231	0.226
$3s^23p^5\,{}^2P^o_{1/2}$ - $3s^23p^4({}^3P)3d\,{}^2P_{1/2}$	0.554	0.468	0.520	
$3s^23p^5\,{}^2P^o_{1/2}$ - $3s^23p^4({}^3P)4s\,{}^2S_{1/2}$	0.099	0.097		
$3s^23p^5\,{}^2P^o_{1/2}$ - $3s^23p^4({}^1D)3d\,{}^2D_{3/2}$	0.003	0.003		
$3s^23p^5\,{}^2P^o_{1/2}$ - $3s^23p^4({}^3P)3d\,{}^2P_{3/2}$	0.078	0.062		
$3s^23p^5\,{}^2P^o_{1/2}$ - $3s^23p^4({}^3P)4s\,{}^2P_{3/2}$	0.036	0.035		
$3s^23p^5\,{}^2P^o_{3/2}$ - $3s3p^6\,{}^2S_{1/2}$	0.028	0.026		
$3s^23p^5\,{}^2P^o_{3/2}$ - $3s^23p^4({}^1D)3d\,{}^2S_{1/2}$	0.327	0.288	0.304	0.300
$3s^23p^5\,{}^2P^o_{3/2}$ - $3s^23p^4({}^3P)3d\,{}^2P_{1/2}$	0.098	0.079	0.099	
$3s^23p^5\,{}^2P^o_{3/2}$ - $3s^23p^4({}^3P)4s\,{}^2S_{1/2}$	0.025	0.024		
$3s^23p^5\,{}^2P^o_{3/2}$ - $3s^23p^4({}^1D)3d\,{}^2D_{3/2}$	0.002	0.002		
$3s^23p^5\,{}^2P^o_{3/2}$ - $3s^23p^4({}^3P)3d\,{}^2P_{3/2}$	0.714	0.589	0.729	
$3s^23p^5\,{}^2P^o_{3/2}$ - $3s^23p^4({}^3P)4s\,{}^2P_{3/2}$	0.090	0.087		

[a]Tayal [112]
[b]Deb et al. [118]
[c]Fuhr et al. [119]

${}^3P$, ${}^1D$, ${}^1S$, $3s3p^3\,{}^3D^o$, ${}^3P^o$, ${}^1D^o$, ${}^3S^o$, ${}^1P^o$, $3s^23p3d\,{}^3P^o$, ${}^1D^o$, ${}^3D^o$, ${}^3F^o$, ${}^1F^o$ and ${}^1P^o$ states of Fe XIII which give rise to 26 fine-structure levels. A brief description of the earlier works for excitation of Fe XIII can also be found in their paper. The 10 orthogonal 1s, 2s, 2p, 3s, 3p, 3d, 4s, 4p, 4d and 4f one-electron radial functions were used in the construction of their target states for scattering calculation. They chose a boundary radius of a = 4.84 au and included 24 continuum orbitals for each angular momentum to obtain convergence in the energy range up to 60 Ryd. They noted significant differences with earlier calculations for some transitions. More recently, Aggarwal and Keenan [127] considered fine-structure levels of the $3s^23p^2$, $3s3p^3$, $3s^23p3d$, $3p^4$, $3s3p^23d$ and $3s^23d^2$ configurations of Fe XIII in their Dirac Atomic R-matrix Code calculations. They noted generally a good agreement with the calculation of Gupta and Tayal [126].

Several theoretical calculations for electron impact excitation collision strengths of fine-structure transitions in Fe XIV have been reported in the literature [128, 129, 130, 131, 132]. Bhatia and Kastner [132] reported a distorted wave calculation for electron impact excitation collision strengths of fine-structure transitions in Fe XIV. Dufton and Kingston [131] presented a simple R-matrix calculation and Storey et al. [129, 130] reported a somewhat detailed scattering calculation using the R-matrix method. Storey et al. [129, 130] included 18 target states of the $3s^23p$, $3s3p^2$, $3s^23d$, $3p^3$ and $3s3p3d$ configurations in the close-coupling expansion and performed R-matrix calculation in the LS-coupling by including the Darwin and mass relativistic corrections. The LS-coupled reactance matrices were then transformed to intermediate coupling using the term-coupling coefficients [3]. Electron impact excitation rates for transitions between fine-structure levels of the $3s^23p$, $3s3p^2$ and $3s^23d$ configurations and from these levels to the fine-structure levels of the $3p^3$, 3s3p3d and $3p^23d$ configurations in Fe XIV have been calculated by Tayal [128]. The 135 target levels were included in the close-coupling expansion in electron excitation calculations which were performed by the use of the Breit-Pauli R-matrix approach. The lowest 135 Fe XIV energy levels belong to the $3s^23p$, $3s3p^2$, $3s^23d$, $3p^3$, 3s3p3d, $3p^23d$, $3s3d^2$, $3p3d^2$, $3s^24s$, $3s^24p$, 3s3p4s and $3s^24d$ configurations. An accurate representation of the target levels was obtained using spectroscopic and correlation radial functions. The atomic wave functions were found to give excitation energies in close agreement with experiment, and oscillator strengths and transition probabilities for Fe XIV lines normally compared very well with other calculations. Absolute direct electron impact excitation cross sections of the forbidden $3s^23p\ ^2P^o_{1/2}$ - $3s^23p\ ^2P^o_{3/2}$ transition have been measured by Hossain et al. [133] at low incident electron energies in the near-threshold region. The experiment was performed using electron energy-loss merged-beams approach. Theoretical cross sections for excitation of the $3s^23p\ ^2P^o_{1/2}$ - $3s^23p\ ^2P^o_{3/2}$ fine-structure transition from the full Breit-Pauli R-matrix calculation were also reported [133]. The relativistic effects were included through the Darwin, mass and spin-orbit Breit-Pauli operators in the scattering equations. Storey et al. [129] found that their collision rates for electron excitation of the $^2P^o_{1/2} - ^2P^o_{3/2}$ transition are a factor of 2-3 higher than the earlier R-matrix calculation of Dufton and Kingston [131]. The larger values of collision rates reported by Storey et al. [129] were not supported by the observations from Seyfert galaxies [134]. The 18-state LS R-matrix plus term-coupling coefficients [129] cross sections for the $^2P^o_{1/2} - ^2P^o_{3/2}$ transition were found to be larger than both experiment and theory of Hossain et al. [133].

Tayal [128] constructed the target wave functions with ten orthogonal one-electron orbitals 1s, 2s, 2p, 3s, 3p, 3d, 4s, 4p, 4d and 4f. The lowest 135 fine-structure levels in Fe XIV belonged to the 59 LS terms of the $3s^23p$, $3s3p^2$, $3s^23d$, $3p^3$, 3s3p3d, $3p^23d$, $3s3d^2$, $3p3d^2$, $3s^24s$, $3s^24p$, $3s^24d$ and 3s3p4s configurations. First the 1s, 2s, 2p, 3s and 3p orbitals for the ground $3s^23p\ ^2P^o$ state were obtained in a Hartree-Fock calculation and then the spectroscopic 3d, 4s, 4p and 4d radial functions were determined by optimization on the $3s^23d$, $3s^24s$, $3s^24p$ and $3s^24d$ states respectively. A correlation 4f function was obtained on the ground $3s^23p\ ^2P^o$ state. The spectroscopic and correlation functions were used to construct configuration-interaction expansions for different atomic states by allowing one-electron and two-electron excitations from all the $3s^23p$, $3s3p^2$, $3s^23d$, $3p^3$, 3s3p3d, $3p^23d$, $3s3d^2$, $3p3d^2$, $3s^24s$, $3s^24p$ and $3s^24d$ basic configurations. In the construction of expansions for the fine-structure levels with various J and π the configurations generated in this excitation scheme were used for the atomic LS states and configurations with coefficients less than 0.02 omitted from the expansions. It was necessary to keep the scattering calculation manageable. However, excitation energies calculated with the extensive expansions and with the reduced expansions agree to within 1%.

The resonant collision strengths in the thresholds energy region has been plotted as a function of electron energy for the green line at 5303 $\mathring{A}$ due to the forbidden $3s^23p\ ^2P^o_{1/2}$ - $3s^23p\ ^2P^o_{3/2}$ transition in Fig. 24. The collision strengths from the calculation of Tayal [128] have been compared with the results of Storey et al. [129]. It is clear from this figure

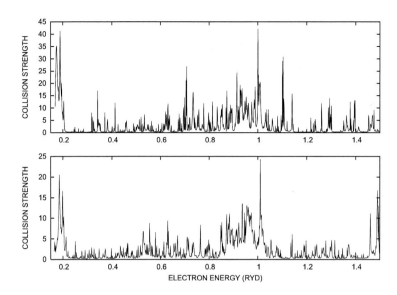

Figure 23. The collision strength for the $3s^23p\ ^2P^o_{1/2}$ - $3s23p\ ^2P^o_{3/2}$ transition in Fe XIV as a function of electron energy. Lower panel, 135-level BPRM [128]; upper panel, 18-state R-matrix [129].

that the resonance structures are complex and may make significant enhancements in the collision strengths. The resonance enhancement in collision strengths for the forbidden and semi-forbidden transitions is normally larger compared to allowed transitions. A fine energy mesh for collision strength calculation in the thresholds energy region needs to be used to delineate the major resonance structures. Tayal [128] chose an energy mesh of 0.001 Ryd in the region of dense resonances up to the $3p^3\ ^2D^o_{3/2}$ around 5.2 Ryd and an energy mesh of 0.002 Ryd in the closed-channel threshold region above 5.2 Ryd. Storey et al. [130] checked the convergence of the average collision strengths as the number of mesh points is increased and found the results to be converged to better than 1%. In this adopted scheme Tayal [128] might have missed some very narrow resonances, but the average collision strength is expected to be converged to about 1%. In the energy region of all open channels where there are no resonances, collision strengths show smooth variation. There is significant contribution made by resonance series converging to the additional excited levels which were not included in the 18-state R-matrix calculations [129, 130].

Resonance structures are quite dense in the energy region up to the $3s^23d\ ^2D$ thresholds around 4.30 Ryd. The background collision strength away from resonances for the forbidden $3s^23p\ ^2P^o_{1/2}$ - $^2P^o_{3/2}$ transition appears to be similar, but there are significant differences in the peaks of resonance structures in the two sets of calculations. The peak resonant collision strengths in the calculation of Hossain et al. [133] are substantially smaller for the forbidden $3s^23p\ ^2P^o_{1/2}$ - $^2P^o_{3/2}$ transition than the calculation of Storey et al. [129]. There are also differences in the position and width of resonances from the two calculations caused by the differences in wave functions and details of physical model and scattering calculations.

In the collision strength for the forbidden $3s^23p\ ^2P^o_{1/2}$ - $^2P^o_{3/2}$ transition large resonance features appear in the near threshold energy region. The position and other parameters of these resonances are sensitive to the details of scattering model. There are serious discrepancies in the position and peak of these resonances between the two calculations. A major part of the discrepancies appears to be caused by the neglect of fine-structure effects in the calculation of Storey et al. [129, 130].

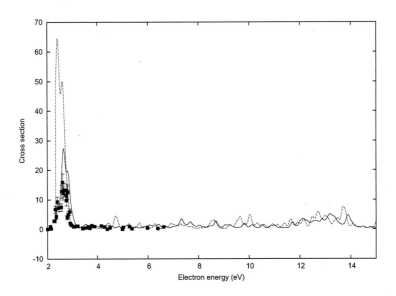

Figure 24. The excitation cross section $(10^{-16} cm^2)$ for the $3s^23p\ ^2P^o_{1/2}$ - $3s^23p\ ^2P^o_{3/2}$ transition in Fe XIV as a function of electron energy. Solid curve, 135-level BPRM [133]; dashed curve, 18-state R-matrix [129]; solid rectangles, measured data [133].

The cross sections for the forbidden $3s^23p\ ^2P^o_{1/2}$ - $3s^23p\ ^2P^o_{3/2}$ transition in Fe XIV from the 135-level Breit-Pauli R-matrix calculation [133] have been compared in Fig. 25 with the 18-state R-matrix calculation [129] and absolute measured cross sections [133]. Both theories have been convoluted to measured electron energy resolution of 125 meV. The experiment confirms the predicted resonance structure in the near-threshold energy region. The peak cross section from the 18-state R-matrix calculation is a factor of 2.4 above the 135-level Breit-Pauli R-matrix calculation and a factor of 4.2 above experiment. The measured peak position of 2.7 ± 0.1 is in good agreement with the theoretical range of 2.4 - 2.6 eV. A good agreement with the measured absolute direct excitation cross sections for the forbidden $3s^23p\ ^2P^o_{1/2}$ - $3s^23p\ ^2P^o_{3/2}$ transition [133] provides some indication that the positions of other resonances in the calculation of Tayal [128] are likely to be accurate. The approach used by Storey et al. [129, 130] has two major limitations compared to the full Breit-Pauli R-matrix calculation. The fine-structure splitting of the target terms is neglected, and in the resonance region where there are both open and closed channels, only those components of the term coupling coefficients are used for which channels are open. Both of these approximations caused substantial inaccuracies in the 18-state calculation because of the importance of spin-orbit interaction for Fe XIV.

5. Conclusion

In this chapter, we have discussed status of the electron excitation of atoms and ions results by giving some examples of the accurate cross sections in the low energy region obtained in the theoretical and experimental efforts in the past about 10 years. The availability of the computer resources has made possible to carry out large scale quantum mechanical calculations. A vast amount of atomic cross section data has been produced for numerous ions and transitions in these calculations. However, there are still serious gaps in the available results for electron excitation cross sections and rates for applications to the astrophysical and fusion plasmas.

The accuracy of the predicted cross sections in the low energy region is sensitive to the scattering model and the accuracy of the target description. Both short-range correlations and polarization effects are very important for the open-shell atomic systems. The non-orthogonal orbitals technique allows for term-dependence optimization of the bound orbitals which generally leads to more accurate target descriptions than the orthogonal orbitals technique. The use of non-orthogonal orbitals for the representation of the scattering functions provides a consistent treatment of the N-electron target and the (N+1)-electron scattering problems. The relaxation of the orthogonality constraints is important for the cases where numerous pseudo-orbitals and large configuration expansions are used for an accurate target description.

The measured results in the near-threshold region obtained from the electron energy-loss merged-beams experiments for selected ions and transitions can provide a check on the accuracy of the theoretical predictions. The joint theoretical and experimental efforts in recent years have been proved to be very fruitful. Once the theoretical results and scattering models in the low energy have been verified for selected transitions in the low energy region by experiments, the theory can be extended to many more transitions in a wide energy range with greater confidence.

References

[1] O. Zatsarinny, O. and Tayal, S. S. *J. Phys. B* 2001, 34, 1299.

[2] Berrington, K. A., Eissner, W. B., and Norrington, P. H. *Comput. Phys. Commun.* 1995, 92, 290.

[3] Saraph, H. E. *Comput. Phys. Commun.* 1978, 15, 247.

[4] Griffin, D. C., Badnell, N. R., and Pindzola, M. S. *J. Phys. B* 1998, 31, 3713.

[5] Zatsarinny, O. and Froese Fischer, C. *Comput. Phys. Commun.* 2000, 124, 247.

[6] Zatsarinny, O. and Bartschat, K. *J. Phys. B* 2004, 37, 2173.

[7] Zatsarinny, O. *Comput. Phys. Commun.* 2006, 174, 273.

[8] Burgess, A. and Sheorey, V. B. *J. Phys. B* 1974, 7, 2403.

[9] Burgess, A. and Tully, J. A. *Astron. Astrophys.* 1992, 254, 436.

[10] Johnson, P. V., McConkey, J. W., Tayal, S. S., and Kanik, I. *Can. J. Phys.* 2005, 83, 589.

[11] Williams, J. F. and Allen, L. J. *J. Phys. B* 1989, 22, 3529.

[12] Tayal, S. S. *J. Phys. B* 1992, 25, 2639.

[13] Thomas, M. R. J, Bell, K. L., and Berrington, K. A. *J. Phys. B* 1997, 30, 4599.

[14] Berrington, K. A. and Burke, P. G. *Planet. Space Sci.* 1981, 29, 377.

[15] Plummer, M., Noble, C. J., and Dourneuf, M. Le *J. Phys. B* 2004, 37, 2979.

[16] Doering, J. P. *J. Geophys. Res. Space Phys.* 1992, 97, 19531.

[17] Shyn, T. W., Cho, S. Y., and Sharp, W. E. *J. Geophys. Res. Space Phys.* 1986, 91, 3751.

[18] Shyn, T. W. and Sharp, W. E. *J. Geophys. Res. Space Phys.* 1986, 91, 1691.

[19] Doering, J. P. and Gulcicek, E. E. *J. Geophys. Res. Space Phys.* 1989, 94, 1541.

[20] Bell, K. L., Berrington, K. A., and Thomas, M. R. *J. Mon. Not. R. Astron. Soc.* 1998, 293, L83.

[21] Doering, J. P. and Yang, J. *J. Geophys. Res. Space Phys.* 2001, 106, 203.

[22] Kanik, I., Johnson, P. V., Das, M. B., Khakoo, M. A., and Tayal, S. S. *J. Phys. B* 2001, 34, 2647.

[23] Johnson, P. V., Kanik, I., Khakoo, M. A., McConckey, J. W., and Tayal, S. S. *J. Phys. B* 2003, 36, 4289.

[24] Noren, C., Kanik, I., Johnson, P. V., McCartney, P., James, G. K., and Ajello, J. M. *J. Phys. B* 2001, 34, 2667.

[25] Johnson, P. V., Kanik, I., Shemansky, D. E., and Liu, X. *J. Phys. B* 2003, 36, 3203.

[26] Tayal, S. S. *Phys. Rev. A* 2002, 66, 030701.

[27] Tayal, S. S. *J. Geophys. Res. Space Phys.* 2004, 109, A08301.

[28] Zatsarinny, O. and Tayal, S. S. *J. Phys. B* 2001, 34, 3383.

[29] Hibbert, A., Biémont, E., Godefroid, M., and Vaeck, N. *J. Phys. B* 1991, 24, 3943.

[30] Doering, J. P., Gulcicek, E. E., and Vaughan, S. O. *J. Geophys. Res. Space Phys.* 1985, 90, 5279.

[31] Jenkins, D. B *J. Quant. Spectrosc. Rdiat. Transfer* 1985, 34, 55.

[32] Goldbach, C. and Nollez, G. *Astron. Astrophys.* 1994, 284, 307.

[33] Brooks, N. H., Rohrlich, D., and Smith, W. H. *Astrophys. J.* 1977, 214, 328.

[34] Bell, K. L. and Hibbert, A. *J. Phys. B* 1990, 23, 2673.

[35] Vaughan, S. O. and Doering, J. P. *J. Geophys. Res. Space Phys.* 1987, 92, 7749.

[36] Gulcicek, E. E. and Doering, J. P. *J. Geophys. Res. Space Phys.* 1988, 93, 5879.

[37] Vaughan, S. O. and Doering, J. P. *J. Geophys. Res. Space Phys.* 1988, 93, 289.

[38] Zatsarinny, O. and Tayal, S. S. *J. Phys. B* 2002, 35, 241.

[39] Wu, J.-H. and Yuan, *J.-M. Chin. Phys.* 2003 12, 1391.

[40] Zatsarinny, O. ,Bartschat, K., and Tayal, S. S. *J. Phys. B* 2006, 39, 1237.

[41] Robinson, D. J. R. and Hibbert, A. *J. Phys. B* 1997, 30, 4813.

[42] Hibbert, A., Biémont, E., Godefroid, M., and Vaeck, N. Astron. *Astrophys. Suppl. Ser.* 1991, 88, 505.

[43] Tayal, S. S. and Beatty, C. A. *Phys. Rev. A* 1999, 59, 3622.

[44] Tong, M., Froese Fischer, C., and Sturesson, L. *J. Phys. B* 1994, 27, 4819.

[45] Ramsbottom, C. A. and Bell, K. L. *Phys. Scr.* 1994, 50, 666.

[46] Berrington, K. A., Burke, P. G., and Robb, W. D. *J. Phys. B* 1975, 8, 2500.

[47] Tayal, S. S. *At. Data Nucl. Data Tables* 2000, 76, 191.

[48] Tayal, S. S. and Zatsarinny, O. *J. Phys. B* 2005, 38, 3631.

[49] Tayal, S. S. Astrophys. *J. Suppl. Ser.* 2006, 163, 206.

[50] Yang, J. and Doering, J. P. *J. Geophys. Res.* 1996, 101, 765.

[51] Doering, J. P. and Goembel, L. *J. Geophys. Res.* 1991, 96, 16021.

[52] Doering, J. P. and Goembel, L. *J. Geophys. Res.* 1992, 97, 4295.

[53] Tayal, S. S. Astrophys. *J. Suppl. Ser.* 2006, 163, 206.

[54] Goldbach, C., Martin, M., Nollez, G., Plombeur, P., Zimmermann, J. -P., and Babic, D. *Astron. Astrophys.* 1986, 161, 47.

[55] Smith, H. W., Bromander, J., Curtis, L. J, and Buchta, *R. Phys. Scr.* 1970, 2, 211.

[56] Lugger, P. M., York, D. G., and Blanchard, T. *Astrophys. J.* 1978, 224, 1059.

[57] Goldbach, C., Lüdtke, T., Martin, M., and Nollez, G. 1992, *Proc. 4th Int. Coll. on Atomic Spectra and Oscillator Strengths for Astrophysical and Laboratory Plasmas* (Gaithersburg) ed. J. Sugar and D. Leckrone.

[58] Zatsarinny, O. and Tayal, S. S. *J. Phys. B* 2002, 35, 2493.

[59] Tayal, S. S. *J. Phys. B* 1997, 30, L551.

[60] Tayal, S. S. *Astrophys. J. Suppl. Ser.* 2004, 153, 581.

[61] Lennon, D. J., Dufton, P. L., Hibbert, A., and Kingston, A. E. *Astrophys. J.* 1985, 294, 200.

[62] Hayes, M. A. and Nussbaumer, H. *Astron. Astrophys.* 1984, 134, 193.

[63] Keenan, F. P., Lennon, D. J., Johnson, C. T., and Kingston, A. E. *Mon. Not. R. Astron. Soc.* 1986, 220, 571.

[64] Luo, D. and Pradhan, A. K. *J. Phys. B* 1989, 22, 3377.

[65] Luo, D. and Pradhan, A. K. *Phys. Rev. A* 1990, 41, 165.

[66] Blum, R. D. and Pradhan, A. K. *Phys. Rev. A* 1991, 44, 6123.

[67] Blum, R. D. and Pradhan, A. K. *Astrophys. J. Suppl. Ser.* 1992, 80, 425.

[68] Wilson, N. J. and Bell, K. L. *Mon. Not. R. Astron. Soc.* 2002, 337, 1027.

[69] Wilson, N. J., Bell, K. L., and Hudson, C. E. *Astron. Astrophys.* 2005, 432, 731.

[70] Lafyatis, G. P. and Kohl, J. L. *Phys. Rev. A* 1987, 36, 59.

[71] Williams, I. D., Greenwood, J. B., Srigengan, B., O'Neill, R. W., and Hughes, I. G. *Mass. Sci. Technol.* 1998, 9, 930.

[72] Tayal, S. S. *Astron. Astrophys.* 2008 (to be sumitted).

[73] Smith, S. J., Zuo, M., Chutjian, A., Tayal, S. S., and Williams, I. D. Astrophys. J. 1996, 440, 421.

[74] Tayal, S. S. *J. Phys. B* 2006, 39, 4393.

[75] Montenegro, M., Eissner, W., Nahar, S. N, and Pradhan, A. K. *J. Phys. B* 2006, 39, 1863.

[76] McLaughlin, B. M. and Bell, K. L. *J. Phys. B* 1998, 31, 4317.

[77] Zuo, M., Smith, S. J., Chutjian, A., Williams, I. D., Tayal, S. S., and McLaughlin, B. M. *Astrophys. J.* 1995, 440, 421.

[78] Bell, K. L., Hibbert, A., Stafford, R. P., and McLaughlin, B. M. *Phys. Scr.* 1994, 50, 343.

[79] Tayal, S. S. Astrophys. *J. Suppl. Ser.* 2007, 171, 331.

[80] Hayes, M. A. and Nussbaumer, H. *Astron. Astrophys.* 1983, 124, 279.

[81] Hayes, M. A. *J. Phys. B* 1983, 16, 285.

[82] Tayal, S. S. *Astrophys. J. Suppl. Ser.* 2006, 166, 634.

[83] Smith, S. J., Lozano, J. A., Tayal, S. S., and Chutjian, A. *Phys. Rev. A* 2003, 68, 062708.

[84] Correge, G., and Hibbert, A. *J. Phys. B* 2002, 35, 1211.

[85] Tayal, S. S. *Astrophys. J.* 2003, 582, 550.

[86] Griffin, D. C., Badnell, N. R., and Pindzola, M. S. *J. Phys. B* 2003, 33, 1013.

[87] Lozano, J. A., Niimura, M., Smith, S. J., Chutjian, A., and Tayal, S. S. *Phys. Rev. A* 2001, 63, 042713.

[88] Bell, E. W., et al. *Phys. Rev. A* 1994, 49, 4585.

[89] Mohan, A., Landi, E., and Dwivedi, B. *Astrophys. J.* 2003, 582, 1162.

[90] Keenan, F. P., Katsiyannis, A. C., and Widing, K. G. *Astrophys. J.* 2004, 601, 565.

[91] Tayal, S. S. *Astrophys. J. Suppl. Ser.* 1997, 111, 459.

[92] Ramsbottom, C. A., Bell, K. L., and Stafford, R. P. *At. Data Nucl. Data Tables* 1996, 63, 57.

[93] Tayal, S. S. and Gupta, G. P. *Astrophys. J.* 1999, 526, 544.

[94] Tayal, S. S. *Astrophys. J.* 2000, 530, 1091.

[95] Tayal, S. S. *Astrophys. J. Suppl. Ser.* 2005, 159, 167.

[96] Johansson, A. E., Magnusson, C. E., Joelsson, I., and Zetterberg, P. O. *Phys. Scr.* 1992, 46, 221.

[97] Saraph, H. E. and Storey, P. *J. Astron. Astrophys.* 1999, 134, 369.

[98] Bhatia, A. K. and Mason, H. E. *Mon. Not. R. Astron. Soc.* 1980, 190, 925.

[99] Bhatia, A. K. and Landi, E. *Astrophys. J. Suppl. Ser.* 2003, 147, 409.

[100] Zhang, H. L. and Sampson, D. H. *At. Data Nucl. Data Tables* 1999, 72, 153.

[101] Bell, K. L. and Ramsbottom, C. A. *Mon. Not. R. Astron. Soc.* 1999, 308, 677.

[102] Bell, K. L. and Ramsbottom, C. A. *At. Data Nucl. Data Tables* 2000, 76, 176.

[103] Liao, C., Smith, S. J., Hitz, H., Chutjian, A., Tayal, S. S. *Astrohphys. J.* 1997, 484, 979.

[104] Smith, S. J., Greenwood, J. B., Chutjian, A., Tayal, S. S. *Astrophys. J.* 2000, 541, 501.

[105] Thomas, R. J. and Neupert, W. M. *Astrophys. J. Suppl. Ser.* 1994, 91, 461.

[106] Brosius, J. W., Davilla, J. M., and Thomas, R. *J. Astrophys. J.* 1998, 497, L113.

[107] Drake, J. J., Laming, J. M., and Widing, K. G. *Astrophys. J.* 1995a, 443, 393.

[108] Drake, J. J., Laming, J. M., and Widing, K. G. *Astrophys. J.* 1995b, 443, 416.

[109] Brickhouse, N. S., Raymond, J. C., and Smith, B. W. *Astrophys. J. Suppl. Ser.* 1995, 97, 551.

[110] Young, P. R., Landi, E., and Thomas, R. *J. Astron. Astrophys.* 1998, 329, 291

[111] Dere, K. P., Landi, E., Mason, H. E., Monsignori-Fossi, B. F., and Young, P. R. *Astron. Astrophys. Suppl. Ser.* 1997, 125, 149.

[112] Tayal, S. S. *Astrophys. J. Suppl. Ser.* 2001, 132, 117.

[113] Tayal, S. S. *Astrophys. J.* 2000, 544, 581.

[114] Mohan, M., Hibbert, A., and Kingston, A. E. *Astrophys. J.* 1994, 434, 389.

[115] Pelan, J. and Berrington, K. A. *Astron. Astrophys. Suppl. Ser.* 1995, 110, 209.

[116] Aggarwal, K. M. and Keenan, F. P. *Astron. Astrophys.* 2005, 439, 1215.

[117] Niimura, M., Cadez, I., Smith, S. J., and Chutjian, A. *Phys. Rev. Lett.* 2002, 88, 103201.

[118] Deb, N. C., Gupta, G. P., and Msezane, A. *Z. Phys. Rev. A* 1999, 60, 2569.

[119] Fuhr, J. R., Martin, G. A., and Weise, W. L. *J. Phys. Chem. Ref. Data* 1988, 17, 180.

[120] Gupta, G. P. and Tayal, S. S. *Astrophys. J. Suppl. Ser.* 1999a, 123, 295.

[121] Gupta, G. P. and Tayal, S. S. *Astrophys. J.* 1999b, 510, 1078.

[122] Dufton, P. L. and Kingston, A. E. *Phys. Scr.* 1991, 43, 386.

[123] Bhatia, K. A. and Kastner, S. O. *J. Quant. Spectrosc. Radiat. Transf.* 1993, 49, 609

[124] Bhatia, A. K. and Doschek, G. A. *At. Data Nucl. Data Tables* 1995, 60, 97.

[125] Aggarwal, K. M. and Keenan, F. P. *Astron. Astrophys.* 2003, 399, 799.

[126] Gupta, G. P. and Tayal, S. S. *Astrophys. J.* 1998, 506, 464.

[127] Aggarwal, K. M. and Keenan, F. P. *Astron. Astrophys.* 2004, 418, 371.

[128] Tayal, S. S. *Astrophys. J. Suppl. Ser.* 2008 (submitted).

[129] Storey, P. J., Mason, H. E., and Saraph, H. E. *Astron. Astrophys.* 1996, 309, 672.

[130] Storey, P. J., Mason, H. E., and Young, P. R. *Astron. Astrophys. Suppl. Ser.* 2000, 141, 296.

[131] Dufton, P. L. and Kingston, A. E. *Phys. Scr.* 991, 43, 386.

[132] Bhatia, K. A. and Kastner, S. O. *J. Quant. Spectrosc. Radiat. Transf.* 1993, 49, 609.

[133] Hossain, S., Tayal, S. S., Smith, S. J., Raymond, J. C., and Chutjian, A. *Phys. Rev. A* 2007, 75, 022709.

[134] Ferguson, J. W., Korista, K. T., and Ferland, G. J. *Astron. Astrophys. Suppl. Ser.* 1997, 110, 287.

In: Atomic, Molecular and Optical Physics...
Editor: L.T. Chen, pp. 205-234

ISBN: 978-1-60456-907-0
© 2009 Nova Science Publishers, Inc.

Chapter 5

POLARIZATION DEPENDENT EFFECTS IN OPTICAL WAVEGUIDES CONTAINING BRAGG GRATING STRUCTURES

Ping Lu[a], Liang Chen[b], Xiaoli Dai[a], Stephen J. Mihailov[a] and Xiaoyi Bao[b]

[a]Communications Research Centre Canada, Ottawa, ON K2H 8S2, Canada
[b]Department of Physics, University of Ottawa, Ottawa, ON K1N 6N5, Canada

Abstract

Since the discovery of photosensitivity in optical fibers by Hill et al. in 1978 [1], Bragg gratings fabricated in optical fibers and planar waveguides have been extensively investigated over the last three decades and have been widely used in optical communication and sensor applications. In this chapter, the properties of birefringence in optical fibers and planar waveguides with Bragg grating structures are investigated experimentally, and the birefringence-induced impairments in communication systems with Bragg-grating–based components are evaluated by simulation.

1. Introduction

When the attenuation and chromatic dispersion in optical fibers are compensated, polarization mode dispersion (PMD) is the major factor limiting the capacity of fiber optic communication systems at high bit rates (>10Gb/second) [2-6]. The presence of PMD in optical networks introduces differential group delay (DGD), resulting in signal distortion and high bit error ratio (BER). The system impact of PMD has been investigated quantitatively by examining the PMD induced system outage and eye penalty [2,3]. The fiber optic networks are becoming more complex because of the introduction of new types of photonic components that have more functionality and can operate at higher data rates. Optical components, such as reconfigurable optical add-drop multiplexers (ROADMs), fiber Bragg gratings (FBGs), planar waveguide Bragg gratings (WBGs), etc. have both PMD and polarization-dependent loss (PDL) [4, 5]. It has been shown both theoretically and experimentally that the system impact due to the combined effect of PMD and PDL is more severe than that due to PMD alone [4, 6]. As one of the key optical components in modern networks, Bragg gratings in fibers and planar waveguides are widely used in today's communication and sensing systems. They are usually fabricated by exposure of the fiber or waveguide to the ultraviolet (UV) or ultrafast infrared (IR) fringes resulting from two-beam interference or generated by a zero-order nulled phase mask [7, 8]. The fibers and waveguides are typically hydrogen (H_2)-loaded to increase photosensitivity to UV light [9]. Due to the sensitive nature of this fabrication technique, laser exposure is only on one side of the fiber or waveguide, resulting in grating spectra that exhibit polarization dependence (birefringence) [10, 11]. Compared with optical fibers, planar waveguides produced by plasma-enhanced chemical vapor deposition (PECVD) or flame hydrolysis deposition (FHD) have a very strong intrinsic birefringence that makes the Bragg wavelength and peak reflectivity sensitive to the polarization state of the propagation signal.

In this chapter, the properties and system impairments of birefringence in FBGs and WBGs are presented and discussed in detail. In section 2, the sources of birefringence in optical fibers and planar waveguides that incorporate a grating structure will be briefly introduced, and the relationship between birefringence and spectral PMD and PDL of FBGs and WBGs will be presented. This relationship will be used in the following sections to characterize the birefringence in gratings and to evaluate the birefringence-induced system impairments. In section 3, the induced birefringence in fibers by UV and femtosecond pulse duration IR lasers during the grating inscription process is examined, and its dependence on H_2-loadening in fibers, the temperature while the grating is inscribed, and the polarization of the laser beam will be presented. The annealing curves of induced birefringence in FBGs are presented as well. Section 4 shows the technique for compensating the intrinsic birefringence of planar waveguide by using UV laser exposure. Finally, in section 5, the birefringence-induced penalties in optical communication systems consisting of Bragg grating based OADMs are simulated by examining the distortion of eye diagram of 10 Gb/s signal.

2. Birefringence in FBGS and WBGS

2.1. Sources of Birefringence

The birefringence of FBGs and WBGs consists of two parts: 1) the intrinsic birefringence of the fiber and waveguide, and 2) the induced birefringence by UV or ultrafast IR lasers during grating inscription. The intrinsic birefringence may come from the asymmetric geometry of the waveguide, the stress induced during waveguide fabrication, and environmental perturbations (such as bending, twisting, etc.). The amount of birefringence in single mode fiber, such as *Corning SMF-28* used in communications, is usually low (of the order of 10^{-7}–10^{-6}), and both the local birefringence value and the local principal axes (the fast and slow axes) vary along the fiber length. The birefringence in a planar waveguide is usually higher than that in cylindrical single mode fiber and can be of the order of up to 10^{-4}. This high birefringence is due to the asymmetric geometry and material structures that result in larger geometrical birefringence and stress birefringence than would occur in regular single mode fibers.

Bragg gratings in fibers and waveguide are usually fabricated by exposure of the fibers and waveguides to the UV or ultrafast IR fringes resulting from two-beam interference or generated by a zero-order nulled phase mask. The fibers and waveguides are typically Hydrogen (H_2) loaded to increase photosensitivity to UV light, but this is not necessary when high peak power IR irradiation is used. Due to the sensitive nature of this fabrication technique, only one side of the fiber and waveguide is exposed to the laser radiation, resulting in the induction of birefringence in addition to the intrinsic birefringence. This laser-induced birefringence has been studied both experimentally and theoretically [12-14]. The birefringent effects in FBGs, and long period gratings has been measured [15-18], and the annealing properties of the UV and ultrafast IR laser-induced birefringence have also been examined [19-21].

2.2. Relationship between Birefringence and Spectral PDL of Bragg Gratings in Fibers and Waveguides

In FBGs and WBGs inscribed by laser exposure, both the effective mode index and the index modulation can show a polarization dependence [18, 21]. The birefringence in effective mode index and the birefringence in index modulation of a grating may be characterized by measuring the spectral PDL (or PMD) at wavelengths around the grating's resonance and using the coupled-mode theory. Here we briefly introduce the relationship between a grating's spectral PDL and birefringence. The transmission and reflection of a uniform grating with grating length L and period Λ can be expressed as:

$$\begin{bmatrix} A_{out} \\ B_{out} \end{bmatrix} = \begin{bmatrix} p & q \\ q* & p* \end{bmatrix} \begin{bmatrix} A_{in} \\ B_{in} \end{bmatrix}, \tag{1}$$

where A and B are the amplitudes of the forward and backward propagating waves, the subscripts of *in* and *out* refer to the waves at the grating's input and output ends respectively. In Eq. (1), $p = \cosh(\sigma L) - i\delta \sinh(\sigma L)/\sigma$ and $q = i\kappa \sinh(\sigma L)/\sigma$, where $\delta = 2\pi(n_{eff}/\lambda - 1/\Lambda)$ is the detuning from Bragg resonance, n_{eff} is the effective mode index and λ is the wavelength in free space. Parameter σ is related to the detuning δ, and mode coupling constant κ, as $\sigma = \sqrt{\kappa^2 - \delta^2}$. The coupling constant, κ, is proportional to the index modulation, Δn, in the fiber core as $|\kappa| = \pi\Delta n / n_{eff}\lambda$. A grating with an apodized index profile can be treated as a concatenation of many uniform sub-gratings. The 2×2 transform matrix in Eq. (1) can then be replaced by the product of the corresponding sub-matrices. The UV exposure on one side of fiber makes both n_{eff} and Δn polarization dependent. The induced PMD and PDL in a grating can then be obtained by calculating the polarization dependent phase delays and magnitudes of A_{out} (in transmission) and B_{in} (in reflection) as follow:

$$PDL_{tran} = \left| 20\log_{10}\left(\left| \frac{A_{out}^{fast}}{A_{out}^{slow}} \right| \right) \right| \qquad \text{(PDL in transmission)}$$

$$DGD_{tran} = \left| \tau_{tran}^{fast} - \tau_{tran}^{slow} \right| \qquad \text{(PMD in transmission)}$$

$$PDL_{ref} = \left| 20\log_{10}\left(\left| \frac{B_{in}^{fast}}{B_{in}^{slow}} \right| \right) \right| \qquad \text{(PDL in reflection)}$$

$$DGD_{ref} = \left| \tau_{ref}^{fast} - \tau_{ref}^{slow} \right| \qquad \text{(PMD in reflection)}$$

$$(2)$$

Where A_{out}^{fast} and A_{out}^{slow} are the amplitudes of the forward waves at the grating's output end when the probe light is launched on to the fast and slow axes respectively. τ_{tran}^{fast} and τ_{tran}^{slow} are the time delays of the forward propagating wave along the fast and slow axes respectively. B_{in}^{fast} and A_{in}^{slow} are the amplitudes of backward waves at the grating's input end when the probe light is launched on to the fast and slow axes respectively. τ_{ref}^{fast} and τ_{ref}^{slow} are the time delays of the backward propagating wave along the fast and slow axes respectively.

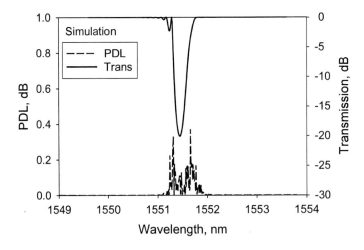

Figure 1. The simulated spectral PDL of a FBG induced by the intrinsic birefringence of the fiber.

The intrinsic birefringence in *SMF-28* fiber is on the order of 10^{-7}–10^{-6} and varies along the fiber length in both amplitude and orientation (the fast or slow axis). When the intrinsic fiber birefringence in a FBG is the major part of the total birefringence, which is likely for weakly inscribed gratings in single mode fiber, the spectral PDL shows noisy peaks due to the random variation of the intrinsic birefringence along the grating length. Using Eqs. (1) and (2), a simulated spectral PDL induced by the intrinsic birefringence is shown in Fig. 1 where the intrinsic birefringence is randomly distributed from 0 to 10^{-6} along the grating length. The grating length used for the simulation results shown in Fig. 1 is 6 mm with an apodization of the Gaussian index profile with maximum index modulation of 4×10^{-4} at the grating center. The birefringence in planar waveguide is usually higher than that in single mode fiber and the fast and slow axes are usually fixed, the spectral PDL of gratings in planar waveguide is similar to that when the major part of birefringence is induced by laser irradiation that will be discussed below.

When a grating is inscribed in fiber or waveguide by exposing to UV or IR laser fringes, the "side-writing" technique produces birefringence creating fast axis and slow axes that are wavelength independent. The orientation of the fast/slow axes is dependent upon the direction of the incident beam used for the grating inscription. In the case when the laser-induced birefringence is much higher than the fiber's intrinsic birefringence, the spectral PDL of a grating can be modeled using Eqs. (1) and (2). Figs. 2(a) and 2(b) show the simulated spectral PDLs of two gratings with the induced birefringence of 5×10^{-6}. Usually within the bandwidth of one transmission (or reflection) peak (from λ_1 to λ_2 in Figs. 2(a) and 2(b)) in a grating spectrum, there are two PDL peaks. When the induced index profile is apodized with the highest index change in the middle of the grating, such as Gaussian or $\cos^2$ profile, the PDL peak on the shorter wavelength side will be narrower than that of the PDL peak in the longer wavelength side if the laser induced index is positive (Fig. 2(a)). If the bandwidth of the PDL peak on the short wavelength side is broader than that of the PDL peak in the longer wavelength side, then the laser-induced index change is negative (Fig. 2(b)). In the simulations of Figs. 2(a) and 2(b), the FBG total length is 6 mm with an apodization of the Gaussian index profile with maximum index modulation of 5×10^{-4} at the center of the grating. Figs. 1 and 2 clearly show that by modeling the grating spectral PDL, the birefringence of a grating can be obtained.

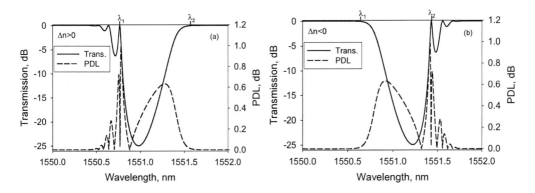

Figure 2. The simulated spectral PDLs of a FBG induced by the laser during FBG inscription. (a) the induced index is positive, (b) the induced index is negative.

3. Characterization of Grating Birefringence

3.1. Experiment Setup

The inscription of a grating in optical fibers and waveguides usually takes a few minutes to a few tens of minutes depending on the requirements of grating bandwidth, strength and apodization profile etc. For real time monitoring of birefringence variation during grating inscription, the measurement set-up has to be fast with reasonable wavelength resolution. The set-up to inscribe FBG and to measure birefringence used in this work is shown in Fig. 3. The laser beam (either UV or ultrafast IR) is focused on the fiber by a cylindrical lens (not shown in Fig. 3) through a phase mask of pitch Λ_{Bragg} that is optimized to the laser wavelength producing a Bragg resonance of $\lambda_{Bragg} = n_{eff}\Lambda_{Bragg}$. A tunable laser beam (*8164A, Agilent*) is launched into the polarization test set (*A2000 Components Analyzer, Adaptif Photonics GmbH*) into which the two ends of *SMF-28* fiber are connected in order to measure the grating's spectral PMD and PDL in transmission during grating inscription. The whole measurement set-up is controlled by a personal computer through a GPIB interface. The set-up has the ability of measuring a set of data, including grating spectrum, PMD, PDL, principal state of polarization (PSP), etc., every 2 to 3 seconds for a wavelength window of a few nanometers with a resolution of 10 picometers. The time needed for the measurement at single wavelength is less than 1 ms, which makes it possible for real-time measurements of the variations of birefringence during grating inscription. The grating's birefringence can also change very quickly when the grating is annealed at certain temperatures [19-21] as will be shown later in this section, therefore a fast measurement set-up is preferred.

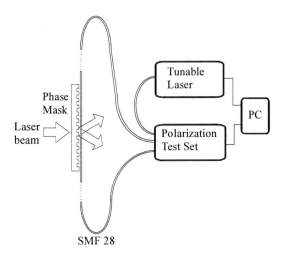

Figure 3. Set-up of writing fiber Bragg gratings and measuring the laser-induced spectral PMD and PDL.

3.2. UV Induced Birefringence in Fiber Bragg Gratings

3.2.1. Growth of Birefringence during Grating Inscription

The experimental set-up for writing FBGs and measuring the UV-induced birefringence is shown in Fig. 3. Gratings are inscribed on *SMF-28* fibers by spot writing a focused 244 nm beam from a frequency doubled Ar+ laser through a phase mask. The focused beam spot size is 9 mm. The focal length of the cylindrical lens is 150 mm and the pitch of the phase mask is $\Lambda_{mask} = 1.07$ μm, which produces a grating period of $\Lambda = 0.535$ μm in the fiber core. The polarization state of the UV beam is parallel to the fiber axis, which is preferred in order to produce minimum birefringence [12]. The fibers used in the experiment were photosensitized by hydrogen loading at room temperature and 2500 psi pressure for one week. Fig. 4 and Fig. 5 show the experimental results of spectral DGD and PDL of the grating with fitting curves by using the coupled mode theory. In the simulations of DGD and PDL, both the birefringence in n_{eff} and the birefringence in Δn need to be considered, especially in the fitting of spectral PDL. This is clearly shown in Fig. 5 where the experimental data are compared to the simulated results (dashed curves) in which only n_{eff} is polarization dependent. Using the set-up shown in Fig. 3, the UV-induced spectral DGDs and PDLs in the grating are measured every 2~3 seconds with the tunable laser power of 140 mW. Due to the dynamic range of the polarization test set, the UV-induced DGD and PDL are measured up to the grating strength of -25 dB in transmission. By fitting the spectral DGDs and PDLs, the changes in

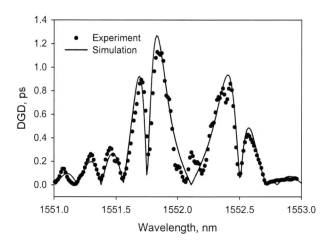

Figure 4. Measurement results of the UV-induced DGD in a 2 mm long grating. The solid curve is the simulated result.

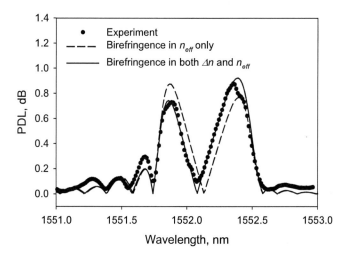

Figure 5. Measurement (dots) and simulated (solid and dashed curves) results of the UV-induced PDL in a 2 mm long grating. Solid curve: both the mode effective index and the index modulation are polarization dependent; dashed curve: the mode effective index is polarization dependent only.

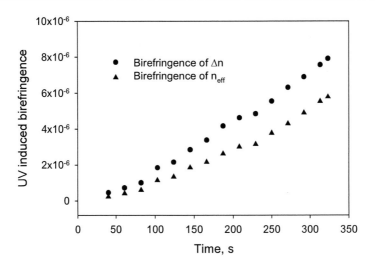

Figure 6. The growths of the birefringence in effective mode index and the birefringence in index modulation.

birefringence for both the n_{eff} and Δn are obtained and presented in Fig. 6. The birefringence in Δn and the birefringence in n_{eff}, shown in Fig. 6 are 1.4% to 1.7% and 0.7% to 1.2% of the Δn value, respectively. The variation in induced birefringence with exposure time is not linear, as is shown in Fig. 6. For various UV laser powers (from 100 to 300 mW) and various distances (from 0 to 10 mm) between the fiber and the focus, the growths of PDL and DGD are measured as well. No obvious change in birefringence is found for the same grating strength.

3.2.2. Annealing of UV-Induced Birefringence

In H_2-loaded *SMF-28* fibers, Canning et al. [19, 20] showed that the UV-induced birefringence can be annealed out at temperatures below 150 °C without significant decay of the induced index modulation. This is shown in Fig. 7 where the reductions of UV-induced Δn and birefringence in H_2-loaded *SMF-28* fibers during various annealing temperatures are plotted. It can be seen clearly that for temperatures above 120 °C, the birefringence rapidly decreases to that of the fiber's intrinsic birefringence ($\sim 10^{-6}$).

The grating with its annealing curve shown in Fig. 7 was inscribed when the UV laser beam was focused using a lens with focal length of 150 mm. During the inscription, the temperature of the fiber section being exposed to UV laser is almost the same as room temperature (after the UV beam was turn off, no obvious Bragg wavelength shift was found). When the UV laser was tightly focused by using a 80 mm focal length lens, the temperature of the fiber section that was exposed to UV beam can be higher than room temperature. Fig. 8 shows the annealing curve of the birefringence up to 260 °C when the inscription temperature is around 140 °C during UV exposure. From Fig. 8 we can see that the UV-induced birefringence cannot be annealed out at low temperature. The induced birefringence is still on the order of 10^{-5} after the grating is annealed up to 260 °C.

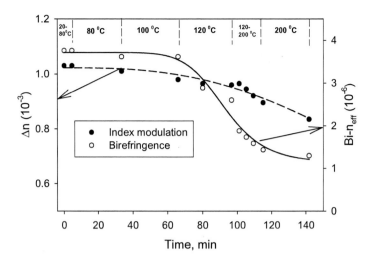

Figure 7. Annealing of UV-induced birefringence of FBGs inscribed in H_2-loaded fibers at room temperature.

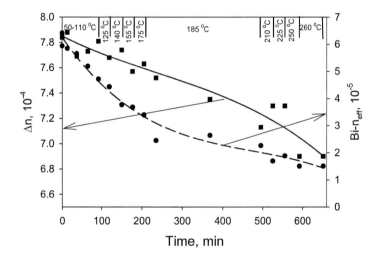

Figure 8. Annealing of UV-induced birefringence of FBGs inscribed in H_2-loaded fibers at 140 °C.

3.3. High Peak Power IR Laser-Induced Birefringence in FBGs

Recently high peak power, regeneratively amplified Ti:Sapphire ultrafast infrared (ultrafast-IR) lasers ($\lambda = 800 nm$, pulse durations from $120\ fs$ to $2\ ps$) have been used to fabricate high quality FBGs in single mode fibers with or without phase masks [8, 22-27]. Depending on the pulse energy and pulse width of the ultrafast-IR laser, two types of gratings, type I-IR and type II-IR, have been observed in single mode fibers [22]. The formation of type I-IR gratings is associated with nonlinear absorption of the IR light and the grating annealing property is similar to that of type I UV gratings. Type II-IR gratings are usually fabricated with high pulse intensity and they are associated with highly localized damage in the fiber and exhibits high temperature stability [22]

3.3.1. Grating Inscription and Birefringence Monitoring

The Type I-IR and Type II-IR gratings under test are spot written in *SMF-28* fibers (both H_2-loading and non H_2-loading) by focusing the IR laser beam through a phase mask with the pitch of 1.07 μm. The focal length of the cylindrical lens is 19 mm and the focused beam spot size is approximately ~3 μm × 6.4 mm. The focused beam is scanned across the fiber core using a piezo-electric actuated stage [23]. The distance between the phase mask and the fiber for Type II-IR grating inscription is 300 μm in order to obtain high peak intensity and it is 1.3 mm for Type I-IR inscription to get a pure two-beam interference fringe [27]. The pulse durations for Type I-IR and Type II-IR grating inscriptions are measured with an autocorrelator to be 125 fs and 2.5 ps respectively. The set-up of FBG inscription and birefringence measurement is shown in Fig. 3.

3.3.2. Birefringence in Type I-IR Gratings

The ultrafast-IR laser induced birefringence during the inscription of Type I-IR grating in H_2-loaded SMF-28 fibers is examined with the polarization of the inscription laser beam either normal or parallel to the fiber axis (S-polarized or P-polarized respectively). No obvious dependence of the induced birefringence on the polarization of the IR laser beam is observed. A typical PDL spectrum of a Type I-IR grating inscribed in H_2-loaded fiber is shown in Fig. 9. The strength of the grating is -21 dB in transmission with an Δn of 5.2×10^{-4} at the center of the grating. Comparing Fig. 9 with the simulation result shown in Fig.1, we can see clearly that the major part of birefringence in the Type I-IR grating in H_2-loaded fibers is randomly distributed along the grating length, which means that the birefringence induced by the IR laser is smaller than the fiber's intrinsic birefringence. The growth of the birefringence with increasing laser induced index change is plotted in Fig. 10. For comparison, the UV-induced birefringence in H_2-loaded SMF-28 fibers is plotted as well in Fig. 10 with the polarization of the UV beam P-polarized, which minimizes the induced birefringence [12]. We can see that for the same induced index in H_2-loaded SMF-28 fibers, the birefringence induced by ultrafast-IR laser is much lower than that induced by the UV laser. After the Type I-IR gratings in H_2-loaded fibers are inscribed, the gratings are left at room temperature for two days. The birefringence is measured again and it is found that the birefringence induced by ultrafast-IR laser is annealed out at room temperature during the two days, which is different from the annealing of UV-induced birefringence in H_2-loaded SMF-28 fibers [19-21].

The birefringence of Type I-IR gratings induced by the ultrafast-IR laser in unloaded *SMF-28* fibers is examined and a typical spectral PDL is shown in Fig. 11. It can be seen that there are clearly characteristic peaks in the spectral PDL curve. Compared to the curves in Figs. 1 and 2 we can see that the laser-induced birefringence is much larger than the fiber's intrinsic birefringence and the laser-induced index is positive (similar to Fig. 2(a), the PDL peak with the narrower bandwidth is on the short wavelength side). The growths of birefringence in effective index, $B_i\text{-}n_{eff}$ with increasing index modulation, *Δn*, are shown in Fig. 12 with the polarization of the laser beam either *P* or *S*-polarized. We can see that the IR laser with S-polarization induces a much higher birefringence than the P polarized laser beam, which is similar to the induced birefringence in H_2-loaded *SMF-28* fibers by UV laser [12]. Comparing Fig. 12 with Fig. 10, for the same IR laser induced Δn in Type I-IR gratings, the induced birefringence in H_2-loaded fibers is much lower than that in unloaded fibers (10^{-6} vs.10^{-5}).

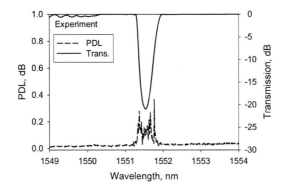

Figure 9. Spectral PDL of type I-IR grating inscribed in H_2-loaded fiber.

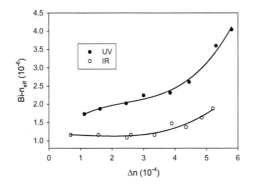

Figure 10. Comparison of the birefringence growth in H_2-loaded SMF28 fibers exposed by UV and IR lasers, respectively.

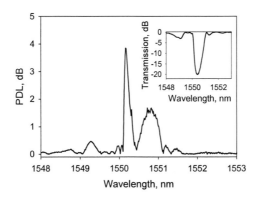

Figure 11. Spectral PDL of type I-IR grating inscribed in non H_2-loading fiber.

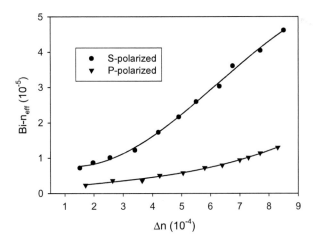

Figure 12. Growth of the IR laser induced birefringence with the polarization of the laser beam parallel and perpendicular to the fiber axis respectively.

3.3.3. Birefringence in Type II-IR Gratings

The inscription of Type II-IR gratings is usually associated with the introduction of wavelength independent scattering loss and high cladding mode coupling. The time inscription of a Type II-IR grating (Δn of up to 10^{-2}) requires far fewer laser pulses making it difficult to measure the birefringence growth during grating inscription.

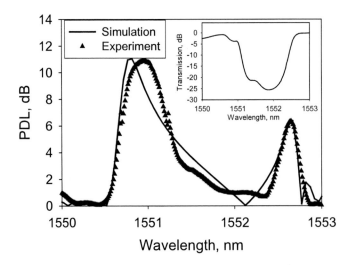

Figure 13. The spectral PDLs of Type II-IR grating inscribed in H_2-loaded fiber. Triangle dots: experiment, solid curve: simulation.

In H_2-loaded *SMF-28* fibers, the ultrafast-IR laser induces high birefringence and PDL for both *P* and *S*-polarized. The spectral PDL changes from sample to sample that is due to the large variation of the grating spectra between samples even for the same exposure condition. Fig. 13 shows a spectral PDL of Type II-IR grating inscribed in H_2-loaded fiber with laser beam S-polarized. The triangle dots present the experimental results and the solid curve is the simulation. From Fig. 13 we can see that the induced Δn is negative when compared to Fig. 2(b), which is different from the positive index change in Type I-IR gratings inscribed by ultrafast-IR and CW UV lasers. For the three samples of Type II-IR gratings in H_2-loaded *SMF-28* fibers, the induced index changes are all negative. In the simulation curve in Fig. 13, the induced Δn and birefringence at the center of the apodized grating are -1.7×10^{-3} and 3×10^{-4}, respectively, i.e., the birefringence is more than 15% of the Δn, which is much higher than that of the UV laser induced birefringence in H_2-loaded *SMF-28* fibers. When the IR laser beam is P-polarized, the induced birefringence is ~6% of the induced Δn, which is still much higher than that induced by UV laser (<2% [18]).

For non H_2-loaded fibers, four Type II-IR grating samples have been tested with two samples for each polarization of the laser beam (S- or P-polarized). It was found that, similar to the case of Type I-IR gratings, the S-polarized laser beam induced a higher birefringence than that of the P-polarized. For the induced Δn of 10^{-3}, the birefringence induced by P-polarized beam is around 10^{-5} and it is around 10^{-4} for the S-polarized. For all the four samples in non H_2-loaded fibers, the induced refractive indices are negative, which is the same as the Type-II gratings in H_2-loaded fibers. Negative induced index changes and generation of birefringence has been observed in bulk silica irradiated with 800 nm 200 fs pulses with intensities similar to the Type II-IR gratings presented here [28].

3.3.4. Annealing Properties of Birefringence in Type I-IR Gratings

The IR induced birefringence of Type I-IR gratings in H_2-loaded fiber anneals out at room temperature after ~48 hours with only the intrinsic birefringence of the fiber remaining and the index modulation drops by ~10%.

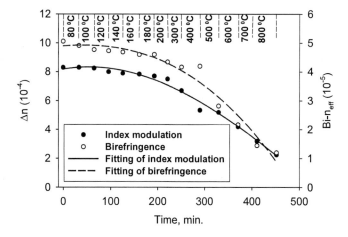

Figure 14. The annealing curves of the induced index modulation (solid dots) and birefringence (hollow dots) of Type I-IR gratings inscribed in unloaded fibers.

The annealing of birefringence of Type I-IR gratings inscribed in unloaded fibers is shown in Fig. 14 (the laser beam is *S*-polarized, the annealing of birefringence induced by *P*-polarized IR beam is the same). The induced birefringence has the same annealing resistance as the index modulation even for temperatures up to 800 °C, which is different from the annealing property of the induced birefringence in H_2-loaded *SMF-28* fibers both by UV and IR lasers. In Fig. 14, after 450 minutes annealing at the increased temperature up to 800 °C, the induced Δn and birefringence drop by 73% and 76% respectively.

3.3.5. Annealing Properties of the Birefringence in Type II-IR Gratings

The annealing properties of Type II-IR gratings are more complicated than the Type I-IR gratings, partially due to the observation that a weak Type I-IR grating is usually inscribed initially before a Type II-IR grating is fabricated, especially in H_2-loaded fibers. Due to the different signs of induced index in the fibers (positive/negative) and different annealing properties of Type I-IR and Type II-IR gratings, the annealing property of a Type II-IR grating has a strong dependence on the grating sample, fiber (H_2-loading or non H_2-loading) and the polarization of the laser beam.

For low birefringence induced by *P*-polarized laser beam in non H_2-loading fibers ($\sim 10^{-5}$), the birefringence is relatively stable during annealing at increased temperature from room temperature to 200 °C (total 3 hours) and it is still stable during a long term annealing at 200 °C (16 hours). The spectral PDLs at 50, 100, 140 and 200 °C during the annealing process are shown in Fig. 15. By modelling of the spectral PDLs, the variation of the birefringence over the whole annealing period (19 hours) is on the order of the fiber's intrinsic birefringence ($\sim 10^{-6}$).

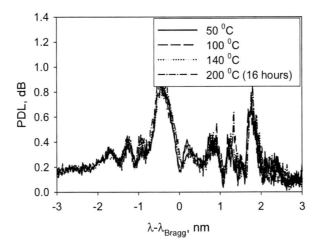

Figure 15. The type II-IR grating spectral PDLs induced by P-polarized laser beam in non H_2-loading fibers at 50, 100, 140 and 200 °C during the annealing process.

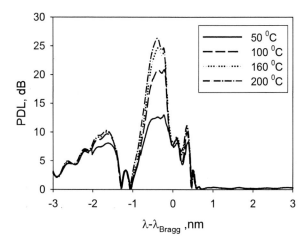

Figure 16. The type II-IR grating spectral PDLs induced by S-polarized laser beam in non H_2-loaded fibers during annealing at various increased temperatures.

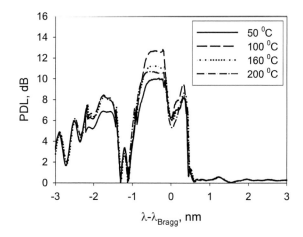

Figure 17. Another example of the changes of type II-IR grating spectral PDLs during annealing process, same as grating whose annealing properties shown in Fig. 16, the type II-IR grating is written in non H_2-loaded fiber with the laser beam S-polarized.

For Type II-IR gratings inscribed by using P-polarized laser in H_2-loaded fibers and by using S-polarized laser in H_2- and non H_2-loaded fibers, the high birefringence ($>10^{-4}$) shows a much larger variation during the annealing process than the low birefringence ($\sim 10^{-5}$) induced by P-polarized beam in non H_2-loaded fibers. Even for the same exposure conditions (laser beam polarization, laser pulse energy, pulse width, with/without H_2-loading in fibers and exposure time etc.), different samples show different annealing behaviours. Fig. 16 shows the changes of spectral PDL in non H_2-loaded fibers induced by S-polarized laser beam during annealing at various increased temperatures. The spectral PDL increases with increasing annealing temperature, which is opposite to the annealing of Type I-IR gratings. By using the same laser exposure condition and the same fiber, another Type II-IR grating with the same grating strength (~ 23 dB in transmission) is inscribed and its spectral PDL during various annealing temperature is shown in Fig. 17. Different from the annealing property shown in Fig. 16, the spectral PDL increases first when the temperature increases from 50 to 100 °C, and then decreases with increasing temperature. The grating with its spectral PDL shown in Fig. 16 was annealed with high temperature up to 800 °C. The spectral PDL still showed strong temperature dependence and no obvious decay of the birefringence was observed (see Fig. 18). The variation of spectral PDL during annealing at different temperatures is due to the change of the grating strength. Through close examination of the grating spectra when the probe laser is launched into the fast and slow axes, it is found that both of grating strengths along fast and slow axes showed strong but different variations with temperature. The reason for the different variations of the grating strength with different polarizations needs to be investigated further.

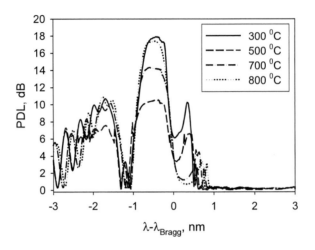

Figure 18. The spectral PDLs of type II-IR gratings at various temperatures (above 200 °C) after being annealed up to 200 °C.

4. Compensation of the Planar Waveguide Birefringence Using UV Laser Exposure

Compared with optical fibers, planar waveguides produced by plasma enhanced chemical vapor deposition (PECVD) or flame hydrolysis deposition (FHD) have a very strong intrinsic birefringence that makes the Bragg wavelength and peak reflectivity sensitive to the polarization state of the signal propagating through the waveguide. In this section we present the technique of compensating the intrinsic birefringence of planar waveguide by UV exposure. In silica-on-silicon planar technology, the ridge waveguide has both asymmetrical geometrical and material structures, and thus has larger geometrical birefringence B_g and stress birefringence B_s than that in an optical fiber. The waveguide intrinsic birefringence B_i is given by

$$B_i = B_s + B_g \qquad (3)$$

where B_s is the stress-induced birefringence and B_g is the geometrical birefringence. The stress-induced birefringence is from the stress anisotropy that is induced mainly by the thermal-expansion mismatch among dissimilar materials. In practice, some novel fabrication methods have been used to minimize waveguide birefringence, for example, trenching on the both sides of waveguide, depositing additional cladding layers, and laser processing waveguide with mid-infrared or UV, etc [29-31]. UV irradiation is a direct and effective method to obtain polarization insensitive Bragg gratings by compensating the intrinsic birefringence B_i in the waveguide with the UV induced birefringence B_{uv}, which is produced automatically during the Bragg grating formation [32-33]. The mechanism of the UV controlled birefringence can be attributed to the stress relief of the waveguide core and the surrounding area. It has also been suggested, that the mechanism could be a refractive index change in the waveguide core exposed by UV, which in turn changes the waveguide geometrical birefringence. In the paper [34], we addressed the relative contribution of these mechanisms and provided some clarification with respect to which process is dominant during the UV trimming of a planar waveguide's birefringence. A theoretical model is developed which reveals the relationships of the waveguide intrinsic birefringence B_i with the core refractive index, the core dimension, and the stress change of the waveguide, allowing us to separately examine the contributions of these different parameters. In the experiments, the method to measure the intrinsic birefringence B_i and UV induced birefringence B_{uv} are established by inducing a weak Bragg grating and monitoring the polarization-dependent Bragg wavelength change under the UV irradiation. The weak Bragg grating with the modulated index $\Delta n \sim 1.49 \times 10^{-4}$ are induced by using a zero-order nulled phase mask and ArF excimer laser operating under the following conditions: 40 mJ/pulse and 50 Hz. The polarization-dependent Bragg wavelength shift is given as

$$\lambda_{TM} - \lambda_{TE} = 2\Lambda\left(n_{TM} - n_{TE}\right) \tag{4}$$

where λ_{TM}, λ_{TE}, n_{TM} and n_{TE} are the Bragg wavelengths and effective indexes for TM and TE modes, respectively. The birefringence B of the waveguide and the initial birefringence B_i of the waveguide are defined as n_{TM} - n_{TE} and n_{0TM} - n_{0TE}, respectively. n_{0TM} and n_{0TE} are the initial effective indexes of the waveguide unexposed by UV for TM and TE modes, respectively. The Bragg grating is subsequently exposed to an un-modulated blanket UV irradiation by removing the phase mask. By observing the changes of two Bragg grating wavelengths for TM and TE modes, B and B_i are available. To test the change of Bragg wavelength with exposure time, the UV irradiation was interrupted periodically in order to measure the TM and TE modes of the Bragg wavelengths. As shown in Fig. 19, the results were fitted with a polynomial regression, and expressed with a solid line. The waveguide birefringence B is estimated with λ_b (TE) - λ_b (TM). The initial Bragg wavelengths λ_{b0} (TE), λ_{b0} (TM) are then determined by extrapolating the curves to zero exposure time. It is clear that the birefringence of the waveguides can be reduced, in some cases to zero, by long UV exposures. A, B, C, D and E denote waveguides of dimensions 8.8μm×5.6μm, 7.7μm×5.6μm, 6.6μm×5.6μm, 5.7μm×5.6μm and 4.6μm ×5.8μm respectively. The birefringence in the 8.8μm×5.6μm large core size waveguide is tuned to 1×10^{-4} from -1×10^{-4} after 26 minutes UV irradiation. With the same UV irradiation condition, the birefringence in the 4.6μm×5.8μm small core size waveguide is tuned to - 4×10^{-4} from - 7×10^{-4}. This technique has been used successfully to fabricate polarization independent Bragg gratings in FHD waveguides with a small intrinsic birefringence of $\sim 2 \times 10^{-4}$. The conclusion is that changes of the birefringence in silica ridge waveguide structures with UV irradiation can be attributed mainly to stress changes in the waveguide core and its surrounding area, and not to the core refractive index change of 10^{-3} induced by UV irradiation. The experimental results are in agreement with this analysis.

Often high UV induced index changes and high UV exposure dosages are required in order to induce enough complementary UV-induced birefringence to compensate for the intrinsic birefringence of the waveguide. Such extreme conditions require extremely long

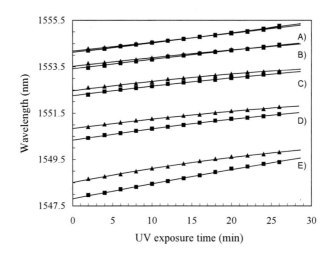

Figure 19. Experimental results of Bragg wavelength as a function of UV exposure time for different waveguide dimensions. The TE mode is denoted by squares while the TM mode is denoted by triangles. A, B, C, D and E denote waveguides of dimensions 8.8 μm X 5.6 μm, 7.7 μm X 5.6 μm, 6.6 μm X 5.6 μm, 5.7 μm X 5.6 μm and 4.6 μm X 5.8 μm respectively. The solid line is a polynomial regression through each data set.

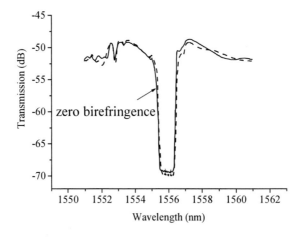

Figure 20. Transmission spectra for TM (solid line) and TE (dashed line) modes of the Bragg grating trimmed after low fluence blanket UV s-polarization exposure.

and intense exposures, which can result in damage to the surface of waveguides. By enhancing the photosensitivity of the planar waveguides through hydrogen loading, a larger UV induced birefringence is available in addition to a strong index change [35]; however, the induced birefringence is limited by the saturation of the index change. In the paper [36], a method for rapid compensation of high intrinsic birefringence in plasma-enhanced chemical vapor deposition (PECVD) based planar waveguides is presented. Compensation of the intrinsic waveguide birefringence is achieved by the induction of a large UV induced birefringence opposite that of the intrinsic birefringence. The UV induced birefringence is created with polarized UV ArF excimer laser irradiation (oriented normal to the waveguide axis) and enhanced further through hydrogen loading. The fabrication of polarization insensitive Bragg gratings (PIBG) in PECVD planar waveguides with a large intrinsic birefringence of 6.3×10^{-4} is demonstrated in Fig. 20 by trimming it to zero birefringence with a low fluence blanket UV polarized exposure. After annealing for removal of the hydrogen, the polarization insensitive nature of the Bragg grating is unchanged. The big reduction in the waveguide birefringence with UV polarized irradiation is due to the change in the anisotropic densification at the core caused by UV irradiation. The birefringence of two waveguides irradiated with either UV S-polarization or UV P-polarization is different, indicating that the anisotropy at the core and its surrounding area is modified differently. The core along the direction of the TE mode is more compressed by the irradiation polarized along the same direction as the TE mode. It is clear that UV polarized irradiation with a wide beam could tune the core densification anisotropy by changing the compressive stress in the core and its surrounding area along the polarization direction of the irradiation through dichroic absorption. With the technique of the polarization independent Bragg grating by UV irradiation, the quality of planar waveguide Bragg gratings used as the telecommunication devices and sensors is improved [37].

5. The Impairments of Birefringence in FBGs and WBGs in Communication Systems

5.1. Simulation Model

The model to simulate system impairments of birefringence in FBGs and WBGs is shown in Fig. 21(a) that consists of a transmitter, receiver and cascaded optical add-drop multiplexers (OADMs) with each OADM consisting a FBG (or WBG) and two optical circulators, see Fig. 21(b). The FBG birefringence-induced power penalty in the network is calculated by examining the eye opening of 10 Gb/s non-return-to-zero (NRZ) signals modulated by a pseudorandom binary sequence (PRBS) with length of 2^7-1. The pulse shape of an isolated pulse in the sequence is Gaussian with 3-dB pulse duration of 60 ps.

We assume in the simulation that all gratings in Fig. 21 are apodized with a 3-dB bandwidth of 60 GHz and the grating lengths are 10 mm. By using the coupled mode theory, the grating spectrum, spectral PDL, dispersion, and spectral DGD can be calculated, see Figs. 22(a) and 22(b) where the index apodization profile of the FBG is Gaussian. The birefringence in mode effective index and the birefringence in index modulation used in the calculation of spectral PDL and DGD curves in Figs. 22(a) and 22(b) are 5×10^{-6} and

8×10^{-6} respectively. These two values were observed previously when H₂-loaded *SMF-28* fiber was exposed to CW frequency doubled Ar⁺ laser radiation through a phase mask for a few minutes [18, 38]. The birefringence in FBGs inscribed in non H₂-loaded *SMF-28* fiber by using ultrafast IR laser and the birefringence in planar waveguide can be an order of magnitude higher than these values [21], resulting in much larger spectral DGD and PDL. We assume that the FBG-based OADMs are linked with optical fibers and the polarization of the signal in the fiber links changes randomly with time due to variations of temperature, strain, vibration, etc. The polarization independent insertion losses from OADMs, optical fibers and connectors are not included in the simulation.

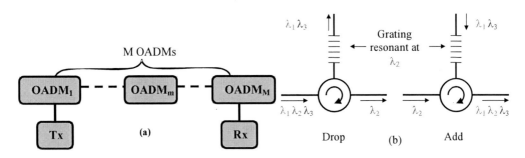

Figure 21. The simulation model: (a) the optical network consists of a transmitter, receiver and *M* FBG-based OADMs; (b) the diagram of OADM made of one FBG and one circulator.

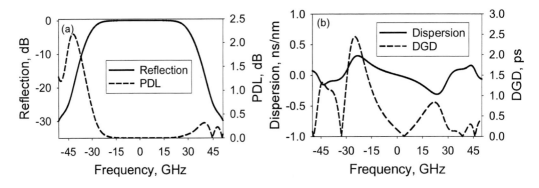

Figure 22. The properties of a FBG: (a) FBG spectrum and spectral PDL in reflection; (b) FBG dispersion and spectral DGD in reflection.

Pulse broadening due to the first order PMD can be simply expressed as $\Delta\sigma = \sigma_{out} - \sigma_{in} = \sqrt{\sigma_{in}^2 + \tau^2 \gamma (1-\gamma)} - \sigma_{in}$ [39], where σ_{in} and σ_{out} are the input and output pulse widths, τ is the DGD of the system and γ $(0 \leq \gamma \leq 1)$ is the relative power launched to the fast (or slow) principal state of polarization. In the presence of both PMD and PDL, especially when higher order PMD and PDL are considered as shown in Figs. 22(a) and 22(b), the calculation of pulse distortion is much more complicated so that it cannot be expressed by a simple equation [40, 41]. In this work, the signal distortion induced by FBGs in the presence of birefringence is obtained by examining the evolutions of the Jones vectors at all Fourier frequencies by using the following equation,

$$\vec{E}_{out}(f_n) = T(f_n)\vec{E}_{in}(f_n) \tag{5}$$

where f_n (n=1, 2, …, N) is the discrete frequency for the Fourier transform of the signal, $\vec{E}_{in}(f_n)$ and $\vec{E}_{out}(f_n)$ are the Jones vectors (electric fields) at frequency f_n at the network's input and output respectively. The polarization state of the input signal $\vec{E}_{in}(f_n)\big/\big|\vec{E}_{in}(f_n)\big|$ is set to $\begin{pmatrix}1\\1\end{pmatrix}\big/\sqrt{2}$. $T(f_n)$ in Eq. (5) is the transmission matrix at frequency f_n of the network consisting of M OADMs and fiber links and it can be written as

$$T(f_n) = T_1(\theta_1)\cdot T_1^{FBG}(f_n)\cdot T_2(\theta_2)\cdot T_2^{FBG}(f_n)\cdots T_m(\theta_m)\cdot T_m^{FBG}(f_n)\cdots T_M(\theta_M)\cdot T_M^{FBG}(f_n)$$
$$(m = 1, 2, …M) \tag{6}$$

where $T_m(\theta_m)$ represents the rotation matrix of the mth fiber link and $T_m^{FBG}(f_n)$ represents the transmission matrix of the mth FBG. They can be written as

$$T_m(\theta_m) = \begin{pmatrix} \cos(\theta_m) & \sin(\theta_m) \\ -\sin(\theta_m) & \cos(\theta_m) \end{pmatrix} \tag{7}$$

$$T_m^{FBG}(f_n) = \begin{pmatrix} r_m^{fast}(f_n) & 0 \\ 0 & r_m^{slow}(f_n) \end{pmatrix} \tag{8}$$

In Eq. (7), θ_m ($0 \le \theta_m \le \pi$) is the rotation angle and it can be randomly distributed between 0 and π to represent the random polarization variation of the signal in the fiber link. $r_m^{fast}(f_n)$ and $r_m^{slow}(f_n)$ in Eq. (8) are the reflection coefficients of the mth FBG when the signal is launched onto the fast and slow axes respectively. The output electrical field signal in time domain, $\vec{E}_{out}(t)$, can then be obtained through the inverse Fourier transform of $\vec{E}_{out}(f_n)$. The distorted NRZ output signal, $P(t) = \big|\vec{E}_{out}(t)\big|^2$, is compared to NRZ input signal and the power penalty of eye opening can be calculated.

5.2. Simulation Result and Discussion

5.2.1. Definition and Representation of FBG Birefringence-Induced Power Penalty

The power penalty of the cascaded Bragg grating filters in the network is examined by studying the eye-diagram and is compared to the result of a back-to-back connection. Fig. 23

shows the eye diagrams for the cases of a back-to-back connection (23a), a 10 OADM cascade considering grating spectral bandwidth and dispersion but no birefringence effect (23b) and a 10 OADM cascade in the presence of grating spectral bandwidth, dispersion and birefringence (23c). The power penalty resulting from the spectral bandwidth and dispersion of FBGs in an optical network has been investigated previously [42-45]. In the presence of grating birefringence, the eye openings are reduced further and are changing with time (Fig. 23c). In order to quantitatively study the system impact of FBG birefringence, the statistical characteristics of the eye-opening diagram in Fig. 23(c) needs to be examined. In this work, the power penalty induced by grating birefringence is defined as $-10\log_{10}\left(H/H_0\right)$ in the unit of dB, where H represents the eye opening in the presence of grating birefringence and H_0 is the minimum eye opening in Fig. 23b. By using this definition, the presented power penalty is induced by the birefringence only; other parameters, such as grating spectral bandwidth, dispersion, index apodization profile, and the wavelength misalignment that will be discussed later, are all included in the calculation of H_0. Due to the random variations of the polarization state of the signal, the statistical property of the power penalty is presented as a percentage value of when the power penalty is larger than a certain amount for a statistical ensemble consisting of 50,000 eye-opening values.

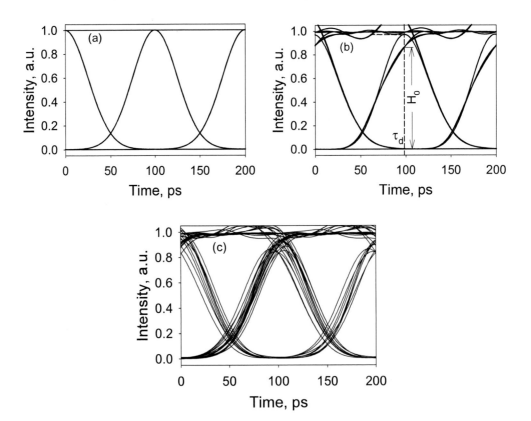

Figure 23. Eye diagrams of NRZ signal: (a) back-to-back; (b) with grating dispersion; and (c) with grating dispersion and birefringence.

5.2.2. The Induced Power Penalty of Birefringence in FBGs with Various Index Apodization Profiles

The index profiles of FBG filters used in wavelength-division multiplexing (WDM) systems are usually apodized in order to promote side lobe suppression and high channel isolation. Depending on the application, various apodization profiles are often used such as Gaussian, raised cosine, Blackman, sine, sinc, tanh, super Gaussian etc. With grating lengths of 10 mm and 3-dB bandwidths of 60 GHz, the reflection spectra and dispersion of FBGs with various index apodization profiles along with the NRZ 10Gb/s signal are shown in Figs. 24(a) and 24(b) respectively. In the presence of birefringence of the effective index of 1×10^{-5} and the birefringence in index modulation of 1.3×10^{-5} (i.e., the ratio between the birefringence in effective index and the birefringence in Δn as shown experimentally [18, 21]), the spectral PDL and DGD of FBGs with various index apodization profiles are shown in Figs. 24(c) and

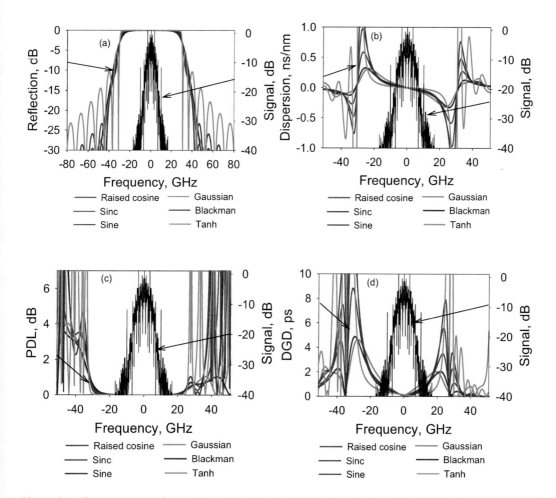

Figure 24. The responses of FBG with various index apodization profiles along with 10 Gb/s NRZ signal: (a) FBG reflection spectra; (b) FBG dispersion curves; (c) FBG spectral PDLs; and d) FBG spectral DGDs.

24(d), respectively, along with the 10 Gb/s NRZ signal. From Fig. 24(c), while within the bandwidth of the 10 Gb/s NRZ signal (-20 to 20 GHz), the spectral PDLs in FBGs with various apodization profiles are almost zero when the birefringence in effective index is $<10^{-5}$. The spectral DGD shown in Fig. 24(d) however, has a strong dependence on the index apodization profile. Among these FBGs, the Gaussian apodization function produces the highest spectral DGD within the signal bandwidth, which is consistent with the dispersion curve associated with Gaussian apodization profile in Fig. 24(b). Figs. 24(b) and 24(d) also show that the FBG with the TANH index apodization profile has the lowest dispersion and spectral DGD within the signal bandwidth. From the point of view of having lower system impairment in the presence of birefringence, TANH apodization is preferred. However, this apodization function produces higher out-of-band side lobe reflection resulting in poor channel isolation which is undesirable in WDM systems, see Fig. 24a. It is obvious that higher spectral DGD in FBGs would produce higher system power penalty. To quantify this effect, Fig. 25 presents outage probabilities for various power penalty values calculated for a network with 10 FBG-based OADMs in the presence of grating birefringence in effective index of 5×10^{-5} and birefringence in Δn of 6×10^{-5}. For the same grating birefringence values, Gaussian apodization profiles of the FBGs are more likely to induce high power penalty when compared to Raised cosine and Blackman apodization profiles. This is consistent with results shown in Fig. 24(d) where within the spectral bandwidth (-20 to 20 GHz) of the 10 Gb/s NRZ signal, the spectral DGD is highest for the grating with a Gaussian apodization. The birefringence induced power penalties of FBGs with other apodization profiles, such as sine, sinc, etc., are also examined resulting in similar penalties to that for gratings with raised cosine apodization functions.

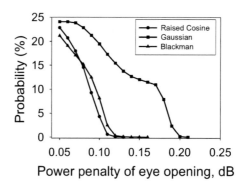

Figure 25. Outage probabilities for various power penalties for a network consisting of 10 FBG-based OADMs with various FBG apodization profiles.

The 3-dB bandwidth of the FBGs used in above simulations is 60 GHz, which is chosen for the 10 Gb/s WDM system with channel spacing of 100 GHz. In the case where FBGs with narrower bandwidths are used for dense WDM systems, the spectral profiles of PDL and DGD shown in Figs. 24c and 24d respectively will be narrower as well. As a result, the Fourier components of the signal will experience higher PDL and DGD, which induces a larger system power penalty. Similarly, for higher bit rate systems, the signal spectrum shown in Fig. 24 will be broader, the frequency components then will experience higher PDL and DGD values resulting in higher power penalty. To quantify these effects, another two sets of simulations are performed. In the first statistical ensemble, the 3-dB bandwidth of the FBG filter (with Gaussian apodization) is reduced from 60 GHz to 30 GHz (appropriate for a 50 GHz channel spacing). In the second statistical ensemble, the FBG 3-dB bandwidth is still 60 GHz but the bit rate is increased from 10 Gb/s to 20 Gb/s. Simulation results showed that in a system consisting of 10 FBGs with each of the FBG having a birefringence value of 10^{-5}, the maximum birefringence induced power penalty is 0.034 dB (based on 50,000 simulations) for the FBG 3-dB bandwidth of 60 GHz and the bit rate of 10 Gb/s. When the 3-dB bandwidth is reduced to 30 GHz and other parameters remain the same, the outage probability for power penalty of 0.1 dB is 1.4%. When the bit rate is increased to 20 Gb/s, the outage probability for power penalty of 0.1 dB is 5.1%.

5.2.3. FBG Birefringence Induced Power Penalty in the Presence of Wavelength Misalignment

During the process of FBG fabrication, due to the inscription alignment, laser power fluctuations, and the speed accuracy of the moving stage, the Bragg wavelength of the grating can shift up to ± 0.1 nm from the designed value for a 30 dB grating. In industry, FBGs that are manufactured for network applications are typically written at a wavelength that is lower than the specified wavelength and then tension tuned to the desired value by special packaging, resulting in errors of the FBG central wavelength of ± 0.05 nm. When these gratings are used in an optical network with a laser source at a designed operating wavelength, this wavelength misalignment may introduce more signal distortion in addition to distortions arising from grating spectral band width, dispersion etc [42]. In the presence of grating birefringence, Figs. 26a and 26b show the spectral PDL and spectral DGD of Gaussian apodized FBGs with 0, 0.05, and 0.1 nm wavelength misalignments along with the NRZ signal. Fig. 26 shows that with the increase of wavelength misalignment, the spectral PDL and DGD at frequencies close to the center of the signal are higher when compared to the case of zero wavelength misalignment. Obviously this increase of PDL and DGD within the signal bandwidth would introduce higher power penalty.

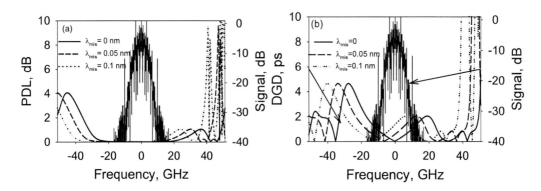

Figure 26. The spectral PDL and DGD of FBGs with wavelength misalignment of 0, 0.05 and 0.1 nm: (a) spectral PDL; (b) spectral DGD.

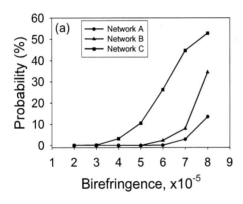 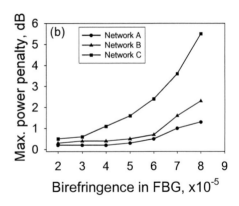

Figure 27. Birefringence induced power penalty in optical networks with various wavelength misalignment values: (a) The outage probabilities for power penalty of 0.5 dB; (b) the maximum induced power penalties based on 50,000 simulations in each statistical ensemble.

Considering the fact that the amount of wavelength misalignment of an individual FBG is usually random within a certain range, in the simulation of this subsection the wavelength misalignments of FBGs in each network are assumed to vary from zero to a maximum value. More specifically, the power penalties of three networks with 10 FBG-based OADMs with different wavelength misalignments are examined for various birefringence values. In the first network (network A), the Bragg resonances of the 10 FBGs are perfectly matched with the wavelength of the laser source (for the purpose of comparison). In the second network (network B) there are two FBGs with each of following misalignments: 0.05, 0.025, 0, -0.025 and –0.05 nm. In the third network (network C) there are two FBGs with each of the following wavelength misalignments: 0.1, 0.05, 0, -0.05, and -0.1 nm. Fig. 27(a) shows the outage probabilities for power penalty of 0.5 dB in the three networks at various birefringence values and Fig. 27(b) shows the maximum power penalties introduced in the three networks at various birefringence values (based on 50,000 simulation in each statistical ensemble). Figs. 27(a) and 27(b) clearly show that with wavelength misalignment in the networks, the birefringence induced power penalty is much higher when high birefringence is present in the FBGs. The power penalties induced by low birefringence ($<10^{-5}$) for various wavelength misalignments up to 0.1 nm are also examined and they are close to the result when there is no wavelength misalignment. The induced power penalty resulting from FBG birefringence is a combined effect of PMD and PDL. In order to examine whether PMD or PDL introduces more system impairment, the birefringence induced power penalties in two systems, either with PMD only or PDL only, are studied with the birefringence value of 8×10^{-5} in each of the FBGs in network C. It should be noted however, that the treatment PMD or PDL in isolation when considering birefringence induced power penalties by FBGs in a system is not possible in reality as the two effects are coupled together [21]. Simulation results show that the outage probability for power penalty of 0.5 dB is 20% when PMD only is present and it is 27% when PDL only is present. From these simulation results and the results shown in Fig. 27, we may conclude that: 1) PDL produces slightly higher system impairment than PMD in a system with a cascade of FBGs, and 2) the combination of PMD and PDL introduces more system impairment than the sum of those induced by PMD and PDL separately (the outage probability for power penalty of 0.5 dB is 53% when both PMD and PDL are considered with birefringence of 8×10^{-5}, see the result of network C in Fig. 27a). The later conclusion is consistent with previous studies [6, 46].

6. Conclusion

UV and ultrafast IR lasers-induced birefringence in FBGs and WBGs is characterized experimentally. The dependence of spectral PMD and PDL on the induced birefringence in gratings is analyzed by using the coupled mode theory. It is seen that the birefringence of the index modulation is between 1.4% and 1.7%, while the birefringence of the effective index is between 0.7% and 1.2% of the index modulation value for the one-side writing technique when UV laser is used. The changes of birefringence in two types of FBGs, Type I-IR and II-IR, inscribed in *SMF-28* fibers using an ultrafast-IR laser, are studied. For Type I-IR gratings, the birefringence growth during grating inscription and its decay during grating annealing at high temperature are examined and compared to those created using a UV laser. It is shown that, in Type I-IR gratings, ultrafast IR laser-induced birefringence in H_2-loaded *SMF-28* fibers is low, can be annealed at room temperature and does not show obvious dependence on the polarization of the laser beam. The IR laser-induced birefringence in Type I-IR gratings in unloaded *SMF-28* fibers is much higher than in H_2-loaded *SMF-28* fibers and exhibits strong polarization dependence on the laser beam. It has similar stability as the induced index for annealing temperatures up to 800 °C. The *P*-polarized ultrafast IR laser-beam–induced birefringence in Type II-IR gratings in non H_2-loaded fibers is low ($\sim 10^{-5}$) and it is stable during annealing. However, the birefringence of Type II-IR gratings inscribed in H_2-loaded fibers with both *P*- and *S*-polarized laser beams and in non H_2-loaded fiber with *S*-polarized laser beam is high ($\sim 10^{-4}$) and shown strong variation during annealing. The technique for compensation of the intrinsic birefringence in planar waveguide using UV laser exposure is presented.

Considering the presence of birefringence in Bragg grating based optical components in an all-optical network, the system power penalty induced by PMD and PDL is examined along with grating spectral bandwidth and dispersion by simulation. The eye diagram with spectral DGD and PDL in FBG-based OADMs is presented, and the distributions of the induced power penalties for various numbers of OADMs in the network are calculated. It is shown that when the birefringence in the Bragg grating is less than 10^{-5}, the birefringence-induced power penalty of eye opening is less than 0.1 dB with up to 30 Bragg gratings in the network. For the same birefringence values, FBGs with Gaussian apodization profile have higher spectral DGD resulting in higher induced power penalties. The presence of wavelength misalignment in optical networks significantly increases the birefringence-induced power penalty, especially when high FBG birefringence is present.

References

[1] K. O. Hill, Y. Fujii, D. C. Johnson, and B. S. Kawasaki. (1978). Photosensitivity in optical fiber waveguides: Application to reflection filter fabrication. *Applied Physics Letters*, **32**, 647–649.

[2] H. Bulow. (1998). System outage probability due to first- and second-order PMD. *IEEE Photonics Technology letters*, **10**, 696-698.

[3] J. Cameron, L. Chen and X. Bao. (2000). Impact of chromatic dispersion on the system Limitation due to polarization mode dispersion. *IEEE Photonics Technology Letters*, **12**, 47-49.

[4] N. Gisin, B. Huttner. (1997). Combined effects of polarization mode dispersion and polarization dependent losses in optical fibers. *Optics Communications*, **142**, 119-125.

[5] A. El Amari, N. Gisin, B. Perny, et al. (1998). Statistical prediction and experimental verification of concatenation of fiber optic components with polarization dependent loss. *Journal of Lightwave Technology*, **16**, 332-339.

[6] B. Huttner, C. Geiser, and N. Gisin. (2000). Polarization-induced distortion in optical fiber networks with polarization-mode dispersion and polarization-dependent losses. *IEEE Journal of Selected Topics in Quantum Electronics*, **6**, 317-329.

[7] K. O. Hill, G. Meltz. (1997). Fiber Bragg grating technology fundamentals and overview. *Journal of Lightwave Technology*, **15**, 1263-1276.

[8] S. J. Mihailov, C. W. Smelser, D. Grobnic, R. B. Walker, P. Lu, H. Ding, and J. Unruh. (2004). Bragg gratings written in all-SiO_2 and Ge-doped core fibers with 800-nm femtosecond radiation and a phase mask. *Journal of Lightwave Technology*, **22**, 94-100.

[9] P. J. Lemaire, R. M. Atkins, V. Mizrahi, and W. A. Reed. (1993). High pressure H_2 loading as a technique for achieving ultrahigh UV photosensitivity and thermal sensitivity in GeO_2 doped optical fibers. *Electronics Letters*, **29**, 1191-1193.

[10] A. M. Vengsarkar, Q. Zhong, D. Inniss, W. A. Reed, P. J. Lemaire, and S. G. Kosinski. (1994). Birefringence reduction in side-written photoinduced fiber devices by a dual-exposure method. *Optics Letters*, **19**, 1260-1262.

[11] H. Renner, D. Johlen, and E. Brinkmeyer. (2000). Modal field deformation and transition losses in UV side-written optical fibers. *Applied Optics*, **39**, 933-940.

[12] T. Erdogan and V. Mizrahi. (1994). Characterization of UV-induced birefringence in photosensitive Ge-doped silica optical fibers. *Journal of the Optical Society of America B*, **11**, 2100-2105.

[13] F. Ouellette, D. Gagnon, and M. Poirier. (1991). Permanent photoinduced birefringence in a Ge-doped fiber. *Applied Physics Letters*, **58**, 1813-1815.

[14] K. Dossou, S. LaRochelle, and M. Fontaine. (2002). Numerical analysis of the contribution of the transverse asymmetry in the photo-induced index change profile to the birefringence of optical fiber. *Journal of Lightwave Technology*, **20**, 1463-1470.

[15] E. Simova, P. Berini, and C. P. Grover, (1999). Characterization of chromatic dispersion and polarization sensitivity in fiber gratings. *IEEE Transactions on Instrumentation and Measurement*, **48**, 939-943.

[16] Y. Zhu, E. Simova, P. Berini, and C. P. Grover. (2000). A Comparison of wavelength dependent polarization dependent loss measurements in fiber gratings. *IEEE Transactions on Instrumentation and Measurement*, **49**, 1231-1239.

[17] B. L. Bachim and T. K. Gaylord. (2003). Polarization-dependent loss and birefringence in long-period fiber gratings. *Applied Optics*, **42**, 6816-6823.

[18] P. Lu, D. S. Waddy, S. J. Mihailov, and H. Ding. (2005). Characterization of the growths of UV-Induced birefringence in effective mode index and index modulation in fiber Bragg gratings. *IEEE Photonics Technology Letters*, **17**, 2337-2339.

[19] J. Canning, H.J. Deyerl, H.R. Sørensen, and M. Kristensen. (2005). Annealing of UV-induced birefringence in hydrogen loaded germanosilicate fibres. *Proceeding of Bragg Gratings, Poling & Photosensitivity*, Star City, Sydney, Australia.

[20] J. Canning, H. J. Deyerl, H. R. Sørensen, and M. Kristensen. (2005). Ultraviolet-induced birefringence in hydrogen-loaded optical fiber. *Journal of Applied Physics*, **97**, paper 053104.

[21] P. Lu, D. Grobnic, and S. J. Mihailov. (2007). Characterization of the birefringence in fiber Bragg gratings fabricated with an ultrafast-infrared laser. *Journal of Lightwave Technology*, **25**, 779-786.

[22] C. W. Smelser, S. J. Mihailov, and D. Grobnic. (2005). Formation of type I-IR and type II-IR gratings with an ultrafast IR laser and a phase mask. *Optics Express*, **13**, 5377-5386.

[23] D. Grobnic, C. W. Smelser, S. J. Mihailov, R. B. Walker, and P. Lu. (2004). Fiber Bragg gratings with suppressed cladding modes made in SMF-28 with a femtosecond IR laser and a phase mask. *IEEE Photonics Technology Letters*, **16**, 1864-1866.

[24] A. Martinez, M. Dubov, I. Khrushchev and I. Bennion. (2004). Direct writing of fibre Bragg gratings by femtosecond laser. *Electronics Letters*, **40**, 1170-1172.

[25] Y. Lai, A. Martinez, I. Khrushchev, and I. Bennion. (2006). Distributed Bragg reflector fiber laser fabricated by femtosecond laser inscription. *Optics Letters*, **31**, 1672-1674.

[26] A. Martinez, M. Dubov, I. Khrushchev, and I. Bennion, (2006). Photoinduced modifications in fiber gratings inscribed directly by infrared femtosecond irradiation. *IEEE Photonics Technology Letters*, **18**, 2266-2268.

[27] C. W. Smelser, S. J. Mihailov, D. Grobnic, P. Lu, R. B. Walker, H. Ding, and X.Dai. (2004). Multiple-beam interference patterns in optical fiber generated with ultrafast pulses and a phase mask. *Optics Letters*, **29**, 1458-1460.

[28] E. Bricchi, B. G. Klappauf, and P. G. Kazansky. (2004). Form birefringence and negative index change created by femtosecond direct writing in transparent materials. *Optics Letters*, **29**, 119-121.

[29] M. Huang. (2003). Thermal stress in optical waveguides. *Optics Letters*, **28**, 2327-2329.

[30] H. Takahashi, Y.Hibino, Y.Ohmori, and M.Kawachi. (1993). Polarization-insensitive arrayed-waveguide wavelength multiplexer with birefringence compensation film, *IEEE Photonics Technology Letters*, **5**, 707-708.

[31] M. Okuno, A. Sugita, K.Jinguji, and M. Kawachi. (1994). Birefringence control of silica waveguides on Si and its application to a polarization-beam splitter/switch. *Journal of Lightwave Technology*, **12**, 625-633.

[32] S. Suzuki,Y.Inoue, and Y.Ohmori. (1994). Polarization-insensitive arrayed-waveguide grating multiplexer with SiO2-on SiO2 structure. *Electronics Letters*, **30** 642-643.

[33] J. Albert, F.Bilodeau, D.C.Johnson, K.O. Hill, S.J. Mihailov, D. Stryckman, T. Kitagawa and Y. Hibino. (1998). Polarization-independent strong Bragg gratings in planar lightwave circuits. *Electronics Letters*, **34**, 485-486.

[34] X. Dai, S. J. Mihailov, C. L. Callender, R. B. Walker, C. Blanchetière, J. Jiang. (2005). Measurement and control of birefringence and dimension of ridge waveguides with Bragg grating and ultraviolet irradiation. *Optical Engineering*, **44**, 124602.

[35] J. Canning, M. Aslund, A. Ankiewicz, M. Dainese, H. Fernando, J. K. Sahu,and L. Wosinski. (2000). Birefringence control in plasma- enhanced chemical vapor deposition planar waveguide by ultraviolet irradiation. *Applied Optics*, **39**, 4296-4299.

[36] X. Dai, S. J. Mihailov, C. Blanchetière, C. L. Callender, R.B.Walker. (2005). High birefringence control and polarization insensitive Bragg grating fabricated in PECVD planar waveguide with UV polarized irradiation. *Optics Communications,* **248**, 123-130.

[37] X. Dai, S. J. Mihailov, C. L. Callender, C. Blanchetiere and R. B. Walker. (2006). Ridge-waveguide-based polarization insensitive Bragg grating refractometer. *Measurement Science and Technology*, **17**, 1752-1756.

[38] P. Lu, S. J. Mihailov, D. Grobnic, and R. B. Walker. (2006). Comparison of the induced birefringence in fiber Bragg gratings fabricated with ultrafast-IR and CW-UV lasers. *Proc. of ECOC 2006*, paper Th.3.3.5.

[39] C. D. Poole, R. W. Tkach, A. R. Chraplyvy, and D. A. Fishman.(1991). Fading in lightwave systems due to polarization-mode dispersion. *IEEE Photonics Technology Letters*, **3**, 68-70.

[40] L. Guo, Y. Zhoiu and Z. Fang. (2003). Pulse broadening in optical fiber with polarization mode dispersion and polarization dependent loss. *Optics Communications,* **227**, 83-87.

[41] M. Wang, T. Li, and S. Jian. (2003). Analytical theory of pulse broadening due to polarization mode dispersion and polarization dependent loss. *Optics Communications.* **223**, 75-80.

[42] N. N. Khrais, A. F. Elrefaie, R. E. Wagner, and S. Ahmed. (1996). Performance of cascaded misaligned optical (De)multiplexers in multiwavelength optical networks. *IEEE Photonics Technology Letters*, **8**, 1073-1075.

[43] B. J. Eggleton, G. Lenz, N. Litchinitser, D. B. Patterson, and R. E. Slusher. (1997). Implications of fiber Grating dispersion for WDM communication systems. *IEEE Photonics Technology Letters*, **9**, 1403-1405.

[44] G. Nykolak, B. J. Eggleton, G. Lenz, and T. A. Strasser. (1998). Dispersion penalty measurements of narrow fiber Bragg gratings at 10 Gb/s. *IEEE Photonics Technology Letters*, **10**, 1319-1321.

[45] M. Kuznetsov, N. M. Froberg, S. R. Henion, and K. A. Rauschenbach. (1999). Power penalty for optical signals due to dispersion slope in WDM filter cascades. *IEEE Photonics Technology Letters*, **11**, 1411-1413.

[46] P. Lu, L. Chen, and X. Bao. (2002). System outage probability due to the combined effect of PMD and PDL. *Journal of Lightwave Technology.* **20**, 1805-1808.

In: Atomic, Molecular and Optical Physics...
Editor: L.T. Chen, pp. 235-260

ISBN: 978-1-60456-907-0
© 2009 Nova Science Publishers, Inc.

Chapter 6

FIBER BRAGG GRATINGS AND THEIR APPLICATIONS AS TEMPERATURE AND HUMIDITY SENSORS

Qiying Chen[*] and Ping Lu

Department of Physics and Physical Oceanography,
Memorial University of Newfoundland
St. John's, Newfoundland A1B 3X7, Canada

Abstract

As an important waveguiding medium, optical fiber plays significant roles in optical communications, optoelectronics, and sensors. A new type of microstructure inscribed in the optical fibers, i.e., fiber Bragg gratings (FBGs), has received considerable attention in recent years. A FBG is a type of distributed Bragg reflector constructed in a short segment of optical fiber that reflects specific wavelengths of light and transmits all the other components. In this chapter, optical properties of FBGs will be reviewed first with the underlying physical mechanisms. Different techniques to fabricate FBGs will be illustrated with the comparison of their advantages and drawbacks. For their important sensing applications, FBGs as temperature and humidity sensors will be discussed. The FBG sensors exceed other conventional electric sensors in many aspects, for instance, immunity to electromagnetic interference, compact size, light weight, flexibility, stability, high temperature tolerance, and resistive to harsh environment. A novel approach to realize separate temperature or humidity measurement, and their simultaneous measurement will be demonstrated by the use of FBGs coated with different polymers. The polymer-coated FBGs indicate linear shifts in the Bragg resonance wavelengths of the gratings with the temperature changes. A polyimide-coated FBG is sensitive to humidity due to the unique hygroscopic properties of polyimide while an acrylate-coated FBG shows insensitivity to humidity. The experimental results are in good agreement with the theoretical analysis.

Keywords: Fiber Bragg grating (FBG), humidity sensing, temperature sensing, polymers.

[*] E-mail address: qiyingc@mun.ca; Phone: 1 709 737-8878; Fax: 1 709 737-8739. Corresponding author.

1. Introduction

Guided wave optics is the forefront of research in optics nowadays. For modern applications, such as optical telecommunications and optical sensors, optical waveguides are the key components in which generation, modulation, propagation, and detection of light are governed by the principles of guided wave optics. An optical waveguide is a dielectric structure that transports energy at a wavelength in the infrared or visible portions of the electromagnetic spectrum [1]. Common waveguides used for optical communications are highly flexible fibers composed of nearly transparent dielectric materials. Charles Kao first suggested the possibility that low-loss optical fibers could be competitive with coaxial cable and metal waveguides for telecommunications applications in 1966 [2]. The commercial applications of optical fibers were not possible until 1970 when Corning Glass Work discovered an optical fiber with a loss less than the benchmark level of 10 dB/km [3,4].

With the increasing interests in the studies of all-fiber systems, fiber Bragg gratings (FBGs) have been applied in many photonic devices. A FBG is a type of distributed Bragg reflector constructed in a short segment of optical fiber that reflects specific wavelengths of light and transmits all the other components. A FBG can also be regarded as a fiber device with a periodic variation of the refractive index of the fiber core along the length of the fiber. The physical mechanism of inscribing the Bragg grating in a fiber is the photosensitivity of the fiber core. When a germanium-doped (GeO$_2$-doped) fiber is exposed to a high-intensity ultraviolet light, the refractive index of the fiber core is permanently changed [5]. The amount of change in the refractive index varies from 10^{-5} to 10^{-3}. The phenomenon of the change in the refractive index of the fiber upon light irradiation is usually referred as photosensitivity. In 1978, Hill *et al.* first reported the photosensitivity phenomenon in optical fibers achieved by the interference between counter-propagating waves inside the fiber core [6]. A decade later, Meltz *et al.* obtained the first FBG which was imprinted in Ge-doped silica single mode fiber by transverse coherent 244 nm UV beams from a tunable excimer-pumped dye laser with a frequency-doubled crystal [7]. Since then, FBGs have been applied in optical communications and optical fiber sensor networks owing to their unique advantages and versatility as in-fiber devices. FBGs are now widely used in a variety of lightwave communication applications such as fiber laser [8, 9], fiber amplifier [10], fiber Bragg filter [11,12], wavelength division multiplexers/demultiplexers [13,14], and dispersion compensation [15]. On the other hand, FBGs have been used as excellent sensor elements for measuring many environmental parameters, including temperature, strain, bending, refractive index, pressure, and flow rate [16].

In the following part of this chapter, basic optical properties of FBGs will be briefly reviewed first together with their fabrication techniques followed by an overview of FBG sensors. The third part of this chapter will discuss FBG sensors for temperature and humidity measurements.

2. Properties of Fiber Bragg Gratings

2.1. Optical Properties

The fundamental properties of optical fibers and FBGs inscribed on the fibers have been discussed in several books and monographs [5, 17-21]. In this section, a brief review on the optical properties of FBGs will be given and details can be found from these references.

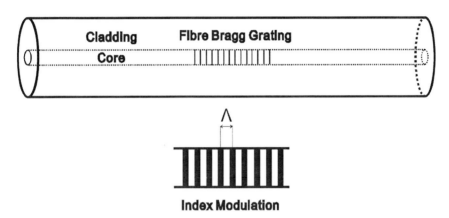

Figure 1. Illustration of a uniform fiber Bragg grating.

For a FBG consisting of a periodic modulation of the refractive index in the core of an optical fiber, the phase fronts are perpendicular to the fiber longitudinal axis and the grating planes are of a constant period, as illustrated in Fig. 1. Light propagating along the fiber core will interact with each grating plane, in which the Bragg condition is used for the discussion of the light propagation,

$$2\Lambda \sin \theta = n\lambda, \tag{1}$$

where Λ is the spacing between the grating planes, θ is the angle between the incident light and the scattering planes, λ is the light wavelength, and n is an integer. In the case when the Bragg condition is not satisfied, the light reflected from each of the subsequent planes becomes progressively out of phase and will eventually disappear. Only when the Bragg condition is met, the contributions of reflected light from each grating plane add constructively in the backward direction to form a back-reflected peak with a center wavelength defined by the grating parameters, i.e., the Bragg wavelength.

The Bragg grating resonance condition is the requirement to satisfy both energy and momentum conservation, in which the energy conservation ($\hbar\omega_f = \hbar\omega_i$) requests that the frequency of the reflected radiation should be the same as that of the incident radiation while the momentum conservation implies that the incident wave vector, $\vec{k}_i$, plus the grating wave vector, $\vec{K}$, equals the wave vector of the scattered radiation $\vec{k}_f$,

$$\vec{k}_i + \vec{K} = \vec{k}_f \tag{2}$$

where the grating wave vector $\vec{K}$ has a direction normal to the grating plane with a magnitude $2\pi/\Lambda$.

The diffracted wave vector is equal in magnitude with opposite direction with regard to the incident wave vector. The momentum conservation condition can be simplified as the first –order Bragg condition

$$\lambda_B = 2n_{eff}\Lambda \tag{3}$$

where λ_B is the Bragg grating wavelength, which is the center wavelength of the input light in the free space that will be back-reflected from the Bragg grating. n_{eff} is the effective refractive index of the fiber core at the free-space centre wavelength.

For a uniform FBG inscribed in the core of an optical fiber with a refractive index n_0. The refractive index profile is [22]

$$n(x) = n_0 + \Delta n \cos\left(\frac{2\pi x}{\Lambda}\right) \tag{4}$$

where Δn and x are the amplitude of the induced refractive index perturbation and the distance along the fiber longitudinal axis, respectively. The reflectivity of a grating with constant modulation amplitude and period can be described as [22]

$$R(l,\lambda) = \frac{\Omega^2 \sinh^2(sl)}{\Delta k^2 \sinh^2(sl) + s^2 \cosh^2(sl)} \tag{5}$$

where $R(l,\lambda)$ is the reflectivity which is a function of the grating length l and wavelength λ, Ω is the coupling coefficient, $\Delta k = k - \pi/\lambda$ is the detuning wave vector, $k = 2\pi n_0/\lambda$ is the propagation constant, and $s = \sqrt{\Omega^2 - \Delta k^2}$. The coupling coefficient Ω for the sinusoidal variation of index perturbation along the fiber axis has been found as

$$\Omega = \frac{\pi \Delta n \eta(V)}{\lambda} \tag{6}$$

where $\eta(V)$ is a function of the normalized frequency V of the fiber that represents the fraction of the fiber mode power contained in the core, $\eta(V) \approx 1 - V^{-2}$. The normalized frequency V can be expressed as [22]

$$V = \left(\frac{2\pi}{\lambda}\right) a \left(n_{co}^2 - n_{cl}^2\right)^{1/2} \tag{7}$$

where a is the core radius, n_{co} and n_{cl} are the refractive indices of the core and cladding, respectively.

At the Bragg grating center wavelength, there is no wave vector detuning and Δk equals zero, the expression for the reflectivity becomes

$$R(l,\lambda) = \tanh^2(\Omega l). \tag{8}$$

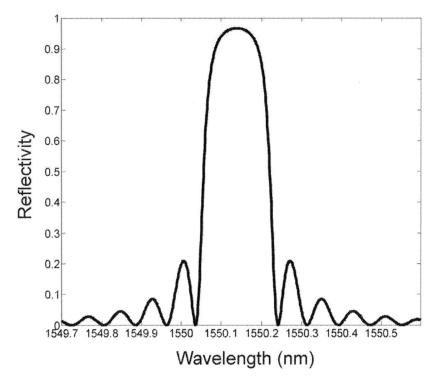

Figure 2. Simulation of the reflection spectrum of a FBG as a function of wavelength detuning.

The reflectivity increases with the change in the induced refractive index. It can also be found that the reflectivity increases with the increase in the length of the grating. Figure 2 gives a simulated reflection spectrum as a function of the wavelength detuning, in which the side lobes of the resonance are due to multiple reflections to and from opposite ends of the grating region.

A Bragg grating written in a highly photosensitive fiber exhibits a pronounced transmission feature on the short-wavelength side of the Bragg peak (Fig. 3). This feature is only observable in the transmission spectrum and only the main peak is visible in the reflection spectrum, which is due to the light leaving the fiber from the side. Radiation-mode coupling as a loss mechanism on core-mode transmission has been suggested to explain this phenomenon [23, 24]. For the cylindrical cladding-air interface, the transmission spectrum of the Bragg grating consists of multiple sharp peaks that modulate the radiation-mode coupling. The light energy in the grating will be coupled to other modes at shorter wavelengths: some will be reflected or absorbed, and the others will be radiated away from the fiber. These interactions are seen as a series of many transmission dips in the spectrum at wavelengths that are shorter than the Bragg wavelength.

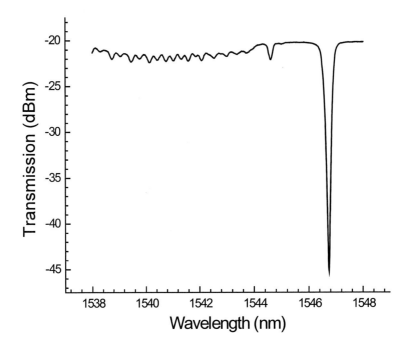

Figure 3. Transmission spectrum of a typical fiber Bragg grating exhibiting loss to radiation modes at the short-wavelength side.

2.2. Photosensitivity and Fabrication Techniques

As discussed in the previous section, photosensitivity in an optical fiber refers to a permanent change in the index of refraction of the fiber core when exposed to light of specific wavelength and intensity. Photosensitivity in optical fibers has significant importance in the fabrication of FBGs for applications in telecommunications and sensors.

After the first observation of weak changes ($\sim 10^{-6}$) in the index of refraction in germanosilica fibers by Hill *et al.* [6] in 1978, Meltz *et al.* showed that a strong index of refraction change occurred when a germanium-doped fiber was exposed to UV light close to the absorption peak of a germanium-related defect at a wavelength range of 240–250 nm [7]. Photosensitivity of optical fibers can be regarded as a measure of the amount of refractive index change in a fiber core. It is desirable to fabricate photoinduced devices in standard optical fibers for compatibility with existing systems, however, the standard single mode telecommunication fibers with a doping concentration of germanium around 3%, typically display weak index changes of 10^{-5}, far below the expected value on the order of 10^{-4}. Hydrogen loading (hydrogenation) [25-27], flame brushing [28], boron co-doping [29,30], and short wavelength light source [31,32] have been used for enhancing the photosensitivity in silica optical fibers.

Bragg gratings have been written in many types of optical fibers using various methods. However, the mechanism of index change is not fully understood. Several models have been proposed for these photoinduced changes in the refractive index, for instance, the color center model [33,34], the dipole model [35], the compaction model [36], the stress-relief model [37-39]. The general idea from these theories is that the germanium–oxygen vacancy defects, Ge–Si or Ge–Ge, are responsible for the photoinduced index changes.

Direct optical inscription of high quality gratings into the cores of optical fibers has been actively pursued by many research laboratories and various techniques have been reported. Different FBG fabrication techniques can be classified as internal inscription and external inscription techniques. Adopted primarily during the earlier years [6], internally writing technique uses relatively simple experimental setup in which the standing wave inside the optical fiber photoimprints a Bragg grating with the same pattern as the standing wave. In recent years, the internal inscription technique has been superseded by the external inscription technique due to the intrinsic limitation of the internally written gratings. The reported externally written fabrication techniques can be categorized into four groups, i.e., interferometric technique, point-by-point exposure technique, phase mask technique, and most recently ultrashort pulse lasers technique.

The interferometric fabrication technique, the first external writing technique to inscribe Bragg gratings in photosensitive fibers, was demonstrated by Meltz *et al.* [7], in which two beams split from one UV light recombined to form an interference pattern and expose a photosensitive fiber inducing a refractive index modulation in the fiber core. Either amplitude-splitting or the wave-front-splitting interferometer has been used in the fabrication of FBGs.

The point-by-point exposure technique [40], which produces grating fringe one-by-one, is accomplished through UV light irradiation over consecutive steps. After each irradiation, either the fiber or the laser beam itself is translated in the axial direction of the fiber by a computer controlled translation stage. The advantage of this technique is that a laser beam of high intensity is obtained due to the focus of the laser beam with a flexibility to alter the Bragg grating parameters. The drawback of the point-by-point technique is that it is a time-consuming process. During the slow fabrication process, errors in the grating spacing due to thermal effects and/or small variations in the fiber's strain can occur, which limits the gratings to a very short length.

One of the most effective methods for inscribing Bragg gratings in photosensitive fiber is the phase-mask technique [41, 42]. This method employs a diffractive optical element, i.e., phase mask, to spatially modulate the UV writing beam. The phase-mask grating has a one-dimension surface-relief structure fabricated in a high-quality fused silica block transparent to the UV writing beam. The profile of the periodic gratings is chosen such that, when an UV beam is incident on the phase mask, the zero-order diffracted beam is suppressed to less than a few percent of the transmitted power. In addition, the diffracted plus and minus first orders are maximized to contain, typically, more than 35% of the transmitted power. A near-field fringe pattern is produced by the interference of the plus and minus first-order diffracted beams. The period of the fringes is one-half that of the mask. The interference pattern photoimprints a refractive-index modulation in the core of a photosensitive optical fiber placed in contact with or in close proximity immediately behind the phase mask. A cylindrical lens may be used to focus the fringe pattern along the fiber core. The phase mask greatly reduces the complexity of the fiber grating fabrication system. The simplicity of using only one phase mask provides a robust and an inherently stable method for reproducing FBGs.

Recently, with the development of ultrafast laser technology, change of refractive index in transparent dielectrics with ultrashort laser pulses for photonic devices has been an interesting topic. Some published papers reported the use of femtosecond lasers to fabricate FBGs in commercial, non-photosensitized, and unhydrogenated fibers. These reported techniques include FBG inscription with standard or special phase masks [43,44] and direct point-by-point writing [45]. The wavelengths in the near-infrared region around 800 nm and ultraviolet (UV) around 248 nm have been used for FBG fabrication [45,46]. On the other hand, femtosecond lasers have also been demonstrated as a powerful tool in trimming FBGs to realize continuously tuning of the grating performance [47]. The ultrafast laser enhancement of photosensitivity response and modification of anisotropic index profile in silica fiber is a powerful technique to precise control of the performance of FBG devices for applications in optical filtering and polarization mode dispersion management.

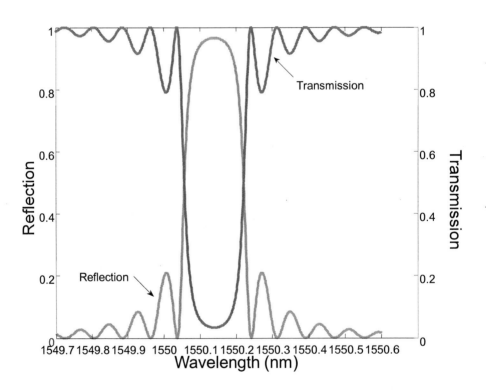

Figure 4. Simulated spectral response of a typical Bragg reflector with a uniform period.

According to the different grating pitch or tilt, fiber Bragg gratings can be classified into the common Bragg reflector, the blazed Bragg grating, and the chirped Bragg grating. The simplest and most frequently used FBG is the Bragg reflector, which has a constant pitch. A typical spectral response of a uniform period is shown in Fig. 4. The Bragg reflector can function as a narrowband transmission or reflection filter and be used in sensor applications

[48-51]. The blazed grating has phase fronts tilted with respect to the fiber axis. The applications of blazed gratings are in mode conversion [52] and sensing field [53]. The chirped grating has an periodic pitch, displaying a monotonous increase in the spacing between grating planes. The applications of the chirped Bragg gratings are for dispersion compensation in high-bit-rate transmission systems [54, 55] and sensing elements [56-59].

3. Fiber Bragg Grating Sensors

Sensors based on FBGs have attracted considerable attention since the early stage of the discovery. FBG sensors exceed other conventional electric sensors in many aspects, for instance, immunity to electromagnetic interference, compact size, light-weight, flexibility, stability, high temperature tolerance, and resistance to harsh environment. Additional advantages of FBG sensors include very low insertion loss, narrowband wavelength reflection, linearity in response over many orders of magnitude and compatibility with the existing fiber optics system, especially their absolute wavelength-encoding of measurand information, making FBG sensors interrupt immune [60].

FBG sensors can measure many physical parameters, in which strain [61-64] and temperature [65, 66] measurements are the major fields of interest. Meltz and Morey propounded that the shift in the Bragg grating wavelength was mainly due to strain and temperature changes [67]. The strain response is induced due to both the fractional change in a grating period due to the physical elongation of the optical fiber and the change in fiber index due to photo-elastic effects. The thermal response is induced due to both the inherent thermal expansion of the fiber material and the temperature dependence of the refractive index. It is apparent that any shift in the Bragg wavelength is the sum of the strain and temperature factors. The sensing measurements of other measurands can be realized by transforming to the strain and temperature factors. FBG sensors used to measure humidity [68, 69], vibration [70], pressure [71-73], and refractive index [74-77], and salinity [78] have also been reported recently.

Various kinds of methods for temperature measurement have been developed by using optical fibers, which include blackbody radiation and pyrometry, absorption, intrinsic scattering, various forms of interferometry, and fluorescence based techniques [79], in addition to the newly emerged FBG techniques. Excellent reviews of optical fiber-based temperature sensing techniques can be found in Refs. [80,81]. Current research on optical fiber temperature sensing is focused on improving accuracy and/or the measurement range, and also reducing the cost. One particularly important topic is on identifying approaches to realize simultaneous measurement of multiple parameters, i.e., temperature and other environmental parameters. In many practical applications, it is required to monitor the temperature in order to achieve temperature compensation while measuring the environmental parameters, for example, flow rate, pressure, refractive index, and salinity.

Humidity is another important physical parameter for environmental control, industrial applications, and daily life. Lee *et al.* reviewed different types of optical humidity sensors [82]. One type of fiber-optic humidity sensors responses to the colorimetric interaction of materials immobilized on the surface of the fiber core or its cladding in the humidity-sensing section, in which the transmitted optical power changes as a function of the relative humidity due to the humidity-induced changes in the refractive indices of the materials. Kharaz *et al.*

reported an optical fiber humidity sensing system based on the colorimetric interaction of cobalt chloride with water molecules [83]. The use of a hydrophilic gel (agarose) deposited on the thinner zone of a biconically tapered single-mode optical fiber to achieve humidity measurement has been reported [84]. Gupta *et al.* discussed a fiber-optic humidity sensor based on the moisture-dependent absorption of light by using a phenol red doped polymethylmethacrylate (PMMA) film over a small portion of the core of the plastic clad silica fiber with a low refractive index [85]. Yeo *et al.* reported the use of a FBG together with a moisture sensitive polymer coating for humidity sensing [69]. Some polymer materials, such as polyimide, undergo volume expansion as a result of humidity absorption and desorption [86]. Once these polymers are coated on FBGs, the strain effects occur due to the volume changes of these coatings. Mrotek *et al.* discussed the diffusion of moisture through optical fiber coatings [87]. Other optical techniques have also been reported for humidity sensors, for example, changes in the incident light due to absorption [88], cooled mirror principles [89], and photo-acoustic technique [90]. Although optical sensors possess an advantage of high sensitivity, their drawbacks include hysteresis, large size, high power consumption for some optical fiber sensors other than in the forms of optical fibers, and especially missing of temperature compensation. The superiority of optical humidity sensors in tracking low moisture level has been noted [82].

As discussed above, FBGs respond to changes in both strain and temperature. It is necessary to discriminate these effects in order to reveal each physical parameter and various methods have been proposed. A practical approach is to use a reference grating or grating pairs [91,92]. The reference grating or grating pairs, which is isolated from one parameter, e.g., strain, is placed near the sensor FBG. The reference grating can be on the same fiber as the sensor FBG [93]. Another method is to use two FBGs with a large difference in their Bragg wavelengths, which show different responses to the same measurands [94]. FBGs written on fibers of different diameters have also been proposed, which give different strain responses, while the temperature responses are the same [95,96]. A sensing head for simultaneous measurement of strain and temperature is demonstrated based on two Bragg gratings arranged in a twisted configuration [97]. By writing FBG with close wavelengths in undoped and boron doped fibers, different temperature sensitivities are obtained while the strain sensitivities remain the same [98]. In these reported techniques, it is necessary to use special fibers (specialty fibers with different doping elements, microstructured fibers, and photonic crystal fibers), complicated configurations (external lasers), or special spectroscopic techniques (fluorescence and interferometry) in order to distinguish temperature and strain, which result in bulky sensor systems targeting for the sole purpose of simultaneous measurement of temperature and strain only. Lu *et al.* reported an approach to resolve the cross-sensitivity between temperature and strain of FBGs [99], in which acrylate and polyimide polymers were used as the coating materials for different FBGs to achieve simultaneous measurement of axial strain and temperature. Since the standard FBG fabrication technique needs to strip the protective plastic coating off the fiber before FBG inscription and recoat a polymeric layer afterwards to protect the grating, the coating of FBGs with different polymeric materials proposed does not complicate the procedures or add extra cost in the FBG fabrication, which can be easily realized by fiber recoaters or dip coating technique. Furthermore, without additional optical devices or spectroscopic techniques, this approach provides one-fiber solution for multiple applications of FBGs in addition to resolve the cross-sensitivity of axial strain and temperature.

A large number of FBG sensors may be integrated at different locations along a single optical fiber or several fibers to form a quasi-distributed sensor [100-102]. The advantage of such kind of FBG array is that each FBG has a unique Bragg wavelength and can be individually interrogated. These FBG arrays can be fabricated with an arbitrary linear distance between the gratings, while the wavelength separation between the adjacent Bragg wavelengths is determined only by the wavelength shift of the neighboring higher and lower Bragg wavelengths.

Field applications of FBG sensors have also been reported [103-107]. FBG sensors possess a number of advantages for applications in spacecraft with embedded sensors that monitor the performance of reinforced carbon fiber composites as well as advanced testing of gas turbine engines [103]. There are numerous applications of FBG sensors for structural health monitoring in civil engineering, including monitoring of bridges [104], crack detection [105], and power transmission lines [106]. The FBG sensors can also be used in harsh environments [107].

4. Fiber Bragg Grating Temperature and Humidity Sensors

4.1. Theory

4.1.1. Fiber Bragg Grating Temperature Sensitivity

The Bragg grating resonance wavelength, λ_B, which is the center wavelength of light back reflected from a Bragg grating, depends on the effective index of refraction of the core (n_{eff}) and the periodicity of the grating (Λ) through the relation $\lambda_B = 2n_{eff}\Lambda$. The effective index of refraction, as well as the periodic spacing between the grating planes, will be affected by changes in temperature and strain. The shift in the Bragg grating wavelength $\Delta\lambda_B$ due to temperature and strain changes is given by [21]

$$\Delta\lambda_B = 2\left(\Lambda\frac{\partial n}{\partial T} + n\frac{\partial\Lambda}{\partial T}\right)\Delta T + 2\left(\Lambda\frac{\partial n}{\partial l} + n\frac{\partial\Lambda}{\partial l}\right)\Delta l. \tag{9}$$

The first term in Eqn. (9) represents the temperature effect on an optical fiber. The changes in the grating spacing and the index of refraction caused by thermal expansion result in a shift in the Bragg wavelength. This fractional wavelength shift for a temperature change ΔT may be written as [108]

$$\Delta\lambda_{B,T} = \lambda_B(\alpha + \zeta)\Delta T, \tag{10}$$

where $\alpha = (1/\Lambda)(\partial\Lambda/\partial T)$ is the thermal expansion coefficient of the fiber ($\sim 0.55\times10^{-6}$ °C^{-1} for silica). The quantity $\zeta = (1/n)(\partial n/\partial T)$ represents the thermo-optic coefficient, which is 8.6×10^{-6} °C^{-1} for a germanium doped silica-core fiber. Clearly, the index change is by far the dominant effect. From Eqn. (9), the expected temperature sensitivity of the FBG at 1550 nm is 0.0142 nm/°C.

4.1.2. Temperature and Humidity Sensitivity of Polyimide-coated Bragg Gratings

Using equation $\lambda_B = 2n_{eff}\Lambda$, the shift in the Bragg wavelength due to thermal expansion changes in the grating spacing and changes in the index of refraction can also be expressed as [108]

$$\frac{\Delta\lambda_B}{\lambda_B} = S_{RH}\Delta RH + S_T\Delta T = \left[\beta_{cf} - \hat{P}_e\left(\beta_{cf} - \beta_f\right)\right]\Delta RH + \left[\alpha_{cf} - \hat{P}_e\left(\alpha_{cf} - \alpha_f\right) + \zeta\right]\Delta T$$

(11)

where S_{RH} and S_T are the sensitivities to relative humidity and temperature, respectively. ΔRH and ΔT are the changes in the relative humidity and temperature accordingly. β_i is the hygroscopic longitudinal expansion coefficient, which is zero for a bare fiber, and α_i is the thermal longitudinal expansion coefficient. The subscript stands for bare fiber ($i = f$) and coated fiber ($i = cf$). ζ is the thermo-optic coefficient of the fiber core, and $\hat{P}_e$ is the effective photoelastic coefficient of the coated fiber [107]

$$\hat{P}_e = \frac{n^2}{2}\left[P_{12} - \upsilon\left(P_{11} + P_{12}\right)\right] = 0.213$$

(12)

where P_{11} and P_{12} are the components of the strain-optic tensor, n is the index of refraction of the core, and υ is the Poisson's ratio. For a typical optical fiber $P_{11} = 0.113$, $P_{12} = 0.252$, $n = 1.482$, $\upsilon = 0.16$, and $\upsilon = -\varepsilon_{f,r}/\varepsilon_{f,z}$ where $\varepsilon_{f,r}$ and $\varepsilon_{f,z}$ are the radial and axial elastic fiber strains, respectively.

The temperature sensitivity S_T can be expressed as

$$S_T = \alpha_{cf} - \hat{P}_e\left(\alpha_{cf} - \alpha_f\right) + \zeta$$

(13)

Table 1 lists some thermo-optic parameters of the fused silica fiber and the polyimide coating [78]. According to the table, the temperature sensitivity of the grating can be calculated to be $S_T = (6.70\pm0.48)\times10^{-6}$ K^{-1} °C^{-1}. The temperature coefficient K_T, defined as $K_T = \Delta\lambda_B/\Delta T$, is $K_T = 0.0104$ nm/°C at 1550 nm.

As we know that the hygroscopic longitudinal expansion coefficient of the bare fiber β_f is zero, the relative humidity sensitivity S_{RH} can be expressed as:

$$S_{RH} = \left(1 - \hat{P}_e\right)\beta_{cf}.$$

(14)

Following the parameters in Table 1, the relative humidity sensitivity S_{RH} is 1.676×10^{-6} (%RH)$^{-1}$. The humidity coefficient K_{RH}, defined as $K_{RH} = \Delta\lambda_B / (\Delta RH)$, is 2.6×10^{-3} nm/(%RH) at 1550 nm.

Table 1. Properties of the fused silica fiber and the polyimide coating.

Parameter	Value
Thermal expansion coefficient α_f (K^{-1})	5×10^{-7}
Thermal expansion coefficient α_c (K^{-1})	4×10^{-5}
Thermo-optic coefficient ξ (K^{-1})	$(55 \pm 4.8) \times 10^{-7}$
Young's modulus, E (Fiber) (GPa)	72
Young's modulus, E (coating) (GPa)	2.45
Poisson's ratio, ν (fiber)	0.17
Poisson's ratio, ν (coating)	0.41
Hygroscopic expansion coefficient β_f (%RH^{-1})	0
Hygroscopic expansion coefficient β_c (%RH^{-1})	7×10^{-5}
Thermal longitudinal expansion coefficient of coated fiber α_{cf} (K^{-1})	1.39×10^{-6}
Hygroscopic longitudinal expansion coefficient of coated fiber β_{cf} (%RH^{-1})	1.58×10^{-6}

4.2. Experimental Details

A standard telecommunication optical fiber (SMF-28, Corning Inc.) was stored in a high hydrogen pressure environment (1900 psi) at room temperature for two weeks. Afterwards, the fiber was stored in a freezer at -70 °C before use. This process can prevent the hydrogen from diffusing out of the optical fiber. Hydrogen loading of a fiber satisfactorily enhances the photosensitivity of a fiber. A phase mask of 10 mm in length was used to write 1 cm FBG gratings onto the fiber with the laser irradiation from an ArF excimer laser (193 nm). The transmission spectrum of the FBG was monitored *in situ* during the laser exposure with a white light source and an optical spectrum analyzer. Since the photosensitivity of an optical fiber also resulted in further changes of the central Bragg wavelength and the peak loss of the attenuation bands generated after fabrication when the hydrogen diffused out of the fiber, the

FBG was baked at 150 °C overnight to eliminate the residual hydrogen and to stabilize the UV-induced index changes. The experimental setup for inscribing gratings is shown in Fig. 5.

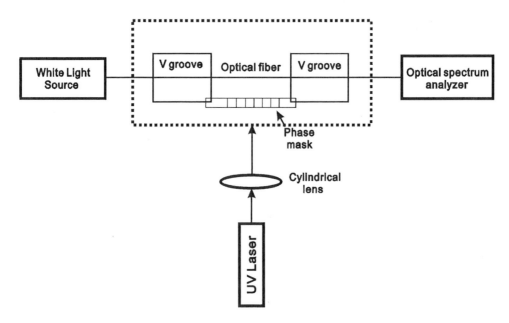

Figure 5. Schematic diagram of the fabrication of a fiber Bragg grating.

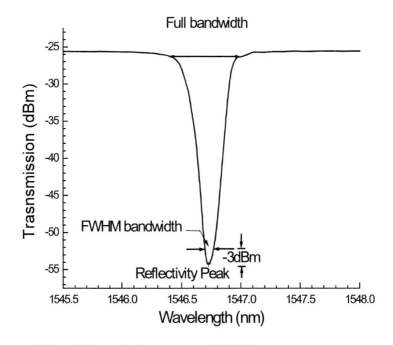

Figure 6. Transmission spectrum of a FBG at room temperature.

The transmission spectrum of a typical FBG is shown in Fig. 6. A broadband light travels through the optical fiber and enters into the FBG, from which one specific wavelength is reflected back by the FBG. Full width at half maximum (FWHM) is a parameter to characterize the bandwidths of laser beams or optical devices. For a strong FBG, it is normally measured at 3 dBm from the reflectivity peak.

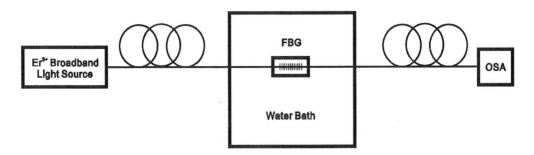

Figure 7. Schematic diagram of a temperature sensing measurement system.

Figure 7 illustrates the setup used to determine the temperature-induced shift in the resonance band of the FBG transmission spectrum. Light from a broadband light source (EBS-7210, MPB Communication, Inc.) was launched into one end of the fiber containing the grating and the transmission spectrum was recorded by an optical spectrum analyzer (ANDO AQ-6315E, Yokogawa Co.). The emission spectrum of the Er^{3+} broadband source is shown in Fig. 8, which indicates an output in the wavelength range of 1525-1600 nm. The temperature of the grating was controlled by employing a microprocessor-controlled water bath (Precision 281). The FBG sample was immersed in the water bath for which the temperature has been kept constant for at least half an hour before measurement in order to ensure that the transmission spectrum of the FBG sample was immovable and the system was in a stable condition.

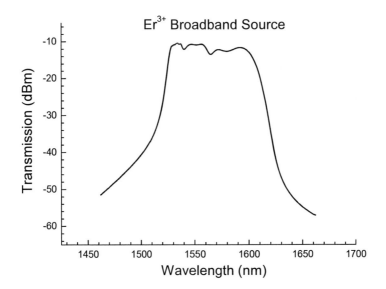

Figure 8. Emission spectrum of the Er^{3+} broadband light source used in this study.

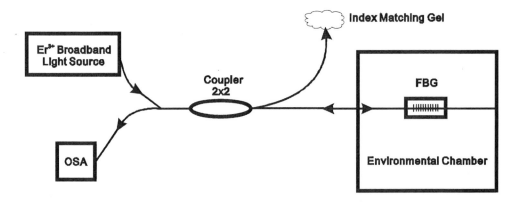

Figure 9. Schematic diagram of the humidity measurement system.

Once the FBG is inscribed onto the SMF-28 fiber, either acrylate or polyimide can be re-coated on the grating section using fiber recoaters or dip-coating method to protect the grating. The experimental setup used to characterize the humidity response of the sensor is shown in Fig. 9. The broadband light source was connected to the FBG sensor using a 2×2 fiber coupler. The reflected signal from the sensor was monitored using the optical spectrum analyzer. The polyimide-coated FBG sensor was stored inside an environmental chamber with controllable relative humidity. All experiments performed in this work were carried out at a room temperature of 20 °C. Sufficient time was given to allow the test environment to stabilize before readings of the wavelength from the optical spectrum analyzer were recorded.

4.3. Temperature Sensing Response of Fiber Bragg Gratings

For testing the strong FBG used here, the water temperature in the water bath was first changed from 5 °C to 90 °C with a step of 5 °C, the temperature was then decreased from 90 °C to 35 °C with a step of 5 °C. At a temperature of 25 °C, the transmission spectrum is shown in Fig. 10.

For the heating process, the observed transmission spectra and the change of the Bragg resonance wavelength as a function of increasing temperature are shown in Figs. 11 and 12, respectively. The dependence of the red-shift of the resonance wavelength of the FBG on the increasing temperature was measured to be 0.0091 nm/ °C.

For the cooling process, the measured change of the Bragg resonance wavelength as a function of decreasing temperature is shown in Fig. 13. The shift in the Bragg wavelength with respect to the decreasing temperature was measured to be 0.0098 nm/ °C. The experimental results of the shift in the Bragg wavelength as a function of temperature are close to the expected value of 0.0142 nm/ °C for a FBG recoated with an acrylate coating. The difference of the temperature sensitivities between the experimental result and theoretical value is mainly due to a discrepancy between different coating materials induced thermal expansion coefficients of the cladding materials and the thermo-optic coefficients of different core materials. Results obtained from the heating and cooling processes demonstrated high repeatability of the temperature sensing measurement.

A general expression for the approximate FWHM bandwidth of a grating is given by [21]

$$\Delta\lambda = \lambda_B s \sqrt{\left(\frac{\Delta n}{2n_0}\right)^2 + \left(\frac{1}{N}\right)^2} \qquad (15)$$

where N is the number of the grating planes. The parameter s is ~1 for strong gratings (for grating with near 100% reflection) whereas s ~0.5 for weak gratings. The FWHM bandwidth of the FBG loss peak in the heating process is shown in Fig. 14 with a standard deviation of 4.3%, which indicates that the FWHM bandwidth is independent of the changing temperature.

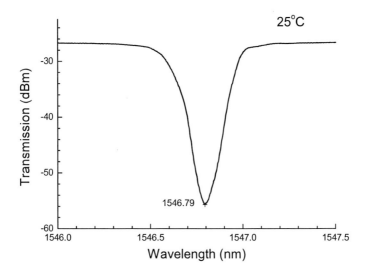

Figure 10. Transmission spectrum of a strong FBG at room temperature.

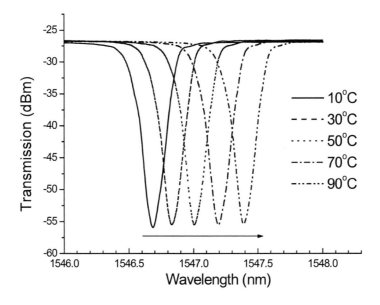

Figure 11. Transmission spectra of the FBG as a function of increasing temperature.

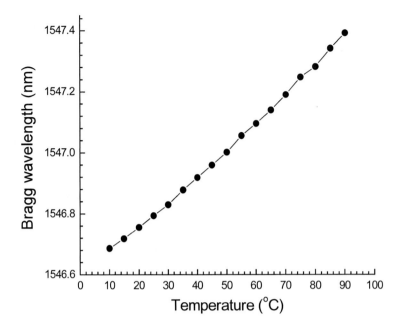

Figure 12. Bragg wavelength of the FBG as a function of increasing temperature.

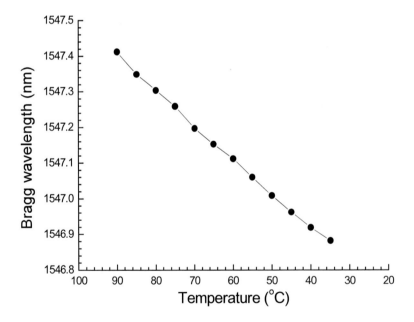

Figure 13. Bragg wavelength of the FBG as a function of decreasing temperature.

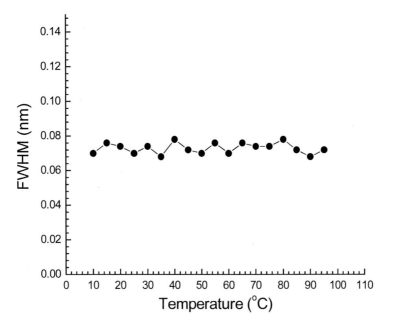

Figure 14. Change in the FWHM bandwidth of the FBG during heating process.

4.4. Humidity Sensing Response of Fiber Bragg Gratings

4.4.1. Humidity Sensing Response of a Polyimide-coated Fiber Bragg Grating

The polyimide-coated FBG was placed in the environmental chamber where the temperature was kept at 20 °C. At each fixed humidity setting, a reading was taken after the output signal from the optical spectrum analyzer had stabilized for 1 hour. The process was carried out three times to ensure reproducibility. As shown in Fig. 15, the data shows the behavior of the FBG spectral characteristics when the grating was exposed to different humidity conditions. From the figure, the Bragg wavelength shift was observed when the humidity in the test environment was varied from 13 %RH to 91 %RH. As the humidity level increased, the wavelength was found to shift toward the longer wavelength, which is consistent with the elongation of the FBG caused by the expansion of the polymer coating. This humidity change resulted in a wavelength shift of 0.22 nm.

Figure 16 shows the shifts of the Bragg wavelength versus relative humidity during three repeated tests. A linear regression was performed to establish the relationship between the wavelength shift due to the material expansion and the humidity level in the test chamber. The linear fitting reveals a linear dependence between the two parameters, which agrees well with the observations made by Kronenberg et al. [68] and Sager et al. [86], where the volume expansion of the polyimide film varies linearly with humidity. The humidity sensitivities of the FBG at 1550 nm were calculated to be 2.85, 2.62, and 2.60 pm/(%RH) respectively. The discrepancy among these test results may be ascribed to the incomplete adhesion between the silica surface and the polyimide layer due to the weak physical bonding at the interface between the fiber and the coating.

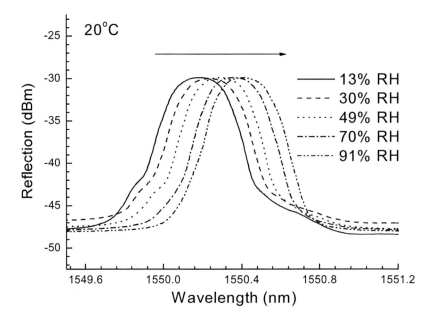

Figure 15. Reflection spectra of the polyimide-coated FBG at different relative humidities at room temperature.

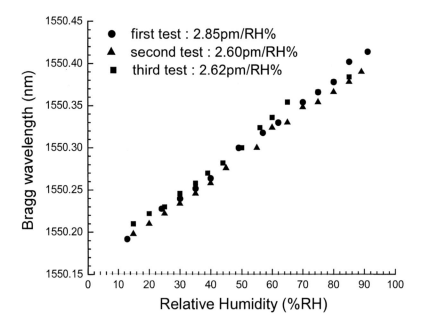

Figure 16. Bragg wavelength of the polyimide-coated FBG as a function of relative humidity in the environmental chamber.

Bare silica fibers are not sensitive to humidity. Polyimide polymers, however, are hygroscopic and will swell in aqueous media as the water molecules migrate into them. The swelling of the polyimide coating strains the fiber, which modifies the Bragg condition of the FBG and thus serves as the basis as the humidity sensor.

4.4.2. Humidity Sensing Response of an Acrylate-coated Fiber Bragg Grating

Following the same procedures mentioned above, FBG wavelength of an acrylate-coated FBG as a function of increasing and decreasing relative humidity was measured with the results shown in Figs. 17 and 18, respectively. The results in these two figures clearly indicate that the change of relative humidity will not change the center wavelength of the FBG at room temperature. The acrylate-coated FBG is therefore not sensitive to the change in relative humidity.

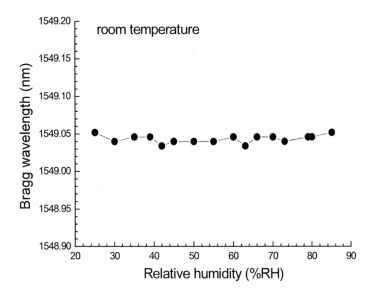

Figure 17. Bragg wavelength of the acrylate-coated FBG as a function of increasing relative humidity.

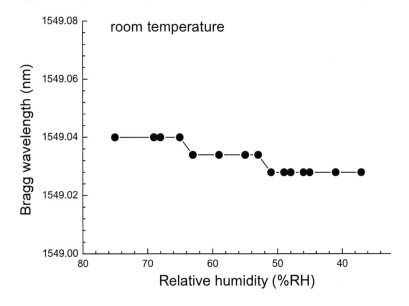

Figure 18. Bragg wavelength of the acrylate-coated FBG as a function of decreasing relative humidity.

5. Conclusions

Fiber Bragg gratings as temperature and humidity sensors have been reviewed and demonstrated, which possess many unique advantages over conventional techniques. Due to the nature of the fiber grating, it is possible to realize simultaneous measurement of these two parameters or to achieve quasi-distribution monitoring. Investigation on FBGs for applications in high temperature or harsh environment has been one of the research foci nowadays. For FBG humidity sensors, further research is needed for a systematic study of different polymer materials in order to identify materials with better performance as well as the sensing mechanisms including diffusion processes and response time.

Acknowledgments

This work has been supported by the Natural Sciences and Engineering Research Council of Canada (NSERC), Canada Research Chairs Program, Canada Foundation for Innovation, the Province of Newfoundland and Labrador, and the Memorial University of Newfoundland.

Reviewer:
Yanfei Ding, Department of Physics and Physical Oceanography, Memorial University of Newfoundland, St. John's, NL A1B 3X7, Canada.

References

[1]   Snyder, A. W.; Love, J. D. *Optical Waveguide Theory*; Chapman and Hall: London, UK, 1983, 3-5.
[2]   Kao, K. C.; Hockham, G. A. *Proc. IEE*, 1966, 113, 1151-1158.
[3]   Keck, D. B.; Schultz, P. C.; Zimar, F. W. U. S. Patent 3, 737, 393.
[4]   Kapron, F. P.; Keck, D. B.; Maurer, R. D. *Appl. Phys. Lett.* 1970, 17, 423-425.
[5]   Iizuka, K. Elements of Photonics, Vol. II *For Fiber and Integrated Optics*; Wiley-Interscience, New York, NY, 2002, 741.
[6]   Hill, O.; Fujii, Y.; Johnson, D. C.; Kawasaki, B. S. *Appl. Phys. Lett.* 1978, 32, 647-649.
[7]   Meltz, G.; Morey, W. W.; Glenn, W. H. *Opt. Lett.* 1989, 14, 823-825.
[8]   Feng, X.; Tam, H. Y.; Wai, P. K. A. *Photon. Technol. Lett.* 2006, 18, 1088-1090.
[9]   Liu, Y.; Chiang, K. S.; Chu, P. L. *Appl. Opt.* 2005, 44, 4822-4829.
[10]   Liaw, S. K.; Dou, L.; Xu, A. *Opt. Exp.* 2007, 15, 12356-12361.
[11]   Lausten, R.; Rochon, P.; Ivanov, M.; Cheben, P.; Janz, S.; Desjardins, P.; Ripmeester, J.; Siebert, T.; Stolow, A. *Appl. Opt.* 2005, 44, 7039-7042.
[12]   Littler, I.; Rochette, M.; Eggleton, B. *Opt. Exp.* 2005, 13, 3397-3407.
[13]   Bilodeau, F.; Johnson, D. C.; Thenault, S.; Malo, B.; Albert, J.; Hill, K. O. *IEEE Photon. Technol. Lett.* 1995, 7, 388-390.
[14]   Dong, L.; Hua, P.; Birks, T. A.; Reekie, L.; Russell, P. St. J. *IEEE Photon Technol. Lett.* 1996, 8, 1656-1658.
[15]   Sheu, L. G.; Chuang, K. P.; Lai, Y. *IEEE Photon. Technol. Lett.* 2003, 15, 939-941.

[16] López-Higuera, J. M. ed. *Handbook of Optical Fibre Sensing Technology*; John Wiley & Sons, Ltd: West Susses, England, 2002.

[17] Buck, J. A. *Fundamentals of Optical Fibers*; Wiley-Interscience: Hoboken, NJ, 2004.

[18] Bass, M. ed. Handbook of Optics Vol. IV. *Fiber Optics & Nonlinear Optics*, Second Edition; McGraw-Hill: New York, NY, 2001.

[19] Truesdale, C. M. Optical Fibers, in Waynant, R. W. and Ediger, M. N. ed. *Electro-Optics Handbook*, Second Edition, McGraw-Hill: New York, NY, 2000.

[20] Kashyap, R. *Fiber Bragg Gratings*; Academic Press: San Diego, CA, 1999.

[21] Othonos, A.; Kalli, K. *Fiber Bragg Gratings*; Artech House: Norwood, MA, 1999.

[22] Lam, D. K. W.; Garside, B. K. *Appl. Opt.* 1981, 20, 440-445.

[23] Mizrahi, V.; Sipe, J. E. *IEEE J. Light. Technol.* 1993, 11, 1513-1517.

[24] Erdogan, T.; Sipe, J. E. *J. Opt. Soc. Am. A* 1996, 13, 296-313.

[25] Lemaire, P. J.; Atkins, R. M.; Mizrahi, V.; Reed, W. A. *Electron. Lett.* 1993, 29, 1191-1193.

[26] Atkins, R. M.; Lemaire, P. J.; Erdogan, T.; Mizrahi, V. *Electron. Lett.* 1993, 29, 1234-1235

[27] Awazu, K.; Kawazoe, H.; Yamane, M. *J. Appl. Phys.* 1990, 68, 2713-2718.

[28] Bilodeau, F.; Malo, B.; Albert, J.; Johnson, D. C.; Hill, K. O. *Opt. Lett.* 1993, 18, 953-955.

[29] Williams, D. L.; Ainslie, B. J.; Armitage, R.; Kashyap, R.; Campbell, R. *Electron. Lett.* 1993, 29, 45-47.

[30] Williams, D. L.; Ainslie, B. J.; Kashyap, R.; Maxwell, G. D.; Armitage, J. R.; Campbell, R. J.; Wyatt, R. *Proc. SPIE*, 1993, 2044, 55-68.

[31] Albert, J.; Malo, B.; Bilodeau, F.; Johnson, D. C.; Hill, K. O.; Hibino, Y.; Kawachi, M., *Opt. Lett.* 1994, 19, 387-389.

[32] Dyer, P. E.; Farley, R. J.; Gied, R.; Byron, K. C.; Reid, D. *Electron. Lett.* 1994, 30, 860-862.

[33] Atkins, R. M.; Mizrahi, V. *Electron. Lett.* 1992, 28, 1743-1744.

[34] Hand, D. P.; Russell, P. St. J. *Opt. Lett.* 1990, 15, 102-104.

[35] Bernandin, J. P.; Lawandy, N. M. *Opt. Commun.* 1990, 79, 194-199.

[36] Douay, M.; Xie, W. X.; Taunay, T.; Bernage, P.; Niay, P.; Cordier, P.; Poumellec, B.; Dong, L.; Bayon, J. F.; Poignant, H.; Delevaque, E. *IEEE J. Light. Technol.* 1997, 15, 1329-1342.

[37] Fonjallaz, P. Y.; Limberge, H. G.; Salathe, R. P.; Cochet, F.; Leuenberger, B. *Opt. Lett.* 1995, 20, 1346-1348.

[38] Ky, N. H.; Limberge, H. G.; Salathe, R. P.; Cochet, F.; Dong, L. *Phys. Lett.* 1999, 74, 516-518.

[39] Limberge, H. G.; Fonjallaz, P. Y.; Salathe, R. P.; Cochet, F. *Appl. Phys. Lett.* 1996, 68, 3069-3071.

[40] Malo, B.; Hill, K. O.; Bilodeau, F.; Johnson, D. C.; Albert, J. *Electron. Lett.* 1993, 29, 1668-1669.

[41] Hill, K. O.; Malo, B.; Bilodeau, F.; Johnson, D. C.; Albert, J. *Appl. Phys. Lett.* 1993, 62, 1035-1037.

[42] Anderson, D. Z.; Mizrahi, V.; Erdogan, T.; White, A. E. *Proc. of the Conf. on Optical Fibre Communication, OFC'93,* Tech. Dig., 1993, 68.

[43]   Dragomir, A.; Nikogosyan, D. N.; Zagorulko, K. A.; Kryukov, P. G.; Dianov, E. M.; *Opt. Lett.* 2003, 28, 2171-2173.

[44]   Mihailov, S. J.; Smelser, C. W.; Grobnic, D.; Walker, R. B.; Lu, P.; Ding, H.; Unruh, J. *J. Light. Technol.* 2004, 22, 94-100.

[45]   Martinez, A.; Dubov, M.; Khrushchev, I.; Bennion, I. *Electron. Lett.* 2004, 40, 1170-1172.

[46]   Violakis, G.; Konstantaki, M.; Pissadakis, S. *IEEE Photon. Technol. Lett.* 2006, 18, 1182-1184.

[47]   Chen, Q.; Chen, K. P.; Xu, W.; Nikumb, S. *Opt. Commun.* 2006, 259, 123-126.

[48]   Kersey, A. D.; Davis, M. A.; Patrick, H. J.; LeBlanc, M.; Koo, K. P.; Askins, C. G.; Putnam, M. A.; Friebele, E. J. *IEEE J. Light. Technol.* 1997, 15, 1442-1463.

[49]   Ball, B.; More, W. W. *Opt. Lett.* 1992, 17, 420-422.

[50]   Othonnos, A.; Alavie, A. T.; Serge, S. M.; Karr, S. E.; Measures, R. M. *Opt. Eng.* 1993, 32, 2841-2846.

[51]   Alavie, A. T.; Karr, S. E.; Othonos, A.; Measures, R. M., *IEEE Photon. Technol. Lett.* 1993, 5, 1112-1114.

[52]   Hill, K. O.; Bilodeau, F.; Faucher, S.; Malo, B.; Johnson, D. C. *Electron. Lett.* 1991, 27, 1548-1550.

[53]   Laffont, G.; Ferdinand, P. *Meas. Sci. Technol.* 2001, 12, 765–770.

[54]   Williams, J. A. R.; Bennion, I.; Sugden, K.; Doran, N. J. *Electron. Lett.* 1994, 30, 985-987.

[55]   Kashyap, R.; Chernikov, S. V.; Mckee, P. F.; Taylor, J. R. *Electron. Lett.* 1994, 30, 1078-1080.

[56]   Fallon, R. W.; Zhang, L.; Gloag, A.; Bennion, I. *Electron. Lett.* 1997, 33, 705-706.

[57]   Putnam, M. A.; Williams, G. M.; Friebele, E. J. *Electron. Lett.* 1995, 31, 309-311.

[58]   Leblanc, M.; Huang, S. Y.; Ohn, M.; Measures, R. M.; Guemes, A.; Othonos, A. *Opt. Lett.* 1996, 21, 1405-1407.

[59]   Huang, S.; Leblanc, M.; Ohn, M. M.; Measures, R. M. *Appl. Opt.* 1995, 34, 5003-5009.

[60]   Othonos, A.; Kalli, K. Fibre Bragg gratings Fundamentals and Applications in Telecommunications and Sensing, 1999, (Boston, MA: Artech House), 98-99

[61]   Betz, D. C.; Thursby, G.; Culshaw, B.; Staszewski, W. J. *IEEE J. Light. Technol.* 2006, 24, 1019-1026.

[62]   Botsis, J.; Humbert, L.; Colpo, F.; Giaccari, P. *Opt. Laser. Eng.* 2005, 43, 491-510.

[63]   Kersey, A. D.; Berkoff, T. A.; Morey, W. W. *Opt. Lett.* 1993, 18, 1370-1372.

[64]   Fernandez, A.; Berghmans, F.; Brichard, B.; Megret, P.; Decreton, M.; Blondel, M.; Delchambre, A. *Meas. Sci. Technol.* 2001, 12, 1-4.

[65]   Jung, J.; Nam, H.; Lee, B.; Byun, J. O.; Kim, N. S. *Appl. Opt.* 1999, 38, 2752-2754.

[66]   Flockhart, G. M. H.; Maier, R. R. J.; Barton, J. S.; MacPherson, W. N; Jones, J. D. C.; Chisholm, K. E.; Zhang, L.; Bennion, I.; Read, I.; Foote, P. D. *Appl. Opt.* 2004, 43, 2744-2751.

[67]   Meltz, G.; Morey, W. W. *Proc. SPIE* 1991, 1516, 185-199.

[68]   Kronenberg, P.; Rastogi, P. K.; Giaccari, P.; Limberger, H. G. *Opt. Lett.* 2002, 27, 1385-1387.

[69]   Yeo, T. L.; Sun, T.; Grattan, K. T. V.; Parry, D.; Lade, R.; Powell, B. D. *IEEE Sensors J.* 2005, 5, 1082-1089.

[70]   Dong, X.; Huang, Y.; Lang, K.; Zhang, W.; Kai, G.; Dong, X. *Microw. Opt. Tech. Lett.* 2004, 42, 474-476.

[71]   Bock, W. J.; Urbańczyk, W. *Appl. Opt.* 1998, 37, 3897-3901.

[72]   Hsu, Y. S.; Wang, L.; Liu, W.; Chiang, Y. J. *IEEE Photon. Technol. Lett.* 2006, 18, 874-876.

[73]   Peng, B.; Zhao, Y.; Yang, J.; Zhao, M. *Measurement* 2005, 38, 176-180.

[74]   Zhao, C.; Yang, X.; Demokan, M. S.; Jin, W. *IEEE J. Light. Technol.* 2006, 24, 879-883.

[75]   Iadiciccoa, A.; Campopianoa, S.; Cutoloa, A.; Giordanob, M.; Cusanoa, A. *Sens. Actuators B,* 2006, 120, 231-237.

[76]   Shu, X.; Gwandu, B. A. L.; Liu, Y.; Zhang, L.; Bennion, I. *Opt. Lett.* 2001, 26, 774-776.

[77]   Iadicicco, A.; Campopiano, S.; Cutolo, A.; Giordano, M.; Cusano, A. *IEEE Photon. Technol. Lett.* 2005, 17, 1250-1252.

[78]   Men, L.; Lu, P.; Chen, Q. *J. Appl. Phys.* 2008, 103, 053107.

[79]   Wade, S. A.; Collins, S. F.; Baxter, G. W. *Appl. Phys. Lett.* 2003, 94, 4743-4756.

[80]   Grattan, K. T. V.; Meggitt, B. T. Optical Fiber Sensor Technology; Chapman & Hall, London & Kluwer Academic, Dordrecht Netherlands, 1994/98/99, Vols. 1-4.

[81]   Dakin, J.; Culshaw, B. Optical Fiber Sensors; Artech House, Boston, 1988/89/96, Vols. 1-3.

[82]   Lee, C. Y.; Lee, G. B. *Sensors Lett.* 2005, 3, 1-14.

[83]   Kharaz, A.; Jones, B. E. *Sens. Actuators A* 1995, 46, 491-493.

[84]   Bariáin, C.; Matías, I. R.; Arregui, F. J.; López-Amo, M. *Sens. Actuators B* 2000, 69, 127-131.

[85]   Gupta, B. D.; Ratnanjali *Sens. Actuators B* 2001, 80, 132-135.

[86]   Sager, K.; Schroth, A.; Nakladal, A.; Gerlach, G. Sens. Actuators A 1996, 53, 330-334.

[87]   Mrotek, J. L.; Matthewson, M. J.; Kurkjian, C. R. *IEEE J. Light. Technol.* 2001, 19, 988-993.

[88]   Schirmer, B.; Venzke, H.; Melling, A.; Edwards, C. S.; Barwood, G. P.; Gill, P.; Stevens, M.; Benyon, R.; Mackrodt, P. *Meas. Sci. Technol.* 2000, 11, 382-391.

[89]   Sorli, B.; Pascal-Delannoy, F.; Giani, A.; Foucaran, A.; Boyer, A. *Sens. Actuators A* 2002, 100, 24-31.

[90]   Bozóki, A.; Szakáll, M.; Mohácsi, Á.; Szabó, G.; Bor, Z., *Sens. Actuators B* 2003, 91, 219-226.

[91]   Tang, J. L.; Wang, J. N. *Sens. Transducers J.* 2006, 68, 597-605.

[92]   Xu, M. G.; Reekie, L.; Chow, Y. T.; Dakin, J. P. *Electron. Lett.* 1993, 29, 389-399.

[93]   Song, M.; Lee, S. B.; Choi, S. S.; Lee, B. *Opt. Fiber Technol.* 1997, 3, 194-196.

[94]   Xu, M. G., Archambault, J. L.; Reekie, L.; Dakin, J. P. *Electron. Lett.* 1994, 30, 1085-1087.

[95]   James, S. W.; Dockney, M. L.; Tatam, R. P. *Electron. Lett.* 1996, 32, 1133-1134.

[96]   Song, M.; Lee, B.; Lee, S. B.; Choi, S. S. *Opt. Lett.* 1997, 22, 790-792.

[97]   Frazao, O.; Ferreira, L. J. Opt. A: *Pure Appl. Opt.* 2005, 7, 427-430.

[98]   Cavaleiro, P. M.; Araujo, F. M.; Ferreira, L. A.; Santos, J. L.; Farahi, F. *IEEE Photon. Technol. Lett.* 1999, 11, 1635-1637.

[99]   Lu, P.; Men, L.; and Chen, Q. 2008, *Appl. Phys. Lett. 92,* 2008, 171112.

[100] Kersey, A. D.; Dandridge, A. *Proc. SPIE* 1988, 985, 113-116.

[101] Berkoff, T. A.; Davis, M. A.; Bellemore, D. G.; Kersey, A. D. *Proc. SPIE* 1995, 2444, 288-294.

[102] Askins, C. G.; Putnam, M. A.; Friebele, E. J. *Proc. SPIE,* 1995, 2444, 257-266.

[103] Friebele, E. J.; Askins, C. G.; Bosse, A. B.; Kersey, A. D.; Patrick, H. J.; Pogue, W. R.; Putnam, M. A.; Simon, W. R.; Tasker, F. A.; Vincent, W. S.; Vohra, S. T. *Smart Mater. Struct.* 1999, 8, 813-838.

[104] Zhou, Z.; Graver, T. W.; Hsu, L.; Ou, J. *Pacific Science Review* 2003, 5, 116-121

[105] Okabe, Y.; Tanaka, N.; Takeda, N. *Smart Mater. Struct.* 2002, 11, 892–898.

[106] Bjerkan, L. *Appl. Opt.* 2000, 39, 554-560

[107] Wnuk, V. P.; Mendez, A.; Ferguson, S.; Graver, T. *Proc. SPIE* 2005, 5758, 46-53.

[108] Meltz, G.; Morey, W. W. *Proc. SPIE* 1991, 1516, 185-199.

In: Atomic, Molecular and Optical Physics…
Editor: L.T. Chen, pp. 261-311

ISBN: 978-1-60456-907-0
© 2009 Nova Science Publishers, Inc.

Chapter 7

TIME-VARIANT SIGNAL ENCRYPTION BY DUAL RANDOM PHASE ENCODING SETUPS APPLIED TO FIBER OPTIC LINKS

Christian Cuadrado-Laborde[*]

Centro de Investigaciones Ópticas (CONICET-CIC)
P.O. Box 3, Gonnet 1897, La Plata, Buenos Aires, Argentina

Abstract

Spatial optical techniques have shown great potential in the field of information security to encode high-security images. Among them, the dual random phase encoding method has received much attention since it was proposed by Réfrégier and Javidi in the middle 1990s. Since then, a number of works in the field were proposed introducing different variations of this technique. On the other hand, the space-time duality refers to the close mathematical analogy that exists between the equations describing the paraxial diffraction of beams in space and the first-order temporal dispersion of optical pulses in dielectric media. It is generally used for extending to the temporal domain well-known properties of spatial optical configurations. In this work a new approach is developed for the secure data transmission problem in fiber optic links. We propose the encoding of time-varying optical signals, mainly for short-haul applications, with encryption methods that can be considered the time domain counterparts of the dual random phase encoding process and two of its more frequent variations: the fractional Fourier transform dual random phase encoding and the Fresnel transform dual random phase encoding. Further, as performance is a very relevant subject in fiber optic links, we will analyze mechanisms to produce time limited, as well as bandwidth limited, encoded signals. To this end, the different signal broadenings produced by each stage of the encoding process, in both time and frequency domains, are analyzed by using the Wigner distribution function formalism, and general expressions for the time width, as well as the bandwidth, in every encryption stage is obtained. The numerical simulations show good system performances, and a comparison between the different encryption processes is made. Furthermore, the robustness of the proposed methods is analyzed against the variation of

[*] E-mail address: claborde@ciop.unlp.edu.ar. Permanent Address: Centro de Investigaciones Ópticas (CONICET-CIC) P.O. Box 3, Gonnet 1897, La Plata, Buenos Aires, Argentina
E-mail address: Christian.Cuadrado@uv.es. Temporary Address: Departamento de Física Aplicada y Electromagnetismo, Universidad de Valencia, C/ Dr. Moliner 50, Burjassot 46100, Valencia, Spain

typical parameters of the encryption-decryption setup. Finally, the implementation of this proposal with current photonic technology is discussed.

1. Introduction

With the rapid development of Internet and computer techniques, information security is becoming more important than ever and, as a consequence, has become a very active research field. The branch of science that studies how to protect data from unauthorized uses and/or eavesdrop intrusions is cryptography, whereas encryption is the process by which plaintext (source data to protect) is transformed into ciphertext (protected data). This is done using a key, which becomes the cornerstone of the process. In cryptography, a key is a piece of information (a parameter) that controls the operation of encryption; together with the encryption system itself, it completely specifies the particular transformation from plaintext into ciphertext. The recovery of the source data from the protected data is done throughout decryption, i.e., the inverse operation of encryption. In a good enough encryption system, it becomes virtually impossible to make a successful decryption without the key. In designing security systems, it is commonplace to assume that the coarse details of the encryption system are already known to the attacker; in this way, the proper design of the keys relies almost entirely on the security of the encryption. On the other side, cryptanalysis is the branch of science that studies the methods for decrypting information without a priori knowledge of the system parameters. Typically, this involves finding the secret key circumventing the system security, in other words, cryptanalysis is the practice of code breaking. This work mainly focuses in cryptography, although we will discuss also some results corresponding to the cryptanalysis field.

On the other hand, whereas encoding optical time-varying signals by quantum-cryptographic schemes require the use of fragile quantum states and ultra sensitive detection equipment, non-quantum encryption setups are more tolerant to optical power losses, they allow the use of off-the-shelf components, and generally they can include optical amplifier steps. The methods proposed here can be considered the temporal analogues of the original Réfrégier and Javidi proposal in the spatial domain for encoding high-security images [1], and improvements or variations appeared subsequently. These spatial encoding techniques are grouped under the acronym *dual random phase encoding* (DRPE). Basically, the method allows one to encode a source image into a stationary white noise using two random phase masks (RPMs) in the input and Fourier planes, respectively, of a $4f$ system. It is impossible to recover the encrypted images without knowing the RPM, i.e., the key that has been used in the encoding step. Since then, different versions have arisen; to avoid confusion among them, we will add the acronym of the basic mathematical principle operation after DRPE. In that way we will refer to the seminal proposition of Réfrégier and Javidi by DRPE-FT, where the last acronym (FT) stands for Fourier transform. Later, Towghi et al. proposed a full-phase version of this setup in which the images are recorded as phase-only versions in a double random phase encoding encryption system [2,3]. Thus, it works opposed to the DRPE-FT technique, in which the images are recorded as intensity variations. Furthermore, they numerically compare the two encoding techniques, demonstrating a slight robustness of the full-phase DRPE-FT processor against noise addition [3]. By that time, Matoba et al. proposed an encrypted optical memory system using three-dimensional keys in the Fresnel domain [4]. This work was the starting point at which Unnikrishnan et al. proposed an optical encryption system through double-random phase encoding, by using the analogy between Fresnel diffraction patterns and the fractional Fourier transform (DRPE-FrFT) [5,6]. This encryption setup was also studied later by Hennelly and Sheridan [7]. In this case additional keys are obtained, namely, the fractional order ps of the fractional Fourier transform (FrFT). Another approach closely related to these techniques (and somewhat simpler) is the lensless DRPE [8,9], and subsequent developments [10,11]. It replaces the $4f$ systems by simple free-space propagation of the light amplitude between RPMs. From a mathematical point of view, the Fresnel transform (FsT) behaves here as the basic principle operation. In this way, we will refer to this encryption setup by DRPE-FsT. More recently, Chen and Zhao proposed an encryption setup based on the Hartley transform (DRPE-HT), where a Michelson-type interferometer was used for the optical implementation [12]. Multiplexing is another option explored in the literature in connection with encrypting techniques [10,11,13]. It has been shown previously that the decryption performance is quite sensitive to the operation wavelength—when the decryption wavelength is sufficiently different from the encryption wavelength, only noise results. As a consequence, it was possible to adopt a wavelength division multiplexing (WDM) technique to achieve encryption of multiple images, in which each decrypted image is obtained by decoding a single encrypted image with its corresponding wavelength.

Our purpose here is to develop temporal encryption techniques that allows encode-decode optical time-varying signals in an efficient way. From the aforementioned setups, we do not focus our attention in the full-phase DRPE-FT approach, since it becomes more cumbersome to manipulate phase signal fluctuations as compared with optical power variations, and neither in the DRPE-HT, due to its inherent complexity. Thus, we propose the temporal analogue devices of the spatial systems based on the DRPE-FT, DRPE-FrFT, and DRPE-FsT, by making use of the space-time duality [14–17]. The multiplexing capabilities of the presented photonic devices are discussed also, and analytical expressions are found to evaluate the system performance. This work is organized as follows. In Section 2 we briefly sketch the DRPE-FT, DRPE-FrFT, and the DRPE-FsT, as they were originally proposed in the space domain, Subsections 2.1 to 2.3, respectively. The space-time analogy is described in Subsection 3.1, whereas the time domain single channel and WDM encryption-decryption setups are presented in Subsection 3.2 (DRPE-FT), Subsection 3.3 (DRPE-FrFT), and Subsection 3.4 (DRPE-FsT). Because the three transform operations analyzed here are special cases of the more widespread Linear Canonical Transform (LCT), we will describe it in Section 4, together with the Wigner distribution function (WDF), to find analytical expressions for both time width and bandwidth extents at each stage of the encryption process for every encryption setup. In this way, in Subsection 4.1 we analyze the time-frequency spreading in the DRPE-FT setup, in Subsection 4.2 we analyze the DRPE-FrFT setup, and in Subsection 4.3 we conclude the theoretical study with the DRPE-FsT setup. Numerical examples are shown in Section 5; in this way Subsections 5.1 to 5.3 show the corresponding performances for the DRPE-FT, DRPE-FrFT and DRPE-FsT, respectively. The robustness against blind decryption is numerically shown also. The signal-to-noise ratio (SNR) is used for measuring the degree of resemblance between original and decrypted signals, as different parameters' setups are varied. Also, we discuss some practical implementation features of this technique (Section 6), and our conclusions are presented in Section 7. Finally, a list of common acronyms used throughout this work is summarized in Table I.

Table 1. Nomenclature

DRPE	Dual Random Phase Encoding
FT	Fourier Transform
FrFT	Fractional Fourier Transform
FsT	Fresnel Transform
RPM	Random Phase Mask
TO	Transform Operation
PM	Phase Modulator
WDF	Wigner Distribution Function
LCT	Linear Canonical Transform
CP	Corner Point

2. The DRPE in the Space Domain

It is appropriate to start by reviewing the main features of each encryption process in the space domain as they were originally proposed. We start in Subsection 2.1 with the DRPE-FT system proposed by Réfrégier and Javidi. In Subsection 2.2 the DRPE-FrFT is presented, together with the mathematical definition of the fractional Fourier transform, plus a short review of the bulk optics implementation of the fractional Fourier transform. Finally in Subsection 2.3 we present the DRPE-FsT setup, together with the connection between the free space propagation and the Fresnel transform. Further details can be checked at the appropriate cited references. Although it is clear that images are two-dimensional, in what follows a one dimensional notation is used for the sake of simplicity, for two reasons, first, because for extrapolating these results to the time domain one-dimensional notation is, of course, enough, and second, because in most cases the extrapolation to two dimensions is straightforward. Finally, throughout this work, a capital letter stands for the Fourier transform of the corresponding signal represented in lower case.

2.1. The DRPE-FT in the Space Domain

Figure 1 shows the classical DRPE-FT setup as it was originally proposed in Ref. 1. Let $f(x)$ and $\psi(x)$ denote the image to be encoded and the encoded image, respectively, being x the spatial coordinate transversal to the paraxial distance z. To encode $f(x)$ into a white stationary sequence, we perform two operations. First, we multiply this image by a random phase mask, which we mathematically express as $a(x) = \exp[j2\pi m(x)]$, i.e. a phase only transfer function, where $m(x)$ represents an independent white noise sequence uniformly distributed in [0; 1]. Then we convolve this image by the impulse response $b(x)$, which same as with $a(x)$, has a phase-only transfer function. The impulse response $b(x)$ is defined by its Fourier transform $B(\nu) = \exp[j2\pi n(\nu)]$, where ν represents a spatial frequency within this

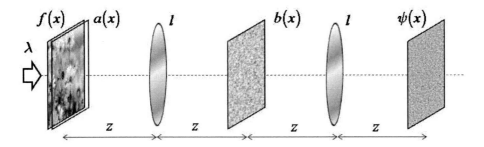

Figure 1. Optical setup for encryption in the space domain DRPE-FT. $a(x)$ and $b(x)$ are random phase masks, l represents the Fourier transforming lens, z is a free space propagation distance, $f(x)$ is the input image, and $\psi(x)$ is the encrypted image.

context; same as with $m(x)$, $n(\nu)$ is an independent white noise sequence uniformly distributed in [0; 1]. In this way the encoded image can be expressed as:

$$\psi(x) = [f(x)a(x)] * b(x) ,$$ (1)

where $*$ denotes the convolution operation. The encoding method can be implemented optically through a conventional $4f$ system [1,3,18]. In either case the complex amplitude encoded image $\psi(x)$ must be represented with both amplitude and phase. For decoding, $\psi(x)$ must be read out by coherent light. An optical method can be implemented for decoding, partially reversing the encoding process. To this end, $\psi(x)$ is optically Fourier transformed and multiplied by the random phase mask $B^*(\nu) = \exp[-j2\pi n(\nu)]$, and then inverse Fourier transformed to produce the output $f(x)a(x)$. The decryption generally stops here, and the final phase correction through $a^*(x)$, is not generally needed, because if a charge-coupled device array is used, simply the desired image $|f(x)|^2$ is obtained at the output. Regarding the optical implementation of the Fourier transform, a $2f$ system performs the desired operation. The input image is free space propagated a distance z, then pass through a lens l with focal length f, and finally free space propagates the same distance z again, see Figure 1 [18]. The relationship $f = z$, must be fulfilled. If x (x') represents the transversal coordinate at the input (transform) plane, then at the transform plane, and immediately before the second random phase mask; see Fig. 1, we have:

$$h(x') = C \int_{-\infty}^{\infty} dx f(x)a(x)\exp\left(\frac{-j2\pi}{\lambda f}xx'\right) .$$ (2)

The operation involved in Eq. (2) can be shortly rewritten as $h(x') = FT[f(x)a(x)]$, and in this way, the full encryption process of Fig. 1 can be described by:

$$\psi(x) = FT\left(\left\{FT\left[f(x)a(x)\right]\right\}B(\nu)\right) , \tag{3}$$

where the different primes over the spatial variables has been omitted for simplicity. Between Eqs. (1) and (3) there is a change of sign that provokes an inversion in the transversal coordinate.

2.2. The DRPE-FrFT in the Space Domain

Again $f(x)$ and $\psi(x)$ denote the image to be encoded and the encoded image, respectively. To encode $f(x)$ into a white stationary sequence, we perform two operations, see Fig. 2. First, we multiply this image by the random phase mask $a(x)$ at the input plane, which has a phase only transfer function as defined before. Then we fractionally Fourier transform this signal to the p_1-th order. At the encryption plane the signal is phase modulated with $b(x)$, which same as with $a(x)$, has a phase-only transfer function. The impulse response $b(x)$ is defined by its Fourier transform $B(\nu) = \exp\left[j2\pi n(\nu)\right]$; where $n(\nu)$ represents an independent white sequence uniformly distributed in [0; 1]. Finally the signal is fractionally Fourier transformed to the p_2-th order. At the output the encoded image can be expressed as [5–7]:

$$\psi(x) = FrFT_{p_2}\left(\left\{FrFT_{p_1}\left[f(x)a(x)\right]\right\}B(\nu)\right) , \tag{4}$$

where $FrFT_{p_i}(.)$ denotes the fractional Fourier transform operation of p_i-th order, and the different primes over the variables have been omitted for simplicity. As we will explain bellow, the encoding method can be implemented optically. For decoding, $\psi(x)$ must be read

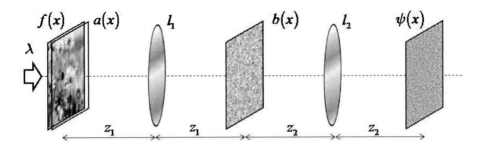

Figure 2. Optical setup for encryption in the space domain DRPE-FrFT. $a(x)$ and $b(x)$ are random phase masks, l_i represents the Fourier transforming lens, z_i is a free space propagation distance, $f(x)$ is the input image, and $\psi(x)$ is the encrypted image.

out by coherent light. To this end, $\psi(x)$ is optically fractionally Fourier transformed to the $-p_2$-th order and multiplied by the phase mask $B^*(\nu) = \exp\left[-j2\pi n(\nu)\right]$ and then fractionally Fourier transformed to the $-p_1$-th order to produce at the output the recovered image, plus a phase factor $f(x)a(x)$. Same as in the preceding subsection the final correction through $a^*(x)$, is not generally needed.

Regarding the fractional Fourier transform, there are several ways to define it, some of them equivalent between. For our purposes, the integral definition is enough, which is defined as follows [5,19–22]:

$$
f_{p_i}(x) = C \int_{-\infty}^{\infty} dx_0 f(x_0) \exp\left[\frac{j\pi}{\lambda f_0 \tan\theta}(x_0^2 + x^2)\right] \exp\left[\frac{-j2\pi}{\lambda f_0 \sin\theta} xx_0\right] ,
$$

$$
C = \frac{1}{|\sin\theta|^{1/2}} \exp\left\{-j\left[\frac{1}{4}\pi \operatorname{sgn}(\sin\theta) - \frac{\theta}{2}\right]\right\} , \qquad \theta = p_i\frac{\pi}{2} .
$$

$$(5)$$

Here $f_{p_i}(x) = FrFT_{p_i}\left[f(x)\right]$, where p_i is the fractional order, λ is the wavelength, and f_0 is a scaling factor of the transformed function. As it was mentioned above, the fractional Fourier transform can be implemented with bulk optics in two equivalent ways, widely known as setups types I and II, after Lohmann [7,23,24]. In the type I setup, the image to be fractionally Fourier transformed, i.e. the input signal $f(x_0)$, is illuminated by a monochromatic plane wave of wavelength λ. The diffracted field propagates a distance z_I, interacts with a lens l_I of focal length f_I, and further propagates by the same distance z_I. The amplitude distribution at the output plane becomes the fractional Fourier transform of fractional order p_i associated with the input object. In this optical arrangement, the geometrical parameters z_I, and f_I should be chosen to satisfy the following relationships

$$
z_I = f_0 \tan\left(p_i\,\pi/4\right) ,
$$

$$
f_I = f_0/\sin\left(p_i\,\pi/2\right) ,
$$

$$(6)$$

where f_0 is the same scale parameter used in Eq. (5). Of course, if the fractional order $p = 0$, the fractional Fourier transform becomes identical to the input object $f(x_0)$, whereas if $p = 1$ the fractional Fourier transform becomes the normal Fourier transform of $f(x_0)$. At another values of p, the fractional Fourier transform exhibits both space and spatial frequency information about the signal. In Fig. 2, for performing the fractional Fourier transform, this setup was chosen. On the other hand, in the type II setup, the field transmitted by $f(x_0)$ first interacts with a lens of focal length f_{II}, then propagates by a distance z_{II}, and interacts again with a second lens having the same focal length. In this case, the geometrical parameters are related with p_i as:

$$z_{\mathrm{II}} = f_0 \sin\left(p_i \pi/2\right) \ ,$$
$$f_{\mathrm{II}} = f_0 / \tan\left(p_i \pi/4\right) \ , \tag{7}$$

Several relevant properties of the fractional Fourier transform can be found in the literature [20,23].

2.3. The DRPE-FsT in the Space Domain

The double random-phase encoding technique can be applied in a more compact way, an optical security system much simpler proposed in Refs. 8 and 9. Compared with previous techniques, this system is lensless, which minimizes the hardware requirement, becoming easier to implement. The optical setup of the system is shown in Fig. 3. Three planes are defined as the input plane, the transform plane, and the output plane. The distances between adjacent planes are z_1 and z_2. The first random phase mask $a(x)$, attached to the input image $f(x)$, is located in the input plane. The second random phase mask $b(x)$, is located in the transform plane. If $h(x)$ represents the image after the free space propagation of distance z_1, i.e. in the first transform plane (but immediately before the second random phase mask), then it is related through a Fresnel transform with $f(x)a(x)$:

$$h_i\left(x'\right) = \int \mathrm{d}x f\left(x\right) a\left(x\right) \eta\left(x, x', z_i, \lambda\right) \quad ,$$
$$\eta\left(x, x', z_i, \lambda\right) = \frac{\exp\left(j2\pi z_i/\lambda\right)}{j\lambda z_i} \exp\left[\frac{j\pi}{\lambda z_i}\left(x' - x\right)^2\right] \ , \tag{8}$$

where $\eta\left(x, x', z_i, \lambda\right)$ can be considered the impulse response of the first stage of the system. The operation involved in Eq. (8) can be shortly rewritten as $h_i\left(x'\right) = FsT_{z_i}\left[f\left(x\right)a\left(x\right)\right]$, and in this way, the full encryption process can be described by:

$$\psi(x) = FsT_{z_2}\left(\left\{FsT_{z_1}\left[f\left(x\right)a\left(x\right)\right]\right\} B(\nu)\right) \ , \tag{9}$$

where the different primes over the spatial variables has been omitted for simplicity. In this way, when the system is perpendicularly illuminated with a plane wave of wavelength λ, the encrypted image $\psi(x)$ is obtained at the output. Distance parameters z_1 and z_2 are determined according to the size of the aperture to satisfy the Fresnel approximation. Since the inverse Fresnel transform does not optically exist, the complex conjugation of $\psi(x)$ should be used for decryption. The optical setup for decryption is the same as that for encryption but in the reverse direction, i.e. the input of the decryption system is the output of its counterpart. The complex conjugation of the encrypted image $\psi^*(x)$ is generated through phase conjugation in a photorefractive crystal in which the encrypted data was stored [8].

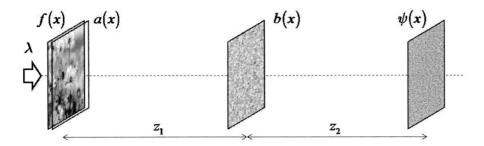

Figure 3. Optical setup for encryption in the space domain DRPE-FsT. $a(x)$ and $b(x)$ are random phase masks, z_i is a free space propagation distance, $f(x)$ is the input image, and $\psi(x)$ is the encrypted image.

3. The DRPE in the Time Domain

The space-time duality theory allows the transference of ideas relatives to spatial signal processing to the temporal domain. This theory is based on the close mathematical analogy that exists between the impulse response functions associated with certain spatial and temporal optical components. From this analogy, several devices can be developed, where the time-varying pulse amplitude has the same properties that those exhibited by the amplitude distributions produced by the "analogue", or spatial, counterparts [14-17]. The space-time duality is reviewed in Subsection 3.1, whereas the time domain counterparts of the DRPE-FT, DRPE-FrFT, and DRPE-FsT setups are presented in Subsections 3.2, 3.3, and 3.4, respectively.

3.1. The Space-Time Duality

One of the main operations involved in the spatial optical processors is the Fresnel diffraction. For free-space propagation of a monochromatic light beam of wavelength λ, there is associated an amplitude impulse response which is given by the kernel of the Eq. (8), which we rewrite here more simply:

$$\eta(z,x) = \eta_0 \exp\left(-j\frac{\pi}{\lambda z}x^2\right) , \tag{10}$$

where x is the spatial coordinate perpendicular to the propagation direction z, and $\eta_0 = (j/\lambda z)\exp(-j2\pi z/\lambda)$ is a constant with respect to the x variable [25]. In the case of narrowband modulated signals propagated in a dispersive medium, there are two photonic devices which can realize an operation temporally analogous to the given by Eq. (10), which are: i) pulse transmission in an optical fiber link, and ii) pulse reflection in a linearly chirped fiber Bragg grating (LCFG).

Most of the effects involving fiber pulse propagation with a temporal width > 1 ps can be explained by using the simplified generalized nonlinear Schrödinger equation, that gives the solution for the slowly varying amplitude A of a pulse envelope [26]:

$$j\frac{\partial A}{\partial z} = -\frac{j}{2}\alpha A + \frac{1}{2}\beta_2\frac{\partial^2 A}{\partial t'^2} - \gamma|A|^2 A \ , \tag{11}$$

where $t' = t - z/v_g = t - \beta_1 z$ is the time scale in a frame of reference moving with the pulse at the group velocity v_g. The three terms on the right-hand side of Eq. (11) describe the effects of absorption (α), dispersion (β_2), and nonlinearity (γ), respectively, inside optical fibers. When $\alpha = 0$, Eq. (11) is merely referred as the nonlinear Schrödinger equation. Next, if we introduce the normalized amplitude U by using the definition

$$A(z,\tau) = \sqrt{P_0}\exp(-\alpha z/2)U(z,\tau) \ , \tag{12}$$

where we have changed to a normalized time scale $\tau = t'/T_0$, being T_0 the initial pulse width, and P_0 the initial optical peak power. If we are only interested in the effect of group velocity dispersion (GVD) on pulse propagation, we must set $\alpha = 0$ and $\gamma = 0$ in Eq. (11). Then, by replacing Eq. (12) in (11), reduces this to

$$j\frac{\partial U}{\partial z} = \frac{1}{2}\beta_2\frac{\partial^2 U}{\partial t^2} \ , \tag{13}$$

besides this, for simplicity of notation, the prime over t' was dropped. It should be noted that Eq. (13) is equivalent to the paraxial wave equation when diffraction occurs in one transversal direction. Equation (13) can be solved by Fourier analysis, if $\tilde{U}(z,\omega)$ represents the Fourier transform of $U(z,\tau)$, then it satisfies:

$$j\frac{\partial \tilde{U}}{\partial z} = -\frac{1}{2}\beta_2\omega^2\tilde{U} \ , \tag{14}$$

whose solution is:

$$\tilde{U}(z,\omega) = \tilde{U}(0,\omega)\exp\left(\frac{j}{2}\beta_2 z\omega^2\right) \ . \tag{15}$$

The exponential term in Eq. (15) can be considered as the transfer function of the optical fiber, under the aforementioned restrictions. The impulse response $h(z,t)$ can be obtained by inverse Fourier transforming the exponential term of Eq. (15):

$$h(z,t) = h_0' \exp\left(\frac{-j}{2\beta_2 z}t^2\right) \quad , \tag{16}$$

being h_0' a constant. By comparing Eqs. (10) and (16), the same behavior is accomplished by the free-space diffracted fields and the pulse amplitudes propagated in an optical fiber, whenever the following equivalences are done

$$\lambda z \leftrightarrow 2\pi |\beta_2| z \quad , $$
$$x \leftrightarrow t \quad , \tag{17}$$

where the symbol $\leftrightarrow$ means that the magnitudes of the left side (spatial parameters) should be replaced by the magnitudes of the right side (time parameters), in order to produce the same impulse response.

In the case of pulse reflection in a LCFG, the linearly varying period of the effective mode index along the fiber axis produces a propagation delay in the reflected signal, which is a linear function of frequency. Therefore, the properties of the pulse amplitude reflected from the LCFG are determined by its complex reflection coefficient $r(\Delta\omega) = |r(\Delta\omega)|\exp[-j\Phi(\Delta\omega)]$, which can be written as [22,25]:

$$r(\Delta\omega) = r_0 \exp\left[-j\left(\Phi_{10}\Delta\omega + \frac{\Phi_{20}}{2}\Delta\omega^2\right)\right] \quad , \tag{18}$$

being $\Delta\omega = \omega - \omega_0$ the frequency difference variable, with ω_0 the Bragg central frequency, and $|r(\Delta\omega)| = r_0$ is the reflectivity for $\omega = \omega_0$. Besides, $\Phi_{k0} = \partial^k\Phi/\partial\omega^k\big|_{\omega=\omega_0}$, with $k = 1,2$, the k-th derivative of the phase function $\Phi(\Delta\omega)$ associated with the LCFG, evaluated at $\omega = \omega_0$. Inside a certain limited spectral bandwidth around ω_0, the coefficients Φ_{k0} can be considered as nearly constant. The spectral behavior of the reflected light pulse is characterized by $r(\Delta\omega)$, therefore, it can be considered as the transfer function for the LCFG pulse reflection. By taking its inverse Fourier transform, the temporal impulse response is obtained as

$$h(t') = h_0'' \exp\left(j\frac{1}{2\Phi_{20}}t'^2\right) \quad , \tag{19}$$

where $t' = t - \Phi_{10}$, h_0'' is a constant, and the effect of the group delay ripple has been neglected. Therefore, looking at Eqs. (10) and (19), it yields to the equivalences

$$\lambda z \leftrightarrow 2\pi |\Phi_{20}| \quad , $$
$$x \leftrightarrow t \quad , \tag{20}$$

where again the prime over t' was omitted, which is equivalent to ignoring the average pulse delay Φ_{10}.

The other relevant operation carried out by the spatial optical processors is the passage of the field amplitude through a lens, which has associated the following transmission function [18,22]

$$l(x) = P_0(x) \exp\left(-j\frac{\pi}{\lambda f}x^2\right) , \tag{21}$$

where $P_0(x)$ is the pupil function that takes into account the limited spatial extent of the lens. In the temporal domain, a conventional optical modulator which introduces a quadratic-phase modulation $l(t)$ is generally implemented by an electro-optic phase modulator, fed with an electrical sinusoidal signal [14]. The temporal pupil action is implemented with a consecutive electro-optic intensity modulator. Thus, the time lens, can be written as [27-33]

$$l(t) = l_0(t) \exp[-j\phi(t)] = l_0(t) \exp\left(-j\frac{\phi_{20}}{2}t^2\right) , \tag{22}$$

where it was considered that $\phi(t) = \cos(\omega_m t)$, and $\phi_{20} = \partial^2\phi/\partial t^2\big|_{t=0}$ is the second term in the Taylor series expansion of the modulation phase $\phi(t)$. The finite duration of the phase modulation, determined by the function $l_0(t)$, is denoted as the time aperture (TA) of the time lens. Under this assumption, by comparing Eqs. (21) and (22), the analogy, or equivalence, between spatial and temporal lenses can be established as

$$\frac{2\pi}{\lambda f} \leftrightarrow \phi_{20} , \tag{23}$$

$$x \leftrightarrow t .$$

Therefore, by properly using the duality relationships given by Eq. (17) −or equivalently Eq. (20)− and Eq. (23), the encryption setups presented in Subsections 2.1 to 2.3 can be transferred to the time domain. Further, in Ref. 27, an enlightening table can be consulted were several space-time dualities have been ordered.

3.2. The DRPE-FT in the Time Domain, Single Channel and WDM

From now on, for this and subsequent encryption setups, we denote as $f(t)$ and $\psi(t)$ the time-varying amplitudes of the original signal and the encrypted signal, respectively, see Fig. 4; whereas $m(t)$ and $n(t)$ denote two independent electrical white noise sequences uniformly distributed in [0; 1]. We perform two operations to encode $f(t)$ into a white noise stationary sequence $\psi(t)$. First, we transmit the input signal $f(t)$ through a phase modulator, i.e. $g(t) = f(t)a(t)$, with $a(t) = \exp[j2\pi m(t)]$. At the output, we convolve this signal with the impulse response $b(t)$, which has a phase-only transfer function. The impulse response $b(t)$ is defined by its Fourier transform: $B(\nu) = \exp[j2\pi n(\nu)]$. Thus, the encoded signal $\psi(t)$ could be represented as:

$$\psi(t) = [f(t)a(t)] * b(t) \ , \tag{24}$$

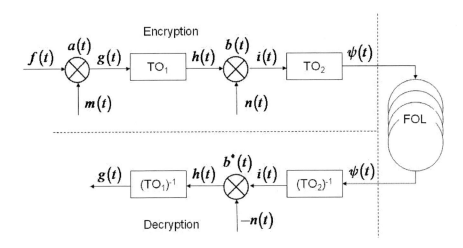

Figure 4. Single channel setup for encryption (above) and decryption (below), where $f(t)$ is an optical input signal, and $\psi(t)$ its encrypted counterpart. $a(t) = \exp[j2\pi m(t)]$ and $b(t) = \exp[j2\pi n(t)]$ are phase modulators (symbol $\otimes$), $m(t)$ and $n(t)$ are random electric noise signals, TO_i stands for a transform operation, e.g., a Fourier transform, whereas FOL is a fiber optic link.

where $*$ denotes the convolution operation. By Fourier transform analysis, it is known that this operation can be performed by first translating the phase modulator output $g(t)$ to the frequency domain through a Fourier transform, next phase modulating this signal with $B(\nu)$, and finally returning the signal information to the time domain through another Fourier transform. With the only purpose to be general, we have grouped under the acronym of transform operation (TO) the j-th Fourier transform stage in Fig. 4. Because it is precisely in the type of the transform operation performed, in which the different encryption setups studied here differs. The signal decryption is obtained partially reversing the encoding process, namely, the received signal $\psi(t)$ is convolved with $b^*(t)$, with this purpose $n(t)$ should be replaced by $-n(t)$ at the decryption stage. In this way, the decrypted signal results as $g(t) = f(t)a(t)$, see Fig. 4. As the detectors are phase insensitive, the final phase correction through $a^*(t)$ is not generally needed. Concerning the j-th transform operation, which in this case consists in a Fourier transform, same as in the spatial case (see Subsection 2.1), a temporal setup can be implemented [34], see Fig. 5. To this end, by using the space-time analogies developed in the preceding subsection, the input signal is transmitted through a dispersive component $r_j(t)$. If a LCFG was chosen for performing the dispersion, then it will be characterized by its second order dispersion coefficient $\Phi_{20}^{(j)}$, see Eq. (18). Otherwise, if a fiber optic of length z was chosen to perform the dispersive effect, then it will be characterized with $\beta_2 z$, see Eq. (15). At the output, this signal is phase modulated (time lens) with a quadratic phase modulator, having a phase modulation factor $\phi_{20}^{(j)} = \omega_j^2$, being ω_j the angular frequency variable in baseband of the electric signal driving the phase modulator, see Eq. (22). Finally, it is transmitted through another dispersive component characterized by the same dispersion coefficient $\Phi_{20}^{(j)}$, Fig. 5. Finally, the relation $\phi_{20}^{(j)} = 1/\Phi_{20}^{(j)}$ must be fulfilled in order to successfully perform the Fourier transform operation, which could be considered the temporal analogue of the relation $z = f$ of a $2f$ system to perform a Fourier transform in the space domain [18]. Regarding with the dispersion effects on the encoded signal transmitted through the fiber optic link, they can be completely discarded if we include the link dispersion in the value of the dispersion parameter required to perform any of the two Fourier transformations involved in the presented setup [34].

Since optical communications involve a large data transmission capability, signal multiplexing is a standard way of increasing the bit-rate requirements. A variety of multiplexing schemes can be implemented in order to increase the system capacity by generating the possibility to transfer data streams to different end users by employing the same fiber. As an example, WDM increases the link capacity by transmitting through the same optical fiber many signals having different wavelengths, so becoming non-interfering signals [35]. As the proposed setup can be quickly upgraded for multiplexing applications, it becomes relevant to analyse the multiplexing behaviour of the methods discussed here. With this purpose, we extend the single DRPE-FT configuration to n channels, see Fig. 6. In what follows, for every encryption setup studied here in this, and subsequent subsections, in each channel there is an original signal to encrypt $f(t,\lambda_i)$, and an encrypted signal $\psi(t,\lambda_i)$ at the output. The encryption is performed through $\kappa(t,\lambda_i)$ that represents the impulse response of the whole single channel encryption setup, as depicted in the top of Fig. 4, where the subscript stands for the wavelength channel λ_i. It is important noting that each encryption

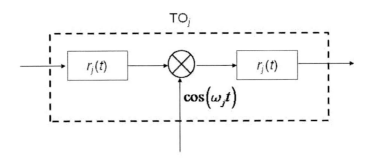

Figure 5. Scheme of the photonic device used to produce two particular kinds of transform operations: the Fourier transform and fractional Fourier transform. Two dispersive elements $r_j(t)$ are separated by a phase modulator driven with a sinusoidal voltage $\cos(\omega_j t)$ in order to act as a temporal lens.

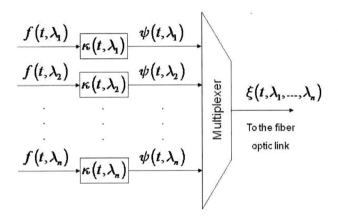

Figure 6. WDM setup for encryption, where $f(t,\lambda)$ is an optical input signal, $\psi(t,\lambda)$ its encrypted counterpart, $\kappa(t,\lambda)$ the impulse response of a single channel encryption stage, and $\xi(t,\lambda_1,...,\lambda_n)$ the WDM encrypted signal at the input of the fiber optic link.

channel has associated its own couple of keys $m_i(t)$ and $n_i(t)$. The several encrypted channels are transmitted together by the same fiber optic link by recombining them through a multiplexer, i.e.

$$\xi(t,\lambda_1,...,\lambda_n) = \sum_{i=1}^{n} \left[f_i(t) a_i(t) \right] * b_i(t) = \sum_{i=1}^{n} \psi(t,\lambda_i) \ , \tag{25}$$

where $\xi(t,\lambda_1,...,\lambda_n)$ represents a multiplexed signal composed of n encrypted signals at different wavelengths. This expression means that the multiplexed transmitted signal is composed of a linear addition of white noise-like signals. During decryption, first $\xi(t,\lambda_1,...,\lambda_n)$ is demultiplexed, and then each encrypted signal $\psi(t,\lambda)$ is decoded by its particular key, see Fig. 7, where $\eta(t,\lambda_i)$ represents the impulse response of the whole single channel decryption setup, as depicted at the bottom of Fig. 4. If the frequency spacing between neighboring channels is adequately chosen, they do not interfere, and as a consequence crosstalk effects become negligible. However it is important noting that the system also works if the same wavelength is used for a limited number of channels. That could be considered the temporal analogue of the recent work of Hennelly et al., in which a number of images were sequentially encoded and recorded on the same material (at the same position) by using the same reference beam, but different pairs of RPMs for each image [13]. In that case the output signal at the decryption stage, corresponding to the m-th channel can be written as:

$$g_D(t) = f_m(t) a_m(t) + \sum_{j=1}^{n, j \neq m} \left[f_j(t) a_j(t) \right] * b_j(t) \ , \tag{26}$$

where the subscript indicating wavelength dependence was omitted for obvious reasons. In this particular case, the m-th decoded signal is composed of two terms: i) the effectively decoded signal $g_m(t) = f_m(t)a_m(t)$ and *ii*) a summation term which is composed of a liner addition of white-noise like signals. This result represents the main difference between a WDM and a not WDM encryption setup. The first case offers, at least theoretically, a perfect decoding procedure, whereas the second comprises, even theoretically, a degree of imperfection (i.e. noise) which linearly increases with the channel numbers. As it was shown in Ref. 34, a stronger degradation in the decrypted signal is expected as the channel number increases. For this reason we will focus our attention just to the WDM approach.

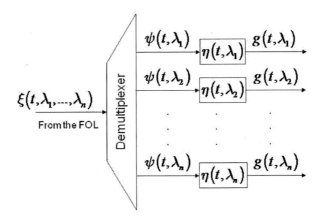

Figure 7. WDM setup for decryption, where $\xi(t,\lambda_1,...,\lambda_n)$ is the WDM encrypted signal at the output of the fiber optic link, $\psi(t,\lambda_i)$ a single channel encrypted signal, $g(t,\lambda_i)$ its decrypted counterpart, and $\eta(t,\lambda_i)$ the impulse response of a single channel decryption stage.

3.3. The DRPE-FrFT in the Time Domain, Single Channel and WDM

As it was mentioned above, Fig. 4 represents every encryption setup analyzed in this work. For the particular case of DRPE-FrFT, TO$_i$ stands for a fractional Fourier transform of arbitrary order p_i. For decryption, same as in the spatial case, each encryption component must be replaced by its complex conjugate, i.e. the $(TO_i)^{-1}$ at the decryption stage stands for a fractional Fourier transform of order $-p_i$. Concerning the temporal fractional Fourier transform production, a setup combining dispersive transmission and quadratic phase modulation was recently proposed [22,36,37]. To this end, Fig. 5 serves to our purpose as well, there the input signal is transmitted through a dispersive component represented by its temporal impulse responses $r_j(t)$ (with dispersion coefficient $\Phi_{20}^{(j)}$). At the output, this signal is phase modulated with a quadratic phase modulator having a phase modulation factor $\phi_{20}^{(j)}$, and transmitted through another dispersive component identical to the first. Finally, in order to appropriately perform the fractional Fourier transform of order p_i, the following relationships should be fulfilled, which could be considered the duals of Eq. (6) [22,36]:

$$\Phi_{20}^{(1)} = \frac{s}{2\pi}\tan\left(\frac{p_1\pi}{4}\right) , \tag{27}$$

$$\phi_{20}^{(1)} = \frac{2\pi}{s} \sin\left(\frac{p_1 \pi}{2}\right) , \tag{28}$$

where s is a scaling factor having dimensions of ps^2/rad, and an analogous expression also holds for performing the second transform operation TO$_2$, i.e. the second fractional Fourier transform by replacing index 1 by 2 in the preceding equations. Same as before (Subsection 3.2) the total dispersion between the quadratic phase modulators of the fractional Fourier transforms of order p_2 (encryption stage) and $-p_2$ (decryption stage) should reduce to zero for a successful decryption [37]. Again, for mathematical simplicity, and without losing any generality, every dispersive effects of the fiber optic link could be included in $r_2(t)$ or $[r_2^*(t)]$. Then by a simple Fourier transform analysis it can be concluded that $|g(t)| = |f(t)a(t)| = |f(t)|$, so being the output of the setup the decrypted signal $g(t)$.

The WDM DRPE-FrFT setup for encryption can be represented as well by Fig. 6, so for signals and systems we will refer to the preceding Subsection. The only difference between this and the preceding encryption setup (the WDM DRPE-FT) is in the temporal impulse response associated with $\kappa(t,\lambda_i)$, which now refers to a DRPE-FrFT. Figure 7 shows as well the WDM DRPE-FrFT setup for decryption, where now $\eta(t,\lambda_i)$ refers to the temporal impulse response associated with a single channel DRPE-FrFT. It should be reminded that the fractional Fourier transform orders p_1 and p_2 also acts as keys in this setup, besides the electrical signal $n_i(t)$ which fed each phase modulator of the i-th decrypting stage.

3.4. The DRPE-FsT in the Time Domain, Single Channel and WDM

Finally, we discuss the time domain single channel DRPE-FsT setup. For transferring the DRPE-FsT setup to the time domain, we will use the space-time duality as before. Again, Fig. 4 represents as well this encryption setup, and, for the particular case of a DRPE-FsT, TO$_i$ stands for a Fresnel transform. By making use of the space-time duality, the Fresnel transform can be performed in the time domain by simple propagating the optical signal through a dispersive element characterized by its impulse response $r_i(t)$, see Fig. 8. For decryption, as opposed to the spatial case, each encryption component must be replaced by its complex conjugate, i.e., the $(TO_i)^{-1}$ at the decryption stage stands for a Fresnel transform performed with dispersion $-\Phi_{20}^{(i)}$. The fact that positive or negative dispersion can be obtained, adds an extra degree of versatility in the time domain, something that is not present in the spatial case, where the complex conjugate of the encrypted signals should be taken for decoding, see Subsection 2.3 [8,38]. Again, the amount of dispersion between the last phase modulator at the encryption stage $b(t)$, and the first at the decryption stage $b^*(t)$, should reduce to zero for a successful decryption, and for mathematical simplicity, and without losing any generality, we assume that every dispersive effects of the fiber optic link are already included at the last dispersive stage $r_2(t)$ of the encryption stage. Under this assumption the fiber optic link behaves as an ideal transmission medium with unity transfer function. Then, through a simple Fourier transform based analysis it could be easily demonstrated that $\left|f(t,\lambda_i)\right|=\left|g(t,\lambda_i)\right|$. Because detectors are phase insensitive, the final phase correction at the decryption stage through $a^*(t)$ is not generally needed. Finally it should be kept in mind that in these encryption systems, as the encrypted data propagates through the encryption process, the signal temporally broadens. Therefore, the time aperture of the $(k+1)$-th phase modulator of the encryption stage should be slightly larger than the k-th phase modulator, in order to correctly phase modulate the partially encrypted signal completely. Below, in Section 4, we will be able to theoretically determine how much the signal spreads in the time (and frequency) domain at every encryption stage.

The WDM DRPE-FsT setup for encryption can be represented as well by Fig. 6, so for signals and systems we will refer to Subsection 3.2. The only difference between this and the WDM DRPE-FT encryption setup is in the temporal impulse response associated with $\kappa(t,\lambda_i)$, which now refers to a DRPE-FsT. Figure 7 shows as well the WDM DRPE-FsT setup for decryption, where now $\eta(t,\lambda_i)$ refers to the temporal impulse response associated with a single channel DRPE-FsT. It should be reminded that the dispersions chosen to perform the Fresnel transform also acts as keys in this setup, besides the electrical signal $n_i(t)$ which fed each phase modulator of the i-th decrypting stage.

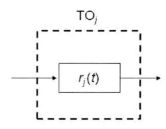

Figure 8. A single dispersive elements $r_j(t)$ performs the remaining TO$_j$: the Fresnel transform.

4. Signal Spreading Analysis in the Time and Frequency Domains

Quantifying the spreading in the time-frequency domains as the signal is encrypted, becomes a relevant subject for analyzing the multiplexing possibilities of the proposed setup. In what follows we assume that signals, e.g., $f(t)$, are bounded within some finite region in the time-frequency phase space domain. Of course, this really means that we only take into account (for the analysis purposes) the time-frequency phase space domain (t,ν) where the optical power of the signal itself as well as its spectrum is significantly a non-zero function [20,38,39]. This means that the following relations should be satisfied:

$$\{f(t),F(\nu)\}\approx 0 \qquad \forall \qquad \{|t|,|\nu|\}>\{\Delta t_f/2,\Delta\nu_f/2\}\ , \tag{29}$$

it should be remembered that through this work a capital letter stands for the Fourier transform of the corresponding signal, i.e. $F(\nu) = \Im[f(t)]$, whereas Δt_f and $\Delta \nu_f$ are the total temporal and frequency extents, respectively. If E represents the total signal energy, i.e.

$$E = \int_{-\infty}^{\infty} dt |f(t)|^2 = \int_{-\infty}^{\infty} d\nu |F(\nu)|^2 \ . \tag{30}$$

Then, the condition expressed in Eq. (29), means that:

$$\int_{-\Delta t_f/2}^{\Delta t_f/2} dt |f(t)|^2 = \int_{-\Delta \nu_f/2}^{\Delta \nu_f/2} d\nu |F(\nu)|^2 \approx E \ . \tag{31}$$

Besides, arguments or subscripts denoting wavelength dependence will be omitted in this subsection, since the results to be derived are equally applicable to any wavelength. The WDF is specially appropriated for performing this analysis because it gives, roughly speaking, the distribution of the signal energy over time and frequency. The WDF of a one-dimensional signal $f(t)$ is given by [20]

$$W_f(t,\nu) = \int dt' f(t+t'/2) f^*(t-t'/2) \exp(-j2\pi\nu t') \ . \tag{32}$$

Then, the WDF doubles the number of dimensions as can be checked in Eq. (32). There are two processes involved as the input signal is progressively encrypted in the DRPE systems, namely: phase modulation with a random signal and subsequent transform operation (i.e. Fourier transform, fractional Fourier transform or Fresnel transform), see Fig. 4. In the time domain and from a mathematical point of view, the first process is a product between the signal and the complex exponential associated with the phase modulation. Regarding with this, there is a property of the WDF especially useful: the multiplication of two functions in the time domain – e.g., $g(t) = f(t)a(t)$, see Fig. 4 – implies a convolution in the frequency domain of their corresponding WDFs [20,38,39]:

$$W_g(t,\nu) = \int d\nu' W_f(t,\nu-\nu') W_a(t,\nu') \ . \tag{33}$$

From Eq. (33) the time width Δt_g of the product signal $g(t)$ becomes the temporal overlapping of the individual signals (having time widths Δt_f and Δt_a). Now, let us assume that one of the signals starts in time after the other, and ends in time before the other, i.e., one of them it is completely contained in the time extent of the other. Then Δt_g can be expressed as $\Delta t_g = MIN\{\Delta t_f, \Delta t_a\}$, where $MIN\{.\}$ stands for the lesser quantity between the braces. Otherwise if a different combination between the start (and/or end) time of both signals arise, a different expression to evaluate the temporal overlapping should be considered. In what follows we will assume that the time extent of the phase modulator signal is always larger than the signal to be modulated, in this way validating the aforementioned expression. On the other hand, following Eq. (33), the frequency extent of the product signal is the sum of the bandwidths of the individual signals, having bandwidths $\Delta\nu_f$ and $\Delta\nu_a$. Both effects are summarized below as:

$$\Delta t_g = \Delta t_f \ , \tag{34}$$

$$\Delta\nu_g = \Delta\nu_f + \Delta\nu_a \ . \tag{35}$$

Now we turn our attention to the second process, i.e. the transform operation. The Fourier transform, fractional Fourier transform and Fresnel transform, together with the scaling operation and chirp multiplication, are members of the three-parameter class of linear integral transforms, widely known by linear canonical transforms, defined as [20,39]:

$$f_{\alpha,\,\beta,\,\gamma}\left(t'\right) = \exp\left(-j\pi/4\right)\sqrt{\beta}\int \mathrm{d}t f\left(t\right)\exp\left[j\pi\left(\alpha t^2 - 2\beta tt' + \gamma t'^2\right)\right] \ , \tag{36}$$

as we will see in the following, α, β, and γ are real constant parameters that acquires specific values depending of the specific transform operation performed. The effect on the WDF of a LCT can be best represented with the following matrix notation acting in the time-frequency phase space domain:

$$\begin{pmatrix} t' \\ \nu' \end{pmatrix} = \begin{pmatrix} \gamma/\beta & 1/\beta \\ -\beta+\alpha\gamma/\beta & \alpha/\beta \end{pmatrix}\begin{pmatrix} t \\ \nu \end{pmatrix} \ , \tag{37}$$

where (t',ν') and (t,ν) denote the transformed and initial points in the time-frequency phase space, respectively. In the following we use the WDF formalism to analyze for each encryption setup the different signal spreading in time and frequency domains at every encryption stage.

4.1. Signal Spreading Analysis for the DRPE-FT in the Time and Frequency Domains

In this specific case the transform operation performed is a Fourier transform. The optical Fourier transform definition in terms of the LCT parameters gives: $\alpha = \gamma = 0$, and $\beta = 1/\lambda f$, where f is the focal length of the Fourier transforming lens [39,40]. Translating this result from the space to the time domain, together with the auxiliary condition for performing a Fourier transform $z = f$ (see Subsection 2.1), results in $\beta = \left(2\pi\Phi_{20}\right)^{-1}$, see Eq. (20). In this way the matrix showed in Eq. (37), transforms in this particular case into:

$$\begin{pmatrix} t' \\ v' \end{pmatrix} = \begin{pmatrix} 0 & 2\pi\Phi_{20} \\ -(2\pi\Phi_{20})^{-1} & 0 \end{pmatrix} \begin{pmatrix} t \\ v \end{pmatrix}. \tag{38}$$

In this way, as can be checked from a proper inspection of Eq. (38), a Fourier transform induces a $\pi/2$ clockwise scaled rotation on each point of the WDF. Figure 9 describes the WDFs time-frequency phase space domains for generic signals involved in the encryption setup, where it can be easily visualized the several changes performed throughout the encryption process on the energy distribution of the signals in time and frequency, simultaneously. This figure should be read together with Table 2 that describes essentially the same operations, but from an analytical point of view. The coordinates of just one corner point (CP 1) of the WDF distribution are enough for this encryption setup to characterize the maximum extents of both the time width and bandwidth.

Table 2. Temporal and Spectral Broadening in the DRPE-FT

Stage	Time Width	Bandwidth
$f(t)$	Δt_f	Δv_f
$a(t)$	Δt_a	Δv_a
$g(t) = f(t)a(t)$	Δt_f	$\Delta v_f + \Delta v_a$
$h(t) = FT[g(t)]$	$2\pi\Phi_{20}\Delta v_g$	$(2\pi\Phi_{20})^{-1}\Delta t_g$
$b(t)$	Δt_b	Δv_b
$i(t) = h(t)b(t)$	Δt_h	$\Delta v_h + \Delta v_b$
$\psi(t) = FT[i(t)]$	see Eq. (39)	see Eq. (40)

The location of this corner point helps visualize the several changes performed in the time-frequency phase space during encryption (especially rotations for this specific case). We start in Fig. 9a with the WDF of the input signal, see also 1st row of Table 2. The input signal $f(t)$ is random phase modulated with $a(t)$ –with time width Δt_a and bandwidth Δv_a; see Fig. 9b and 2nd row of Table 2– so obtaining a new signal $g(t) = f(t)a(t)$. As we have previously analyzed, this operation (product in the time domain) results in a convolution in the frequency domain between the WDFs representing each signal, i.e., $W_f(t,v)$ and $W_a(t,v)$, see Figs. 9a and 9b, respectively. As a consequence, the signal spreads only in the frequency domain, becoming temporally as narrow as the narrower of both, see Fig. 9c, and 3rd row of Table 2. Next, we obtain the new signal $h(t)$, as the Fourier transform of $g(t)$. As we have seen before this induces a $\pi/2$ clockwise scaled rotation on each point of the WDF, see Fig. 9d, accordingly with Eq. (38). Now, for the temporal spreading, see 4th row of Table 2 (2nd column); whereas for the new signal bandwidth, see 4th row of Table 2 (3rd column). Next, the signal $h(t)$ is random phase modulated with $b(t)$ –with time width Δt_b

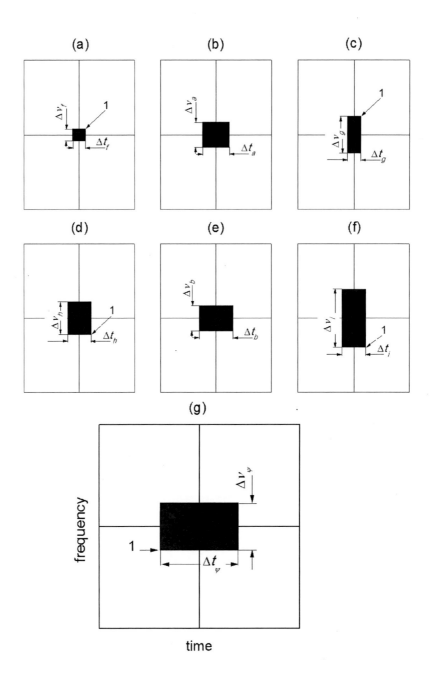

Figure 9. The Wigner distribution function as the signal becomes progressively encrypted in the DRPE-FT. (a) Input signal $W_f(t,\nu)$ (b) the first random phase modulator $W_a(t,\nu)$, (c) $W_{g=fa}(t,\nu)$, (d) $W_{h=FT(g)}(t,\nu)$, (e) the second random phase modulator $W_b(t,\nu)$, (f) $W_{i=hb}(t,\nu)$ (g) the finally encrypted signal $W_{\psi=FT(i)}(t,\nu)$. The evolution of the WDF is showed through the location of one corner point (1).

and bandwidth $\Delta\nu_b$; see Fig. 9e and 5th row of Table 2– so obtaining a new signal $i(t) = h(t)b(t)$. As a result of this process both WDFs –i.e. $W_h(t,\nu)$ and $W_b(t,\nu)$, see Figs. 9d and 9e, respectively– convolve in the frequency domain, where the result of this process can be observed in Fig. 9f, for the time extent, see 6th row of Table 2 (2nd column); whereas for the new bandwidth, see 6th row of Table 2 (3rd column). Finally, the signal at the output of the second phase modulator is Fourier transformed again, accordingly with Eq. (38) this induces another $\pi/2$ clockwise scaled rotation on each point of the WDF of the original signal; see Fig. 9g. As a consequence, the encrypted signal becomes spread in both domains in the following way:

$$\Delta t_\psi = \Delta t_f + 2\pi\Phi_{20}\Delta\nu_b \ , \tag{39}$$

$$\Delta\nu_\psi = \Delta\nu_f + \Delta\nu_a \ . \tag{40}$$

A similar result was obtained before by Hennelly et al. for the spatial case [13], and here we have transferred to the time domain those results under the most widespread formalism of the LCTs. As can be observed in Eq. (39), white noise cannot be used for encryption because the time width of the encoded signal broadens excessively thereby slowing the data transmission speed; whereas the bandwidth of the encrypted signal enlarges, preventing a WDM implementation, see Eq. (40).

4.2. Signal Spreading Analysis for the DRPE-FrFT in the Time and Frequency Domains

In this encryption setup the transform operation performed is an fractional Fourier transform of arbitrary order p. Then, defining the optical fractional Fourier transform in terms of the LCT parameters gives $\alpha = \gamma = [\lambda f \tan(p\pi/2)]^{-1}$ and $\beta = [\lambda f \sin(p\pi/2)]^{-1}$, where f is the focal length of the Fourier transforming lens [39,40]. Translating this result from the space to the time domain, see Section 3.1, results in the following transformation matrix:

$$\begin{pmatrix} t' \\ \nu' \end{pmatrix} = \begin{pmatrix} \cos(p\pi/2) & s\sin(p\pi/2) \\ -s^{-1}\sin(p\pi/2) & \cos(p\pi/2) \end{pmatrix}\begin{pmatrix} t \\ \nu \end{pmatrix} \ , \tag{41}$$

where s has been defined before, see Eqs. (27) and (28). In this way, as can be checked from a proper inspection of Eq. (41), a fractional Fourier transform of order p, induces a $p\pi/2$ clockwise scaled rotation on each point of the WDF. This relationship fulfills for every point in the time-frequency phase space domain, however, as we are analyzing the signal spreading, we will use the temporal and frequency extents. Now we will analyze the spreading in both domains for the DRPE-FrFT setup. Same as in the preceding subsection, Fig. 10 describes the WDFs time-frequency phase space domains for generic signals involved in the encryption setup, one can easily visualize the several changes performed throughout the encryption process on the energy distribution of the signals in both time and frequency. This figure should be analyzed together with Table 3 that describes essentially the same operations, but from an analytical point of view. We start in Fig. 10a with the Wigner distribution function of the input signal; corner points 1 and 2 will serve to characterize the maximum extents of both the time width and bandwidth. At this initial stage both coordinates of corner points 1 or 2, either of them, serves for that purpose as well, see Fig. 10a and 1st row of Table 3. Next, this input signal $f(t)$ is random phase modulated with $a(t)$ –with time width Δt_a and bandwidth $\Delta \nu_a$, see Fig. 10b and 2nd row of Table 3– so obtaining a new signal $g(t) = f(t)a(t)$. As we have previously analyzed, this operation (product in the time domain) results in a convolution in the frequency domain between the Wigner distribution functions representing each signal, i.e., $W_f(t,\nu)$ (see Fig. 10a) and $W_a(t,\nu)$ (see Fig. 10b).

Table 3. Temporal and Spectral Broadening in the DRPE-FrFT.

Stage	Time Width	Bandwidth
$f(t)$	Δt_f	$\Delta \nu_f$
$a(t)$	Δt_a	$\Delta \nu_a$
$g(t) = f(t)a(t)$	Δt_f	$\Delta \nu_f + \Delta \nu_a$
$h(t) = FrFT[p_1, g(t)]$	$\cos(p_1 \pi/2)\Delta t_g +$ $s\sin(p_1 \pi/2)\Delta \nu_g$	$s^{-1}\sin(p_1 \pi/2)\Delta t_g +$ $\cos(p_1 \pi/2)\Delta \nu_g$
$b(t)$	Δt_b	$\Delta \nu_b$
$i(t) = h(t)b(t)$	Δt_h	$\Delta \nu_h + \Delta \nu_b$
$\psi(t) = FrFT[p_2, i(t)]$	see Eq. (45)	see Eq. (45)

As a consequence, the signal spreads only in the frequency domain, becoming temporally as narrow as the narrower of both, see Fig. 10c. Again either of them, corner point 1 or 2 serves to characterize the spreading, see 3rd row of Table 3. Next, we obtain the new signal $h(t)$, as the fractional Fourier transform of $g(t)$ having fractional order p_1. As we have seen before, this induces a $\theta_1 = p_1 \pi/2$ clockwise scaled rotation on each point of the WDF, see Fig. 10d, accordingly with Eq. (41). If θ_1 is a 1st or 3rd quadrant angle, i.e. $0 \leq p_1 \leq 1$ or $2 \leq p_1 \leq 3$, respectively, then corner point 1 characterize the temporal spreading, whereas corner point 2 characterize the new signal bandwidth. On the other hand, if θ_1 is a 2nd or 4th quadrant angle (i.e., $1 \leq p_1 \leq 2$ or $3 \leq p_1 \leq 4$, respectively), then exactly the opposite should be considered, i.e., corner point 2 for the time extent and corner point 1 for the new signal bandwidth. In the 4th row of Table 3 is expressed the time extent (2nd column) and the bandwidth (3rd column) at this stage, when θ_1 belongs to the 1st quadrant. However, due to the involved symmetry, the same expressions of 4th row can still be used for any other θ_1 value, with a convenient reduction to the first quadrant, which is equivalent to replace in the 4th row the following expressions: $p_1 \rightarrow 2 - p_1$ (for $1 \leq p_1 \leq 2$), $p_1 \rightarrow p_1 - 2$ (for $2 \leq p_1 \leq 3$), and $p_1 \rightarrow 4 - p_1$ (for $3 \leq p_1 \leq 4$). Next, the signal $h(t)$ is random phase modulated with $b(t)$ –with time

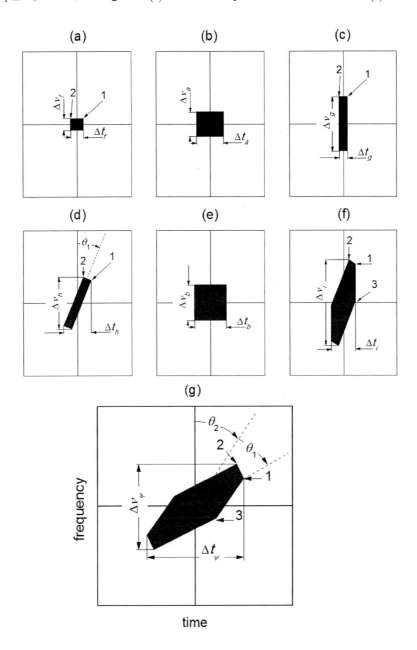

Figure 10. The Wigner distribution function as the signal becomes progressively encrypted in the DRPE-FrFT. (a) Input signal $W_f(t,\nu)$ (b) the first random phase modulator $W_a(t,\nu)$, (c) $W_{g=fa}(t,\nu)$, (d) $W_{h=FrFT_1(g)}(t,\nu)$, (e) the second random phase modulator $W_b(t,\nu)$, (f) $W_{i=hb}(t,\nu)$ (g) the finally encrypted signal $W_{\psi=FrFT_2(i)}(t,\nu)$. The evolution of the WDF is showed through the location of two and three corner points 1, 2, and 3; with $\theta_i = p_i \pi / 2$.

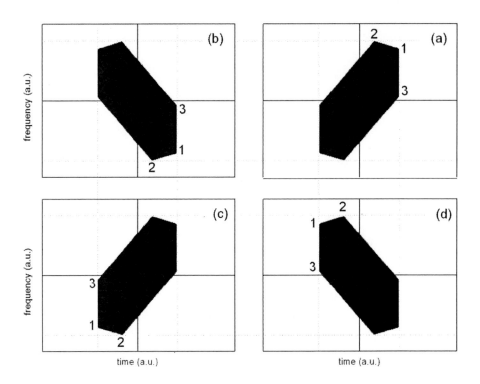

Figure 11. Four different $W_i(t,\nu)$, resulting from the frequency convolution between $W_b(t,\nu)$ (see Fig. 10e) and $W_h(t,\nu)$ (see Fig. 10d): (a) for θ_1 in the 1st quadrant ($p_1 = 0.3$), (b) for θ_1 in the 2nd quadrant ($p_1 = 1.7$), (c) for θ_1 in the 3rd quadrant ($p_1 = 2.3$), and (d) for θ_1 in the 4th quadrant ($p_1 = 3.7$); with $\theta_1 = p_1 \pi/2$.

width Δt_b and bandwidth $\Delta \nu_b$ – see Fig. 10e and 5th row of Table 3), so obtaining a new signal $i(t) = h(t)b(t)$. As a result of this process both WDFs $[W_h(t,\nu)$ and $W_b(t,\nu)$; see Figs. 10d and 10e] convolve in the frequency domain, where the result of this process can be observed in Fig. 10f, when θ_1 belongs to the 1st quadrant. For the new temporal extent, see 6th row of Table 3 (2nd column); whereas for the new signal bandwidth, see 6th row of Table 3 (3rd column). On the other hand, if θ_1 belongs to any other quadrant, i.e., for $1 \le p_1 \le 4$, the same 6th row expressions can still be used by applying the same reduction of θ_1 (i.e. p_1) to the first quadrant, performed in the same way as for the 4th row. The reason for this can be best explained graphically. Figure 11 shows four different $W_i(t,\nu)$ resulting from the proper frequency convolution between $W_b(t,\nu)$ (see Fig. 10e) and $W_h(t,\nu)$ (see Fig. 10d), when θ_1 belongs to the 1st quadrant (see Fig. 11a), the 2nd quadrant (see Fig. 11b), the 3rd quadrant (see Fig. 11c), and the 4th quadrant (see Fig. 11d). In Fig. 11, once p_1 was chosen for the first quadrant, e.g., $p_1 = 0.3$, then every other p for the remaining quadrants was chosen in such a way to be reduced to the first quadrant, as was explained above, i.e., $p_1 = 1.7$, $p_1 = 2.3$, and $p_1 = 3.7$, for the 2nd, 3rd, and 4th quadrant, respectively. It can be observed that when

$1 \leq p_1 \leq 4$, only a change of sign in one or both variables is necessary to return to the 1st quadrant WDF, so becoming the temporal and frequency extents exactly the same. Finally, the signal at the output of the second phase modulator is fractionally Fourier transformed with fractional order p_2. Accordingly with Eq. (41) this induces a $\theta_2 = p_2 \pi/2$ clockwise rotation on each point of the WDF of the signal; see Fig. 10g. From Fig. 10g, it becomes clear that with the location of any 3 consecutive corner points is possible to describe the time extent and bandwidth of the encrypted signal $\psi(t)$, with independence of the value chosen for p_2. As we need the original position (see Fig. 10f) of the corner points before the transformation, below we provide the full expressions for their locations in the time-frequency phase space domain, when θ_1 belongs to the 1st quadrant:

$$
\begin{aligned}
t_i^{(1)} &= \cos(p_1 \pi/2)\Delta t_f/2 + s\sin(p_1 \pi/2)(\Delta \nu_f + \Delta \nu_a)/2 \ , \\
t_i^{(2)} &= -\cos(p_1 \pi/2)\Delta t_f/2 + s\sin(p_1 \pi/2)(\Delta \nu_f + \Delta \nu_a)/2 \ , \\
t_i^{(3)} &= \cos(p_1 \pi/2)\Delta t_f/2 + s\sin(p_1 \pi/2)(\Delta \nu_f + \Delta \nu_a)/2 \ ,
\end{aligned}
\tag{42}
$$

$$
\begin{aligned}
\nu_i^{(1)} &= -s^{-1}\sin(p_1 \pi/2)\Delta t_f/2 + \cos(p_1 \pi/2)(\Delta \nu_f + \Delta \nu_a)/2 + \Delta \nu_b/2 \ , \\
\nu_i^{(2)} &= s^{-1}\sin(p_1 \pi/2)\Delta t_f/2 + \cos(p_1 \pi/2)(\Delta \nu_f + \Delta \nu_a)/2 + \Delta \nu_b/2 \ , \\
\nu_i^{(3)} &= -s^{-1}\sin(p_1 \pi/2)\Delta t_f/2 + \cos(p_1 \pi/2)(\Delta \nu_f + \Delta \nu_a)/2 - \Delta \nu_b/2 \ ,
\end{aligned}
\tag{43}
$$

which has been obtained by applying the usual WDF analysis. Equations (42) and (43) are still valid for any other quadrant (as it can be deduced from Fig. 11), provided the following changes of sign and reductions to the first quadrant be performed: *i*) $p_1 \rightarrow 2 - p_1$ and $\nu_i^{(l)} \rightarrow -\nu_i^{(l)}$ (for $1 \leq p_1 \leq 2$); *ii*) $p_1 \rightarrow p_1 - 2$, $t_i^{(l)} \rightarrow -t_i^{(l)}$, and $\nu_i^{(l)} \rightarrow -\nu_i^{(l)}$ (for $2 \leq p_1 \leq 3$); *iii*) and $p_1 \rightarrow 4 - p_1$ and $t_i^{(l)} \rightarrow -t_i^{(l)}$ (for $3 \leq p_1 \leq 4$), where $l = 1,2,3$ indicates the specific corner point. After the second fractional Fourier transform, each corner point has as new position in the time-frequency phase space domain that can be expressed in matrix notation as:

$$
\begin{pmatrix} t_\psi^{(1)} & t_\psi^{(2)} & t_\psi^{(3)} \\ \nu_\psi^{(1)} & \nu_\psi^{(2)} & \nu_\psi^{(3)} \end{pmatrix} = \begin{pmatrix} \cos(p_2\pi/2) & s\sin(p_2\pi/2) \\ -s^{-1}\sin(p_2\pi/2) & \cos(p_2\pi/2) \end{pmatrix} \begin{pmatrix} t_i^{(1)} & t_i^{(2)} & t_i^{(3)} \\ \nu_i^{(1)} & \nu_i^{(2)} & \nu_i^{(3)} \end{pmatrix} .
\tag{44}
$$

Any of these corner points could serve to describe the time extent or bandwidth, depending on the fractional Fourier transform orders and the scale parameter s. Therefore, the time extent and bandwidth of the encrypted signal can be obtained as twice the maximum absolute value of each row, which can be expressed in matrix notation in the following way:

$$\begin{pmatrix} \Delta t_\psi \\ \Delta \nu_\psi \end{pmatrix} = 2MAX \begin{pmatrix} \left| t_\psi^{(1)} \right| & \left| t_\psi^{(2)} \right| & \left| t_\psi^{(3)} \right| \\ \left| \nu_\psi^{(1)} \right| & \left| \nu_\psi^{(2)} \right| & \left| \nu_\psi^{(3)} \right| \end{pmatrix} , \tag{45}$$

where we have introduced the $MAX(.)$ operator, which selects the maximum quantity of each row. In this way the maximum value at each row is determined giving as a result a 2×1 vector, being the element at the top row the time extent of the encrypted signal Δt_ψ, whereas the element at the bottom row is the total bandwidth $\Delta \nu_\psi$. It is necessary to remark that for WDFs without symmetry, it becomes necessary applying the much broader analysis described by Hennelly and Sheridan for the spatial case [39-41]. In such case, the reductions performed here are no longer possible, and if the WDF is bounded by n sides, then a $2 \times \sum_{i=1}^{n-1} i$ matrix would be necessary in Eq. (45), e.g., 2×15 for $n = 6$.

4.3. Signal Spreading Analysis for the DRPE-FsT in the Time and Frequency Domains

In this specific case transform operation stands for a Fresnel transform, then, defining the optical Fresnel transform in terms of the LCT parameters gives $\alpha = \beta = \gamma = 1/\lambda z$, where z is the propagation distance [39,40]. Translating this result from the space to the time domain by using Eq. (20) results in $\beta = (2\pi \Phi_{20})^{-1}$, which in turn determines the following transformation matrix:

$$\begin{pmatrix} t' \\ \nu' \end{pmatrix} = \begin{pmatrix} 1 & 2\pi \Phi_{20} \\ 0 & 1 \end{pmatrix} \begin{pmatrix} t \\ \nu \end{pmatrix} , \tag{46}$$

where Φ_{20} represents the dispersion used to perform the Fresnel transform. According to Eq. (46) a Fresnel transform induces an horizontal shearing in the time-frequency phase space domain. Now we can analyze the spreading in both domains for the DRPE-FsT. Figure 12 describes the WDFs time-frequency phase space domains for generic signals involved in the encryption setup. This figure should be read together with Table 4 that describes essentially the same operations, but from an analytical point of view. We start in Fig. 12a with the WDF of the input signal; same as in DRPE-FT, the coordinates of just one corner point (CP 1) is enough for this encryption setup to characterize the maximum extents of the time width and bandwidth, see Fig. 12a and 1st row of Table 4. Next, this input signal $f(t)$ is random phase modulated with $a(t)$ –with time width Δt_a and bandwidth $\Delta \nu_a$, see Fig. 12b and 2nd row of Table 4– so obtaining a new signal $g(t) = f(t)a(t)$. As we have previously analyzed, this operation (product in the time domain) results in a convolution in the frequency domain between the WDFs representing each signal, i.e. $W_f(t,\nu)$ and $W_a(t,\nu)$, see Figs. 12a and 12b, respectively. As a consequence, the signal spreads only in the frequency domain, becoming temporally as narrow as the narrower of both, see Fig. 12c and 3rd row of Table 4. Next, we obtain the new signal $h(t)$, as the FsT$_1$ of $g(t)$ with dispersion $\Phi_{20}^{(1)}$. This induces a horizontal shearing on each point of the WDF, which linearly increases with frequency and dispersion; see Fig. 12d, accordingly with Eq. (46). For the new temporal extent, see 4th row of Table 4 (2nd column); whereas for the new signal bandwidth, see 4th row of Table 4 (3rd column). Next, the signal $h(t)$ is one more time random phase modulated with $b(t)$ (with time width Δt_b and bandwidth $\Delta \nu_b$, see Fig. 12e and 5th row of Table 4, so obtaining a new signal $i(t) = h(t)b(t)$. As a result of this process both WDFs $-W_h(t,\nu)$ and $W_b(t,\nu)$; see Figs. 12d and 12e– convolve in the frequency domain, where the result of this process can be observed in Fig. 12f. For the new temporal extent, see 6th row of Table 4 (2nd column); whereas for the new signal bandwidth, see 6th row of Table 4 (3rd column). Finally, the signal at the output of the second phase modulator is once again Fresnel transformed, now with dispersion $\Phi_{20}^{(2)}$. Accordingly with Eq. (46) this induces another horizontal shearing on

Table 4. Temporal and Spectral Broadening in the DRPE-FsT.

Stage	Time Width	Bandwidth
$f(t)$	Δt_f	$\Delta \nu_f$
$a(t)$	Δt_a	$\Delta \nu_a$
$g(t) = f(t)a(t)$	Δt_f	$\Delta \nu_f + \Delta \nu_a$
$h(t) = FsT_1[g(t)]$	$\Delta t_g + 2\pi \Phi_{20}^{(1)} \Delta \nu_g$	$\Delta \nu_g$
$b(t)$	Δt_b	$\Delta \nu_b$
$i(t) = h(t)b(t)$	Δt_h	$\Delta \nu_h + \Delta \nu_b$
$\psi(t) = FsT_2[i(t)]$	see Eq. (47)	see Eq. (48)

each point of the WDF, see Fig. 12g. As a consequence, the encrypted signal has temporal and spectral extents given by:

$$\Delta t_\psi = \Delta t_f + 2\pi \left(\Phi_{20}^{(1)} + \Phi_{20}^{(2)} \right) \left(\Delta \nu_f + \Delta \nu_a \right) + 2\pi \Phi_{20}^{(2)} \Delta \nu_b \;, \tag{47}$$

$$\Delta \nu_\psi = \Delta \nu_f + \Delta \nu_a + \Delta \nu_b \;. \tag{48}$$

In the following section we will show some numerical simulations to illustrate the behavior of the proposed encryption setups and to corroborate the validity of the derived results for the encrypted time widths and bandwidths.

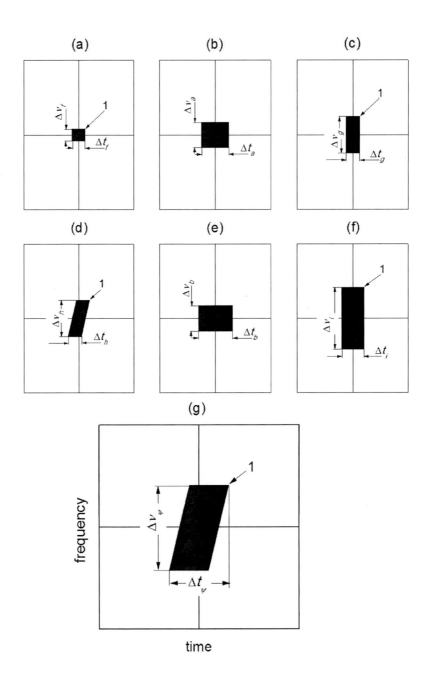

Figure 12. The Wigner distribution function as the signal becomes progressively encrypted in the DRPE-FsT. (a) Input signal $W_f(t,\nu)$ (b) the first random phase modulator $W_a(t,\nu)$, (c) $W_{g=fa}(t,\nu)$, (d) $W_{h=FsT_1(g)}(t,\nu)$, (e) the second random phase modulator $W_b(t,\nu)$, (f) $W_{i=hb}(t,\nu)$ (g) the finally encrypted signal $W_{\psi=FsT_2(i)}(t,\nu)$. The evolution of the WDF is showed through the location of one corner point (CP 1).

5. Results

From the analysis of the preceding section, it becomes clear that white noise cannot be used for encryption because the system performance degrades severely as $\Delta\nu_a$ and $\Delta\nu_b$ increases, i.e. there is an increment in both the time width and bandwidth of the encrypted signal. In spatial DRPE this problem appears due to the finite size of the apertures at the Fourier lenses and RPMs. Thus, only the low spatial frequencies are used for recording, being the higher frequencies discarded. Therefore, in spatial optical systems it is a very common situation that a severe information loss occurs, and as a consequence, the reconstructed image includes speckle noise [42]. Therefore, for exploiting the fiber optic multiplexing capabilities, a quasi-white noise must be used, i.e. a white noise signal but with a well-defined bandwidth. To this end, we follow an iterative process that has been already applied also by Nomura et al. in the context of optical encryption for the design of input phase masks [43]. We will shortly describe it in four steps using $a(t)$ as an example [an analogous procedure must be performed for $b(t)$]: i) It starts with a true random phase modulator function described by $a_0(t) = \exp\left[j2\pi m_0(t)\right]$, being $m_0(t)$ a white noise signal. This signal is Fourier transformed, so obtaining $A_0(\nu)$. ii) We get $A_0(\nu)\,\text{rect}(\nu/\Delta\nu_a)$, in order to limit the spectral content of $A_0(\nu)$ up to $\Delta\nu_a$; where $\text{rect}(.)$ stands for the rectangle function which is 1 for $|\nu| \le \Delta\nu_a/2$ and 0 otherwise. Next, it is inverse Fourier transformed and we have a new input random phase modulator $a_1(t)$. iii) The amplitude of $a_1(t)$ is set to unity through $\exp\left\{j\arg\left[a_1(t)\right]\right\}$. Finally, iv) this new random phase modulator function is the new input to the process which is continued up to the i-th iteration, when the error given by $e = \int_{|\nu|>\Delta\nu_a/2} d\nu\,|A_i(\nu)|^2$ becomes small enough. For computational purposes, we have used random PMs having a spectral content limited to below $\Delta\nu_a \approx 45$ GHz (unless otherwise specified), obtained with 15 iterations, after which e is well below 1%. In the following, we will illustrate the systems behavior through some numerical examples. As the single channel behavior can be inferred from the WDM operation, we will only illustrate the later for every encryption setup analyzed here. Whenever necessary, the signal degradation at the decryption stage will be measured with the SNR, defined by:

$$\text{SNR} = 10\log\left\{\frac{\int |f(t,\lambda_i)|^2\,dt}{\int \left[f(t,\lambda_i) - |g(t,\lambda_i)|\right]^2\,dt}\right\}, \tag{49}$$

where $f(t,\lambda_i)$ is the input signal, and $g(t,\lambda_i)$ its decrypted counterpart at the output of the decrypting stage. In the following we numerically show each encryption setup performance for every encryption setup. The numerical calculations were performed in a time window of

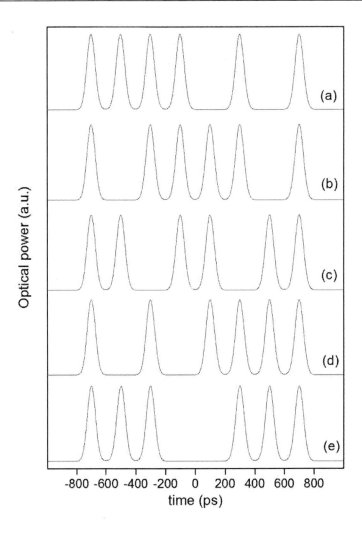

Figure 13. Input signal intensities of: (a) $f(t,\lambda_1 \approx 1548.51 \text{ nm})$, (b) $f(t,\lambda_2 \approx 1549.31 \text{ nm})$, (c) $f(t,\lambda_3 \approx 1550.12 \text{ nm})$, (d) $f(t,\lambda_4 \approx 1550.92 \text{ nm})$, and (e) $f(t,\lambda_5 \approx 1551.72 \text{ nm})$. The data stream is composed of eight time slots of 200 ps each, and a "1" data bit is represented by a Gaussian optical pulse of a 40 ps mean width at $1/e$ intensity point.

~137 ns with 2^{17} equally spaced points (time discretization). These values are large enough to correctly visualize both the signals in the time domain as well as its Fourier spectra. Throughout this section, we will use the same input signals for every encryption system studied. Figure 13 shows the binary optical input signals to the encryption setups in the time domain. Each input is composed of eight slots, each one having a time width of $T_1 = 200$ ps, so becoming the temporal extension of one data stream $\Delta t_f = 1.6$ ns. Inside each slot, a "1" data bit is represented by an optical Gaussian pulse of unitary intensity and mean width $T_0 = 40$ ps, at $1/e$ intensity point; whereas an empty slot represents a "0" bit. Figure 14a shows the spectrum of a typical input signal, the bandwidth of the input signal is $\Delta\nu_f \approx 10$ GHz (i.e., $\Delta\lambda_f \approx 0.1$ nm), and it has been obtained from the calculated spectrum. In order to illustrate

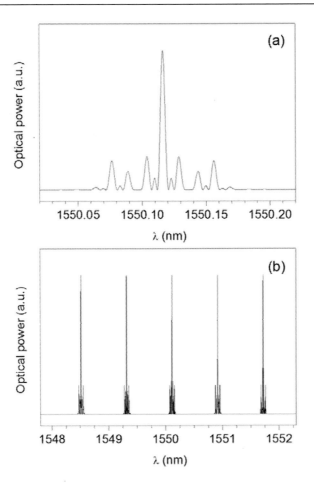

Figure 14. (a) Spectrum power of one of the input signals represented in the previous figure, e.g., $f(t,\lambda_3)$, (b) The spectral power of the five inputs signals together according to WDM, see the caption of Fig. 13 for signal wavelengths.

the interference effects between adjacent channels, we have used the following five input signals (ordered in increasing wavelengths): (a) $f(t,\lambda_1 \approx 1548.51$ nm$)$, (b) $f(t,\lambda_2 \approx 1549.31$ nm$)$, (c) $f(t,\lambda_3 \approx 1550.12$ nm$)$, (d) $f(t,\lambda_4 \approx 1550.92$ nm$)$ and (e) $f(t,\lambda_5 \approx 1551.72$ nm$)$, which are enough representative of a narrow band in a WDM transmission at 100 GHz spacing[1], Figure 14b.

In the following we analyze the performance of each particular encryption setup.

5.1. The DRPE-FT Performance

In the DRPE-FT system, in each channel the only keys used are $m_i(t)$ and $n_i(t)$, whereas for performing the Fourier transform, a value of $\Phi_{20} = 2 \times 10^3$ ps^2/rad was chosen for every

---

[1] ITU-T Recommendation G.694.1 (06/2002) "Spectral grids for WDM applications: DWDM frequency grid".

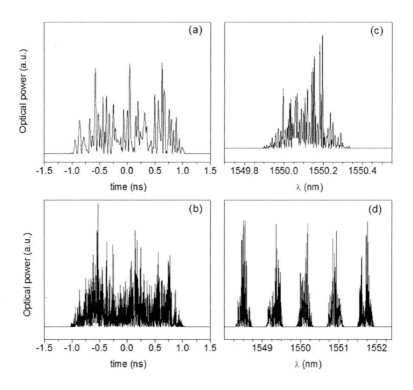

Figure 15. The DRPE-FT encrypting performance. (a) Encrypted signal intensity corresponding to a single channel, e.g., $\psi(t,\lambda_3)$. (b) Encrypted signal intensity corresponding to the WDM transmission of the five data streams, i.e., $\xi(t,\lambda_1,...,\lambda_5)$. (c) and (d) spectral powers of the signals showed in (a) and (b), i.e., $\Psi(\nu,\lambda_3)$ and $\Xi(\nu,\lambda_1,...,\lambda_5)$, respectively. Channel wavelengths are the same as in Fig. 13 caption.

channel. Figure 15a shows the intensity of an encrypted signal as they look in the time domain, e.g., $\psi(t,\lambda_3)$, whereas Fig. 15b shows the total intensity of the simultaneous transmission of the five encrypted data streams, i.e., $\xi(t,\lambda_1,...,\lambda_5)$. Their corresponding spectral powers are shown in Figs. 15c and 15d, i.e., $\Psi(\nu,\lambda_3)$ and $\Xi(\nu,\lambda_1,...,\lambda_5)$, respectively. It can be observed, by comparing Figs. 13c and 15a that the encrypted signal has temporally broadened, i.e., from $\Delta t_f^{(3)} \approx 1.6$ ns up to $\Delta t_\psi^{(3)} \approx 2.2$ ns. This last value compares reasonably well with the predicted value obtained by applying Eq. (39) $\Delta t_\psi^{(3)} \approx 2.16$ ns. Whereas the spectral extent, as a consequence of the encryption process, it has increased from $\Delta\nu_f \approx 10$ GHz up to $\Delta\nu_\psi^{(3)} \approx 50$ GHz (i.e. $\Delta\lambda_\psi^{(3)} \approx 0.4$ nm), compare Fig. 14a with 15c. This value also compares reasonably well with the value predicted by Eq. (40), i.e., $\Delta\nu_\psi^{(3)} \approx 55$ GHz. Finally, it should be taken into account that Eqs. (39) and (40) are approximate, as they were obtained by assuming that signals are completely bounded in both time and frequency, where the time-frequency uncertainty principle was relaxed. This is the main reason of the difference between obtained and calculated values. After passing through the multiplexer, every encrypted signal is recombined, thereby sharing the same time window, which is the distinctive feature of WDM transmission [35]. For this reason the five channels together have the same total temporal extension than one single encrypted signal (see Fig. 15a as compared with 15b). Finally, in Fig. 15d the whole five encrypted signal spectrum is shown, where it can be seen that there is not any appreciable overlapping between single channel spectra. Figure 16 shows the decrypted signals corresponding to the DRPE-FT encryption of the signals showed in Fig. 13. It can be clearly observed the close resemblance between the input and decrypted signals. The small ripple present in the decrypted signal diminishes as the separation between channels enlarges, therefore, a compromise between noise and channel capacity must be reached. The SNR of the decrypted signals are typically $SNR \approx 35$ dB at every channel, which evidences the good overall behaviour. As it was mentioned before, only $n_i(t)$ acts as key in this encryption setup, as a consequence, it becomes interesting investigate the robustness of the proposed method against blind

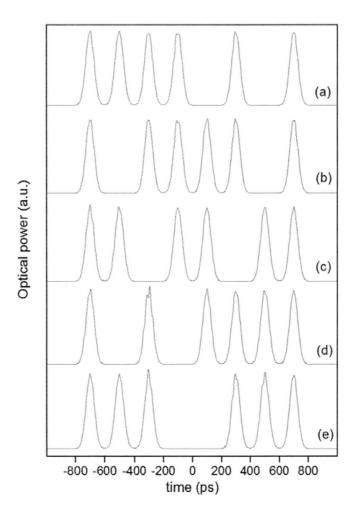

Figure 16. The DRPE-FT decrypting performance. Decrypted signal intensities corresponding to the inputs showed in Fig. 13: (a) $g(t,\lambda_1)$, (b) $g(t,\lambda_2)$, (c) $g(t,\lambda_3)$, (d) $g(t,\lambda_4)$ and (e) $g(t,\lambda_5)$. Channel wavelengths are the same as in Fig. 13 caption.

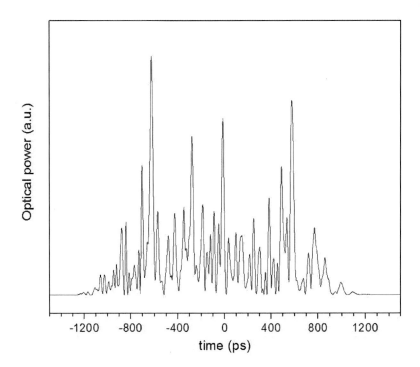

Figure 17. The DRPE-FT performance against blind decryption with $n(t)$ wrong, $f(t,\lambda_3)$ was used as an example.

decryption when an incorrect key $n_i(t)$ has been used in one channel. Figure 17 shows the decrypted signal using $f(t,\lambda_3)$ as an example. As expected, the original signal cannot be recognized when an incorrect key is used, being the $SNR \approx 1$ dB in this case.

5.2. The DRPE-FrFT Performance

As scale parameter for performing the fractional Fourier transform operations a value of $s = 2.14 \times 10^4$ ps^2/rad was chosen, see Eqs. (27) and (28). In each channel, besides the keys $m_i(t)$ and $n_i(t)$, the fractional Fourier transform orders can be used also as keys, with this purpose in mind, it have been chosen the following combinations of fractional Fourier transform orders: $p_1^{(1)} = 0.2$, $p_2^{(1)} = 0.5$; $p_1^{(2)} = 0.4$, $p_2^{(2)} = 0.3$; $p_1^{(3)} = 0.3$, $p_2^{(3)} = 0.4$; $p_1^{(4)} = 0.5$, $p_2^{(4)} = 0.2$; and $p_1^{(5)} = 0.35$, $p_2^{(5)} = 0.35$, where the superscript indicates channel wavelength. Additionally, we have restricted the preceding fractional Fourier transform orders to the condition $p_1^{(i)} + p_2^{(i)} = 0.7$, to avoid excessive disparity between time extents and bandwidths of adjacent channels. Figure 18a shows a typical encrypted signal, e.g., $\psi(t,\lambda_3)$, whereas Fig. 18b shows the total intensity of the simultaneous transmission of the five encrypted data streams, i.e., $\xi(t,\lambda_1,...,\lambda_5)$. It can be clearly observed that its temporal extension is confined to a well-defined temporal region instead of being largely spread, being this a direct consequence of using quasi-white noise signals in the encryption process. The encrypted signal has temporally broadened from $\Delta t_f^{(3)} \approx 1.6$ ns up to $\Delta t_\psi^{(3)} = 2.34$ ns, which has been obtained with Eq. (45) that reasonably matches with the situation depicted by Fig. 18a, i.e., $\Delta t_\psi^{(3)} \approx 2$ ns. Besides, $\Delta t_\psi^{(3)}$ could be made even smaller by selecting a lower $\Delta \nu_a$ or $\Delta \nu_b$, or both, as it was analyzed in Section 4. As a consequence of the encryption process, its corresponding spectrum has largely broadened $\Delta \nu_\psi^{(3)} = 110$ GHz (i.e., around ≈ 0.7 nm if the mean wavelength is around 1550 nm), see Fig. 18c, as compared with the original signal spectrum $\Delta \nu_f^{(3)} \approx 10$ GHz (see Fig. 14a). Whereas the bandwidth predicted by using Eq. (45) results in $\Delta \nu_\psi^{(3)} = 128$ GHz. Again, it should be kept in mind the inherent approximate nature of the derived equations, as it was mentioned in the preceding subsection. After passing through the multiplexer, every encrypted signal is recombined; thereby sharing the

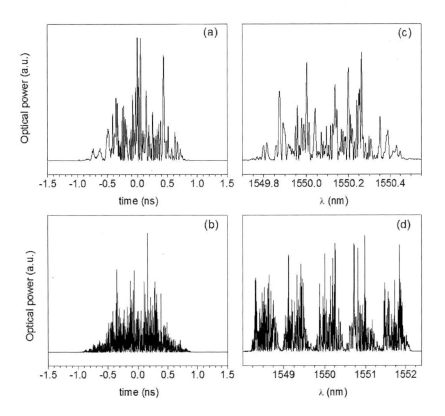

Figure 18. The DRPE-FrFT encrypting performance with $p_1^{(i)} + p_2^{(i)} = 0.7$. (a) Encrypted signal intensity corresponding to a single channel, e.g., $\psi(t, \lambda_3)$. (b) Encrypted signal intensity corresponding to the WDM transmission of the five data streams, i.e. $\xi(t, \lambda_1, ..., \lambda_5)$. (c) and (d) spectral powers of the signals showed in (a) and (b), i.e., $\Psi(\nu, \lambda_3)$ and $\Xi(\nu, \lambda_1, ..., \lambda_5)$, respectively. Channel wavelengths are the same as in Fig. 13 caption.

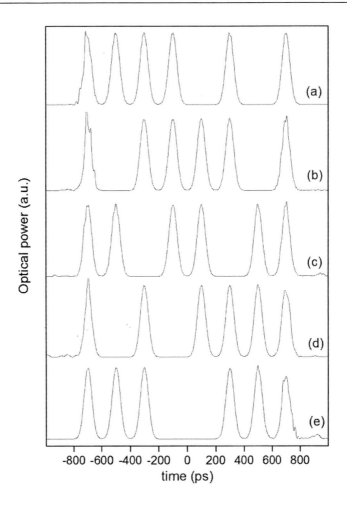

Figure 19. The DRPE-FrFT decrypting performance with $p_1^{(i)} + p_2^{(i)} = 0.7$. Decrypted signal intensities corresponding to the inputs showed in Fig. 13: (a) $g(t,\lambda_1)$, (b) $g(t,\lambda_2)$, (c) $g(t,\lambda_3)$, (d) $g(t,\lambda_4)$ and (e) $g(t,\lambda_5)$. Channel wavelengths are the same as in Fig. 13 caption.

same time window, see Fig. 18b. Finally, in Fig. 18d the whole five encrypted signal spectrum is shown, where it can be seen a certain amount of overlapping between single channel spectra. The decoded signals are shown in Fig. 19, although it exist close resemblance between the original and decrypted signals (compare Figs. 13 and 19), it is a fact that decrypted signals have some kind of distortion, mainly due to the existing overlapping between neighboring spectra. This is present also in the SNR that has dropped to 19 dB as compared to the 35 dB obtained in the preceding subsection with the DRPE-FT setup. From Eq. (45), it is clear that there will be fractional Fourier transform orders for which the time and frequency extents will be different, and if the fractional Fourier transform orders are properly chosen, the overlapping between neighboring channels will be negligible. For this reason, results for another combinations of fractional Fourier transform orders are shown:

$$p_1^{(1)} = 0.85, \quad p_2^{(1)} = 0.75; \quad p_1^{(2)} = 0.9, \quad p_2^{(2)} = 0.7; \quad p_1^{(3)} = 0.7, \quad p_2^{(3)} = 0.9; \quad p_1^{(4)} = 0.8,$$

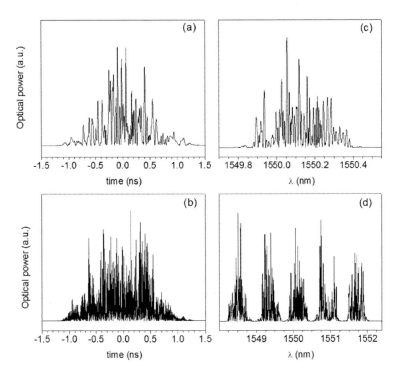

Figure 20. The DRPE-FrFT encrypting performance with $p_1^{(i)} + p_2^{(i)} = 1.6$. (a) Encrypted signal intensity corresponding to a single channel, e.g., $\psi(t, \lambda_3)$. (b) Encrypted signal intensity corresponding to the WDM transmission of the five data streams, i.e., $\xi(t, \lambda_1, ..., \lambda_5)$. (c) and (d) spectral powers of the signals showed in (a) and (b), i.e., $\Psi(\nu, \lambda_3)$ and $\Xi(\nu, \lambda_1, ..., \lambda_5)$, respectively. Channel wavelengths are the same as in Fig. 13 caption.

$p_2^{(4)} = 0.8$; and $p_1^{(5)} = 0.75$, $p_2^{(5)} = 0.85$. Same as with the preceding example, we have restricted the preceding fractional Fourier transform orders to the condition $p_1^{(i)} + p_2^{(i)} = 1.6$, to avoid excessive disparity between time extents and bandwidths of adjacent channels. Figure 20a shows a typical encoded signal, e.g., $\psi(t, \lambda_3)$, whereas Fig. 20b shows the total intensity of the simultaneous transmission of the five encrypted data streams, i.e. $\xi(t, \lambda_1, ..., \lambda_5)$. As a consequence of changing the fractional Fourier transform orders, the temporal and spectral extents have changed. In this case, the predicted value by using Eq. (45) is $\Delta t_\psi^{(3)} \approx 2.94$ ns that reasonably matches with the value obtained through Fig. 20c, i.e., $\Delta t_\psi^{(3)} \approx 2.7$ ns. Its corresponding spectrum has become narrower, see Fig. 20c as compared with Fig. 18c. The bandwidth predicted by using Eq. (45) results in $\Delta \nu_\psi^{(3)} = 95.4$ GHz that can be compared reasonably well with the bandwidth observed at Fig. 20c, i.e., $\Delta \nu_\psi^{(3)} \approx 80$ GHz (i.e., around ≈ 0.6 nm if the mean wavelength is around 1550 nm). Finally, in Fig. 20d the whole five encrypted signal spectrum is shown, it can be observed that the amount of overlapping between single channel spectra has diminished as compared to the preceding example in which another fractional Fourier transform orders where used, i.e. compare Figs. 20d and 18d. The decoded signals are shown in Fig. 21, here the resemblance between original and decrypted signals (compare Figs. 19 and 21) has improved. This behavior can be numerically corroborated through the SNR that has raised up to 32 dB, as compared to the 19 dB obtained with the preceding fractional Fourier transform orders. It is clear also that this better decryption performance has been obtained at expenses of increasing the time extent of the encrypted signal, as can be checked by comparing Figs. 18a and 20a. Following Eq. (45), Figs. 22 and 23 illustrates grayscale two-dimensional plots of the time width and bandwidth of the encrypted signal, respectively, as functions of the fractional Fourier transform orders: p_1 (horizontal axis) and p_2 (vertical axis). As explained before (Subsection 4.2), it is enough to limit the p value to the interval $[0,1]$, because for higher p values a convenient reduction to the first quadrant can be performed. It can be concluded that, no matter which fractional Fourier transform orders have been chosen as keys, the encrypted

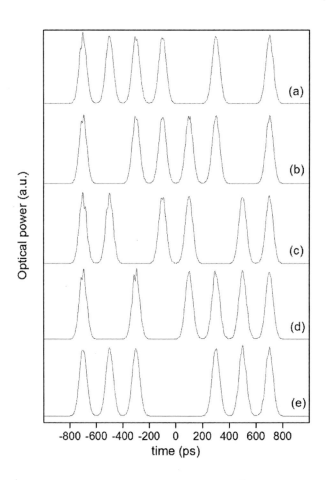

Figure 21. The DRPE-FrFT decrypting performance with $p_1^{(i)} + p_2^{(i)} = 1.6$. Decrypted signal intensities corresponding to the inputs showed in Fig. 13: (a) $g(t, \lambda_1)$, (b) $g(t, \lambda_2)$, (c) $g(t, \lambda_3)$, (d) $g(t, \lambda_4)$ and (e) $g(t, \lambda_5)$. Channel wavelengths are the same as in Fig. 13 caption.

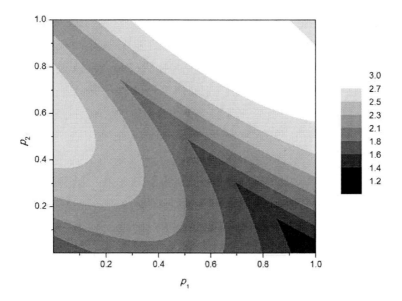

Figure 22. The DRPE-FrFT encrypting performance. Encrypted optical signal time width Δt_ψ as a function of both fractional Fourier transform orders p_1 and p_2, with $0 \leq p_i \leq 1$ ($i = 1, 2$). The grey tones box at the right shows the Δt_ψ scale in ns.

signal extents in both domains remain limited to narrow temporal and spectral zones, i.e. $1 \text{ ns} \leq \Delta t_\psi \leq 3 \text{ ns}$, and $50 \text{ GHz} \leq \Delta \nu_\psi \leq 140 \text{ GHz}$. This behavior becomes an additional feature for employing the fractional orders of the fractional Fourier transform as flexible keys in the encryption-decryption procedure. Finally, it should not be forgotten that the location of the maximums and minimums in the p_1–p_2 plane of the time width and bandwidth depends on the scale parameter s, the bandwidth of the key signals $m(t)$ and $n(t)$, and the input signal characteristics, see Eq. (45). For another values of these parameters the location of these maximums are, generally, in distinct positions. Now, the robustness of the proposed method against blind decryption will be numerically investigated by using $f(t, \lambda_3)$ as an example. Figure 24 shows the decrypted signals when different combinations of incorrect keys are used: *i*) $p_1 = 0.7$ and $p_2 = 0.9$ right, and $n(t)$ wrong (Fig. 24a), *ii*) $p_1 = 0.7$ and $n(t)$ right, and $p_2 = 0.8$ wrong (Fig. 24b), *iii*) $n(t)$ right, $p_1 = 0.8$ and $p_2 = 0.8$ wrong (Fig. 24c), and finally *iv*) $p_2 = 0.9$ and $n(t)$ right, and $p_1 = 0.8$ wrong (Fig. 24d). The obtained SNRs have been $SNR \approx 0.9 \text{ dB}$, 1.7 dB, 2.2 dB, and 5.7 dB, respectively. Finally, Fig. 25 shows the SNR when a detuning from the right fractional order p is present: *i*) p_2 fixed to its right value (i.e. $p_2 = 0.9$) and $0.2 \leq p_1 \leq 1.2$ (right $p_1 = 0.7$), solid line; *ii*) p_1 fixed to its right value (i.e. $p_1 = 0.7$) and $0.4 \leq p_2 \leq 1.4$ (right $p_2 = 0.9$), dashed line. There, it can be observed that p_2 (dashed line) is more robust than p_1 (solid line) against blind decryption. The reason for this is that an incorrect p_2 will result in multiplication by the random phase modulator $b^*(t)$ in the wrong fractional domain [7].

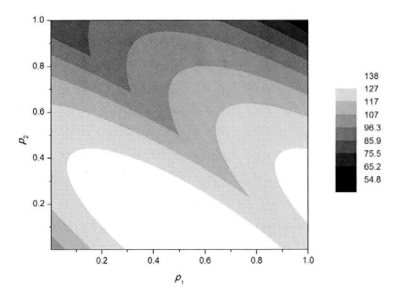

Figure 23. The DRPE-FrFT encrypting performance. Encrypted optical signal bandwidth $\Delta\nu_\psi$ as a function of both fractional Fourier transform orders p_1 and p_2, with $0 \leq p_i \leq 1$ $(i = 1, 2)$. The grey tones box at the right shows the $\Delta\nu_\psi$ scale in GHz.

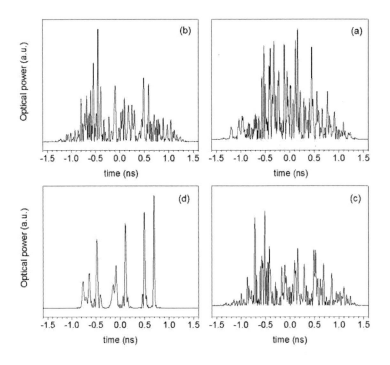

Figure 24. The DRPE-FrFT performance against blind decryption. (a) $p_1 = 0.7$ and $p_2 = 0.9$ right, and $n(t)$ wrong, (b) p_1 and $n(t)$ right, and $p_2 = 0.8$ wrong, (c) $n(t)$ right, $p_1 = 0.8$ and $p_2 = 0.8$ wrong, and (d) p_2 and $n(t)$ right, and $p_1 = 0.8$ wrong, $f(t, \lambda_3)$ was used as example.

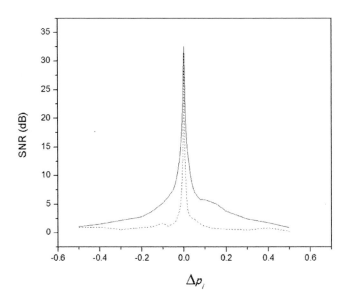

Figure 25. The DRPE-FrFT performance against blind decryption. Signal-to-noise ratio when a detuning from the right fractional order p is present: i) p_2 fixed to its right value (i.e., $p_2 = 0.9$) and $0.2 \leq p_1 \leq 1.2$ (right $p_1 = 0.7$), solid line; ii) p_1 fixed to its right value (i.e., $p_1 = 0.7$) and $0.4 \leq p_2 \leq 1.4$ (right $p_2 = 0.9$), dashed line. $f(t, \lambda_3)$ was used as example.

5.3. The DRPE-FsT Performance

The same total dispersion was chosen for each channel: $\Phi_{20}^{(T)} = \Phi_{20}^{(1)} + \Phi_{20}^{(2)} = -4 \times 10^3 \ \text{ps}^2/\text{rad}$, where $\Phi_{20}^{(i)} = \chi_i \Phi_{20}^{(T)}$, with $0 \leq \chi_i \leq 1$. Further, in order to use different keys for each channel, we have changed the dispersion distribution between encryption stages, i.e., between $\Phi_{20}^{(1, i)}$ and $\Phi_{20}^{(2, i)}$. In this way, the chosen keys χ_1 and χ_2 from $f(t, \lambda_1)$ to $f(t, \lambda_5)$ were chosen as: $\{0.5, 0.5\}$, $\{0.55, 0.45\}$, $\{0.45, 0.55\}$, $\{0.4, 0.6\}$, and finally $\{0.6, 0.4\}$. Figure 26a shows the intensity of an encrypted signal, e.g. $\psi(t, \lambda_3)$, whereas Fig. 26b shows the total intensity of the simultaneous transmission of the five encrypted data streams, i.e., $\xi(t, \lambda_1, ..., \lambda_5)$. Their corresponding spectral powers are shown in Figs. 26c and 26d, i.e., $\Psi(\nu, \lambda_3)$ and $\Xi(\nu, \lambda_1, ..., \lambda_5)$, respectively. It can be observed, by comparing Figs. 26a and 13c, that the encrypted signal has temporally broadened, i.e. from $\Delta t_f^{(3)} = 1.6 \ \text{ns}$ up to $\Delta t_\psi^{(3)} \approx 3 \ \text{ns}$. This last value compares reasonably well with the predicted value obtained by applying Eq. (47) $\Delta t_\psi^{(3)} \approx 3.5 \ \text{ns}$. As a consequence of the encryption process, the spectral extent has increased from $\Delta \nu_f^{(3)} \approx 10 \ \text{GHz}$ (i.e., $\Delta \lambda_f^{(3)} \approx 0.1 \ \text{nm}$) up to $\Delta \nu_\psi^{(3)} \approx 90 \ \text{GHz}$ (i.e., $\Delta \lambda_\psi^{(3)} \approx 0.6 \ \text{nm}$), which also compares reasonably well with the value predicted by Eq. (48),

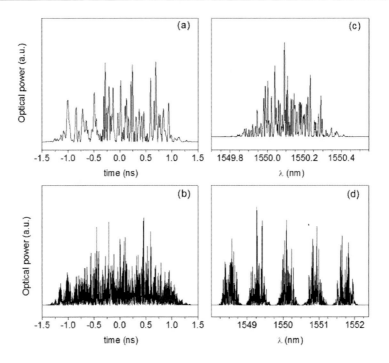

Figure 26. The DRPE-FsT encrypting performance. (a) Encrypted signal intensity corresponding to a single channel, e.g., $\psi(t,\lambda_3)$. (b) Encrypted signal intensity corresponding to the WDM transmission of the five data streams, i.e. $\xi(t,\lambda_1,...,\lambda_5)$. (c) and (d) spectral powers of the signals showed in (a) and (b), i.e., $\Psi(\nu,\lambda_3)$ and $\Xi(\nu,\lambda_1,...,\lambda_5)$, respectively. Channel wavelengths are the same as in Fig. 13 caption.

i.e. $\Delta\nu_\psi^{(3)} \approx 100\ \text{GHz}$. The five channels together have the same total temporal extension than one single encrypted signal, as a consequence of the WDM transmission process; see Fig. 26b as compared with Fig. 26a. Finally, in Fig. 26d the whole five encrypted signal spectrum is shown, where it can be seen that there is not any appreciable overlapping between single channel spectra. Figure 27 shows the decrypted signals corresponding to the Fresnel transform encryption of the signals showed in Fig. 13. It can be clearly observed the close resemblance between the inputs and decrypted signal, with a $SNR \approx 27.5\ \text{dB}$. The small ripple present in the decrypted signal diminishes as the separation between channels enlarges, as before, a compromise between noise and channel capacity must be reached.

In order to make the comparison between different encryption setups possible, we have choose the same dispersion at the final encryption stage ($\Phi_{20}^{(2)}$) in the preceding results. In this way, it is clear by comparing Eqs. (39) and (47) that the DRPE-FsT has a lower performance as compared with the DRPE-FT. This has become in evidence also by comparing the SNRs in both cases, which has dropped from $SNR \approx 35\ \text{dB}$ (DRPE-FT) up to $SNR \approx 27.5\ \text{dB}$ (DRPE-FsT). The DRPE-FsT shows also a higher bandwidth for the encrypted signal as compared with the DRPE-FT, see Eqs. (40) and (48), which also has detrimental consequences in WDM applications. Being the DRPE-FT an especial case of the DRPE-FrFT, the same could be said of a comparison between the later and the DRPE-FsT.

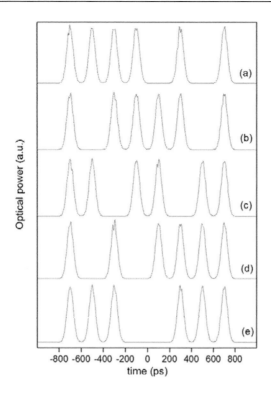

Figure 27. The DRPE-FsT decrypting performance. Decrypted signal intensities corresponding to the inputs showed in Fig. 13: (a) $g(t,\lambda_1)$, (b) $g(t,\lambda_2)$, (c) $g(t,\lambda_3)$, (d) $g(t,\lambda_4)$ and (e) $g(t,\lambda_5)$. Channel wavelengths are the same as in Fig. 13 caption.

The robustness of the proposed method against blind decryption was numerically investigated also. Figure 28 shows the decrypted signal when different combinations of incorrect keys were used, where $f(t,\lambda_3)$ was chosen merely as an example: *i*) $\chi_1 = 0.5$ and $\chi_2 = 0.5$ right, and $n(t)$ wrong (Fig. 28a), *ii*) $\chi_1 = 0.5$ and $n(t)$ right, and $\chi_2 = 0.6$ wrong (Fig. 28b), *iii*) $n(t)$ right, $\chi_1 = 0.6$ and $\chi_2 = 0.6$ wrong (Fig. 28c), and finally *iv*) $\chi_2 = 0.5$ and $n(t)$ right, and $\chi_1 = 0.6$ wrong (Fig. 28d). The obtained SNRs have been $SNR \approx 2.1$ dB, 2.4 dB, 2.3 dB, and 5.9 dB, respectively. Finally, Fig. 29 shows the SNR when a detuning from the right dispersion coefficient is present: χ_1 fixed to its right value (i.e., $\chi_1 = 0.5$) and $0 \leq \chi_2 \leq 1$ (right $\chi_2 = 0.5$), dash line; χ_2 fixed to its right value (i.e., $\chi_2 = 0.5$) and $0 \leq \chi_1 \leq 1$ (right $\chi_1 = 0.5$), solid line. There, it can be observed that χ_2 (dash line) is more robust than χ_1 (solid line) against blind decryption. This result could be compared with the DRPE-FrFT behaviour when a detuning of the right fractional orders is present, see Fig. 25. Both encryption setups are more resistant against blind decryption when the second coefficient is detuned: p_2 (in the DRPE-FrFT) and χ_2 (DRPE-FsT), and for identical reason: the multiplication by the random phase modulator $b^*(t)$ in the wrong fractional domain for the DRPE-FrFT, and by the wrong Fresnel transform plane in the DRPE-FsT.

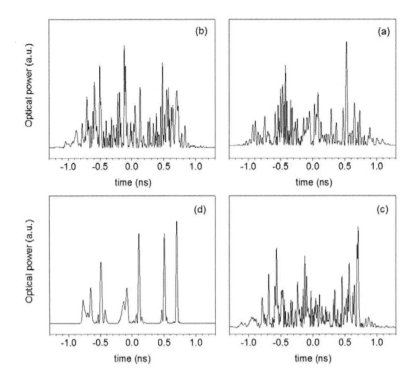

Figure 28. The DRPE-FsT performance against blind decryption. (a) $\chi_1 = 0.5$ and $\chi_2 = 0.5$ right, and $n(t)$ wrong, (b) χ_1 and $n(t)$ right, and $\chi_2 = 0.6$ wrong, (c) $n(t)$ right, $\chi_1 = 0.6$ and $\chi_2 = 0.6$ wrong, and (d) χ_2 and $n(t)$ right, and $\chi_1 = 0.6$ wrong.

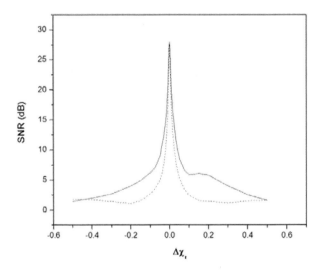

Figure 29. The DRPE-FsT performance against blind decryption. Signal-to-noise ratio when a detuning of the right dispersion coefficient is present: χ_1 fixed to its right value (i.e., $\chi_1 = 0.5$) and $0 \leq \chi_2 \leq 1$ (right $\chi_2 = 0.5$), dashed line; χ_2 fixed to its right value (i.e., $\chi_2 = 0.5$) and $0 \leq \chi_1 \leq 1$ (right $\chi_1 = 0.5$), solid line.

6. Feasibility Considerations

Now we consider the experimental implementation feasibility of the proposed setups. The example studied throughout this work, as regards with the separation between channels, belongs to dense WDM (DWDM). In case of applications using the entire frequency band between second and third transmission window (1310–1550 nm, respectively), OH-free silica fibres could be used in order to fully exploit the WDM capacity of this encryption setup. These fibres span the communication wavelength range from ~1310 nm to ~1625 nm, allowing a full spectrum operation. For instance, a full-spectrum single-mode fibre has typical dispersion values ranging from the zero dispersion value at $\lambda_0 \approx 1317$ nm to $D \approx 20$ ps/nm×km (i.e., $\beta_2 \approx -28.5$ ps^2/km) at $\lambda \approx 1625$ nm.[2] By employing such fibre, and by considering the frequency spacing mainly used throughout this work (i.e., $\Delta\nu^{(c)} = 100$ GHz), ~430 channels could be assigned for secure data transmission purposes.

Otherwise, with an ordinary single mode fibre, ~213 channels would be possible transmitting at both windows (~170 and ~43 channels for the 2nd and 3rd window, respectively). Moreover, the wavelength dependence of dispersion of the fibre optic link should be carefully considered also in the experimental implementation.

Polarization-mode dispersion (PMD) induced by random birefringence into single-mode optical fibres can be the dominant source of pulse distortion in high bit-rate transmission systems at great distances. However PMD coefficients of contemporary fibres can be as low as 0.02 ps/√km. Further, for the aforementioned full-spectrum fibre, the PMD typical value is lower than 0.06 ps/√km.[2] As this encryption method was focused mainly for short-haul fibre optic links (e.g., metropolitan distances); PMD does not constitute a major problem in a WDM context.

There are two encryption setups in which become necessary using time lenses: the DRPE-FT and the DRPE-FrFT. As it is widely known, in a time lens based on electro-optic phase modulation, the time aperture and the phase factor are inverse parameters (i.e., the longer the time aperture, the smaller the phase factor), a fact which sometimes could preclude its utilization for some specific cases, like in temporal imaging [17]. However, by using a time lens based on sum-frequency generation in a nonlinear crystal, one can overcome these limitations, and the time lens can be designed to provide the desired phase factor and time aperture essentially as independent parameters [44]. The remaining encryption setup, the DRPE-FsT, is a time lensless encryption setup, for this reason its implementation becomes easier, although it has a lower performance as compared with the others [38].

Before concluding, a few words are in order about the resistance against eavesdrop attacks of the DRPE, and systems related. Although the DRPE dates back to the middle 1990s only recently has it security strength started to be analyzed from a cryptanalysis point of view [45-49]. Frauel et al. showed that the DRPE-FT is resistant against brute force attacks but susceptible to chosen-plaintext attack (CPA) and known plaintext attacks [45]. CPA is an attack model, commonly used in cryptanalysis, which presumes that the attacker has the capability to choose arbitrary plaintexts to be encrypted and obtain the corresponding ciphertexts. Peng et al. showed that DRPE-FsT scheme is also vulnerable to CPA [46]. They

---

[2] Typical values taken from the technical data sheet of the single-mode full-spectrum optical fiber Corning® SMF-28e+™.

show how an opponent can access both random phase keys in the input and Fresnel domains, respectively, if the wavelength and the propagation distances are known, with the impulse functions as chosen plaintexts. Any cryptographic algorithm could be claimed to be secure if and only if, it were able to endure various attacks from cryptanalysis. Therefore, under the cryptanalysis point of view, the DRPE-FT and the DRPE-FsT have a vulnerability failure. However, CPA becomes more important in other contexts, namely in public key cryptography, where the encryption key is public and attackers can encrypt any plaintext they choose. Moreover, it must not be forgotten that it is not a trivial task for an eavesdropper the knowledge of both propagation distances and the mean wavelength.

7. Conclusion

The implementation of spatial encryption techniques in the time domain was studied to evaluate its potential application for secure data transmission, mainly in metropolitan fiber optic links. Decryption by an eavesdropper becomes unsuccessful because of the large number of involved degrees of freedom, and the vast amount of possible permutations of parameter keys. Special emphasis was taken for theoretically analyzing the optical fiber multiplexing capabilities. We have found general expressions relating the temporal and frequency extent of the encrypted signal transmitted through the fiber optic link with the input signal and key parameters. As a result, a quasi-white noise signal with a well-defined bandwidth is used to enhance the channel number capacity in WDM. With this proposal, a novel alternative addressing the problem of security by handling multiple data in a temporal dual random phase encoding in fiber optic links was introduced, something that has not been deeply explored using these techniques. Summarizing, we settle on a new point of view to a multiplexing transmission mechanism that expands the possible combinations of encrypted data within a given fiber optic link architecture.

Acknowledgments

This work was supported by the Consejo Nacional de Investigaciones Científicas y Técnicas (CONICET-Argentina), the Secretaría de Estado de Universidades e Investigación del Ministerio de Investigación y Ciencia (Spain), and the Facultad de Ingeniería de la Universidad Nacional de La Plata (UNLP-Argentina).

References

[1] P. Réfrégier and B. Javidi, "Optical image encryption based on input plane and Fourier plane random encoding," *Opt. Lett.* **20**, 767-769 (1995).

[2] N. Towghi, B. Javidi, and Z. Luo, "Fully phase encrypted image processor," *J. Opt. Soc. Am. A* **16**, 1915-1927 (1999).

[3] B. Javidi, N. Towghi, N. Maghzi, and C. Verrall, "Error-reduction techniques and error analysis for fully phase- and amplitude-based encryption," *Appl. Opt.* **39**, 4117-4130 (2000).

[4]   O. Matoba and B. Javidi, "Encrypted optical memory system using three-dimensional keys in the Fresnel domain," *Opt. Lett.* **24**, 762-764 (1999).

[5]   G. Unnikrishnan, J. Joseph, and K. Singh, "Optical encryption by double-random encoding in the fractional Fourier domain," *Opt. Lett.* **25**, 887-889 (2000).

[6]   G. Unnikrishnan and K. Singh, "Double random fractional Fourier-domain encoding for optical security," *Opt. Eng.* **39**, 2853-2859 (2000).

[7]   B.M. Hennelly and J.T. Sheridan, "Image encryption and the fractional Fourier transform," *Optik* **114**, 251-265 (2003).

[8]   G. Situ and J. Zhang, "Double random-phase encoding in the Fresnel domain," *Opt. Lett.* **29**, 1584-1586 (2004).

[9]   G. Situ and J. Zhang, "A lensless optical security system based on computer generated phase only," *Opt. Commun.* **232**, 115-122 (2004).

[10] L. Chen and D. Zhao, "Optical color image encryption by wavelength multiplexing and lensless Fresnel transform holograms," *Opt. Exp.* **14**, 8552-8560 (2006).

[11] G. Situ and J. Zhang, "Multiple-image encryption by wavelength multiplexing," *Opt. Lett.* **30**, 1306-1308 (2005).

[12] L. Chen and D. Zhao, "Optical image encryption with Hartley transforms," *Opt. Lett.* **31**, 3438-3440 (2006).

[13] B.M. Hennelly, T.J. Naughton, J. McDonald, J.T. Sheridan, G. Unnikrishnan, D.P. Kelly, and B. Javidi, "Spread-space spread-spectrum technique for secure multiplexing," *Opt. Lett.* **32**, 1060-1062 (2007).

[14] B.H. Kolner, "Space-time duality and the theory of temporal imaging," *IEEE J. Quantum. Electron.* **30**, 1951-1963 (1994).

[15] A. Papoulis, "Pulse compression, fiber communications, and diffraction: a unified approach*," J. Opt. Soc. Am. A* **11**, 3-13 (1994).

[16] J. Azaña, L.R. Chen, M.A. Muriel, and P.W.E. Smith, "Experimental demonstration of real-time Fourier transformation using linearly chirped fibre Bragg gratings," *Electron. Lett.* **35**, 2223-2224 (1999).

[17] J. Azaña and L. Chen, "General temporal self-imaging phenomena," *J. Opt. Soc. Am. B* **20**, 1447-1458 (2003).

[18] J. Goodman, *Introduction to Fourier Optics* (McGraw-Hill, 1996).

[19] D. Mendlovic, R. Dorsch, Z. Zalevsky, A. Lohmann, and C. Ferreira, "Optical illustration of a varied fractional Fourier-transform order and the Radon-Wigner display," *Appl. Opt.* **35**, 3925-3929 (1996).

[20] H.M. Ozaktas, Z. Zalevsky, and M. Alper Kutay, *The Fractional Fourier Transform with Applications in Optics and Signal Processing*, Wiley, 2001.

[21] U. Sümbül and H. Ozaktas, "Fractional free space, fractional lenses, and fractional imaging systems," *J. Opt. Soc. Am. A* **20**, 2033-2040 (2003).

[22] C. Cuadrado-Laborde, P. Costanzo-Caso, R. Duchowicz, and E.E. Sicre, "Ultrafast Optical Temporal Processing using Phase-Space Signal Representations," in: P.V. Gallico (Ed.), Optics Research Trends, Nova Science Publishers, New York, 2007. Chapter 6, pp. 221-270.

[23] A.W. Lohmann, "Image rotation, Wigner rotation, and the fractional Fourier transform," *J. Opt. Soc. Am. A* **10**, 2181-2186 (1993).

[24] I. Moreno, M.M. Sánchez-López, C. Ferreira, and F. Mateos, "Fractional Fourier transforms, symmetrical lens systems, and their cardinal planes," *J. Opt. Soc. Am. A* **24**, 1930-1936 (2007).

[25] J. Azaña and M.A. Muriel, "Temporal Talbot effect in fiber gratings and its applications," *Appl. Opt.* **38**, 6700-6704 (1999).

[26] G. Agrawall, Nonlinear Fiber Optics; Optics and Photonics series; Academic Press, London, UK, 2001; 39-45.

[27] J. van Howe and C. Xu, "Ultrafast optical signal processing based upon space-time dualities," *J. Lightwave Technol.* **24**, 2649-2662 (2006).

[28] A.W. Lohmann and D. Mendlovic, "Temporal filtering with time lenses," *Appl. Opt.* **31**, 6212-6219 (1992).

[29] B.H. Kolner, "Active pulse compression using an integrated electro-optic phase modulator," *Appl. Phys. Lett.* **52**, 1122-1124 (1988).

[30] S. Mookherjea and A. Yariv, "Analysis of optical pulse propagation with two-by-two (ABCD) matrices," *Phys. Rev. E* **64**, 16611 (2001).

[31] N.K. Berger, B. Levit, S. Atkins, and B. Fischer, "Time-lens-based spectral analysis of optical pulses by electro-optic phase modulation," *Electron. Lett.* **36**, 1644-1646 (2000).

[32] X. Yang, "Implementation of time lenses and optical temporal processors," *Opt. Commun.* **116**, 193-207 (1995).

[33] C.V. Bennett and B.H. Kolner, "Aberrations in Temporal Imaging," *IEEE J. Quantum Electron.* **37**, 20-32 (2001).

[34] C. Cuadrado-Laborde, R. Duchowicz, R. Torroba, and E.E. Sicre, "Dual random phase encoding: a temporal approach for fiber optic applications," *Appl. Opt.* **47**, 1940-1946 (2008).

[35] A.E. Willner and Y. Xie, "Wavelength Domain Multiplexed (WDM) Fiber Optic Communication Networks," in *Fiber Optics Handbook*, M. Bass and E. W. Van Stryland, eds. (McGraw-Hill, 2002), 13.1-13.31.

[36] C. Cuadrado-Laborde, P. Costanzo-Caso, R. Duchowicz, and E.E. Sicre, "Pulse propagation analysis based on the temporal Radon Wigner transform" *Opt. Commun.* **266**, 32-38 (2006).

[37] C. Cuadrado-Laborde, R. Duchowicz, R. Torroba, and E.E. Sicre, "Fractional Fourier Transform Dual Random Phase Encoding of Time-Varying Signals" *Opt. Commun.* **281** (2008) 4321-4328.

[38] C. Cuadrado-Laborde, "Time-variant signal encryption by lensless dual random phase encoding applied to fiber optic links," *Opt. Lett.* **32**, 2867-2869 (2007).

[39] B.M. Hennelly and J.T. Sheridan, "Optical encryption and the space bandwidth product", *Opt. Commun.* **247**, 291-305 (2005).

[40] B.M. Hennelly and J.T. Sheridan, "Generalizing, optimizing, and inventing numerical algorithms for the fractional Fourier, Fresnel, and linear canonical transforms," *J. Opt. Soc. Am. A* **22**, 917-927 (2005).

[41] B.M. Hennelly and J.T. Sheridan, "Tracking the space bandwidth product in optical systems," *Proc. of SPIE* **5827**, 334-345 (2005).

[42] L.G. Neto and Y. Sheng, "Optical implementation of image encryption using random phase encoding," *Opt. Eng.* **35**, 2459-2463 (1996).

[43] T. Nomura, E. Nitanai, T. Numata, and B. Javidi, "Design of input phase mask for the space bandwidth of the optical encryption system," *Opt. Eng.* **45,** 17006 (2006).

[44] C. Bennett and B. Kolner, "Upconversion time microscope demonstrating 103 × magnification of femtosecond waveforms," *Opt. Lett.* **24**, 783-785 (1999).

[45] Y. Frauel, A. Castro, T.J. Naughton, and B. Javidi, "Resistance of the double random phase encryption against various attacks," *Opt. Express* **15**, 10253-10265 (2007).

[46] X. Peng, H. Wei, and P. Zhang, "Chosen-plaintext attack on lensless double-random phase encoding in the Fresnel domain," *Opt. Lett.* **31**, 3261-3263 (2006).

[47] A. Carnicer, M. Montes-Usategui, S. Arcos, and I. Juvells, "Vulnerability to chosen-cyphertext attacks of optical encryption schemes based on double random phase keys," *Opt. Lett.* **30**, 1644-1646 (2005).

[48] X. Peng, P. Zhang, H. Wei, and B. Yu, "Known-plaintext attack on optical encryption based on double random phase keys," *Opt. Lett.* **31**, 1044-1046 (2006).

[49] U. Gopinathan, D.S. Monaghan, T.J. Naughton, and J.T. Sheridan, "A known-plaintext heuristic attack on the Fourier plane encryption algorithm," *Opt. Express* **14**, 3181-3186 (2006).

In: Atomic, Molecular and Optical Physics...
Editor: L.T. Chen, pp. 313-343

ISBN 978-1-60456-907-0
© 2009 Nova Science Publishers, Inc.

Chapter 8

MAGNETISM IN PURE AND DOPED MANGANESE CLUSTERS

Mukul Kabir[1],*, *Abhijit Mookerjee*[1] *and D.G. Kanhere*[2]
[1]S. N. Bose National Centre for Basic Sciences, JD Block, Sector III,
Salt Lake City, Kolkata 700 098, India
[2]Department of Physics and Centre for Modeling and Simulation,
University of Pune, Pune - 411 007, India

Abstract

In this chapter, we report electronic and magnetic structure of pure and (As-) doped manganese clusters from density functional theory using generalized gradient approximation for the exchange-correlation energy. Ferromagnetic to ferrimagnetic transition takes place at $n = 5$ for pure manganese clusters, Mn_n, and remarkable lowering of magnetic moment is found for Mn_{13} and Mn_{19} due to their closed icosahedral growth pattern and results show excellent agreement with experiment. On the other hand, for As-doped manganese clusters, Mn_nAs, ferromagnetic coupling is found only in Mn_2As and Mn_4As and inclusion of a single As stabilizes manganese clusters. Exchange coupling in the Mn_nAs clusters are anomalous and behave quite differently from the Ruderman-Kittel-Kasuya-Yosida like predictions. Finally, possible relevance of the observed magnetic behaviour is discussed in the context of Mn-doped GaAs semiconductor ferromagnetism.

Keywords: Cluster, magnetism, semiconductor ferromagnetism

1. Introduction

The terms "clusters" and "microclusters" are usually used to describe aggregates of number of atoms, starting with the diatomic molecule and reaching an upper bound of several hundred thousands of atoms, which are too small to resemble small pieces of crystals. They bridge the gap between the atom and crystals and serve as the dome of basic physics at reduced dimension. The structural, electronic and magnetic properties of these aggregates

---

*E-mail address: mukul@bose.res.in. (Corresponding author)

differ widely from their corresponding bulk material and these properties can change dramatically with the addition of just one or few atoms to it. Surface rearrangements can take place on crystals with adatoms, but these changes are less drastic than the changes occur when one or few atoms are added to smaller clusters.

Particularly, transition metal (TM) clusters and their magnetic properties are of more interest and are unique. How the magnetic properties behave in the reduced dimension and how it evolve with cluster size to the bulk? — are the great fundamental questions with a potential technological importance. Several unexpected magnetic ordering have already been reported in clusters: (1) *Non-zero magnetic moment in the clusters of nonmagnetic bulk material* — Cox and co-workers found that the bare rhodium clusters display non-zero magnetic moment for less than 60 atoms in it [1], which is an indication of either ferromagnetic or ferrimagnetic ordering even though the bulk rhodium is Pauli paramagnet at all temperature. In Rh_n clusters magnetism show strong size dependence: Magnetic moment per atom decreases as the cluster size increases and become non-magnetic above 60 atoms[1]. (2) *Enhancement of magnetic moment in the clusters which is already ferromagnetic in bulk.* For Fe_n clusters, it was found that the magnetic moment per atom oscillates with the size of the cluster, slowly approaching its bulk value[2], and finally (3) *Finite magnetic moment observed in the clusters which is antiferromagnetic as bulk* – The Cr_n [3]and Mn_n [4, 5] have shown such property. It has been found that Cr_n clusters have magnetic moment 0.5-1 μ_B per atom. Knickelbein observed, Mn_5 - Mn_{99} clusters posses finite magnetic moment, which is otherwise antiferromagnetic as bulk. All these magnetic measurements on free clusters have been done by using Stern-Gerlach molecular deflection experiments at "low" temperatures.

As we will be discussing the pure as well as doped manganese clusters in this chapter, we discuss some few aspects of manganese. Among all the $3d$ transition metal elements, manganese is unique as an atom, clusters or crystals. It has $3d^5\ 4s^2$ electronic configuration and has high (2.14 eV) $3d^5\ 4s^2 \to 3d^6\ 4s^1$ promotion energy. This large promotion energy reduces the degree of $s-d$ hybridization as atoms are brought together and leads to weaker bonding. For example, Mn_2 dimer is a weakly bound van der Walls dimer with very low binding energy, ranging $0.1 \pm 0.1 - 0.56 \pm 0.26$ eV per atom[6]. If an additional atom is added and the process keeps on going to form different sized clusters, the binding energy increases monotonically but the improvement is not much and remains the lowest among all the $3d$-transition metal clusters. In the solid phase, manganese exists in four allotropic forms exhibiting a complex phase diagram [7]. The stable form is known as $\alpha-Mn$, which has a very complex lattice structure with as much as 58 atoms in the unit cell and has lowest cohesive energy (2.92 eV). For an isolated manganese atom, according to Hund's rule, the half-filled localized $3d$ electrons give rise to a magnetic moment of 5 μ_B.

We discuss electronic and magnetic properties of Mn_n ($n \leq 20$) and $Mn_n As$ ($n \leq 10$) clusters from the density functional theory. The pure manganese clusters undergo a ferromagnetic $\to$ ferrimagnetic transition at $n = 5$. In the recent Stern-Gerlach experiments a few interesting observations were made [4, 5]. (1) Although the experimental uncertainty in the measurement decreases with the increase in cluster size, Knickelbein found an extraordinarily large uncertainty in the measured magnetic moment of Mn_7, 0.72 ± 0.42 μ_B/atom[5]. (2) A relative decrease in the magnetic moment was observed at $n = 13$ and 19. We have discussed these issues in a recent paper[8]. Parks *et al.* found that free Mn_n

clusters are nonreactive towards H_2 for small sizes $n \leq 15$. However, this reaction rate show an abrupt increase at $n = 16$ [9] and it was argued to be attributed from a nonmetal-metal transition at $n = 16$. All these issues are discussed in a great detail here.

Next we will move on to see how a single dopant affect the electronic and magnetic properties. Recently we observed [10] that if we dope a single As-atom into a Mn_n cluster that binding energy of the resultant Mn_nAs cluster improves substantially due to their strong $p - d$ hybridization, i.e. As-atom stabilizes the Mn_n clusters. Mn_2As and Mn_4As are the only clusters that we found to have ferromagnetic Mn-Mn coupling and for all other sizes this coupling is turned out to be ferrimagnetic. It was also found that the exchange coupling in these doped-clusters are anomalous and behave differently from the Ruderman-Kittel-Kasuya-Yosida like predictions. All these issues will be discussed in the context of semiconductor ferromagnetism. Before going to the results, we briefly discuss the projector augmented wave formalism of pseudopotential what we have been used throughout.

2. The Projector-Augmented-Wave Formalism

The most widely used electronic structure methods can be divided into two classes. First one is the linear method [11] developed from the augmented-plane-wave method [12, 13] and Koringa-Kohn-Rostocar method [14, 15] and the second one is the norm-conserving pseudopotentials developed by Hamann, Schlüter and Chiang [16]. In that scheme, inside some core radius, the all electron (AE) wave function is replaced by a soft nodeless pseudo (PS) wave function, with the restriction to the PS wave function that within the chosen core radius the norm of the PS wave function have to be the same with the AE wave function and outside the core radius both the wave functions are just identical. However, the charge distribution and moments of AE wave function are well reproduced by the PS wave function only when the core radius is taken around the outer most maxima of AE wave function. Therefore, elements with strongly localized orbitals pseudopotentials require a large plane wave basis set. This was improved by Vanderbilt [17], where the norm-conservation constraint was relaxed and a localized atom centered augmentation charges were introduced to make up the charge deficit. But the success is partly hampered by rather difficult construction of the pseudopotential. Later Blöchl [18] has developed the projector-augmented- wave (PAW) method, which combines the linear augmented plane wave method with the plane wave pseudopotential approach, which finally turns computationally elegant, transferable and accurate method for electronic structure calculation. Further Kresse and Joubert [19] modified this PAW method and implemented in their existing Veina ab-initio pseudopotential package (VASP). Here in this section we will discuss briefly the idea of the method.

2.1. Wave Functions

The exact Kohn-Sham density functional is given by,

$$E = \sum_n f_n \langle \Psi_n | -\frac{1}{2}\nabla^2 | \Psi_n \rangle + E_H[n + n_Z] + E_{xc}[n], \quad (1)$$

where $E_H[n + n_Z]$ is the hartree energy of the electronic charge density n and the point charge densities of the nuclei n_Z, $E_{xc}[n]$ is the electronic exchange-correlation energy and

f_n are the orbital occupation number. $|\Psi_n\rangle$ is the all-electron wave function. This physically relevant wave functions $|\Psi_n\rangle$ in the Hilbert space exhibit strong oscillations and make numerical treatment difficult. Transformation of this wave functions $|\Psi_n\rangle$ into a new pseudo wave functions $|\tilde{\Psi}_n\rangle$ in the PS Hilbert space,

$$|\Psi_n\rangle = \tau|\tilde{\Psi}_n\rangle, \tag{2}$$

within the augmentation region Ω_R, then makes PS wave function $|\tilde{\Psi}_n\rangle$ computationally convenient.

Let us now choose a PS partial wave function $|\tilde{\phi}_i\rangle$, which is identical to the corresponding AE partial waves $|\psi_i\rangle$ outside the augmentation region and form a complete set within the augmentation region Ω_R. Within the augmentation region every PS wave function can be expanded into PS partial waves,

$$|\tilde{\Psi}\rangle = \sum_i c_i|\tilde{\phi}_i\rangle, \tag{3}$$

where the coefficients c_i are scalar products,

$$c_i = \langle\tilde{p}_i|\tilde{\Psi}_n\rangle, \tag{4}$$

with some fixed function $\langle\tilde{p}_i|$ of the PS wave function, which is called the projector function. By using the linear transformation[18],

$$\tau = 1 + \sum_i (|\phi_i\rangle - |\tilde{\phi}_i\rangle)\langle\tilde{p}_i|, \tag{5}$$

the corresponding AE wave function $|\Psi_n\rangle$ in the Eq. 1.2 is then,

$$|\Psi_n\rangle = |\tilde{\Psi}_n\rangle + \sum_i (|\psi_i\rangle - |\tilde{\psi}_i\rangle)\langle\tilde{p}_i|\tilde{\Psi}_n\rangle, \tag{6}$$

where i refers to the atomic site $\mathbf{R}$, the angular momentum quantum numbers $L = l, m$ and an additional index k for the reference energy ϵ_{kl} and, with $\tilde{p}_i$ being the projector functions, which within the augmentation region Ω_R satisfy the condition,

$$\langle\tilde{p}_i|\tilde{\phi}_j\rangle = \delta_{ij}. \tag{7}$$

The AE charge density at a given point $\mathbf{r}$ is the expectation value of the real-space projector operator $|r\rangle\langle r|$ and hence given by,

$$n(\mathbf{r}) = \tilde{n}(\mathbf{r}) + n^1(\mathbf{r}) - \tilde{n}^1(\mathbf{r}), \tag{8}$$

where $\tilde{n}(\mathbf{r})$ is the soft PS charge density calculated from the PS wave function

$$\tilde{n}(\mathbf{r}) = \sum_n f_n\langle\tilde{\Psi}_n|\mathbf{r}\rangle\langle\mathbf{r}|\tilde{\Psi}_n|\mathbf{r}\rangle, \tag{9}$$

and onsite charge densities are defined as,

$$n^1(\mathbf{r}) = \sum_{ij} \rho_{ij}\langle\phi_i|\mathbf{r}\rangle\langle\mathbf{r}|\phi_j\rangle, \tag{10}$$

and,

$$\tilde{n}^1(\mathbf{r}) = \sum_{ij} \rho_{ij} \langle \tilde{\phi}_i | \mathbf{r} \rangle \langle \mathbf{r} | \tilde{\phi}_j \rangle. \tag{11}$$

ρ_{ij} are the occupation of each augmentation channel (i, j) and are calculated from the PS wave functions applying the projector function,

$$\rho_{ij} = \sum_{n} f_n \langle \tilde{\Psi}_n | \tilde{p}_i \rangle \langle \tilde{p}_j | \tilde{\Psi}_n \rangle. \tag{12}$$

It should be pointed out here that for a complete set of projector functions the charge density $\tilde{n}^1$ is exactly same as $\tilde{n}$ within Ω_R.

2.2. The Total Energy Functional

The total energy can be written as a sum of three terms,

$$E = \tilde{E} + E^1 - \tilde{E}^1, \tag{13}$$

where the first term,

$$\tilde{E} = \sum_{n} f_n \langle \tilde{\Psi}_n | -\frac{1}{2}\nabla^2 | \tilde{\Psi}_n \rangle + E_{xc}[\tilde{n} + \hat{n} + \tilde{n}_c] + E_H[\tilde{n} + \hat{n}]$$

$$+ \int v_H[\tilde{n}_{Zc}][\tilde{n}(\mathbf{r}) + \hat{n}(\mathbf{r})]d\mathbf{r} + U(\mathbf{R}, Z_{ion}). \tag{14}$$

$n_{Zc}(\tilde{n}_{Zc})$ is the sum of the point charge density of the nuclei $n_Z(\tilde{n}_z)$ and the frozen core charge density $n_c(\tilde{n}_c)$, i.e. $n_{Zc} = n_z + n_c$ and $\tilde{n}_{Zc} = \tilde{n}_z + \tilde{n}_c$ and $\hat{n}$ is the compensation charge, which is added to the soft charge densities $\tilde{n} + \tilde{n}_{Zc}$ and $\tilde{n}^1 + \tilde{n}_{Zc}$ to reproduce the correct multipole moments of the AE charge density $n^1 + n_{Zc}$ located in each augmentation region. As n_{Zc} and $\tilde{n}_{Zc}$ have same monopole $-Z_{ion}$ and vanishing multipoles, the compensation charge $\hat{n}$ is chosen so that $\tilde{n}^1 + \hat{n}$ has the same moments as AE valence charge density n^1 has, within each augmentation region. $U(\mathbf{R}, Z_{ion})$ is the electrostatic interaction potential between the cores.

The second term in the total energy is,

$$E^1 = \sum_{ij} \rho_{ij} \langle \phi_i | -\frac{1}{2}\nabla^2 | \phi_j \rangle + \overline{E_{xc}[n^1 + n_c]} + \overline{E_H[n^1]}$$

$$+ \int_{\Omega_r} v_H[n_{n_{Zc}}]n^1(\mathbf{r})d\mathbf{r}. \tag{15}$$

and the final term is,

$$\tilde{E}^1 = \sum_{ij} \rho_{ij} \langle \tilde{\phi}_i | -\frac{1}{2}\nabla^2 | \tilde{\phi}_j \rangle + \overline{E_{xc}[\tilde{n}^1 + \hat{n} + \tilde{n}_c]} + \overline{E_H[\tilde{n}^1 + \hat{n}]}$$

$$+ \int_{\Omega_r} v_H[\tilde{n}_{Zc}][\tilde{n}^1(\mathbf{r}) + \hat{n}(\mathbf{r})]d\mathbf{r}, \tag{16}$$

In all these three energy terms, the electrostatic potential $v_H[n]$ and electrostatic energy $E_H[n]$ of charge density n is given by:

$$v_H[n](\mathbf{r}) = \int \frac{n(\mathbf{r}')}{|\mathbf{r} - \mathbf{r}'|} d\mathbf{r}', \qquad (17)$$

$$E_H[n] = \frac{1}{2} \int d\mathbf{r} \int d\mathbf{r}' \frac{n(\mathbf{r})n(\mathbf{r}')}{|\mathbf{r} - \mathbf{r}'|}. \qquad (18)$$

In the total energy functional the smooth part $\tilde{E}$ is evaluated on a regular grids in Fourier or real space, and the two one-center contributions E^1 and $\tilde{E}^1$ are evaluated on radial grids for each sphere individually.

2.3. Compensation Charge Density

The compensation charge density $\hat{n}$ is defined such that $\tilde{n}^1 + \hat{n}$ has exactly the same moments as the AE charge density n^1 has, within the augmentation region, which then requires,

$$\int_{\Omega_r} (n^1 - \tilde{n}^1 - \hat{n})|\mathbf{r} - \mathbf{R}|^l Y_L^*(\widehat{\mathbf{r} - \mathbf{R}}) dr = 0. \qquad (19)$$

The charge difference Q_{ij} between the AE and PS partial wave for each channel (i, j) within the augmentation region is defined by,

$$Q_{ij}(\mathbf{r}) = \phi_i^*(\mathbf{r})\phi_j(\mathbf{r}) - \tilde{\phi}_i^*(\mathbf{r})\tilde{\phi}_j(\mathbf{r}), \qquad (20)$$

and their moments q_{ij}^L are,

$$q_{ij}^L = \int_{\Omega_r} Q_{ij}(\mathbf{r})|\mathbf{r} - \mathbf{R}|^l Y_L^*(\widehat{\mathbf{r} - \mathbf{R}}) dr, \qquad (21)$$

which has non zero values only for certain combinations of L, i and j. Then the compensation charge density can be rewritten as,

$$\hat{n} = \sum_{i,j,L} \rho_{ij} \hat{Q}_{ij}^L(\mathbf{r}), \qquad (22)$$

where,

$$\hat{Q}_{ij}^L = q_{ij}^L g_l(|\mathbf{r} - \mathbf{R}|) Y_L(\widehat{\mathbf{r} - \mathbf{R}}), \qquad (23)$$

where $g_l(r)$ are the functions for which the moment is equal to 1.

2.4. Operators

Let us consider some operator $\mathcal{O}$, so its expectation value $\langle \mathcal{O} \rangle = \sum_n f_n \langle \Psi_n | \mathcal{O} | \Psi_n \rangle$, where n is the band index and f_n is the occupation of the state. As in the PAW method we work with the PS wave function, we need to obtain observables as the expectation values of PS wave function. Applying the form of the transformation τ and using $\sum_i |\tilde{\phi}_i\rangle\langle \tilde{p}_i| = 1$ within the augmentation region Ω_R and $|\tilde{\phi}_i\rangle = |\phi_i\rangle$ outside the augmentation region, for quasilocal operators, the PS operator $\tilde{\mathcal{O}}$ has the following form[1]:

$$
\begin{aligned}
\tilde{\mathcal{O}} &= \tau^\dagger \mathcal{O} \tau \\
&= \mathcal{O} + \sum_{ij} |\tilde{p}_i\rangle (\langle \phi_i|\mathcal{O}|\phi_j\rangle - \langle \tilde{\phi}_i|\mathcal{O}|\tilde{\phi}_j\rangle) \langle \tilde{p}_j|.
\end{aligned}
\tag{24}
$$

Overlap operator: The PS wave function obey orthogonality condition,

$$
\langle \tilde{\Psi}_n|\mathcal{S}|\tilde{\Psi}_m\rangle = \delta_{nm},
\tag{25}
$$

where $\mathcal{S}$ is the overlap operator in the PAW approach. The overlap matrix in the AE representation is given by the matrix elements of unitary operator. Therefore, $\mathcal{S}$ has the form given by,

$$
\mathcal{S}[\{\mathbf{R}\}] = 1 + \sum_i |\tilde{p}_i\rangle [\langle \phi_i|\phi_j\rangle - \langle \tilde{\phi}_i|\tilde{\phi}_j\rangle] \langle \tilde{p}_j|.
\tag{26}
$$

Hamiltonian operator: The Hamiltonian operator is defined as the first derivative of the total energy functional with respect to the density operator, $\tilde{\rho} = \sum_n f_n |\tilde{\Psi}_n\rangle\langle\tilde{\Psi}_n|$,

$$
H = \frac{dE}{d\tilde{\rho}},
\tag{27}
$$

where the PS density operator $\tilde{\rho}$ depends on the PS charge density $\tilde{n}$ and on the occupancies of each augmentation channel ρ_{ij}. So the variation of the total energy functional can be rewritten as,

$$
\frac{dE}{d\tilde{\rho}} = \frac{\partial E}{\partial \tilde{\rho}} + \int \frac{\delta E}{\delta \tilde{n}(\mathbf{r})} \frac{\partial \tilde{n}(\mathbf{r})}{\partial \tilde{\rho}} d\mathbf{r} + \sum_{i,j} \frac{\partial E}{\partial \rho_{ij}} \frac{\partial \rho_{ij}}{\partial \tilde{\rho}}.
\tag{28}
$$

The partial derivative of $\tilde{E}$ with respect to the PS density operator is the kinetic energy operator and the variation with respect to $\tilde{n}(\mathbf{r})$ leads to the usual one-electron potential $\tilde{v}_{eff}$,

$$
\frac{\partial \tilde{E}}{\partial \tilde{\rho}} = -\frac{1}{2}\nabla^2,
\tag{29}
$$

and,

$$
\frac{\delta \tilde{E}}{\delta \tilde{n}(\mathbf{r})} = \tilde{v}_{eff} = v_H[\tilde{n} + \hat{n} + \tilde{n}_{Zc}] + v_{xc}[\tilde{n} + \hat{n} + \tilde{n}_c].
\tag{30}
$$

As in the smooth part of the total energy functional, $\tilde{E}$, the occupancies ρ_{ij} enter only through the compensation charge density $\hat{n}$, the variation of $\tilde{E}$ with respect to ρ_{ij} is given by,

$$
\begin{aligned}
\hat{D}_{ij} = \frac{\partial \tilde{E}}{\partial \rho_{ij}} &= \int \frac{\delta \tilde{E}}{\delta \hat{n}(\mathbf{r})} \frac{\partial \hat{n}(\mathbf{r})}{\partial \rho_{ij}} d\mathbf{r} \\
&= \sum_L \int \tilde{v}_{eff}(\mathbf{r}) \hat{Q}_{ij}^L(\mathbf{r}) d\mathbf{r}.
\end{aligned}
\tag{31}
$$

---

[1]The kinetic energy operator $-1/2\nabla^2$ and the real space projector operators $|r\rangle\langle r|$ are quasi local operators. For truely nonlocal operators an extra term must be added to the $\tilde{\mathcal{O}}$ expression.

In the remaining two terms E^1 and $\tilde{E}^1$ in the energy functional, ρ_{ij} enters directly via kinetic energy or through n^1, $\tilde{n}^1$ and $\hat{n}$. Now the variation of E^1 with the occupancies ρ_{ij} is given by,

$$D_{ij}^1 = \frac{\partial E^1}{\partial \rho_{ij}} = \langle \phi_i | -\frac{1}{2}\nabla^2 + v_{eff}^1 | \phi_j \rangle, \tag{32}$$

where,

$$v_{eff}^1[n^1] = v_H[n^1 + n_{Zc}] + v_{xc}[n^1 + n_c],$$

and the variation of $\tilde{E}^1$ is given by,

$$\tilde{D}_{ij}^1 = \frac{\partial \tilde{E}^1}{\partial \rho_{ij}} = \langle \tilde{\phi}_i | -\frac{1}{2}\nabla^2 + \tilde{v}_{eff}^1 | \tilde{\phi}_j \rangle + \sum_L \int_{\Omega_r} d\mathbf{r}\, \tilde{v}_{eff}^1(\mathbf{r}) \hat{Q}_{ij}^L(\mathbf{r}), \tag{33}$$

where,

$$\tilde{v}_{eff}^1[\tilde{n}^1] = v_H[\tilde{n}^1 + \hat{n} + \tilde{n}_{Zc}] + v_{xc}[\tilde{n}^1 + \hat{n} + \tilde{n}_c].$$

The onsite terms D_{ij}^1 and $\tilde{D}_{ij}^1$ are evaluated on radial grid within each augmentation region and they restore the correct shape of the AE wave function within the sphere. The final form of the Hamiltonian operator:

$$H[\rho, \{\mathbf{R}\}] = -\frac{1}{2}\nabla^2 + \tilde{v}_{eff} + \sum_{i,j} |\tilde{p}_i\rangle (\hat{D}_{ij} + D_{ij}^1 - \tilde{D}_{ij}^1) \langle \tilde{p}_j|. \tag{34}$$

2.5. Forces in the PAW Method

Forces are usually defined as the total derivative of the energy with respect to the ionic position $\mathbf{R}$,

$$\mathbf{F} = -\frac{dE}{d\mathbf{R}}. \tag{35}$$

The total derivative consists of two terms, first one is the forces on the ionic core and the second term is the correction due to the change of AE wave functions for fixed PS wave functions when ions are moved. This correction term comes because augmentation depends on the ionic positions and are known as Pulay force [20]. When calculating this Pulay correction one must consider the overlap between the wave functions due to the change in ionic position. According to Goedecker and Maschke[21], the total derivative can be written as,

$$\frac{dE}{d\mathbf{R}} = \frac{\partial U(\mathbf{R}, Z_{ion})}{\partial \mathbf{R}} + \sum_n f_n \left\langle \tilde{\Psi}_n \left| \frac{\partial (H[\rho, \{\mathbf{r}\}] - \epsilon_n \mathcal{S}[\{\mathbf{R}\}])}{\partial \mathbf{R}} \right| \tilde{\Psi}_n \right\rangle, \tag{36}$$

where the first term is the forces between the ionic cores and ϵ_n are the Kohn-Sham eigenvalues, and the PS wave functions assumed to satisfy the orthogonality condition $\langle \tilde{\Psi}_n | \mathcal{S} | \tilde{\Psi}_m \rangle = \delta_{nm}$ and the corresponding Kohn-Sham equation, reads $H | \tilde{\Psi}_n \rangle = \epsilon_n \mathcal{S} | \tilde{\Psi}_n \rangle$.

The first contribution, $\mathbf{F}^1$ to the second term of the total derivative comes from the change in the effective local potential $\tilde{v}_{eff}$ when the ions are moved and $\tilde{v}_{eff}$ depends explicitly on the ionic positions via $\hat{n}_{Zc}$,

$$\mathbf{F}^1 = -\int \left[\frac{\delta Tr[H\tilde{\rho}]}{\delta v_H[\tilde{n}_{Zc}](\mathbf{R})} \frac{\partial v_H[\tilde{n}_{Zc}](\mathbf{R})}{\partial \mathbf{R}} \right] d\mathbf{R}. \tag{37}$$

This equation can be further simplified to,

$$\mathbf{F}^1 = -\int \left(\tilde{n}(\mathbf{r}) + \sum_{i,j,L} \hat{Q}_{ij}^L(\mathbf{r})\rho_{ij} \right) \frac{\partial v_H[\tilde{n}_{Zc}](\mathbf{r})}{\partial \mathbf{R}} d\mathbf{r}$$

$$= -\int [\tilde{n}(\mathbf{r}) + \hat{n}(\mathbf{r})] \frac{\partial v_H[\tilde{n}_{Zc}](\mathbf{r})}{\partial \mathbf{R}} d\mathbf{r}. \tag{38}$$

The second contribution arise from $\hat{D}_{ij}$ due to the changes in the compensation charge density $\hat{n}$, when ions are moved,

$$\mathbf{F}^2 = -\sum_{i,j,L} \int \tilde{v}_{eff}(\mathbf{r})\rho_{ij}q_{ij}^L \frac{\partial g_l(|\mathbf{r}-\mathbf{R}|)Y_L(\widehat{\mathbf{r}-\mathbf{R}})}{\partial \mathbf{R}} d\mathbf{r}. \tag{39}$$

These two terms, $\mathbf{F}^1$ and $\mathbf{F}^2$, together describe the electrostatic contributions to the force. The third term comes due to the change in the projector function $\tilde{p}_i$ as ions are moved,

$$\mathbf{F}^3 = -\sum_{n,i,j} (\hat{D}_{ij} + D_{ij}^1 - \tilde{D}_{ij}^1 - \epsilon_n q_{ij})f_n \left\langle \tilde{\Psi}_n \left| \frac{\partial |\tilde{p}_i\rangle\langle\tilde{p}_j|}{\partial \mathbf{R}} \right| \tilde{\Psi}_n \right\rangle, \tag{40}$$

where $q_{ij} = \langle\phi_i|\phi_j\rangle - \langle\tilde{\phi}_i|\tilde{\phi}_j\rangle$.

As the exchange-correlation potential depends on the nonlinear core corrections $\tilde{n}_c$ gives an additional contribution, which can be obtained from the total energy functional,

$$\mathbf{F}^{\text{nlcc}} = -\int v_{xc}[\tilde{n} + \hat{n} + \tilde{n}_c] \frac{\partial \tilde{n}_c(\mathbf{r})}{\partial \mathbf{R}} d\mathbf{r}. \tag{41}$$

All the differences between the PAW method and ultra-soft pseudopotential are automatically absorbed in the in the definition of $(\hat{D}_{ij} + D_{ij}^1 - \tilde{D}_{ij}^1)$. $D_{ij}^1 - \tilde{D}_{ij}^1$ are constant for US-PP, where they are calculated once and forever whereas in PAW method they vary during the calculation of the electronic ground state.

3.   Computational Method

The electronic and magnetic calculations of the pure and doped- manganese clusters were carried out using density functional theory (DFT), within the pseudopotential plane wave method [16, 17], using projected augmented wave (PAW) formalism [18, 19] and Perdew-Burke-Ernzerhof (PBE) exchange-correlation functional [22] for the spin-polarized generalized gradient approximation (GGA), as it is implanted in VASP code [23]. The wave functions are expanded in a plane wave basis set with the kinetic energy cutoff equal to 337.3 eV and reciprocal space integrations were carried out at the Γ point. The $3d$, $4s$ for Mn-atom and $4s$, $4p$ orbitals for As-atom were treated as valence states. Symmetry unrestricted geometry optimizations were performed using quasi Newtonian and conjugate gradient methods until all the force components are less than 0.005 eV/Å. Simple cubic supercells are used with neighbouring clusters separated by at least 12Å vacuum regions. Several initial structures were studied to ensure that the globally optimized geometry does not correspond to the local minima, as well as, for all clusters, we have explicitly considered *all possible* spin multiplicities to determine the magnetic ground state.

4. Results

4.1. Magnetic Transition: Pure Mn_n ($n \leq 20$) Clusters

We start with the magnetism in the pure manganese clusters. Theory and experimental reports are in contradiction for the Mn_2 dimer. More than 30 years ago, Nesbet [24] calculated binding energy, bond length and magnetic moment of the dimer at the restricted Hartree-Fock (RHF) level and predicted an antiferromagnetic (AFM) ground state with bond length 2.88 Å. Later on, unrestricted Hartree-Fock calculation was done by Shillady et al. [25] and found a ferromagnetic (FM) ground state with total spin 10 μ_B and bond length 3.50 Å. The experiments based on resonant Raman spectroscopy [26] and Electron Spin Resonance (ESR)[27] predicted an AFM ground state with a bond length 3.4 Å. However, all DFT calculations [28, 29, 8, 30] within different levels of approximation and using different exchange-correlation functionals, predict a FM ground state with much smaller bond length, ~ 2.60 Å than the experimental bond length. A comparison among different levels of theory is given in the Table1.

Table 1. Summary of binding energy (E_b), bond length (R_e) and magnetic moment (μ) of Mn_2 reported by various authors.

Authors	Method	E_b (eV)	R_e (Å)	μ (μ_B)
Nesbet [24]	RHF + Heisenberg exchange	0.79	2.88	0
Wolf and Schmidtke [31]	RHF		1.52	0
Shillady et al. [25]	UHF	0.08	3.50	10
Fujima and Yamaguchi [32]	LSDA	0.70	3.40	0
Harris and Jones [33]	LSDA	1.25	2.70	10
Salahub and Baykara [34]	LSDA	0.86	2.52	0
Nayak et al. [28]	All-electron + GGA (BPW91)	0.91	2.60	10
Parvanova et al.[29]	All-electron + GGA (PBE)	0.98	2.60	10
Kabir et al. [8]	Pseudopotential + GGA (PBE)	1.06	2.58	10
Pederson et al. [30]	All-electron + GGA	0.99	2.61	10
Experiment	ESR in rare-gas matrix [27]	0.1 ± 0.1	3.4	0

Due to the filled $4s$ and half-filled $3d$ electronic configuration, as well as high (2.14 eV) $4s^2\,3d^5 \rightarrow 4s\,3d^6$ promotion energy Mn atoms do not bind strongly as two Mn atoms come closer to form Mn_2 dimer, and as a result Mn_2 is very weakly bond van der Walls dimer, which is also evident from the low experimental value, $0.01 \pm 0.1 - 0.56 \pm 0.26$ eV/atom [6]. However, no experimental results available in the gas-phase and all the experiments are done in the gas matrix, and therefore, it is quite possible that the Mn atoms interact with the matrix, which could stretch R_e and could lead to the AFM ground state.

From the Table2, we see that R_e decreases monotonically as the net moment decreases. It is simply because the reduction of the atomic spacing leads to comparatively stronger overlap of the atomic orbitals which reduces the magnetic moment. The manganese trimer and tetramer is found to have FM ground state with 5 μ_B/atom magnetic moment. Resonance Raman spectroscopy study by Bier et al.[26] suggested a distorted D_{3h} structure with

Table 2. The binding energy and equilibrium bond length of Mn$_2$ dimer for all possible spins.

spin	Binding energy (eV)	Bond length
0 (AFM)	0.51	2.57
2	0.44	1.53
4	unbound	1.73
6	unbound	1.94
8	0.47	2.24
10	1.06	2.58

odd-integer magnetic moment for trimer. Manganese tetramer in solid silicon was studied by Ludwig *et al.* [35] and observed a 21-line hyperfine pattern which confirmed that the all four atoms are equivalent with a total magnetic moment 20 μ_B.

Calculated binding energy per atom, relative energies of the isomers, magnetic moments per atom with the experimental comparison are given in the Table3 ($n = 2$-10) and Table4 ($n = 2$-20) and the calculated magnetic moment per atom corresponding to the ground state are plotted in the Figure1. We see a FM $\rightarrow$ ferrimagnetic transition takes place at n=5. and remains the same for larger size clusters [8]. Figure1 shows very good agreement with the recent Stern-Gerlach experiments[4, 5]. However, we find, for all sizes there exists several

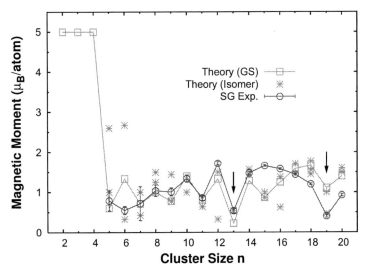

Figure 1. Variation of magnetic moment with cluster size for the size range $n = 2 -20$. A comparison is made with the experimentally measured value [5]. For all sizes, the magnetic moments of closely lying isomers are also plotted.

Table 3. Binding energy per atom, relative energy to the ground state ($\triangle E = E - E_{GS}$) and per atom magnetic moment for Mn_n ($n \leq 10$) clusters. Predicted magnetic moments are compared with the Stern-Gerlach experiment.[4, 5]

Cluster	BE (eV/atom)	$\triangle E$ (eV)	Magnetic Moment (μ_B/atom)	
			Theory[8]	Experiment[4, 5]
Mn_2	0.530	0.000	5.000	—
	0.255	0.531	0.000	
Mn_3	0.823	0.000	5.000	—
	0.808	0.046	1.667	
Mn_4	1.179	0.000	5.000	—
	1.160	0.078	2.500	
	1.130	0.196	0.000	
	1.130	0.195	2.000	
Mn_5	1.413	0.000	0.600	0.79 ± 0.25
	1.401	0.059	2.600	
	1.399	0.069	1.000	
	1.374	0.193	4.600	
Mn_6	1.567	0.000	1.333	0.55 ± 0.10
	1.564	0.017	0.333	
	1.559	0.045	2.667	
	1.543	0.142	4.333	
Mn_7	1.726	0.000	0.714	0.72 ± 0.42
	1.713	0.091	1.000	
	1.699	0.193	0.429	
Mn_8	1.770	0.000	1.000	1.04 ± 0.14
	1.770	0.000	1.500	
	1.765	0.039	1.250	
Mn_9	1.867	0.000	0.778	1.01 ± 0.10
	1.856	0.099	1.444	
	1.850	0.151	0.778	
	1.844	0.211	1.000	
Mn_{10}	1.936	0.000	1.400	1.34 ± 0.09
	1.934	0.015	1.000	
	1.935	0.007	0.400	
	1.935	0.009	0.400	

isomers with different magnetic structure[8].

In the SG experiment [5] it has been seen that the experimental uncertainty in measurement of magnetic moment generally decreases with the cluster size as the cluster production efficiency increases with the size. However, this uncertainty for Mn_7 is quite high, 0.72 ± 0.42 μ_B/atom[5], which is 58% of the measured value. Now we discuss what might be the possible reason for this. We predict [8] the Mn_7 to be a pentagonal bi-pyramidal (PBP) structure in its ground state with magnetic moment 5 μ_B, which is exactly the experimental value. In addition, we found another two PBP isomers, with magnetic moment 7 μ_B and 3 μ_B, to be close in energy to the ground state, which are shown in the Figure2. They lie 0.09

Table 4. Same as Table3 for Mn$_n$ ($n = 11$— 20) clusters.

Cluster	BE (eV/atom)	$\triangle E$ (eV)	Magnetic Moment (μ_B/atom)	
			Theory[8]	Experiment[4, 5]
Mn$_{11}$	1.993	0.000	0.818	0.86 ± 0.07
	1.984	0.107	0.455	
	1.980	0.153	0.636	
Mn$_{12}$	2.081	0.000	1.333	1.72 ± 0.04
	2.077	0.052	0.333	
	2.071	0.114	1.500	
Mn$_{13}$	2.171	0.000	0.231	0.54 ± 0.06
	2.165	0.076	0.538	
Mn$_{14}$	2.171	0.000	1.286	1.48 ± 0.03
	2.170	0.021	1.429	
	2.167	0.055	1.571	
Mn$_{15}$	2.231	0.000	0.867	1.66 ± 0.02
	2.229	0.033	0.333	
	2.228	0.057	0.467	
	2.228	0.049	0.867	
	2.227	0.064	1.000	
	2.213	0.281	0.467	
Mn$_{16}$	2.274	0.000	1.250	1.58 ± 0.02
	2.273	0.017	1.375	
	2.271	0.055	0.625	
	2.268	0.097	0.500	
Mn$_{17}$	2.327	0.000	1.588	1.44 ± 0.02
	2.323	0.069	1.471	
	2.322	0.087	1.706	
Mn$_{18}$	2.351	0.000	1.667	1.20 ± 0.02
	2.350	0.018	1.556	
	2.350	0.020	1.444	
	2.348	0.064	1.778	
Mn$_{19}$	2.373	0.000	1.105	0.41 ± 0.04
	2.373	0.009	1.000	
	2.369	0.076	0.474	
Mn$_{20}$	2.370	0.000	1.400	0.93 ± 0.03
	2.370	0.003	1.500	
	2.368	0.055	1.600	
	2.367	0.067	0.800	

eV and 0.20 eV higher in energy. Therefore, the presence of these two isomers along with the ground state in the SG molecular beam could be a possible explanation to the observed large uncertainty. But it should be pointed out here that this might be a possible argument but never the certain reason, because for all sizes there exist more than one isomer with different magnetic structure, but the corresponding experimental uncertainty is not large.

Now we move to the sudden deep observed in the measured magnetic moment at n=13 and 19 compared to their neighbours[4, 5]. Through *ab initio* calculation, we reproduced

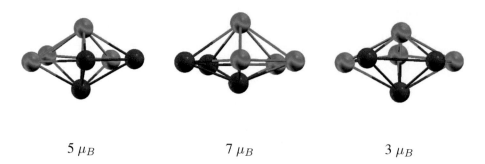

$5\ \mu_B$ $7\ \mu_B$ $3\ \mu_B$

Figure 2. Ground state and isomeric configuration for Mn_7. Note that for all configurations number of spin up atoms and spin down atoms are the same ($N_\uparrow = 4$, $N_\downarrow = 3$). Green(Red) ball represents up(down) atom and we follow the same convention throughout.

this behaviour[8]. For Mn_{13}, we studied icosahedral, cubooctahedral and hexagonal structures for all possible multiplicities and we found the icosahedral structure with magnetic moment 3 μ_B to be the ground state. The optimal cubooctahedral and hexagonal structure have 9 and 11 μ_B magnetic moment, which is much higher than the SG experimental value, $0.54 \pm 0.06\ \mu_B$[4, 5], and they lie much higher in energy than the icosahedral ground state. They are shown in the Figure3. Similarly, the ground state of Mn_{19} is found to be double icosahedral structure with magnetic moment 21 μ_B (Figure4). The optimal FCC structure (Figure4) lie 1.53 eV higher in energy. So in our recent paper we concluded that the "closed" icosahedral geometric structure is responsible for the observed sudden deep.

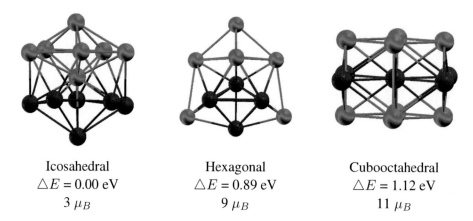

Icosahedral   Hexagonal   Cubooctahedral
$\triangle E = 0.00$ eV $\triangle E = 0.89$ eV $\triangle E = 1.12$ eV
3 μ_B 9 μ_B 11 μ_B

Figure 3. Spin configuration for the icosahedral ground state of Mn_{13}. Lowest energy spin configurations for hexagonal and cubooctahedral structure have also been shown.

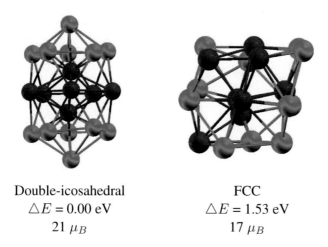

Double-icosahedral	FCC
$\triangle E$ = 0.00 eV	$\triangle E$ = 1.53 eV
21 μ_B	17 μ_B

Figure 4. Ground state configuration for Mn_{19} is a double icosahedron. Optimal FCC structure lies much higher in energy.

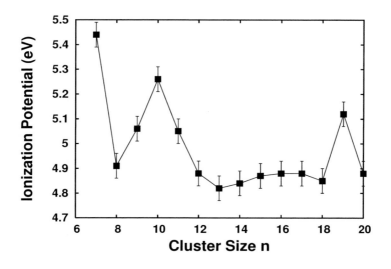

Figure 5. Experimentally measured ionization potential for Mn_n clusters in the size range $n = 2\text{-}20$.(From Ref.[37].)

4.2. Non-metal to Metal Transition?

Parks *et al.* produced Mn_n clusters containing up to 70 atoms by cooling the inert gas condensation source to -160^0C to study their reaction rate with the molecular hydrogen[9]. They found that the clusters with $n \leq 15$ are nonreactive toward H_2, whereas they form stable hydride above $n = 15$ and the reaction rate varies considerably with cluster size. It was thought that this might arise due to the possible abrupt change in the bonding character at $n = 16$: For small clusters $n \leq 15$, the bonding is weak and of van der Walls kind, which perhaps become metallic at $n = 16$ and remains the same for larger clusters. If it is so then once the cluster is metallic, it is energetically possible to transfer charge from metallic

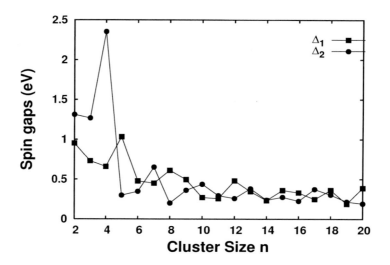

Figure 6. Variation of spin gaps with cluster size in the size range $n = 2$—20. (From Ref.[8].)

cluster to the antibonding state of the H_2 molecule and H-H bond would consequently break and H atom would attach to the metal cluster. If this is indeed the case, it is likely that a significant downward change in the ionization potential would be observed at Mn_{16}. The similar effect has observed in the free mercury clusters[36], where a steady decrease in the ionization potentials attribute to the nonmetal-metal transition. We also expect the closing up of the two spin gaps for this kind of transition. Koretsky et al. measured ionization potentials of Mn_n clusters in the size range $n = 7 - 64$[37]. However, no sudden decrease is observed at $n = 16$ (Figure5).

Spin gaps[2] are defined as follows:

$$\triangle_1 = - \left[\epsilon_{HOMO}^{majority} - \epsilon_{LUMO}^{minority} \right]$$
$$\triangle_2 = - \left[\epsilon_{HOMO}^{minority} - \epsilon_{LUMO}^{majority} \right], \tag{42}$$

and we plot them in the Figure6. We did not find any closing up of these spin gaps. Therefore, there is no evidence of nonmetal-metal transition and the observed sudden change in the reactivity is not due to any kind of structural transition at $n = 16$ either, as all the medium sized clusters adopt icosahedral growth pattern. So the reason for the abrupt change in the reaction rate of Mn_n clusters with H_2 at $n = 16$ is yet unknown.

4.3.   Localization of d−electrons

In finite size clusters coordination number may very drastically for different sites and, therefore, they are very good system to study the fact that how the d−electrons get localized with

---

[2]Any spin arrangement for these Mn_n clusters is magnetically stable only if both the spin gaps are positive. This means that the lowest unoccupied molecular orbital (LUMO) of the minority(majority) spin lies above the highest occupied molecular orbital(HOMO) of the majority(minority) spin.

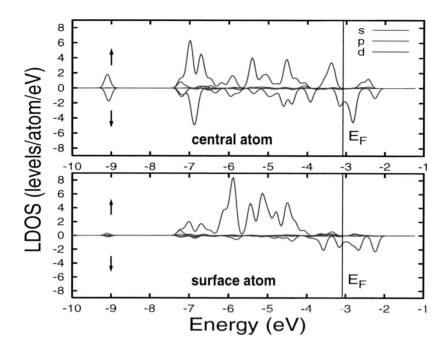

Figure 7. The local density of states (LDOS) for the central and a surface atom for Mn_{13} cluster.

the coordination. Take an example of icosahedral Mn_{13} cluster, where the bonding character of the 12-coordinated central atom is very much different from the the 6-coordinated surface atoms. Here to study the effect of localization, mainly of $d-$ electrons, on the coordination, we calculate the angular momentum projected local density of states, which is plotted in the Figure7 for Mn_{13} cluster. We see that the $d-$ projected local density of states (LDOS) of the central atom are broad for both the majority and minority spin states, whereas the same for the surface atoms are rather localized and the majority spin states are nearly fully occupied. This is also reflected from the local magnetic moments of the central and surface atoms, which is defined as follows:

$$\mathcal{M}_i = \int_0^R \left[\rho^\uparrow(\mathbf{r}) - \rho^\downarrow(\mathbf{r}) \right] d\mathbf{r}, \tag{43}$$

where $\rho^\uparrow(\mathbf{r})$ and $\rho^\downarrow(\mathbf{r})$ are spin up and spin down densities, respectively and R is the radius of the integrating sphere centering each atom. The local magnetic moment of the central atom is small, 1.42 μ_B, compared to the surface atoms $\sim$ 3.60 μ_B due to the above mentioned reason.

5.   Single As-doping in Mn$_n$ Clusters: Mn$_n$As ($n \leq 10$)

We already have discussed the pure Mn$_n$ clusters and have seen that very small clusters containing up to 4 atoms are ferromagnetic and they undergo a magnetic (ferromagnetic to ferrimagnetic) transition at $n=5$. Now it will be interesting to see that what happens to the magnetic structure of those if we dope a single arsenic into it to form Mn$_n$As species. In the previous section, we have also noticed that the binding energy of pure Mn$_n$ clusters are small compared to other $3d-$transition metal clusters due to its $3d^5\ 4s^2$ electronic structure. Here we will also discuss that what happens to the bonding due to single As-doping. Particularly magnetism in these Mn$_n$As clusters are more interesting, as manganese-doped semiconductors, such as (GaMn)N, (GaMn)As and (InMn)As, have attracted considerable attention because of their carrier induced ferromagnetism[38, 39, 40, 41, 42]. The Mn dopants in these III-V dilute magnetic semiconductors serve the dual roles of provision of magnetic moments and acceptor production. A wide range of T_c, 10-940 K, has been reported for Ga$_{1-x}$Mn$_x$N by varying Mn concentration ($x \sim$7-14%) as well as by varying the growth conditions[43, 44, 45, 46]. But still it is unclear at present, whether all these reports of ferromagnetism are indeed intrinsic or arise due to the some kind of 'defects' originated during the growth, as magnetic atoms are not thermodynamically stable in the semiconductor host and tend to form 'defects'. Recent investigation[51] found that ferromagnetism persists up to $\sim$ 980 K for metastable Mn-doped ZnO and further heating transforms the metastable phase and kills the ferromagnetism. Earlier experimental results on Ga$_{1-x}$Mn$_x$As indicate a nonmonotonic behaviour of $T_c(x)$, first increases with the Mn concentration x, reaching a maximum of 110 K for $x \sim$5% and then decreases with the further increase of x[47, 48, 49, 50]. However, recent experimental studies [52, 48, 53], under carefully controlled growth and annealing conditions, suggest that the 'metastable' nature and high 'defect' content (such as clustering of Mn) of low temperature of MBE grown Ga$_{1-x}$Mn$_x$As may be playing an important role in the magnetic properties and could enhance T_c with Mn content. Chiba *et al.* [52] have reported T_c as high as 160 K for 7.4 % Mn concentration, in a layered structure. These studies rule out the earlier suggestion that the T_c in Ga$_{1-x}$Mn$_x$As is fundamentally limited to 110 K and certainly opened up a hope for further possible increase in T_c for such metastable phase under different suitable growth and annealing conditions. Towards the end of the chapter, we will discuss this recently debated clustering phenomena, which at least puts some light on the issue.

5.1.   Geometry and Bonding

We start with the MnAs dimer, which has much higher binding energy of 1.12 eV/atom and much shorter bond length, 2.21 Å compared with the pure Mn$_2$ cluster, which has the values 0.53 eV/atom and 2.58 Å, respectively[10]. The ground state of the Mn$_2$As is an isosceles triangle and the binding energy gets substantially increased to 1.63 eV/atom. The Mn-Mn bond, 2.59 Å, in this Mn$_2$As cluster is nearly equal to that of the Mn$_2$ dimer. As more Mn-atoms are added to the Mn$_2$As cluster, the resultant clusters adopt three dimensional shape and, moreover, the determination of the ground state is far more difficult as it is known that the number of local minima in the potential energy surface is an exponential function of the number of atoms present in that cluster.

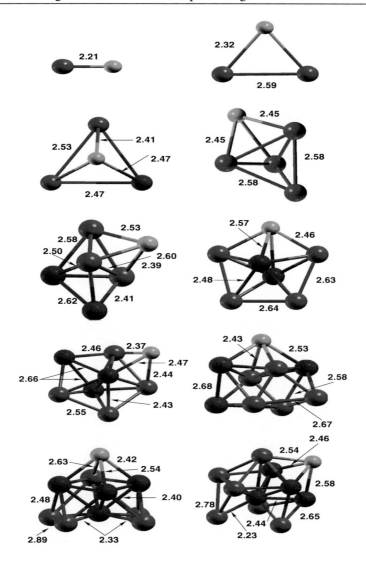

Figure 8. Ground state geometries and spin ordering of Mn_xAs clusters, x=1-10. Blue and red balls represent $Mn_\uparrow$ and $Mn_\downarrow$ atoms, respectively. Green ball represents the As atom. The bond lengths are given in Å. Magnetic polarization of As, is negative for MnAs - Mn_8As and, whereas, positive for Mn_9As and $Mn_{10}As$. Note, Mn atoms are coupled ferromagnetically only in Mn_2As and Mn_4As clusters.

Calculated binding energy, total magnetic moment, shortest Mn-Mn and Mn-As bond lengths and two spin gaps are given in the Table5 for the ground state along with the isomers for Mn_nAs cluster. The presence of an As-atom makes Mn_3As cluster tetrahedral, however, Mn_4As is a Mn_4 tetrahedra with As at a face cap. Generally, the ground state structures of Mn_nAs clusters can be seen as that of the corresponding pure Mn_n cluster with As capped at some face, i.e. As-doping does not give rise to any considerable structural change, but a moderate perturbation to the corresponding pure cluster structure. However, the Mn_6As

cluster is the only cluster whose geometry differs significantly from the pure Mn_6 cluster. Mn_6 is octahedral, whereas Mn_6As is a pentagonal bipyramid, where the As atom is trapped in the pentagonal ring. As n in the Mn_nAs increases, binding energy increases monotonically due to the increase in the coordination number. However, the increase rate is very slow after $n = 2$: Increases slowly from 1.73 eV/atom for Mn_3As to 2.22 eV/atom for $Mn_{10}As$. The shortest Mn-As bond length increases from 2.21 Å for MnAs dimer to 2.46 Å for $Mn_{10}As$ cluster, whereas the shortest Mn-Mn bond length in these clusters decreases with cluster size (2.59 Å for Mn_2As to 2.23 Å for $Mn_{10}As$). Generally, we find, the bonds between Mn atoms of opposite spin to be somewhat shorter (2.20-2.60 Å) than the bonds between Mn atoms of like spin (2.50-2.90 Å), whereas Mn-As distance varies between 2.20-2.60 Å. All the Mn-As-Mn bond angles in these clusters vary in between $\sim$ 60-70^0. All the clusters in the Figure8 and their respective isomers are magnetically stable i.e. both the spin gaps are positive. These two spin gaps, $\triangle_1$ and $\triangle_2$, for Mn_2As (0.83 and 1.34 eV) and Mn_4As (0.89 and 1.14 eV) are the highest among all clusters. As x increases, $\triangle_1$ and $\triangle_2$ decrease to a value 0.47 and 0.35 eV, respectively, for $Mn_{10}As$ (see Table5).

Now we discuss the important issue whether the Mn-clustering around the single As-atom are at all fafourable or not. First we look into the binding energy of pure Mn_n and Mn_nAs clusters, which is plotted in the Figure9. We see that due to single As-doping the binding energy of the Mn_nAs clusters are substantially enhanced from their respective pure counterpart. As it is discussed earlier that due to the half-filled $3d$ and filled $4s$ shell and high $3d^5 4s^2 \rightarrow 3d^6 4s^1$ promotion energy (2.12 eV), Mn atoms do not bind strongly when they are brought together to form clusters. However, as a single As-atom is attached to the pure Mn_n clusters, the $4s^2$ electrons of the Mn-atom interact with the $4p^3$ electrons of As and results enhancement in the binding energy.

To understand this better, one can calculate the energy gains which are of two kind[10]:

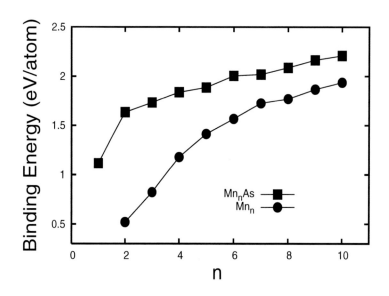

Figure 9. Plot of binding energy per atom for pure and doped manganese clusters, Mn_n and Mn_nAs, in the size range $n \leq 10$.

- The energy gain due to adding an As-atom to a pure Mn_n cluster to form Mn_nAs cluster,

$$\triangle^1 = -[E(Mn_nAs) - E(Mn_n) - E(As)], \qquad (44)$$

and,

- The energy gain due to adding a Mn-atom to a existing $Mn_{n-1}As$ cluster to from Mn_nAs cluster,

$$\triangle^2 = -[E(Mn_nAs) - E(Mn_{n-1}As) - E(Mn)]. \qquad (45)$$

These two energy gains are plotted as a function of Mn concentration in the Figure 10. The $\triangle^1$ increases with n: The increment from $n = 1$ to $n = 3$ is monotonous and much stiff but afterwards it is not monotonic and tends to saturate. The other energy gain, $\triangle^2$, gives the number that how many Mn-atoms can be bonded to a single As-atom, which is significant, 2.65 eV, even for $Mn_{10}As$ cluster. Both the energy gains are positive and the energy gain in adding an As-atom is much greater than that of adding a Mn-atom ($\triangle^1 >> \triangle^2$) and, therefore, we can conclude that the clustering of Mn-atom around a single As is favourable[10]. So, the clustering of Mn is possible around As during the growth in Mn-doped GaAs/InAs. However, the other factors like lattice distortions and available space will play a role in determining the size of Mn clustering in these semiconductors.

5.2. Magnetic Order

Now we ask the next important question that what is the Mn-Mn magnetic coupling in these Mn_nAs clusters? The total magnetic moments of Mn_nAs clusters corresponding to

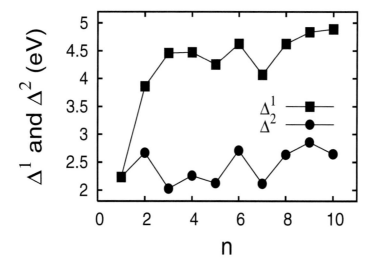

Figure 10. Plot of two different energy gains, $\triangle_1$ and $\triangle_2$ as a function of size.

Table 5. Binding energy, total moment, lowest Mn-Mn (d_{Mn-Mn}) and Mn-As (d_{Mn-As}) bond length and two spin gaps ($\triangle_1$ and $\triangle_2$) for $Mn_n As$ (n = 1-10) clusters for their respective ground and isomeric state.

Cluster	B.E. (eV/atom)	Total moment μ_B	d_{Mn-Mn} Å	d_{Mn-As} Å	$\triangle_1$ eV	$\triangle_2$ eV
MnAs	1.116	4	—	2.208	0.709	1.016
Mn_2As	1.634	9	2.591	2.316	0.826	1.342
	1.597	1	2.428	2.246	0.582	1.345
Mn_3As	1.733	4	2.473	2.413	0.651	0.504
	1.702	12	2.481	2.372	1.295	0.523
Mn_4As	1.838	17	2.573	2.453	0.891	1.144
Mn_5As	1.886	2	2.417	2.389	0.856	0.381
	1.871	12	2.401	2.464	0.620	0.665
Mn_6As	2.004	9	2.469	2.461	0.540	0.669
	1.974	1	2.362	2.435	0.014	0.623
Mn_7As	2.018	6	2.430	2.375	0.409	0.593
	2.014	14	2.405	2.436	0.470	0.611
	2.012	6	2.323	2.472	0.456	0.402
Mn_8As	2.086	7	2.307	2.424	0.405	0.347
	2.075	3	2.305	2.478	0.072	0.552
	2.074	5	2.227	2.468	0.468	0.378
Mn_9As	2.163	10	2.306	2.416	0.418	0.250
	2.158	12	2.308	2.449	0.159	0.826
$Mn_{10}As$	2.207	13	2.225	2.455	0.467	0.347
	2.204	3	2.200	2.501	0.484	0.372

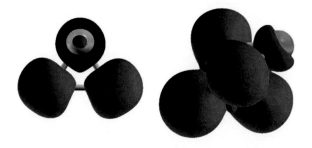

Figure 11. Constant spin density surfaces for Mn_2As and Mn_4As corresponding to 0.04 $e/Å^3$. Red and blue surfaces represent positive and negative spin densities, respectively. Green ball is the As atom, which has negative polarization in both the structures. Note ferromagnetic coupling among Mn atoms in the both Mn_2As and Mn_4As clusters.

the ground state geometries are 4, 9, 4, 17, 2, 9, 6, 7, 10 and 13 μ_B, respectively, for n=1-10 (Table5). These large magnetic moments generally arise form the ferrimagnetic coupling between the moments at Mn sites with exceptions for Mn_2As and Mn_4As, where the magnetic coupling is found to be ferromagnetic. We note, generally, no change in the magnetic coupling between the Mn sites in Mn_nAs clusters from their respective Mn_n, however, for Mn_3As, the Mn-Mn coupling behaviour changes form ferromagnetic to ferrimagnetic, due to As doping in Mn_3. Another tetrahedral isomer with magnetic moment 12 μ_B is 0.12 eV higher, where all the Mn atoms are ferromagnetically ordered. Here we should point out, as mentioned in the Table3, a frustrated antiferromagnetic structure with total magnetic moment 5 μ_B is only 0.05 eV higher in energy than the ferromagnetic Mn_3 cluster. The ground state magnetic moments of Mn_nAs, for n=1, 2, 3, 5, 7, and 10 can be represented as (μ_n-1) μ_B, whereas for n=4, 6, 8, 9, it can be expressed as (μ_n-3) μ_B, where μ_n is the total magnetic moment of the Mn_n cluster corresponding to the ground state or the 'first' isomer. It should be pointed out here that, for n=6 and 7, the ground state magnetic moments of Mn_nAs clusters are higher than that of Mn_n, as due to As doping, Mn_nAs can favour the isomeric magnetic structure of the corresponding Mn_n cluster without changing the overall magnetic behaviour between the Mn sites. For MnAs dimer, the magnetic moment at Mn site, $\mathcal{M}_{Mn}$, is 3.72 μ_B and $\mathcal{M}_{As}$ is -0.26 μ_B for As. This large negative polarization of the anion, As, is due to the strong $p-d$ interaction. Polarized neutron diffraction study found a local magnetic moment -0.23 ± 0.05 μ_B at the As sites for NiAs-type MnAs [54], which is very close to the present value for MnAs dimer. In the Mn_2As cluster, Mn atoms are ferromagnetically coupled with $\mathcal{M}_{Mn}$=3.79 μ_B each, whereas the magnetic polarization is negative for As atom, -0.14 μ_B. Mn atoms show ferrimagnetic ordering with $\mathcal{M}_{Mn}$ 3.1, 3.1 and -3.9 μ_B for Mn_3As, whereas, As has $\mathcal{M}_{As}$=-0.21 μ_B, which gives a total magnetic moment 4 μ_B. For Mn_4As, Mn atoms are ferromagnetically arranged with average $\mathcal{M}_{Mn}$=3.66 μ_B and are coupled antiferromagnetically with As atom, $\mathcal{M}_{As}$=-0.22 μ_B. Nature of magnetic coupling can be visualize through their respective spin iso-density plots, which is shown in the Figure11 for Mn_2As and Mn_4As clusters. For Mn_5As, $\mathcal{M}_{Mn}$ varies between 3.04-3.72 μ_B with a local magnetic moment -0.23 μ_B at As site. We observe, the negative polarization of As, $\mathcal{M}_{As}$, decreases sharply to -0.08 μ_B for Mn_6As and further decreases monotonically to -0.02 μ_B for Mn_8As, however it become positive, 0.04 and 0.02, for Mn_9As and $Mn_{10}As$, respectively. For all Mn_xAs clusters, x =6-10, Mn atoms are coupled ferrimagnetically, where $\mathcal{M}_{Mn}$ varies between 0.8 - 3.7 μ_B. These are unlike the study of Rao and Jena [55], where Mn atoms were found to be coupled ferromagnetically for all sizes, n=1-5, in nitrogen doped Mn_n clusters.

6. Exchange Coupling

To determine the exchange interactions J_{ij}'s between the atoms i and j, we map the magnetic energy onto a classical Heisenberg Hamiltonian:

$$\mathcal{H} = -\sum_{ij} J_{ij}(r)\, \mathbf{S}_i \cdot \mathbf{S}_j, \tag{46}$$

where the sum is over all distinct magnetically coupled atoms, $\mathbf{S}_i(\mathbf{S}_j)$ denotes the localized magnetic moments at $i(j)$-th Mn-site. $r = |\mathbf{R}_i - \mathbf{R}_j|$ is the Mn-Mn spatial separation. Here, rather than the conventional $|\mathbf{S}_i| = 1$ consideration, it is taken as $|\mathcal{M}_i|/2$, where

$$\mathcal{M}_i = \int_0^{R'} [\rho_i^{\uparrow}(\mathbf{r}') - \rho_i^{\downarrow}(\mathbf{r}')]\, d\mathbf{r}'. \tag{47}$$

The reason for this kind of consideration is that, for $\mathrm{Mn}_n\mathrm{As}$ clusters Mn atoms have different environment and so as the bonding, and therefore, it is very much likely that $|\mathbf{S}_i|$ have (slightly or much) different value at different site i. In the Eqn 1.47, R' is the radius of the integrating sphere centering each atom and taken as the half of the lowest Mn-Mn separation in the cluster to make sure that no two spheres overlap. $\rho^{\uparrow}(\mathbf{r}')$ and $\rho^{\downarrow}(\mathbf{r}')$ are the up and down spin densities, respectively, at a point $\mathbf{r}'$. It should be noted that in the construction of the Hamiltonian $\mathcal{H}$ in the Eqn.1.46, the signs (but not the magnitudes) are already absorbed in the definition of J_{ij}: positive for ferromagnetic and negative for antiferromagnetic coupling. We can calculate J_{ij}'s from Eqn.1.46, by computing the total energy for a judicious choice of spin configurations with inequivalent combinations of pair correlation functions $\mathbf{S}_i \cdot \mathbf{S}_j$, which results in a set of linear equations for the J_{ij}'s.

We can then compare the calculated J_{ij}'s with the Ruderman-Kittel-Kasuya-Yosida (RKKY)-like theory. This kind of theory describes the magnetic interaction between the localized Mn moments induced by the free carrier spin polarization. This indirect Mn-Mn exchange interaction arise from the local Zener coupling (or the $p - d$ hybridization) between the holes and the Mn d-levels, which then leads to the effective Mn-Mn RKKY interaction:

$$J_{ij}^{\mathrm{RKKY}}(r) \propto r^{-4}\left[\sin(2k_F r) - 2k_F r \cos(2k_F r)\right], \tag{48}$$

or, in the simplest approximation it reads,

$$J_{ij}^{\mathrm{RKKY}}(r) \propto r^{-3} \cos(2k_F r), \tag{49}$$

k_F is the Fermi wave vector. In dilute magnetic semiconductors, RKKY-like models predict that J_{ij} increases with concentration as $n^{1/3}$ at 0K and for fixed n, are independent of environment. We will check the case for As-doped Manganese ($\mathrm{Mn}_n\mathrm{As}$) clusters.

For the smallest possible cluster $\mathrm{Mn}_2\mathrm{As}$, the Mn-Mn exchange coupling J can be calculated from the energy difference between the two different spin orientations: parallel ($\uparrow\uparrow$) and antiparallel ($\uparrow\downarrow$) as,

$$E_{\uparrow\downarrow}(r) - E_{\uparrow\uparrow}(r) = J(r)\left[|S_1^{\uparrow}|\,|S_2^{\downarrow}| + |S_1^{\uparrow}|\,|S_2^{\uparrow}|\right], \tag{50}$$

which is the energy measure to flip either one of the moments. Spatial ($r_{\mathrm{Mn-Mn}}$) dependence of $E_{\uparrow\downarrow}$, $E_{\uparrow\uparrow}$ and exchange coupling J is plotted in the Figure12. Calculated J is compared with the simplest RKKY analytic form $J^{\mathrm{RKKY}}(r) \propto r^{-3} \cos(2k_F r)$. Exchange coupling J oscillates between positive and negative with r favouring FM and AFM solutions, respectively and dies down as $1/r^3$ — a typical RKKY type behaviour.

The case of $\mathrm{Mn}_3\mathrm{As}$ is very interesting because of the possible magnetic frustration. The Mn moment orientation in the ground state of $\mathrm{Mn}_3\mathrm{As}$ nanostructure is $\uparrow\uparrow\downarrow$ with a total moment 4 μ_B (Figure13). Energies of different Mn-spin orientations are given in the Table6.

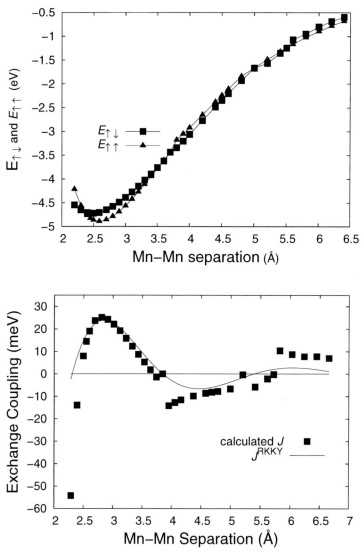

Figure 12. (upper) Spatial dependence of total energy for ↑↑ and ↑↓ configuration of Mn moment in Mn_2As cluster. (lower) Plot of Mn-Mn exchange coupling J with r_{Mn-Mn}. and compared with the simplistic RKKY-like form fitted with $k_F = 1.02$ Å^{-1}.

However, instead of two negative and one positive values, all the computed exchange couplings J_{ij}'s turn out to be negative ($J_{12} = -11$ meV and J_{23}, $J_{13} = -14$ meV) indicating the magnetic frustration of Mn-spins.

Now, it is interesting to see, how exchange couplings behave with the increasing cluster size n in Mn_nAs clusters. We plot averaged exchange coupling $\bar{J}_{ij}$ in the Figure 14, which behaves quite differently form RKKY-like theory: $\bar{J}_{ij}$ decreases as n increases with an exceptional increase for ferromagnetic Mn_4As cluster and it has a strong environment dependency.

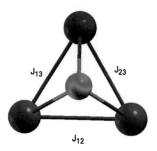

Figure 13. Ground state spin configuration for Mn_3As. Red(Blue) ball(s) represent $Mn(\downarrow)$ and $Mn(\uparrow)$ spin. The structure has a total magnetic moment $4\mu_B$.

7. Summary

In this chapter we discussed the electronic and magnetic properties of pure and doped manganese , Mn_n ($n = 2$ - 20) and Mn_nAs ($n = 1$ - 10), clusters from the density functional theory within the pseudopotential scheme. Here we use projector augmented wave method [19] which has been discussed briefly in the Section2.. The spin polarized generalized gradient approximation has been used to treat the exchange-correlation energy. To determine the ground state structure (both geometrical and magnetic), many different initial geometrical structures have been considered and all the possible spin multiplicities have also been investigated for each geometrical conformation. For both pure and doped clusters, we found many isomers with different magnetic structures are possible for a particular size.

Very small pure clusters, Mn_n, up to 4 atoms show Mn-Mn ferromagnetic coupling with 5 μ_B/atom magnetic moment. An addition of a single Mn atom then makes the magnetic coupling ferrimagnetic for Mn_5 and remains the same for the entire size range $n = 5$–20. These results show very good agreement with the recent experiments[4, 5]. The extraordinarily large experimental uncertainty for Mn_7 cluster[5] is explained due to the possible presence of the isomers in the experimental SG beam. However, the present density functional theory calculation predicts that for almost all the sizes there exist more than one isomer with different magnetic structure, but we do not observe any sufficiently large uncertainty corresponding to the size. The sudden deep in the experimentally measured

Table 6. Total energies of the Mn_3As clusters for different spin orientations.

Mn-spin configuration	Total energy (eV)
↑ ↑ ↑	-6.734
↑ ↑ ↓	-6.926
↑ ↓ ↑	-6.904
↓ ↑ ↑	-6.904

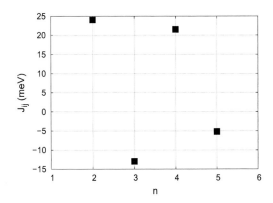

Figure 14. Plot of average exchange coupling $\bar{J}_{ij}$ with the number of Mn-atoms in the cluster n in $Mn_n As$.

magnetic moment at Mn_{13} and Mn_{19} [4, 5] is well reproduced within the density functional calculation. This is attributed to their closed icosahedral structure[8].

Previously it was argued that the the sudden increase in the reaction rate of the Mn_n clusters with molecular hydrogen [9] is attributed to the nonmetal-to-metal transition at $n = 16$. If this is indeed the case then there must be a large decrease in the ionization potential and a closing-up of spin gaps at that particular size $n = 16$. However, such kind of downward discontinuity in the ionization potential [37] or any closing-up of spin gaps [8] are not observed at $n = 16$ and the reason for this abrupt change in the reaction rate remains unknown.

Finite systems like cluster are very good system to study the effect of coordination on the electron localization. For an example we discussed Mn_{13} cluster: The $d-$ electrons of the 12-coordinated central atom are more delocalized than those of the 6-coordinated surface atoms, i.e. the $d-$ electron localization decreases with the increase in coordination[8].

Next we discussed the issue how electronic and magnetic structure of these Mn_n clusters do change due to single As-doping. The geometrical structures do not change substantially due to As-doping, but only a moderate perturbation to the corresponding pure Mn_n cluster, i.e. a $Mn_n As$ cluster, more or less, could be viewed as pure Mn_n cluster with a As-atom caped somewhere. However, the magnetic structure of the $Mn_n As$ cluster change substantially from their pure counterpart. Only in $Mn_2 As$ and $Mn_4 As$ clusters, the Mn-Mn coupling is ferromagnetic, and for all other size range this coupling is ferrimagnetic. It is seen that in $Mn_n As$ clusters the exchange couplings are anomalous and behave quite differently from the RKKY-type predictions[10]. It is known that due to $4s^2$, $3d^5$ electronic structure as well as for high enough $4s^2$, $3d^5 \rightarrow 4s$, $3d^6$ promotion energy of the isolated Mn-atom , they do not bind strongly and as a result Mn_n clusters have lowest binding energies among all other $3d-$transition metal clusters. This scenario has been changed when a single As-atom is attached to them to form a resultant $Mn_n As$ cluster. The binding energy of the resultant $Mn_n As$ clusters are enhanced by the hybridization of the $3d-$electrons of the Mn-atom with the $4p-$electrons of the As-atom. The corresponding energy gain in adding an As-atom to a existing Mn_n cluster ($\triangle^1$) is much larger than that of adding an Mn-atom to an existing $Mn_{1-n} As$ cluster ($\triangle^2$), which finally points out that the clustering of the Mn-atoms around an As-atom is energetically favourable.

This point is very important and should be discussed in the context of the observed ferromagnetism in the Mn-doped GaAs/InAs semiconductors. From the free cluster calculations, one come to the conclusion that As-atom act as a nucleation centre for Mn-atoms and it is very much likely that Mn-clusters might be present in those low temperature molecular beam epitaxy grown (Ga, Mn)As/(In, Mn)As samples. Now how the presence of these clusters influence the observed ferromagnetism and Curie temperature is a debatable question for a long time. The ferromagnetic ordering of Mn-atoms are intrinsic for Mn_2As and Mn_4As clusters and are the source of effective internal magnetic field, which influences the energy structure and transport properties. They provide (high temperature) ferromagnetic contribution to the total magnetization of the GaMnAs samples. Although the Mn_2As and/or Mn_4As phases can be mostly responsible for the observed ferromagnetic behaviour of GaMnAs, it is not excluded that predicted carrier-induced ferromagnetic or other mechanism leading to ferromagnetic behaviour can be effective. The presence of ferromagnetic Mn_2As and/or Mn_4As phases in the GaMnAs samples, which have large magnetic moment and very high exchange coupling would consequently enhance the Curie temperature. On the other hand, the presence of the other sized clusters would eventually lead to the low Curie temperature due to their low magnetic moment and low exchange coupling value.

Finally, it should be pointed out that in this chapter we talked about collinear spin arrangements. However, it will be interesting so see whether these clusters show noncollinear magnetic structure or not. Indeed, for larger clusters both the pure and As-doped manganese clusters with more that 5 manganese atoms into it might have noncollinear magnetic structure with substantially different total magnetic moment [8, 10].

References

[1] A. J. Cox, J. G. Louderback and L. A. Bloomfield, *Phys. Rev. Lett.* **71**, 923 (1993); A. J. Cox, J. G. Louderback, S. E. Apsel and L. A. Bloomfield, *Phys. Rev. B* **49**, 12295 (1994).

[2] D. M. Cox, D. J. Trevor, R. L. Wheetten, E. A. Rohlfing and A. Kaldor, *Phys. Rev. B* **32**, 7290 (1985).

[3] L. A. Bloomfield, J. Deng, H. Zhang and J. W. Emmert, in *Proceedings of the International Symposium on Cluster and Nanostructure Interfaces*, edited by P. Jena, S. N. Khanna and B. K. Rao (World Publishers, Singapore, 2000), p.131

[4] M. B. Knickelbein, *Phys. Rev. Lett.* **86**, 5255 (2001).

[5] M. B. Knickelbein, *Phys. Rev. B* **70**, 14424 (2004).

[6] M. D. Morse, *Chem. Rev.* **86**, 1049 (1986).

[7] D. A. Young, *Phase Diagrams of the Elements* (Berkeley, Los Angeles, CA: University of California Press).

[8] M. Kabir, A. Mookerjee and D. G. Kanhere, (To be published.)

[9] E. K. Parks, G. C. Nieman and S. J. Riley, *J. Chem. Phys.* **104**, 3531 (1996).

[10] M. Kabir, D. G. Kanhere and A. Mookerjee, physics/0503009, (2005).

[11] O.K. Andersen, *Phys. Rev. B* **12**, 3060 (1975).

[12] J.C. Slater, *Phys. Rev.* **51**, 846 (1937). .

[13] P.M. Marcus, *Int. J. Quantum. Chem.* **1S**, 567 (1967).

[14] J. Korringa, *Physica* **13**, 392 (1947).

[15] W. Kohn and N. Rostocker, *Phys. Rev.* **94**, 111 (1954).

[16] D.R. Hamann, M. Schlüter and C. Chiang, Phys. Rev. Lett. **43**, 1494 (1979).

[17] D. Vanderbilt, *Phys. Rev. B* **41**, 7892 (1985) .

[18] P. E. Blöchl, *Phys. Rev. B* **50**, 17953 (1994).

[19] G. Kresse and D. Joubert, *Phys. Rev. B* **59**, 1758 (1999).

[20] P. Pulay, in *Modern Theoretical Chemistry*, edited by H. F. Schaefer (Plenum, New York, 1977); *Mol Phys.* **17**, 197 (1969).

[21] S. Goedecker and K. Maschke, *Phys. Rev. B* **45**, 1597 (1992).

[22] J. P. Perdew, K. Burke and M. Ernzerhof, *Phys. Rev. Lett.* **77**, 3865 (1996).

[23] G. Kresse and J. Furthmuller, *Phys. Rev. B 54*, 11169 (1996).

[24] R. K. Nesbet, *Phys. Rev.* **135**, A460 (1964).

[25] D. D. Shillady, P. Jena, B. K. Rao, and M. R. Press, *Int. J. Quantum Chem.* **22**, 231 (1998).

[26] K. D. Bier, T. L. Haslett, A. D. Krikwood and M. Moskovits, *J. Chem. Phys.* **89**, 6 (1988).

[27] R. J. Van Zee and W. Weltner,Jr., *J. Chem. Phys.* **89**, 4444 (1988); C. A. Bauman, R. J. Van Zee, S. Bhat, and W. Weltner,Jr., *J. Chem. Phys.* **78**, 190 (1983); R. J. Van Zee, C. A. Baumann and W. Weltner, *J. Chem. Phys.* **74**, 6977 (1981).

[28] S. K. Nayak, B. K. Rao and P. Jena, *J. Phys.: Condens Matter* **10**, 10863 (1998).

[29] P. Bobadova-Parvanova, K. A. Jackson, S. Srinivas and M. Horoi, *Phys. Rev. A* **67**, 61202 (2003); *J. Chem. Phys* **122**, 14310 (2005).

[30] M. R. Pederson, F. Ruse and S. N. Khanna, *Phys. Rev. B* **58**, 5632 (1998).

[31] A. Wolf and H-H. Schmidtke, *Int. J. Quantum Chem.* **18**, 1187 (1980).

[32] N. Fujima and T. Yamaguchi, *J. Phys. Sols Japan* **64**, 1251 (1995).

[33] J. Harris and R. O. Jones, *J. Chem. Phys.* **70**, 830 (1979).

[34] D. R. Salahub and N. A. Baykara, *Surf. Sci.* **156**, 605 (1985).

[35] G. W. Ludwig, H. H. Woodbury and R. O. Carlson, *J. Phys. Chem. Solids* **8**, 490 (1959).

[36] K. Rademann, B. Kaiser, U. Even, and F. Hensel, *Phys. Rev. Lett.* **59**, 2319 (1987).

[37] G. M. Koretsky and M. B. Knickelbein, *J. Chem. Phys.* **106**, 9810 (1997).

[38] H. Ohno, *J. Magn. Magn. Mater.* **200**, 110 (1999).

[39] H. Ohno and F. Matsukura, *Solid State Commun.* **117**, 179 (2001).

[40] T. Dietl, H. Ohno, F. Matsukura, J. Cibert, and D. Ferrand, *Science* **287**, 1019 (2000).

[41] H. Ohno, H. Munekata, T. Penney, S. Von Molnar, and L. L. Chang, *Phys. Rev. Lett.* **68**, 2664 (1992).

[42] H. Ohno, A. Shen, F. Matsukura, A. Oiwa, A. Endo, S. Katsumoto, and Y. Iye, *Appl. Phys. Lett.* **69**, 363 (1996).

[43] M. E. Overberg, C. R. Abernathy, and S. J. Pearton, *Appl. Phys. Lett.* **79**, 1312 (2001).

[44] M. L. Reed, N. A. El-Masry, H. H. Stadelmaier, M. K. Ritums, M. J. Reed, C. A. Parker, J. C. Roberts, and S. M. Bedair, *Appl. Phys. Lett.* **79**, 3473 (2001).

[45] S. Sonoda, S. Shimizu, T. Sasaki, Y. Yamamoto, and H. Hori, cond-mat/0108159, (2001).

[46] A. F. Guillermet and G. Grimvall, *Phys. Rev. B* **40**, 10582 (1989).

[47] H. Shimizu, T. Hayashi, T. Nishinaga, and M. Tanaka, *Appl. Phys. Lett.* **74**, 398 (1999).

[48] S. J. Potashnik, K. C. Ku, R. Mahendiran, S. H. Chun, R. F. Wang, N. Samarth, and P. Schier, *Phys. Rev. B* **66**, 012408 (2002).

[49] S. J. Potashnik, K. C. Ku, S. H. Chun, J. J. Berry, N. Samarth, and P. Schier, *Appl. Phys. Lett.* **79**, 1495 (2001).

[50] T. Hayashi, Y. Hashimoto, S. Katsumoto, and Y. Iye, *Appl. Phys. Lett.* **78**, 1691 (2001).

[51] Darshan C. Kundaliya, S. B. Ogale, S. E. Lofland, S. Dhar, C. J. Metting, S. R. Shinde, Z. Ma, B. Varughese, K.V. Ramanujachary, L. Salamanca-Riba, T. Venkatesan, *Nature Materials* **3**, 709 (2004).

[52] D. Chiba, K. Tankamura, F. Matsukura and H. Ohno, *Appl. Phys. Lett.* **82**, 3020 (2003).

[53] K. W. Edmonds, K. Y. Wang, R. P. Campion, A. C. Neumann, N. R. S. Farley, B. L. Gallagher and C. T. Foxon, *Appl. Phys. Lett.* **81**, 4991 (2002); K. C. Ku, S. J. Potashnik, R. F. Wang, S. H. Chun, P. Schiffer, N. Samarth, M. J. Seong, A. Mascarenhas, E. Johnston-Halperin, R. C. Myers, A. C. Gossard, and D. D. Awschalom, ibid 82, 2302 (2003).

[54] Y. Yamaguchi and H. Watanabe, *J. Magn. Magn. Mater.* **31-34**, 619 (1983).

[55] B. K. Rao and P. Jena, *Phys. Rev. Lett.* **89**, 185504 (2002).

In: Atomic, Molecular and Optical Physics…   ISBN: 978-1-60456-907-0
Editor: L.T. Chen, pp. 345-385   © 2009 Nova Science Publishers, Inc.

Chapter 9

POROUS SILICA FOR SOLID-STATE
OPTICAL DEVICES

Carlo M. Carbonaro[1,], Francesca Clemente[2], Stefania Grandi[3] and Alberto Anedda[1,4]*

[1]Dep. of Physics, University of Cagliari, SP n 8 Km 0.700,I-09042 Monserrato (Ca), Italy
[2]IMEC and K.U.Leuven, ESAT-INSYS, B-3001 Leuven, Belgium.
[3]Dep. of Physical Chemistry, University of Pavia, INSTM, and IENI-CNR, V.le Taramelli 16, I-27100 Pavia, Italy
[4]Centro Grandi Strumenti, Dep. of Physics, University of Cagliari, SP n° 8 Km 0.700, I-09042 Monserrato (Ca), Italy

Abstract

The investigation of the optical properties of porous silica samples is presented. Optical spectroscopy measurements, including Raman scattering, steady state and time resolved photoluminescence, optical absorption and excitation of photoluminescence are reported. The chapter reviews the results of the research we carried out upon the emission features of porous silica in the ultraviolet and visible wavelength range and the characterization of the emission properties of dye-doped sol-gel synthesized silica samples. In particular, the study of the emission band recorded at about 3.7 eV and its correlation with the chemical and physical conditions of the surface is discussed. As regards dye-doped silica samples, the analysis of the spectroscopic features of pre- and post-doped hybrid samples is presented and their potential feasibility as solid state dye laser is proposed.

Introduction

Porous glasses are promising materials for several technological applications [1-6]. In particular, owing to its distribution of pore sizes and surface areas, porous silica (PS) has been used as host for materials such as catalysts, polymers, metals and semiconductor nanoparticles and it has potential applications in optoelectronics, chemistry and magnetic

---

[*]E-mail address: cm.carbonaro@dsf.unica.it, phone: +390706754823, fax: +39070510171. (Corresponding author.)

recording [1-6]. The properties of porous silica are related to the surface chemistry of the samples. The surface of sol-gel synthesized PS is generally terminated with OH groups bounded to a silicon atom, SiOH units called silanols [7, 8]. The concentration of OH groups at the surface is about 4-5 OH/nm^2 and it is found to be almost independent on the synthesis conditions of porous silica [7]. Mesoporous silica presents an open and interconnected pore structure with sizes in the range of 2–200 nm and SiO_2 walls of about 1 nm thickness [9, 10]. This is the typical thickness of the oxide layer which covers porous silicon and silicon nanoparticles, promoting porous silica as an ideal material to study chemical and physical properties of silicon/silica based devices with nanometer dimensions. Indeed, in the past few years, the optical features of nanosized silica have been extensively investigated due to the similarities with the PL properties of porous silicon [11-17].

Active organic dopants can add new functionalities to SiO_2 glass [18]. The development of solid-state gain media containing organic laser dyes has been studied in the past decade [18-21]. The encapsulation of organic dyes within inorganic solids offers several advantages with respect to liquid hosts, including ease of use and replacement and a decrease of health and environmental hazards [18-21]. The mild processing temperature of the sol–gel synthesis allows for the easy addition of organic molecules to inorganic hosts [18, 19]. Recently, organic–inorganic complexes have been developed with applications ranging from photonics to optoelectronics and sensing [18, 19].

In this chapter we report on two topics: i) the study of the optical properties of porous silica in the visible and ultraviolet (UV) region; ii) the analysis of the photoluminescence features of silica based organic-inorganic compounds.

Both topics are related to the possibility of developing solid-state devices for optical technology applications. The ultraviolet emission at about 3.7 eV observed in porous silica shares common optical features with UV photoluminescence of oxidized porous silicon, silicon nanostructures and silicon oxide films [11, 12, 22]. Beyond the technological development of silicon/silica-based UV-emitting devices, the analysis of these optical properties as a function of the porous media properties could help clarify the debated origin of the UV and visible luminescence in porous silicon [23-26]. By properly combining the host and guest chemical- and physical properties, the embedding of organic dye molecules into silica glass prepared via sol-gel methods can offer the highest physical and chemical performances of silicon/silica based devices [27-38].

The incorporation of a model organic dye, the very efficient Rhodamine 6G (Rh6G), into sol-gel prepared silica-based material by different routes is presented and the optical features of the realized hybrids are examined in the perspective of solid-state devices for optoelectronics.

The chapter is organized as follows. The first section deals with the optical properties of porous silica and their correlation with the physical and chemical conditions of the sample surface. The analysis of the emission properties as a function of the sample porosity and in samples with different chemical composition or subjected to different chemical and physical treatment is presented. The reported data allow us to propose a model for the emission at about 3.7 eV.

The second section concerns the synthesis and optical spectroscopy characterization of organic-inorganic hybrids, that is Rh6G-silica samples, to exploit their application as photonic devices. The optical features of samples prepared by pre- and post-doping sol-gel method are presented and compared to the properties of a reference liquid solution. The analysis of the sample photostability, that is the light irradiation hardness, indicates the pre-doped samples with covalent bonding between silica host and dye as good candidates for solid state dye laser development.

Emission Properties of Porous Silica on the Ultraviolet Energy Range

The potential applications of PS are widespread in several technological field, from catalysis to gas sensing [1-6]. In this section we discuss PS optical properties, linking the physical and chemical structure of its surface to the detected photoluminescence features and pointing out its possible applications in optoelectronics.

Because of the large surface-to-volume ratio, the surface chemistry determines the properties of PS. It is well known that the surface of sol-gel synthesized PS is naturally terminated with a large number of different surface silanols SiOH (isolated, geminal, and vicinal) where water molecules can be easily adsorbed. In the following section a description of the PS surface is reported and a study of the OH species as a function of pore size is presented. Indeed, in order to engineer high-performing porous SiO_2 devices (e.g. luminescent devices, gas sensors, solid state dye lasers), the control of the physical and chemical interaction of guest molecules within PS is demanded and the knowledge of the surface structure, its morphology and chemistry, represents a technologically relevant task [39-42]. The dependence of PL properties on the conformation of the surface will be described in the second session. Strong evidences for surface emitting centers will be given. PS UV emission at about 3.7 eV shares common optical features with UV photoluminescence (PL) of oxidized porous silicon, silicon nanostructures, and silicon oxide films [11, 12, 22]. Beyond the technological development of silicon-based UV emitting devices, the analysis of these optical properties as a function of the porosity could help to clarify the debated origin of the UV and visible luminescence in porous silicon [23-26]. In the third section PS photoluminescence is further investigated: time resolved PL experiments on samples with different porosity suggest the assignment of the UV emissions to strong and weak interacting silanol species.

Pore Size Distribution and OH-Species

The surface of PS can be represented as the truncation of a random network composed of siloxane (Si-O-Si) rings. Hydroxyl groups bonded to silicon atoms typically terminate the open rings at the surface and form the so-called silanols (SiOH units). Depending on the hydrogen interactions of neighboring silanols isolated, geminal, and vicinal structures can be formed [7, 8].

An isolated silanol is a non-hydrogen bonded OH group located at a Q^3 site; when the distance between OH group sitting at Q^3 sites allows H bonding interaction, the SiOH units are referred as vicinal. Geminal silanols [Si(OH)$_2$] can be described as two OH groups sitting at a Q^2 site [7]. The number of OH covering the PS surface does not depend on the synthesis conditions: its mean value is about 4-5 OH/nm^2 in fully hydrated samples and it is found to be almost independent on the pore size and morphology [7]. Although the density of silanols at the surface is sample independent, the hydrogen-bond interaction of OH groups is determined by a number of factors: the number of hydroxyls per silicon site, the Si-O-Si ring size, the opening degree of the ring, that is the angles of the ring, and the curvature of the surface [7]. In particular, the distance between neighbor hydroxyls is reduced by a small negative curvature radius enhancing the hydrogen-bond interaction with respect to a flat surface. Thus, a larger interaction between silanols is expected in samples with smaller pore diameter. In addition, interacting silanols act as preferential adsorption sites for water molecules as compared to isolated silanols and a larger content of adsorbed water should be expected in samples with smaller pores [7].

Within PS pores, hydroxyls can be found in the form of silanols, physisorbed water or "bulk" water nested into pore puddles. Confined liquids in restricted geometries reside in two different states: a very sluggish adsorbed phase (at the surface-fluid interface) and a relatively free liquid phase with moderately frustrated relaxation (in the inner pore volume). Concerning H_2O, the majority of the studies show that only a small percentage of water confined in PS is affected by confinement [42-53].

Raman and infrared spectroscopy have been extensively used to investigate the vibrational properties of silica at different steps of the sol-gel process. When studying the O-H fundamental stretching vibration, in the 3000-3800 cm^{-1} range, two main features are detected: a broad band assigned to the superimposition of different bonded hydroxyl groups and a narrow peak at 3750 cm^{-1} fingerprint of the isolated silanol vibration [7, 54-57].

In this section we present a comparative study on the vibrational properties of sol-gel synthesized PS. The scientific goal is to determine the distribution of the hydroxyl groups at the porous silica surface and to assess their possible correlation with pore morphology. To fulfill this task, we analyzed the Raman spectra (RS) of samples with different porosity focusing, in particular, on the O-H stretching range. A dependence of the relative concentration of the different hydroxyl species on the mean pore diameter of the samples is detected. By taking into account the contribution of bulk water nested into the pores, we effectively singled out the role of surface SiOH groups.

Room temperature Raman scattering measurements were carried out with a micro-Raman spectrometer (Dilor XY80). The excitation was supplied by an argon ion laser operating at 514.5 nm (Coherent Innova90C-4). The signal, dispersed by a 600 grooves/mm grating, was detected by a 1024x256 LN_2 cooled charge coupled detector (CCD). The spectral resolution was 2.0 cm^{-1}.

Table I. Characteristics of the porous samples, being $\varnothing$ = pore diameter; SSA = Specific Surface Area; ρ = density.

sample	$\varnothing$ (nm)	SSA (m^2/g)	ρ (g/cm^3)
S1	3.2	594	1.2
S2	5.5	540	0.9
S4	18.2	264	0.6

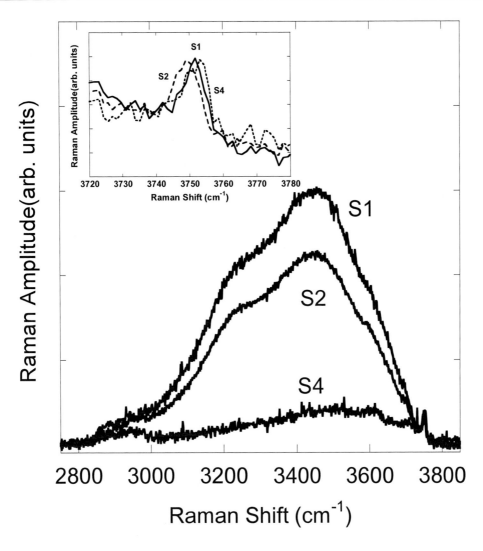

Figure 1. Raman Spectra in the fundamental O-H stretching range of S1, S2 and S4 samples. In the inset an enlarged view of the 3750 cm⁻¹ peak is reported. Adapted with permission from [59]. Copyright 2003 American Chemical Society.

Measurements were performed on sol-gel synthesized porous silica monoliths produced by Geltech Inc. (US). The samples have different pore diameters: 3.2 nm (S1), 5.5 nm (S2) and 18.2 nm (S4), with 5% standard deviation; porosimetric data are reported in Table I [58].

In Figure 1 the Raman spectra of S1, S2 and S4 samples in the 2750-3850 cm⁻¹ range are shown. A composite band in the 2900-3800 cm⁻¹ range is displayed by the three sets of samples with a narrow feature at 3750 cm⁻¹. For a better comparison spectra were arbitrarily normalized to the intensity of the 3750 cm⁻¹ peak. As the pore diameter of the silica samples increases, the contribution of the composite band with respect to the 3750 cm⁻¹ peak decreases.

In order to give an estimate of the relative content of nested water into the pores of the samples, we analyzed the RS of liquid water (Figure 2) through Gaussian best-fit deconvolution: three components peaked at 3250, 3450 and 3610 cm⁻¹ were attributed to the first overtone of the H_2O bending, and to the symmetric and anti-symmetric O-H stretching respectively [60]. Their full width at half maximum (FWHM) are 264, 198 and 141 cm^{-1} respectively.

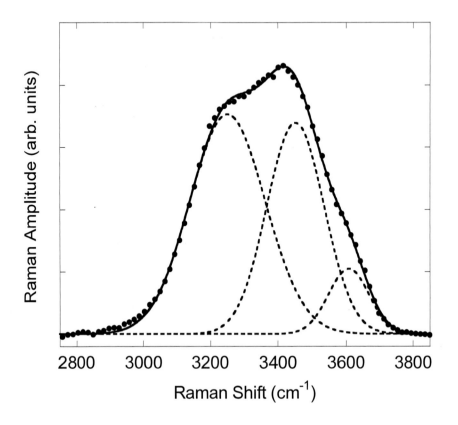

Figure 2. Best fit deconvolution ($r^2 = 0.99$) of liquid water Raman spectrum in the O-H stretching range: circles are the experimental data, dashed lines are the Gaussian components and solid line is the resulting fit. Adapted with permission from [59]. Copyright 2003 American Chemical Society.

Raman spectra of PS samples were tentatively resolved by using Gaussian bands and the reconstructed H_2O contribution. The content of "bulk" water within the pores was a free parameter of the fit, that is a scaling factor, while position and shape of the reconstructed H_2O contribution were constrained. As the properties of molecular water confined inside PS are comparable with those of motion free liquid H_2O, the Raman activity of PS confined water is, in first approximation, very well described by the RS of "bulk" liquid water [42-53]. Thus, the liquid water simulated spectral profile, multiplied by a scaling factor, was used as a function to fit the RS of the three sets of samples in addition to Gaussian components.

The deconvolution of the RS of S1 (Figure 3a), S2 (Fig 3b) and S4 (Figure 3c) samples is reported in Figure 3. Four Gaussian bands are resolved in addition to the sample dependent contribution of water. Peak positions and FWHM of these Gaussian bands are reported in Table II with an estimated uncertainty of 2 cm^{-1}. The integrated percentage area, reported in Table III, allows to estimate the relative contribution of water and different Gaussian bands. It is well known that the hydrogen bonding interaction between neighboring structures and the

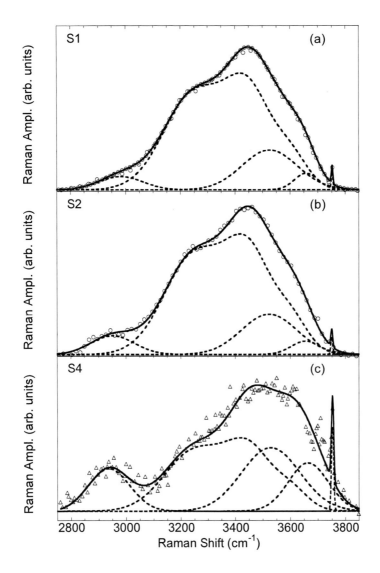

Figure 3. Best fit deconvolution of the Raman spectra in the O-H stretching range of: (a) S1 samples; (b) S2 samples; (c) S4 samples. Markers are the experimental data, dashed lines are the gaussian bands and the reconstructed water contribution according to the fitting analysis and solid line is the resulting fit. Adapted with permission from [59]. Copyright 2003 American Chemical Society.

hydrophilic nature of the samples are increased by decreasing the pore diameter (see [7] and references therein). Figure 1 is a clear confirmation of this effect: by analyzing samples with different pore size we showed that a larger contribution of isolated silanols with respect to the interacting OH-species is observed in samples with larger pore diameter.

As shown in Figure 3 and Table II, the analysis of the RS spectra of the three sets of samples by means of deconvolution in Gaussian bands singled out the contribution of several vibrations. In addition to the composite band of adsorbed water, the vibrations at about 3525, 3660 and 3750 cm^{-1} were resolved. It is worth to note that the spectral features of these bands do not depend on the mean pore diameter of the samples (see Table II). In the literature the vibrations here singled out were attributed to specific O-H stretching modes by assuming that the vibrational frequency of the Raman mode decreases as the H-bonding interaction of the surface OH groups increases. The following attribution were reported: the 3750 cm^{-1} is the fingerprint of the isolated silanol vibration; the vibration at 3660 cm^{-1} was related to O-H stretching in adsorbed water at the surface or to Si-OH stretching of mutually H-bonded surface hydroxyls; the SiO-H stretching of surface silanols bonded to molecular water via hydrogen bonding are located in the 3510-3540 cm^{-1} range. In addition, the vibrations of OH-species interacting via stronger and weaker hydrogen bonding were called out to describe the 3520 and 3660 cm^{-1} bands [7, 60-62].

The analysis of the Raman spectra resolves also a contribution in the 2800-3100 cm^{-1} range whose peak position is slightly dependent on the samples. The attribution of this feature is still open and both the C-H stretching mode and the O-H stretching of water dimer molecules were proposed [63].

Table II. Spectral properties of the gaussian bands of the deconvoluted spectra. Adapted with permission from [59]. Copyright 2003 American Chemical Society.

Sample	Peak –FWHM (cm^{-1})	Peak –FWHM (cm^{-1})	Peak – FWHM (cm^{-1})	Peak – FWHM (cm^{-1})	r^2
S1	2982 - 186	3525 - 235	3658 - 101	3751 - 7	0.99
S2	2950 - 185	3522 - 237	3654 - 123	3749 - 8	0.99
S4	2937 - 176	3527 - 251	3663 - 171	3751 - 10	0.92

The analysis of integrated percentage area, reported in Table III, indicates that sol-gel synthesized PS presents a relative content of adsorbed H$_2$O which is a function of the sample porosity: the relative content of adsorbed water decreases as the pore diameter increases. Also the relative contribution of the SiOH related bands depends on porosity. The relative content of interacting silanol species with respect to the content of isolated silanols decreases as the sample pore diameter increases. The observed trend can be explained by considering the pore curvature effect on the extent of hydrogen bonding interaction between neighboring OH structures. A decrease of the pore diameter increases the H-bonding interaction between neighboring silanol species leading to a larger relative content of interacting silanols with respect to the isolated ones [7]. This is also confirmed by the increase of the relative content of adsorbed water in samples with a smaller pore diameter. This effect is expected since interacting silanols are preferential adsorption sites for water and polar molecules. Samples with smaller pores can be regarded as more hydrophilic system.

The dependence of the SiOH distribution on the pore diameter, effectively proven here by Raman spectroscopy, is a key aspect in achieving an understanding of PS chemical-physical properties. Indeed, owing to the high surface to volume ratio, the role of surface hydroxyls should determine PS properties and in particular surface centers are expected to affect PS optical properties.

Table III. Integrated area percentage of the components of the deconvoluted spectra. Adapted with permission from [59]. Copyright 2003 American Chemical Society.

Sample	IA_{2960} (%)	IA_{3525} (%)	IA_{3658} (%)	IA_{3750} (%)	IA_{water} (%)
S1	4.0	15.9	2.1	0.1	77.9
S2	5.4	14.8	2.5	0.2	77.0
S4	11.7	25.0	13.0	1.3	49.0

Photoluminescence Properties and Hydration Conditions of Porous Silica Surface

In the last decades an increasing number of studies have dealt with the photoluminescence properties of porous silica samples. It was reported that by exciting in the UV energy range a few emissions band centered at about 3.7, 2.8 and 2.0 eV could be observed [12-17]. In the present section we report the investigation we carried out on the PL band centered at about 3.7-4.0 eV. Indeed, the attribution of this UV luminescence was largely discussed and ranges from different kinds of adsorbed water in Si-OH complexes [11, 12, 16, 64] (including isolated and adjacent silanols) to water-carbonyl groups [12], apical-like Si surface centers and surface oxygen deficient centers [13]. In particular, the analysis of the surface reactivity of PS to OH was performed by Yao et al. [16]. The modifications induced to the UV emission by the adsorption of hydroxyl groups from the ambient atmosphere leaded the authors to associate the PL feature to non bridging oxygen hole centers. We investigated the origin of the 3.7 eV PL emission with a detailed analysis of differently synthesized porous silica samples. The PL spectra of both commercial and home-made PS and polyethilenglycole (PEG)-silica hybrids are presented. Data analysis is supported by Raman scattering measurements. The reported emission spectra show common features in the whole set of samples, regardless the nature of the samples (commercial, home-made or hybrids) or the different chemical or physical treatment performed upon the samples. The analysis suggests that the PL feature centered at about 3.7-4.0 eV is related to silanols and OH species adsorbed at the surface.

PL measurements were carried out at room temperature. The excitation was supplied by a MgF_2-sealed Deuterium lamp (Hamamatsu mod. L1835) and dispersed by 0.3 m scanning monochromator (McPherson mod. 218) under computer control The excitation spectral bandwidth was 5.0 nm. The PL signal was dispersed by a spectrograph (ARC-SpectraPro 275) and detected by a gatable intensified array (EG&G 1420). The spectral band-pass was 0.6 nm. The reported spectra are recorded applying a short wavelength cutoff filter (WG280) and corrected for the optical transfer function of the system. Raman measurements were carried out with the set-up described in the former section.

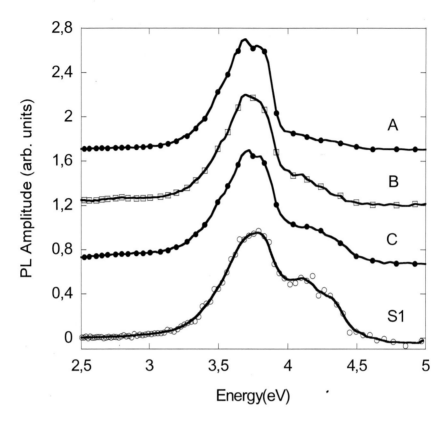

Figure 4. PL spectra of A, B, C and S1 samples excited at 5.6 eV. Adapted from [65], Copyright (2003), with permission from Elsevier.

PL and Raman measurements were performed on a set of commercial and home-prepared sol-gel synthesized porous silica. Disk samples with the pore diameter distribution peaked at around 3.2 nm were produced by Geltech Inc. (S1 samples listed in Table I). In order to investigate the correlation between UV PL band and surface hydration conditions, S1 samples were studied both in their as-grown state and in fully hydrated conditions: the sample was stored in distilled water for about 12 hours, then the inner pore water was removed by clean vacuum exhausting (10^{-2} Torr). Fully hydrated samples can contain clusters of free molecular water nested in puddles on the pore surface [42, 44].

The possible correlation with C-related species was examined by studying the PL properties of fully calcinated samples and C-containing PEG-silica hybrids. A set of samples with a mean pore diameter of about 3 nm and a specific surface of about 600 m^2g^{-1} was synthesized and thermally treated at 500 °C and 600 °C for four hours (hereafter A and B samples, respectively) [66]. In addition porous PEG-SiO$_2$ hybrid materials (hereafter C samples) were prepared via sol-gel route by adding Polyetilenglicol-200 (PEG with an average molecular weight of 200) to the sol in a 35/65 weight ratio. The obtained xerogels have PEG embedded in the porous matrix [66].

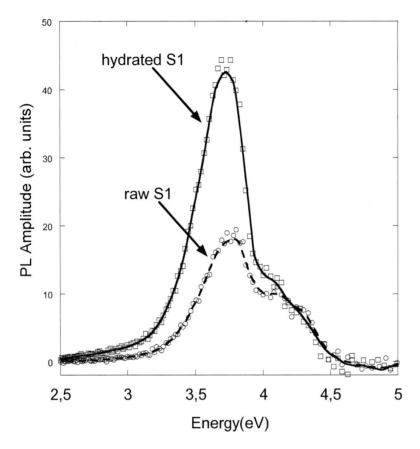

Figure 5. PL spectra of S1 samples before and after hydration. Adapted from [65], Copyright (2003), with permission from Elsevier.

The PL spectra excited at 5.6 eV of A, B, C, and S1 samples are shown in Figure 4. The excitation energy was selected in the 4.6-5.9 eV range as the most efficient in providing the PL feature. For a better comparison, emission intensities are normalized to their maximum and baselines arbitrarily shifted. A structured photoluminescence in the 2.5-4.5 eV range is recorded, the main features of the spectra are a peak at 3.7 eV and a less intense component at about 4.1 eV. No variation of the PL shape and intensity is found with air exposure nor during excitation irradiation.

Figure 5 shows PL spectra of raw and water treated porous silica S1 samples. The 3.7 eV component increases with respect to the 4.1 eV band in fully hydrated samples. As expected Raman activity in the O-H stretching range is observed for both raw and water treated samples (Figure 7) with a composite band in the 3000-3700 cm^{-1}. This broad vibrational band shows a peak at 3450 cm^{-1} and other components at about 3230, 3590 and 3700 cm^{-1}. The sharp peak at 3750 cm^{-1} is detected only in raw samples.

Since the UV PL shares common features in differently synthesized samples, we suggest that the center responsible for the observed emission is a surface center. As an example, samples with a different chemical composition of the silica walls, such as pure and hybrid porous silica, display a comparable PL band, as shown in Figure 5. In addition, the dependence of the intensity of the 3.7 eV on the specific surface of porous silica was recently reported confirming the attribution of the observed PL to a surface center [13].

The analysis of the PL spectra of PS synthesized at different temperatures indicates that the observed luminescence is not related to organic compounds. Since the thermal treatment of B samples at 600 °C produces fully calcinated samples, no organic residual were left in the samples by the sol-gel synthesis and originating from precursors, templates and/or catalysts. Thus the PL band observed in B samples and comparable with the PL emission of the other sets of samples can not be related to organic residues. In addition, the observed emission does not depend on the chemical composition of the matrix (C samples - Figure 5) supporting the attribution of the PL to a non-organic surface center.

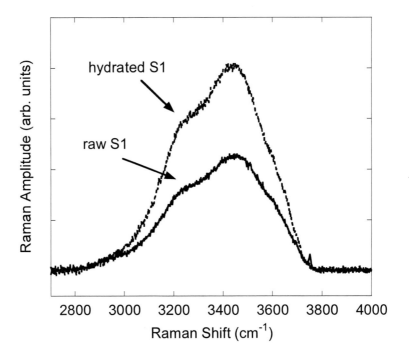

Figure 6. Raman spectra of the S1 samples before (solid line) and after (dashed line) water treatment in the O-H stretching range. Adapted from [65], Copyright (2003), with permission from Elsevier.

The surface morphology of the samples is affected by the hydration procedure, as evidenced by the Raman spectra analysis in the O-H stretching range. The increase of the contributions of interacting silanols (both vicinal and geminal) and adsorbed water molecules [7, 60] was, indeed, expected and shed some light on the attribution of the PL band.

Raman spectra of the fully hydrated samples, with respect of the raw ones, lack of isolated silanol signature (the 3750 cm^{-1} peak in Figure 7).

The hydrogen bonding between superficial hydroxyls is increased by the adsorbed water and causes the bonding of previously isolated silanols. As a consequence, the 3750 cm^{-1} vibration disappears. On the other hand the relative contribution of the PL at about 3.7 eV with respect to the emission at highest energy increases because of the hydration procedure. Thus, we can deduce that the observed emission is related to interacting OH-related centers: the emission spectrum depends on the concentration of hydroxyl groups at the inner pore surface. In addition, the modification induced by the hydration procedure confirms the location of the emitting centers at the surface. Finally one should note the lack of reactivity to the ambient atmosphere observed here, in contrast to previously reported experiments [16]; this can be explained as follows: the as-prepared samples were already in equilibrium with the OH contained in the air or their equilibrium rate with the atmosphere was faster than the time technically required to perform measurements.

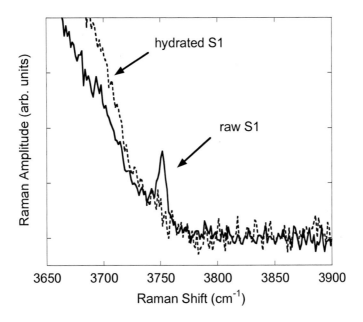

Figure 7. Detail of the Raman spectra in the 3650-3900 cm^{-1} range of S1 samples before (solid line) and after (dashed line) water treatment. Adapted from [65], Copyright (2003), with permission from Elsevier.

Ultraviolet Emission of Silanol Groups

We have shown that the PL observed in porous silica originates from surface centers. In addition PL efficiency and emission range can be modulated by changing the chemical and physical conditions of the PS surface [11-17]. PS displays a main absorption band in the 4.7 – 6.5 eV range [67]. Comparable systems like high surface silica displayed absorption bands at about 5.0 and 5.9 eV which were related to paramagnetic Si-OH and Si-H defects [68]. As shown in the previous section, the 3.7 eV emission excited at 5.6 eV is associated to OH-related surface defects. We carried out a further study of the spectral and temporal characteristics of the UV PL band by means of synchrotron radiation. The performed investigation allows us to spectrally and temporally resolve the UV emission into two different contributions with peak at 3.7 and 3.9 eV and decay time of 2.0 and 20 ns respectively. On the basis of the reported data combined with Raman scattering and PL evidences previously reported, the investigated emissions are assigned to two different kinds of OH related surface defects.

Evidences for the Composite Nature of the UV PL Band

Measurements were performed on commercial sol-gel synthesized porous silica monoliths whose details were reported in Table I. In addition a set of samples with pore diameter of 7.5 nm, SSA of 525 m^2/g and density of 0.7 g/cm^3 (hereafter S3) were investigated (Geltech Inc. US). PL and PLE measurements were carried out with pulsed excitation light provided by the synchrotron radiation (SR) at the SUPERLUMI experimental station on the I beamline of the HASYLAB synchrotron laboratories at Desy (Hamburg). The PL signal, dispersed by a 0.5 m Czerny-Turner monochromator, was detected in the 250-850 nm wavelength range with a photomultiplier (Valvo XP2020Q). The spectral resolution was 16 nm. The PLE measurements were performed in the 125-310 nm wavelength range with 0.3 nm of bandwidth. PL and PLE spectra were recorded under multi-bunch operation and detected with an integral time window of 192 ns correlated to the SR pulses. Time resolved PLE spectra were acquired over time windows of 20 and 150 ns and delay time with respect to the exciting synchrotron radiation pulse of 1.5 and 40 ns respectively. Decay time were gathered in the ns domain under single-bunch operation: the 192 ns interval time between adjacent pulses was scanned with 1024 channels (pulse width of 0.2 ns [69]). Measurements were performed at room temperature (RT) and at 8K; a continuous-flow liquid helium cryostat was used to set the temperature of the sample chamber at 8K.

The PLE spectrum in the UV and vacuum UV range of the emission at 3.65 eV for the S1 sample is reported in Figure 8. A narrow excitation channel at 5.5 eV is detected; its FWHM is 0.55 eV. In addition two less efficient excitation bands are observed at about 5.0 and 6.0 eV.

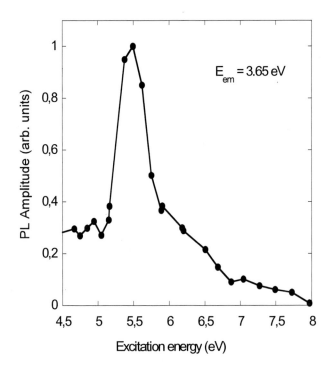

Figure 8. Excitation spectrum of the 3.65 eV PL of S1 samples. Adapted with permission from [70]. Copyright 2003 American Chemical Society.

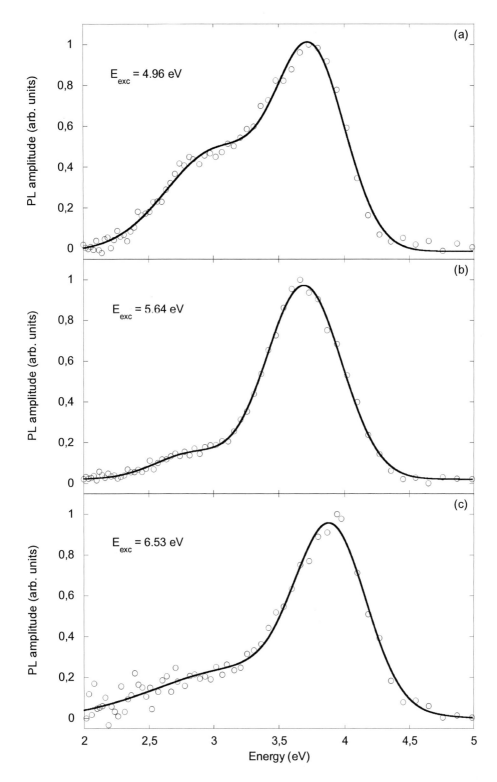

Figure 9. PL of S1 samples at different excitation energies. Adapted with permission from [70]. Copyright 2003 American Chemical Society.

The spectrum of PL of S1 samples excited at a few energies within the excitation spectrum (4.96, 5.64 and 6.53 eV) is reported in Figure 9; PL spectra are arbitrarily normalized to their maximum for the sake of clarity. The spectra are composed of two emission bands centered around 3.7 eV and 2.8 eV respectively. The peak position of the UV PL band depends on the excitation energy. In fact, when the PL spectra were heuristically deconvoluted with two gaussian bands, it could be observed that, when exciting at 4.96 eV, the UV band is peaked at 3.75 eV, then red shifted to 3.70 eV with the excitation energy increased to 5.64 eV and blue shifted to 3.89 eV with the excitation energy set at 6.53 eV. The FWHM keeps almost constant, 0.61, 0.68 and 0.68 eV at 4.96, 5.64 and 6.53 eV respectively. The blue PL peak changes in a similar way: when the excitation is 4.96 eV it is positioned at 3.03 eV, then red shifted to 2.84 eV as excited at 5.64 eV and blue shifted to 3.13 eV at the excitation of 6.53 eV.

One should note that the contribution of the blue band with respect to the UV PL is larger when the excitation is fixed at 4.96 eV than for the other excitation energies. As a consequence, the spectral and temporal properties of the UV PL band can be evidenced in a better way in S1 samples by analyzing the emission spectra excited at 5.64 and 6.53 eV.

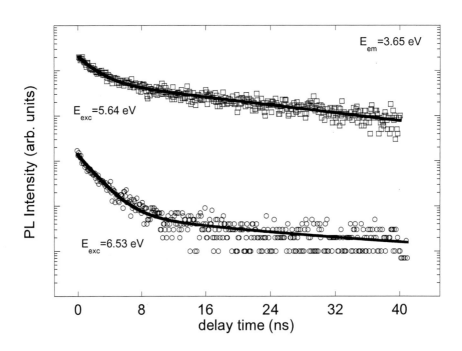

Figure 10. 3.65 eV PL decay time excited at 5.64 and 6.53 eV in S1 samples. Adapted with permission from [70]. Copyright 2003 American Chemical Society.

The decay time of the 3.65 eV PL band is reported in Figure 10 where the time evolution of the emission is recorded under 5.64 eV and 6.53 eV excitations. The semi-logarithmic scale of the plots displays that the time kinetics can not be interpreted as a single exponential decay. The experimental data of the decay time at 5.64 eV excitation can be successfully approximated with two exponential decays of 2.0 and 20 ns (square correlation factor ≥0.97, estimated error 5%). Even thought the signal to noise ratio of PL decay time at 6.53 eV excitation decreases, experimental data can be satisfactorily fitted with the same exponential decays with the addition of a baseline. The relative amplitude of the two exponential decays is different at different excitation energies: is about 3:1 at 5.64 eV excitation and 20:1 at 6.53 eV excitation.

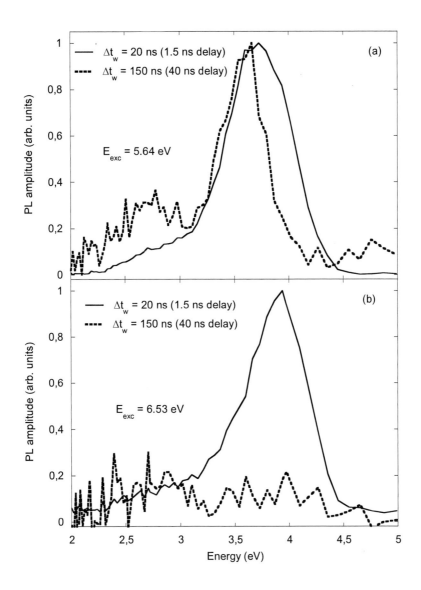

Figure 11. PL excited at 5.64 (a) and 6.53 eV (b) at different delay time and temporal windows (Δt_w) in S1 samples. Adapted with permission from [70]. Copyright 2003 American Chemical Society.

The experimental set-up of the SUPERLUMI station allows to detect the emission spectrum with selected time windows (time widths of acquisition) at different delays from the excitation pulse. In Figure 11 the UV PL band was recorded with a time window of 20 ns and a delay of 1.5 ns after the excitation pulse and with a time window of 150 ns and a delay of 40 ns after the excitation pulse. When excited at 5.64 eV (Figure 11a), the peak position of the UV PL changes at different delay times: with the shortest time window the PL is centered at 3.73 eV with the 1.5 ns delay time and shifts to 3.60 eV with the largest time window and the

40 ns delay time. The PL spectrum excited at 6.53 eV (Figure 11b) indicate that, under this excitation, the UV emission has a decay time of the order of few ns: indeed a very small contribution to the UV PL is recorded with the time window of 150 ns and a delay time of 40 ns after the excitation pulse. The observed spectral and temporal characteristics can be interpreted as the indication of the presence of two different and overlapping emissions.

The reported results can be interpreted by hypothesizing that the PL of two different emitting centers contribute to the spectrum of the UV PL peaked at 3.7 eV. The performed analysis allows the spectral and temporal resolution of the investigated emission features. The first emission is centered at about 3.9 eV, is characterized by a lifetime of about 2.0 ns and is more efficiently excited at 6.53 eV; the second emission is peaked at about 3.7 eV, has a decay time of about 20 ns and is more efficiently excited at 5.64 eV.

In the previous section we have shown that the PL band at 3.7 eV is related to the concentration of OH groups at the PS surface: by increasing the hydroxyls content and the hydrogen interaction between OH species the 3.7 eV PL increases. Thus the data previously reported call for the attribution of the composite emission at 3.7 eV to two interacting silanols. We call them Interacting Silanols 1 (IS-1) and Interacting Silanols 2 (IS-2), indicating the silanols characterized by Raman vibrations peaked at 3525 cm^{-1} and 3658 cm^{-1} respectively.

Dependence of the UV PL Band on the Sample Porosity

Since the relative content of OH-species in PS depends on porosity and a larger relative content of IS-1 with respect to that of IS-2 is detected in samples with smaller pore diameter (see first section), we carried out the investigation of the PL properties as a function of porosity in order to verify the proposed attribution and to associate a specific interacting silanol (IS-1 or IS-2) to one of the reported PL contributions. In Figure 12 the emission spectra of S1 and S4 samples at room temperature (RT) and at 8 K excited at 5.64 eV are reported. The PL spectra of S2 and S3 samples are very similar to those of S1 and S4

Table IV. Spectral characteristics (in eV) of the gaussian bands used to fit the PL spectra of Figure 12: peak position (E_0, maximum standard deviation 5%), full width at half maximum (FWHM, maximum standard deviation 15%), integrated area ratio, A(g2)/A(g3). Adapted with permission from [73]. Copyright 2003 American Chemical Society

Sample		g1 (E_0 - FWHM)	g2 (E_0 - FWHM)	g3 (E_0 - FWHM)	A(g2)/A(g3)
S1	RT	2.94 – 0.82	3.67 – 0.58	4.07 – 0.32	12.0
	8 K	2.77 – 0.60	3.73 – 0.74	4.11 – 0.20	38.5
S2	RT	2.81 – 0.63	3.69 – 0.65	4.17 – 0.52	9.7
	8 K	2.75 – 0.59	3.71 – 0.72	4.07 – 0.28	40.7
S3	RT	3.15 – 1.34	3.74 – 0.60	4.02 – 0.41	3.6
	8 K	2.79 – 0.65	3.69 – 0.68	3.99 – 0.44	2.4
S4	RT	3.05 – 0.84	3.69 – 0.60	4.00 – 0.43	2.4
	8 K	2.83 – 0.69	3.66 – 0.67	3.97 – 0.47	1.50

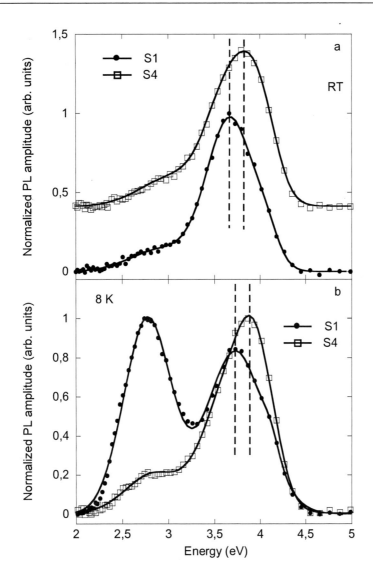

Figure 12. Normalized PL spectra of S1 and S4 samples excited at 5.64 eV at RT (a) and 8 K (b); lines are fitting curves (see text and Table II for details). Adapted with permission from [73]. Copyright 2003 American Chemical Society.

respectively, thus, in the following, only the latter are reported and discussed. The spectra were arbitrarily rescaled to their maximum in order to better compare peak position and shape. As expected, in the 2-5 eV energy range two emission bands are detected: the first one peaked at about 3.7-3.9 eV and the second one centered at about 2.7-2.9 eV. The relative contribution of the blue band increases at low temperature (8 K) and depends on the samples; a detailed analysis of the blue band was reported elsewhere [71, 72]. Concerning the UV band, its peak position at RT is at 3.6 and 3.8 eV in the S1 and S4 samples respectively, while at 8 K is at 3.7 eV in S1 samples and 3.9 eV in S4 samples. One should note, in addition, that a shoulder at higher energy, at about 4.0 eV, is detected in S1 sample at 8 K.

The possible contribution of overlapping emission bands in the 3.2-4.5 eV range was investigated by fitting the spectra with two gaussian bands. A third one was added to heuristically account for the band at about 2.8 eV. Fit results of a least-square fit procedure (square correlation factor $R^2 \geq 0.99$) are reported in Table IV. Two spectrally resolved emissions contribute to the UV band, the first one centered at about 3.7 eV and the second one peaked at about 4.0 eV in all the samples. The relative contribution of the two emissions depends on the sample: the 4.0 eV emission band increases with respect to the weight of the 3.7 eV in samples with larger pore diameter (S3 and S4).

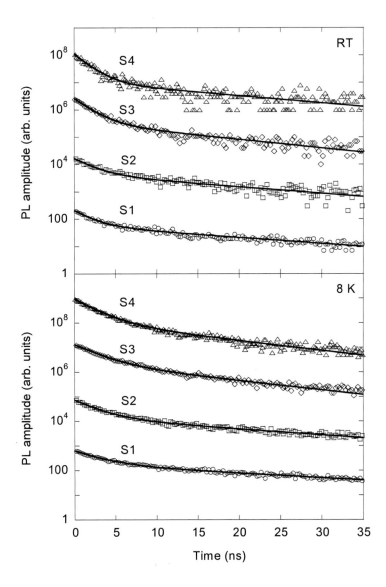

Figure 13. Decay time of 3.7 eV emission excited at 5.64 eV at RT (upper panel) and 8 K (lower panel); lines are fitting curves (see text and Table III for details). Adapted with permission from [73]. Copyright 2003 American Chemical Society.

Table V. Time decays of 3.7 eV PL band excited at 5.64 eV as deduced from best-fit procedure. Adapted with permission from [73]. Copyright 2003 American Chemical Society

sample	RT		8 K	
	τ_1 (ns)	τ_2 (ns)	τ_1 (ns)	τ_2 (ns)
S1	2.1±0.1	20±1	3.0±0.1	24±1
S2	2.2±0.1	18±1	2.7±0.1	18±1
S3	1.9±0.1	14±1	2.5±0.1	12±1
S4	1.7±0.1	16±1	2.3±0.1	11±1

The decay time measurements of the 3.7 eV emission energy are reported in Figure 13. The excitation energy was 5.64 eV at RT (upper panel) and 8 K (lower panel). The semi-log plots indicate that time curves could not be approximated with a single exponential decay. The presence of two temporally resolved decays was hypothesized by fitting the data with two exponential decays. Mean life times of about 2 and 17 ns were estimated by means of a least-square fit procedure (details are reported in Table V, square correlation factor $R^2 \geq 0.97$).

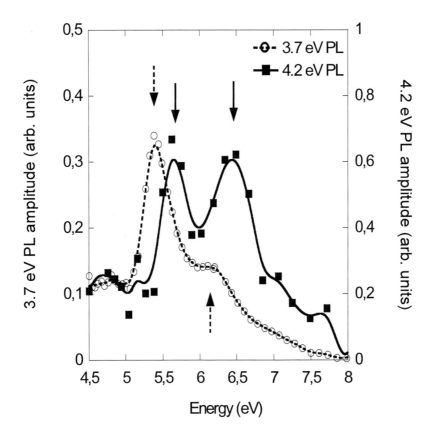

Figure 14. PLE spectra of 3.7 and 4.2 eV emissions in S1 samples (lines are guide for the eyes). Adapted with permission from [73]. Copyright 2003 American Chemical Society.

The PLE spectra for the 3.7 and 4.2 eV emission energies recorded at RT are reported in Fig 14. The two emissions display two main excitation channels but with different relative intensities. The PL at 3.7 eV is excited at 5.4 and 6.2 eV, the first one being the dominant excitation channel. The PL at 4.2 eV is excited with comparable efficiency at 5.7 and 6.5 eV for the 4.2 eV PL. Less efficient excitation channels are detected at 4.8 eV for both the emissions and above 7.0 eV for the 4.2 eV PL.

In order to complete the spectral and temporal resolution of the two emission bands and to correlate all the recorded data, the PLE spectrum of the 3.72 eV band was recorded with different time windows and at different time delays from the excitation channels, as previously explained for the PL emission. The RT PLE spectra are reported in Figure 15. The spectrum recorded 1.5 ns after the SR pulse with a time window of 20 ns displays the excitation channels detected with the integrated time window (Figure 14), that is the main peak at 5.4 eV and two other peaks at 4.8 and 6.2 eV. Once the spectrum is recorded 40 ns after the SR pulse with a time window of 150 ns, the peak at 5.4 eV is the only one detected. The same investigation was carried out upon the 4.2 eV PL emission. A spectrum similar to the one reported in Figure 14 was recorded with the shortest time window while no PL signal was detected with the largest one. The exposed results confirm and complete the picture previously discussed: two different emitting centers participate to the 3.7 eV PL emission. The first one is characterized by an emission centered at 3.7 eV, excitation channel at 5.4 and 6.2 eV and life time of about 17 ns, the second one is characterized by an emission peaked at 4.0 eV, excitation channel at 5.7 and 6.5 eV and life time of about 2.0 ns.

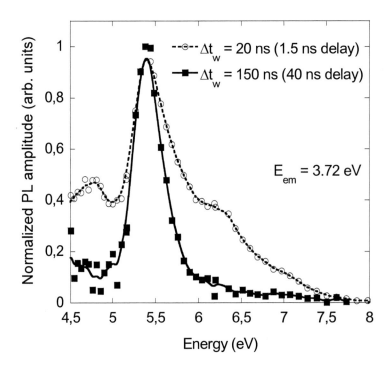

Figure 15. PLE spectra of 3.7 eV PL band at different delays from the excitation pulse and with different time windows (lines are guide for the eyes). Adapted with permission from [73]. Copyright 2003 American Chemical Society.

As previously discussed (2nd section), the 3.7 eV PL band is related to the content of interacting silanols. In the third section we have shown that the UV PL band is a composite band and associated the two spectrally and temporally resolved contributions to different interacting silanols, IS-1 and IS-2 respectively. Since the relative content of IS-1 with respect to that of IS-2 depends on porosity, the relative contributions of the two emitting centers is also expected to depend on it. Indeed, as reported in Table IV, the relative intensity of the PL band at 4.0 eV decreases with respect to the 3.7 eV one in samples with smaller pore diameter. Moreover the reported time decays call for emitting centers with different contributions of non-radiative de-excitation pathways. The detected spectroscopic properties can be summarized with the electronic level schemes depicted in Figure 16 where the emission and excitation channels are reported. The emission at 3.7 eV is ascribed to strongly interacting silanols (IS-1) and the emission at about 4.0 eV is associated to weakly interacting silanols (IS-2) as confirmed by the analysis of the decay time. Indeed, it is worth to note that, according to the following equation: $\tau = (k_r + k_{nr})^{-1}$, k_r and k_{nr} being the radiative and non-radiative decay rate constants, a longer time decay (τ) is expected if the emitting center is characterized by a stronger interaction with the surrounding environments.

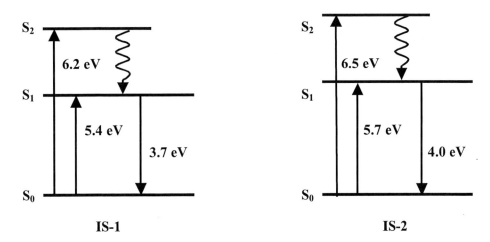

Figure 16. Schematic picture of electronic levels of interacting silanols; vertical arrows represent radiative transitions, curled arrows indicate non-radiative transitions. Adapted with permission from [73]. Copyright 2003 American Chemical Society.

Rhodamine 6G-Porous Silica Hybrids for Solid-State Dye Laser Applications

In the last decades, the effort to produce solid-state tunable lasers by embedding dye molecules into sol-gel matrix has been increasing continuously. It is well known that liquid dye lasers can be tuned over a large emission band (typically about 40 nm) and that by simply changing the dye molecule in solution one can obtain laser emission in the wavelength range from the ultraviolet to the near infrared [18-21]. Since the early 80ies, the research in the field of the dye-laser has been oriented towards the possibility to realize solid state active media by means of the introduction of dye molecules in solid organic, inorganic or hybrid matrices [18-21, 74-77]. Indeed, as compared to liquid solutions, the solid matrix offers a larger mechanical and thermal stability and reduces the risks of environment and operator hazard. In addition, a larger concentration of the dye can be, in principle, achieved, reducing the formation of aggregates which are responsible for the quenching of the luminescence.

Among the investigated possibilities, the embedding of dye molecules into silica glass prepared via sol-gel methods can offer the highest physical and chemical performances [27-38]. In spite of the remarkable progress achieved [18-21, 74-77], the realization of a solid state dye-laser is still limited by the photostability of the confined dye inside of the solid matrix and the consequently fast degradation of the dye molecules and their luminescent properties [75, 76]

Different studies dealt with the challenging task to increase the stability of the solid state dye laser leading to the preparation of samples where a covalent bond between molecules and the structure of the matrix is formed [74, 78, 79]. The presence of the covalent bonds introduces a series of recombination channels for the dye molecule, channels through which it can dissipate the energy excess and therefore the accumulated heat from the excitation wavelength during the pumping process [78, 80].

In this section we present the study of the PL and photostability properties of Rh6G confined in porous silica matrix. PL properties of the samples were investigated via steady-state excitation and compared to the properties of the liquid dye solution. Photostability as a function of the radiation power was investigated by exciting the samples with a pulsed radiation. Two different sol-gel prepared organic/inorganic hybrid categories were studied: type I, where dye molecules is bonded to the structure of the silica matrix by weak bonds (like H-bonding or Van der Walls forces), and type II, where the organic molecules of the dye are tied to the inorganic structure of the silica by means of covalent bonds.

Synthesis Procedures, Samples and Experimental Set-Up

Two different approaches can be followed to prepare dye-doped silica glasses: the impregnation or post-doping method, where the selected dye is incorporated into the sol-gel prepared porous silica by impregnation, [31, 82, 83] and the pre-doping method, where the dye molecule is introduced at the sol stage of the sol-gel procedure synthesis [27, 31, 32, 36, 38]. Because of its processing at low temperatures, the sol-gel method allows for the introduction of organic molecules within the inorganic matrix without thermal degradation. Due to the mixing of the components at a molecular level, pre-doping method results in a more homogeneous distribution of the dye molecules as compared to post-doping. The interaction between dye and silica matrix can be accomplished by electrostatic or covalent bonding: in the first case the samples are classified as class I materials, in the second case as class II. The latter, due to the immobilization of the dye molecules, are expected to resolve or, at least, to reduce the leaching and bleaching effects reported for class I samples [27, 38].

Three groups of samples were synthesized: the first set consisted of type I samples prepared by post-doping method, the other two sets included type I and type II samples prepared by pre-doping method.

Class I post-doped samples (I-post in table VI) were prepared by impregnation of commercial xerogels (Geltech Inc., USA). Mesoporous silica samples (pore diameter 5.4 nm, 5% standard deviation) were loaded with Rh6G ethanol solutions. Sample impregnation was obtained by capillary absorption and drying was performed by leaving hybrids in air at standard pressure and temperature. Ethanol solutions with different dye concentrations in the $5 \times 10^{-5} - 5 \times 10^{-3}$ mol/L ($\pm 2\%$) range were used. Uniformly colored disks were obtained.

Table VI. List of Rh6G-silica hybrid samples. The "pre" or "post" indication in the sample id refers to the doping method (see text for details)

sample id	class	Rh6G concentration (mol/L)
I-post-A	I	5.0×10^{-5}
I-post-B	I	5.0×10^{-4}
I-post-C	I	5×10^{-3}
I-pre-A	I	1.6×10^{-4}
I-pre-B	I	5.0×10^{-4}
I-pre-C	I	1.2×10^{-3}
II-pre-A	II	1.4×10^{-4}
II-pre-B	II	5.0×10^{-4}
II-pre-C	II	1.1×10^{-3}

Pre-doped type I samples (I-pre in table VI) were synthesized by adding dye molecules at the sol stage, according to the following recipe: 0.125 moles of TEOS were added to 50 ml of an aqueous solution of 0.01 N HCl and mixed with 12.5 ml of ethanol as a co-solvent. The mixture was stirred for 15 min and then ethanol was removed by distillation at 35° C by a rotating evaporator to obtain 45 ml of "sol". Then, chloride or perchlorate Rhodamine salts, dissolved in methanol at desired concentration, were added to the sol, together with NH_4OH 0.05N to increase pH from 2 to 4.2.

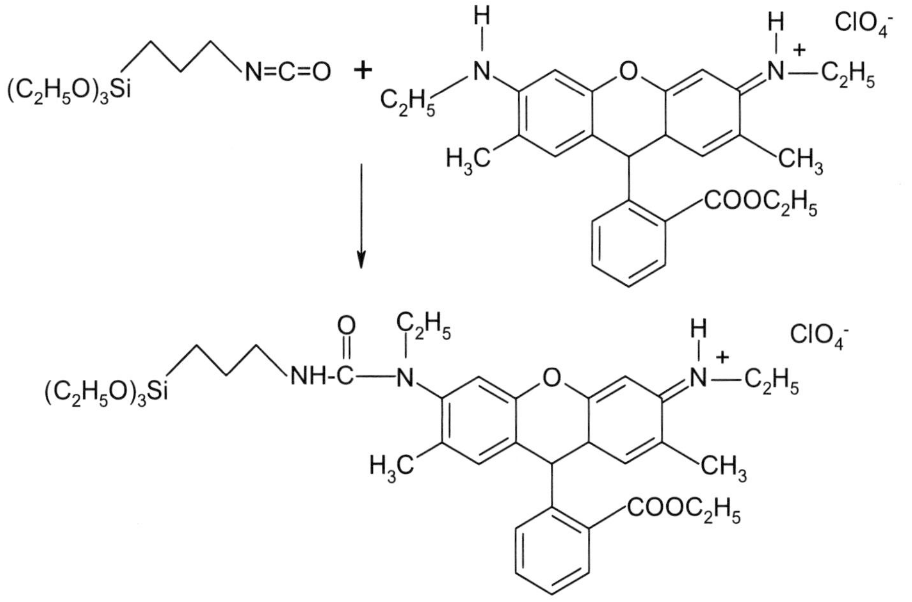

Figure 17. Rh6G precursor reaction (M.W. = 790.39). Adapted from [81], Copyright (2005), with permission from Elsevier.

Pre-doped type II hybrids (II-pre in table VI) were prepared by the synthesis of a tailor-made precursor in a two step process, using Rh6G perchlorate salts. In the first step a grafted Rh6G perchlorate was prepared as follows: 0.3 g of Rhodamine perchlorate were dissolved in 12 ml of acetonitrile then, 0.15 g (with an excess of 5% to take into account the reactivity with moisture in the atmosphere) of 3 isocyanate propyl triethoxysilane were added to the solution. The mixture was stirred for 12 hours and dried at 60 °C. The obtained powder was dried at 60 °C under vacuum for at least 1 hour. The performed reaction is reported in Figure 17. The product was examined by IR spectroscopy at different steps of the reaction. The fraction of the reaction was measured by the decrease of the amplitude of the 2200 cm^{-1} band, the absorption band associated to the NCO group of the isocyanate, and the increase of the amplitude of the 1750 cm^{-1} band, related to the CO ureic group of the grafted Rhodamine.

In the second step, the hybrid synthesis was performed as described for type I hybrids by adding grafted Rh6G molecules dissolved in methanol to the sol. NH4OH 0.05N was then added to the sol to increase the pH from 2 to 4.4. The final molar concentration of the samples ranged between 1.6×10^{-3} and 1.4×10^{-4} mol/L. A 10% error is allowed.

A set of pre-doped type I samples and a set of type II samples were also subjected to different chemical and physical treatment: the first one by heating the samples at 65 °C in order to achieve a fast evaporation of the residual solvent (a T subscript is added to the sample id), the second one by washing the samples with distilled water (three times for at least 1 day each time) in order to eliminate possible residuals of not-grafted Rh6G and then dried at room conditions for about 1 month (a W subscript is added to the sample id).

The absorption spectrum of the Rh6G in solution before and after the grafting reaction was recorded with an UV-VIS spectrophotometer (Kontron Uvikon 941). The Rh6G and the Si-Rh6G were dissolved in methanol and the solutions (Rh6G concentration 0.5×10^{-5} mol/L) were placed in a glass-quartz cuvette and spectra were obtained with a resolution of 1 cm^{-1}.

Low power steady-state PL measurements were carried out by exciting the sample with the 363.8 nm line of an Argon Ion Laser (INNOVA 90C-4). PL signal was collected in a "front face" configuration by an optical fiber probe and recorded with a photonic multi-channel analyzer (PMA11 Hamamatzu). The spectral resolution was 1 nm. All spectra were corrected for the spectral response of the optical systems.

Photostability measurements were performed by means of high power pulsed laser. The excitation was provided by the 3rd harmonic (355 nm) of a Nd:YAG pulsed laser (Spectra Physics Quanta Ray Pro270) with time pulse width at half maximum of 10 ns at 30 Hz repetition rate. The power of the laser was varied over five orders of magnitude, to study the photodegradation features of hybrid samples as a function of the laser excitation power density. PL photostability spectra were recorded in "front face" configuration, focusing the PL emission onto the entrance slit of a triple grating spectrograph (ARC SpectraPro-275), while the light signal detection was achieved through an intensified photodiode array (EG&G mod. 1420). Spectral resolution was 1 nm. All spectra were corrected for the spectral response of the optical systems.

Optical Features of Rh6G-Silica Hybrids

The absorbance spectra of Rh6G molecule and of the precursor are reported in Figure 18. The typical band at about 530 nm and the shoulder at 500 nm due to the absorption of the monomers with its vibronic replica are shown [34, 84].

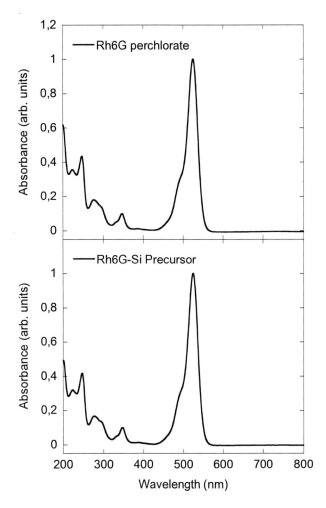

Figure 18. Absorbance spectra of methanolic solution of Rh6G perchlorate and Rh6G-Si precursor (concentration of 0.5×10^{-5} mol/L).

The two spectra do not display differences in wavelength peak position and, concerning the shape, there is only a slight difference due to a larger contribution of the shoulder at 500 nm in the solution containing the precursor molecules. The characteristic UV-VIS absorbance of Rh6G is related to the π-electron delocalized structure in the three conjugated benzene rings. As deduced from the Figure 18, the grafting procedure does not modify the absorbance properties of the dye molecule. On the other hand the larger contribution of the shoulder at 500 nm in the precursor solution could be explained by hypothesizing the presence of dimers. Despite the aggregation of Rh6G molecule in methanol is not expected at the investigated concentration [34, 84], we found that small amounts of grafted Rhodamine tend to form some dimers. In fact, an increase of the contribution at 500 nm of about 2% was estimated by deconvolving the spectra with gaussian bands.

Figure 19 shows the PL spectra of both the Rh6G molecule and the Rh6G-Si precursor in methanolic solution. Both the samples were measured after dissolution in methanol at 0.5×10^{-5} molar concentration. The spectrum of the Rh6G-Si precursor shows the same emission features of the Rh6G commercial salt with an emission peak at 574 nm [85]. Based on absorbance and PL analysis, we can conclude that the grafting procedure does not modify the spectral properties of the dye molecules.

The PL features of the hybrids samples are reported in Figure 20-23. All the investigated samples show an emission spectrum characterized by the PL properties of the dye; no emission signal arising from the silica cage was found, not even when exciting in the excitation channels of porous SiO_2 (see previous sections) [74-77, 86, 87].

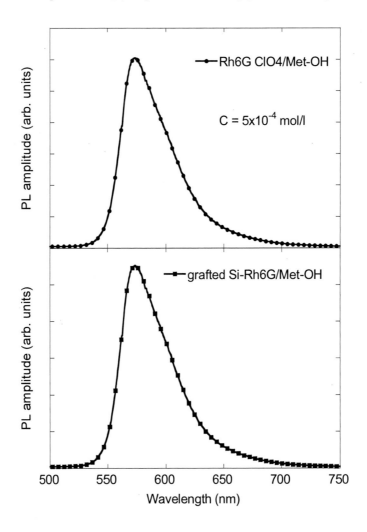

Figure 19. PL spectra of methanolic solution of Rh6G perchlorate and Rh6G-Si precursor (concentration of 0.5×10^{-5} mol/L).

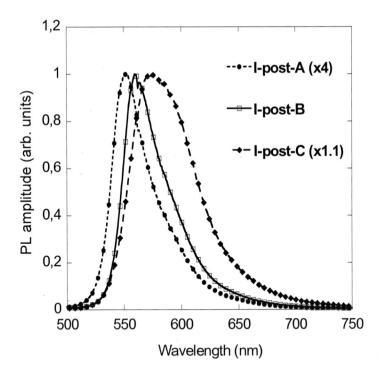

Figure 20. PL emission of impregnated samples as a function of dye concentration. Intensities were arbitrarily normalized to the unity. Adapted from [81], Copyright (2005), with permission from Elsevier.

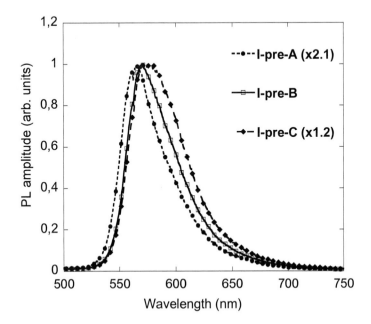

Figure 21. Photoluminescence spectra of sol gel synthesized type I hybrids as a function of dye concentration. Spectra were arbitrarily normalized to the unity. Adapted from [81], Copyright (2005), with permission from Elsevier.

Figure 20 displays the PL spectra of impregnated type I hybrids as a function of dye concentration. For a better comparison, spectra were arbitrarily normalized to one. An increasing red shift of the spectra is observed with increasing Rh6G concentration. Emission peak shifts from 551 nm for I-post-A samples (5×10^{-5} mol/L) to 575 nm for I-post-C samples (5×10^{-3} mol/L). In addition the larger the dye concentration, the greater the FWHM of the detected PL. The most intense emission in the examined range was detected in samples loaded with a solution containing a concentration of dye of 5×10^{-4} mol/L.

PL spectra of pre-doped sol-gel synthesized type I samples are reported in Figure 21; spectra were arbitrarily normalized to one for the sake of comparison. The larger the dye concentration, the smaller the emission energy: PL peak shifts from 564 nm for samples with 1.6×10^{-4} mol/L of Rh6G to 575 nm in hybrids with 1.0×10^{-3} mol/L dye. In I-pre-C samples (1.0×10^{-3} mol/L) a broadening of the emission towards the lower energies is observed.

Figure 22 displays the PL spectra of type II hybrids. Emissions were arbitrarily normalized to one. With increasing Rh6G concentration, the PL maximum shifts from 565 nm for II-pre-A samples (1.4×10^{-4} mol/L) to 570 nm for both II-pre-B (5×10^{-4} mol/L) and II-pre-C samples (1.1×10^{-3} mol/L). No measurable shape variations between the PL spectra of type II samples with different concentration of Rh6G were detected. Both type I and II sol-gel synthesized samples with 5×10^{-4} mol/L Rh6G concentration displayed the most intense photoluminescence in the examined range.

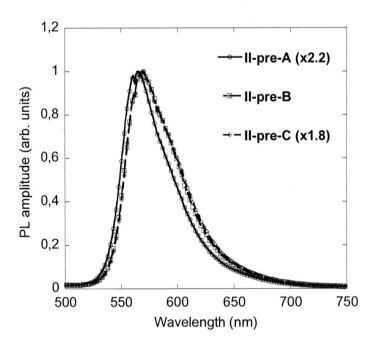

Figure 22. Photoluminescence spectra of sol gel synthesized type II hybrids as a function of dye concentration. Spectra were arbitrarily normalized to the unity. Adapted from [81], Copyright (2005), with permission from Elsevier.

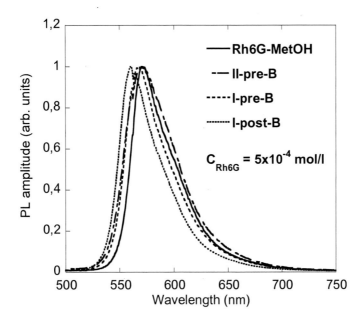

Figure 23. Photoluminescence spectra of samples with 5×10^{-4} M Rh6G concentration. PL of methanolic solution is reported for comparison. Spectra were arbitrarily normalized to the unity. Adapted from [81], Copyright (2005), with permission from Elsevier.

Figure 23 summarizes the PL properties of the three sets of samples with a 5×10^{-4} mol/L concentration of Rh6G. The PL spectrum of a 5×10^{-4} mol/L methanolic solution of Rh6G is reported for comparison. The PL peak position of impregnated xerogels displays a 10 nm blue shift with respect to the peak position of the emission band in the solution. In pre-doped sol-gel synthesized type I hybrids a smaller blue shift in the PL peak position of 3 nm is observed.

These features were already reported for dye molecules hosted in porous glass materials (the so called "cage effect") and are related to surface polarity and to mobility restriction of the molecules [27, 88, 89]. Indeed blue shifts were reported for molecules confined in nanometer pores and related to a reduction of the intermolecular electronic transfer [88]. On the contrary, the PL spectrum of type II hybrids matches the peak position of the solution but with a larger FWHM.

The cage effect can be explained by a solvatochromic mechanism due to the coverage of the SiO_2 walls with hydroxyls (see previous sections) [7]. It is well-known that the peak position of the PL band of Rh6G in solution depends on the polarity of the solvent: the larger the polarity of the solvent, the smaller the PL emission energy for a given concentration [89]-[91]. Since surface hydroxyls at the pore surface act as solvating centers with a larger polarity than alcohol, impregnated samples display a large blue shift of the PL peak position with respect to the reference solution. On the contrary, sol-gel synthesized type I hybrids display minor deviations from the typical PL of Rh6G solutions. The smaller blue shift of the PL peak position can be interpreted in terms of the different entrapment process of the dye molecule within the SiO_2 network. When confining dyes during the sol-gel process, not only a more homogeneous dispersion of organic molecules within the sample is obtained, but a different physical and chemical surrounding of the dye is achieved.

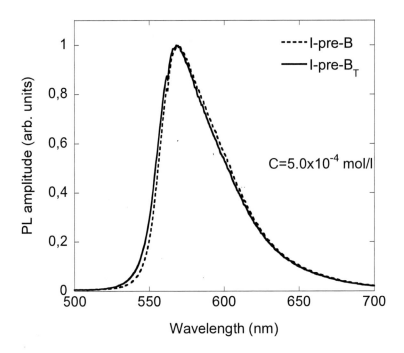

Figure 24. PL spectra of pre-doped class I samples (5.0×10^{-4} mol/L) dried at room temperature or at 65°C.

We can hypothesize that the solvatochromic mechanism is reduced and the blue shift is mainly related to the restriction of the mobility of the molecules. Concerning type II hybrids, the PL properties do not differ from those of solutions of comparable Rh6G concentration with the exception of a larger bandwidth, which calls for a larger distribution of the emitting centers because of the different surroundings. We can conclude that the covalent bonding between dye and silica leads to more stable hybrids by reducing the role of the interaction between the hybrids and the environment.

The effects of thermal treatment are illustrated in Figure 24: the PL properties do not change by drying the samples at 65 °C in order to speed up the evaporation of the residual solvent, at least within the experimental error. Similar results also hold for samples with different concentration (here not shown).

Figure 25 displays the modifications of the PL features of the class II samples caused by the washing procedure. No main differences can be individuated but for a small systematic red-shift of the emission peak (of about 1-2 nm) and a slight enlargement of the band width around 600 nm in the samples with the largest concentration of Rh6G.

The procedure of washing can be regarded as an aging process: the morphology of the sample is modified because of a variation of the porosity of the sample [7, 92]. In addition, the relative content of residual solvent engulfed within the pores changes: while water and Et-OH are entrapped within raw samples, the washed ones contain only water [93]. It is well known that four different processes can be observed during the aging: polycondensation, syneresis, coarsening, and phase transformation.

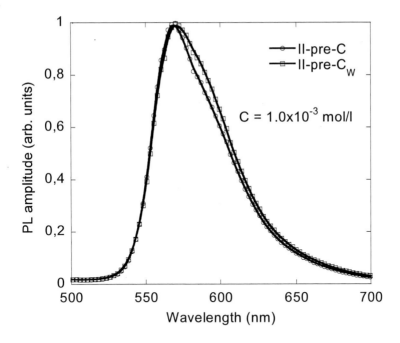

Figure 25. PL spectra of pre-doped class II samples (1.0×10^{-3} mol/L) as prepared and subjected to leaching test.

The effects related to these chemical processes depend on different parameters, such as time, temperature and pH. Typically, due to polycondensation the network connectivity increases while a spontaneous shrinkage of the matrix is caused by the syneresis. The coarsening process modifies the specific surface because of dissolution and reprecipitation phenomena and causes a variation in the mean pore diameter.

Recently, a bimodal distribution of the pore diameter was observed in SiO_2 xerogel samples [93]. In addition, this distribution was modified with the washing procedure: the diameter of the micropores (diameter < 2 nm) increased while the diameter of the mesopores (2<diameter<50 nm) decreased [93]. The observation of two different environments was also reported by studying the photostability of dye-doped sol-gel host, calling for a different mobility of the dye molecules which were divided into non-encapsulated and isolated molecules [78, 79]. In the scenario of a bimodal distribution of the pores, the isolated dye molecules are molecules within the micropores, and the non-encapsulated ones are molecules located within the mesopores. Thus, the observed modification of the photoluminescence features of the class II samples can be interpreted by hypothesizing an increase in the micropore diameter which decreases the isolation of dye molecules: H bonds with –OH groups of SiO_2 or H_2O or dimers can be formed in larger pores causing the emission band to red-shift and widen at larger wavelength [36, 94]. On the other hand, Rh6G molecules in larger pores (mesopores) do not sense a different environment if the pore diameter slightly decreases.

Photostability Measurements

PL photostability measurements were performed at different pulsed laser incident power; it is possible to group the response of the samples of both the types in three different sets: up to 0.1 KW/mm² pulse, from 0.1 to 30 KW/mm² pulse and from 30 to 130 KW/mm² pulse.

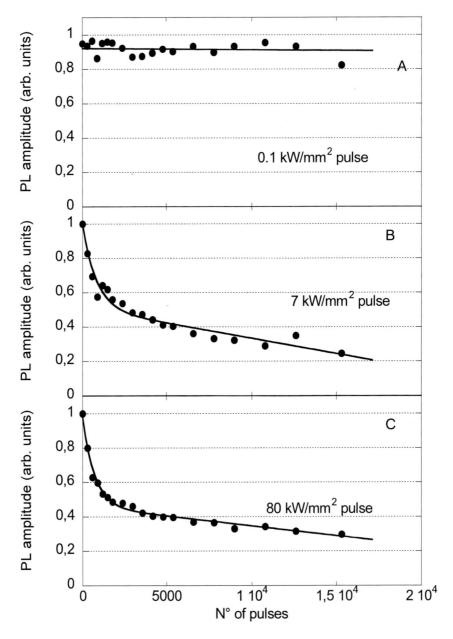

Figure 26. Class I samples (I-pre-C): PL maximum intensity as a function of the number of incident laser pulse at three different powers: (A) 0.1 KW/mm² pulse, (B) 7 KW/mm² pulse, (C) 80 KW/mm² pulse. Dye concentration is 1.0×10^{-3} mol/L (line is fitting curve). Adapted from [95], Copyright (2006), with permission from Elsevier.

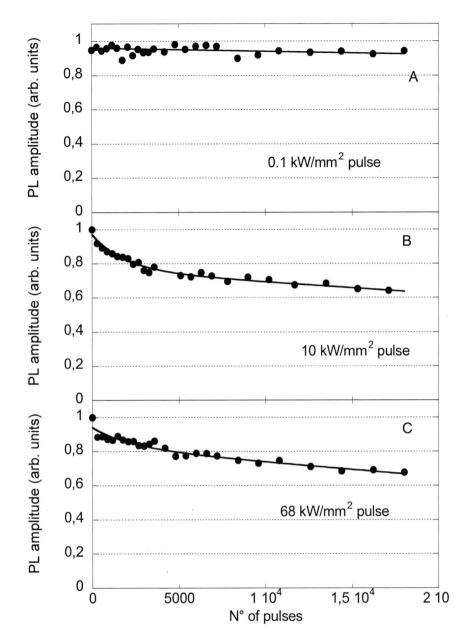

Figure 27. Class II samples (II-pre-C): PL maximum intensity as a function of the number of incident laser pulse at three different powers: (A) 0.1 KW/mm² pulse, (B) 10 KW/mm² pulse, (C) 68 KW/mm² pulse. Dye concentration is 1.0x10-3 mol/L (line is fitting curve). Adapted from [95], Copyright (2006), with permission from Elsevier.

Figures 26 and 27 present the intensity of the PL maximum as a function of the number of laser pulses on the samples in the three ranges of incident laser power, for the two kinds of samples. Data reported refer to dye concentration of 1.0×10^{-3} mol/L where the observed differences in the photostability features are better evidenced. Both the types of hybrid samples did not present PL degradation as the number of pulses increases in the first range of incident laser power (Figure 26A and 27A). The photodegradation starts in the second range of incident laser power (Figure 26B and 27B) with a fast decay of the PL amplitude during the first thousands of laser pulses followed by a slowest decay. The PL spectrum does not change in spectral position and shape for both the sample types. A similar PL bleaching can be observed also in the third range of incident laser power. It is worth to note that the final amplitude decrease is larger in type I samples (Figure 26C and 27C).

As an example of the photobleaching effect, Figure 28 presents a 3D plot of the PL spectra of the 5.0×10^{-4} mol/L type II sample as a function of the number of pulses of incident laser, excitation power being fixed at 125 kW/(mm^2 pulse).

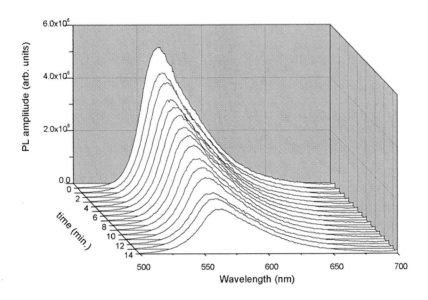

Figure 28. 3D plot of the PL spectra of the 5×10^{-4} mol/L type II samples as a function of the number of pulses of incident laser, excitation power 125 kW/mm^2 pulse. Adapted from [95], Copyright (2006), with permission from Elsevier.

On the contrary, the PL spectrum of a comparable Rh6G methanolic solution (5.0×10^{-4} mol/L) does not present photodegradation effect for all the experimental conditions (here not shown). Even thought the solution was not flown during the measurements, however we can not exclude that spontaneously convective flows in the solution were stimulated by the measurements, avoiding photodegradation effects.

As reported in Figure 26 and 27, photobleaching of the hybrid samples exhibits a composite decrease: PL intensity starts to decay exponentially as the number of pulses increases up to the first thousands of pulses; then the PL intensity decays linearly. Experimental data were successfully fitted with the following theoretical curve:

$$I = a \cdot \exp\left(-\frac{N}{b}\right) + q + m \cdot N$$

where a, b, m and q are fitting parameters and N the number of pulses.

The decrease of the PL amplitude as the number of laser pulses increases can be quantitative analyzed introducing a parameter of photostability. We define a *figure of merit* by the ratio [78]: FM = m/q, where m and q are the angular coefficient (m) and the intercept with the y-axis (q) of the theoretical curve which better interpolate the experimental data in the linear range. FM physically represents the relative variation of intensity with rising number of pulses, normalized to the intensity of the PL at the initial time. Depending on the angular coefficient of the line (m), FM is always negative; FM = 0 corresponds to a perfect photostability of the sample.

In table VII FM values for type I and type II samples with different incident laser power ranges above the photodegradation threshold are collected (dye concentration 1.0×10^{-3} mol/L). Data gathered in the lower power range (up to 0.1 kW/mm^2 pulse) were fitted with a linear curve.

Table VII. Figure of merit of spectra reported in Figure 26 and 27 as deduced from the fitting procedure described in the text. Adapted from [95], Copyright (2006), with permission from Elsevier

Pulse Power (KW/mm^2 pulse)	Figure of Merit (x10^{-6})	
	I-pre-C	II-pre-C
up to 0.1	0.3	0.5
0.1-30	-23	-9
30-130	-30	-11

It is worth to note that the class II samples present a lower FM factor with respect to class I samples: it represents an estimation of their better photostability. We can hypothesize that the photodegradation decays are related to two different sets of bounded dye molecules: the first one has a relative rapid photodegradation (represented by an exponential decay) while the second one is much more stable also under very high energy incident laser pulse. Similar results were reported by Weiss et al. [96] on hybrid silica/Rhodamine sample realized with a different procedure. Moreover, they observed that when the samples are located inside an optical cavity and the laser emission occurs, only the slow photodegradation component is present. The presence of dye molecules surrounded by different environments can explain the reported photobleaching effect. If non-encapsulated molecules are located at the surface of larger pores and isolated molecules are caged by the silica network, different photodegradation patterns can be expected for the two subpopulations of molecules. As reported in comparable hybrid system, the surface molecules may easily react with molecular oxygen or other impurities leading to the photochemical destruction of the molecule: the fast exponential decay is ascribed to this species [78, 79]. On the contrary, isolated molecules are characterized by a longer photostability, represented by the linear decay of the fluorescence signal in the reported spectra. The different photostability of class I and class II samples are expected to be related to the different bonding of the two species to the silica glass: covalent bonding allows the dye molecule to better dissipate the energy excess to the silica host while ionic bonding allows a larger reactivity of the molecules to the environment. The different kind of bonding realized in the two sets of samples, weak-bonding and covalent bonding, could explain the different resistance of both the two species of confined molecules allowing a better photodegradation performance for the class II samples. Indeed covalently bonded samples show a lower contribution of the exponential decay and larger final amplitude of the PL, as confirmed by the FM analysis. We can conclude that type II samples present the same luminescence properties of dye solution commonly employed in tunable laser and that, due to their photostability, are promising candidate to the realization of a solid state dye laser.

Conclusions

The aim of the chapter was to explore the optical properties of porous silica samples for applications as solid-state optical devices. In order to illustrate the enormous potentialities of the material we have highlighted two topics: the optical properties of the porous silica and the photoluminescence and photostability features of the dye doped compounds.

Porous silica exhibits photoluminescence activity in the ultraviolet and visible range. Due to the huge surface to volume ratio, the emissions can be modulated by the physical and chemical conditions of the sample surface. The analysis of the Raman scattering evidenced the presence of different kinds of silanol species whose relative content depends on the porosity of the samples. The emission features are also related to the distribution of silanol species and we demonstrated, in particular, that the photoluminescence band at about 3.7 eV is a composite band related to interacting silanol groups. The reported study indicates porous silica as a photoluminescent material in the technologically relevant ultraviolet-visible window. In addition, from a basic physics point of view, it can shed some light on the topic of porous silicon.

Concerning the dye doped compounds, two different hybrid organic/inorganic materials were studied: type I, where the dye is confined in the structure of the silica by weak bonding, and type II, in which the organic molecules of the dye are tied to the inorganic structure of the silica by covalent bonding. The selected dopant was the standard Rhodamine 6G. The analysis of the photoluminescence properties illustrates that the proposed synthesis routes allow preparing hybrid samples of spectroscopic features comparable to the ones of the dye solutions. Among the different hybrids, the photostability performances of pre-doped class II samples indicate them as the most promising candidate for applications in the solid-state optical filed.

References

[1]   Popov, A.V.; Roizin, Y.O.; Rysiakiewick-Pasek, E.; Marczuk, K.; *Opt. Mater.* 1993, 2, 249.
[2]   Huang, M.H.; Choudrey, M.H.; Yang, P.; *Chem. Commun.* 2000,12, 1063.
[3]   Schlotting, F.; Textor, M.; Georgi, U.; Roewer, G.; *J. Mater. Sci. Lett.* 1999, 18, 599.
[4]   Shafer, M.W.; Awschalom, D.D.; Warnock, J.; Ruben, G.; *J. Appl. Phys.* 1987, 61, 4339.
[5]   Wirnsberg, G.; Stucky, G.D.; *Chem. Phys. Chem.* 2000, 1, 89.
[6]   Casula, M.F.; Corrias, A.; Paschina, G.; J. Non-Cryst. Solids 2001, 293, 25.
[7]   Brinker, J.; Sherer, W. G. Sol-gel science: the physics and chemistry of sol-gel processing; Academic Press: San Diego, CA, 1990.

[8] Klein, L. Sol-gel technology for thin films, fibers, preforms, electronics and specialty shapes; Noyes Publications: Park Ridge, NJ, 1988.

[9] Kresge, C.T.; Leonowicz, M.E.; Roth, W.J.; Vartuli, J.C.; Beck, J.S.; Nature (London) 1992, 359, 10.

[10] Monnier, A.; Schulth, F.; Huo, Q.; Kumar, D.; Margolese, D.; Maxwell, R.S.; Stucky, G.D.; Krishnamurty, M.; Petroff, P.; Firouzi, A.; Janicke, M.; Chmelka, B.F.; *Science* 1993, 261, 1299.

[11] Qin, G. G.; Lin, J.; Duan, J. Q.; Yao, G. Q. *Appl. Phys. Lett.* 1996, 69, 1689.

[12] Glinka, Y. D.; Lin, S.-H.; Hwang, L. P.; Chen, Y.-T. *Appl. Phys. Lett.* 2000, 77, 3968.

[13] Chiodini, N.; Meinardi, F.; Morazzoni, F.; Paleari, A.; Scotti, R.; Di Martino, D. *Appl. Phys. Lett.* 2000, 76, 3209.

[14] Glinka, Y. D.; Naumenko, S. N.; Ogenko, V. M.; Chuiko, A. A. *Opt. Spectrosc. (USSR)* 1992, 71, 250.

[15] Glinka, Y. D.; Lin, S.-H.; Chen, Y.-T. *Phys. ReV. B* 2000, 62, 4733.

[16] Yao, B.; Shi, H.; Zhang, X.; Zhang, L. *Appl. Phys. Lett.* 2001, 78, 174.

[17] Glinka, Y. D.; Lin, S.-H.; Chen, Y.-T. *Appl. Phys. Lett.* 1999, 75, 778.

[18] Sanchez, C. ; Lebeau, B. ; Chaput, F. ; Boilot, J.-P. ; *Adv. Mater.* 2003, 15, 1969.

[19] Proposito, P.; Casalboni, M.; Handbook of Organic–inorganic Hybrid Materials and Nanocomposites, vols. I, XXX, American Scientific Publisher, 2003.

[20] Reisfeld, R.; Yariv, E.; Minti, H.; *Opt. Mater.* 1997, 8, 31.

[21] Reisfeld, R.; *Opt. Mater.* 2001, 16, 1.

[22] Song, H. Z.; Bao, X. M.; Li, N. S.; Wu, X. L. *Appl. Phys. Lett.* 1998, 72, 356.

[23] Cullis, A. G.; Canham, L. T.; Calcott, P. D. *J. Appl. Phys.* 1997, 82, 909.

[24] Koyama, H.; Matsushita, Y.; Koshida, N. *J. Appl. Phys.* 1998, 83, 1776.

[25] Prokes, S. M.; Glembocki, O. J.; Bermudez, V. M.; Kaplan, R.; Friedersorf, L. E.; Searson, P. C. *Phys. ReV. B* 1992, 45, 13788.

[26] Fritsch, E.; Mihut, L.; Baibarac, M.; Baltog, I.; Ostrooumov, M.; Lefrant, S.; Wery, J. J. Appl. Phys. 2001, 90, 4777.

[27] Avnir, D.; Levy, D.; Reisfeld, R. *J. Phys. Chem.* 1984, 88, 5956.

[28] Drexhage, K.H. *Laser Focus*, 1973, 9, 35.

[29] Lo, D. ; Lam, S.K. ; Ye, C. ; Lam, K.S. *Optics Communication* 1998, 156, 316.

[30] Sakka, S. *Sol-Gel Science and Technology, Processing, Characterisation and Applications*, Kluwer Academic Publishers, Boston/Dordrecht/ 2005.

[31] Seddon, A.B.; Illston, T.J.; Cannell, A.C.; Bagnall, C. *Chemtronics* 1991, 5, 117.

[32] Makishima, A.; Tani, T. *J. Am. Ceram. Soc.* 1986, 69, C-72.

[33] Narang, U.; Bright, F.V.; Prasad, P.N. *Applied Spectroscopy* 1993, 47, 229.

[34] Innocenzi, P.; Kozuka, H.; Yoko, T. *J. Non-Cryst. Solids* 1996, 201, 26.

[35] Malashkevich, G.E.; Poddeneznyi, E.N.; Zelnichenko, I.M.; Prokopenko, V.B.; Demyanenko, D.V. *Physics of the Solid State* 1998, 40, 427.

[36] Hungerford, G.; Suhling, K.; Ferriera J.A.; Journal of Photochemistry and Photobiology A: Chemistry 1999, 129, 71.

[37] Zhu, X.-L.; Lam, S.-K.; Lo, D. *Applied Optics* 2000, 39, 3104.

[38] Rao, A.P.; Rao, A.V. *Material Letters* 2003, 57, 3741.

[39] McDonagh, C.; Bowe, P.; Mongey, K.; MacCraith, B. D. *J. Non-Cryst. Solids* 2002, 306, 138.

[40]   Schulz-Ekloff, G.; Wo¨hrle, D.; van Duffel, B.; Schoonheydt, R.A.; *Microporous Mesoporous Mater.* 2002, 51, 91.

[41]   Brinker, C. J.; Kirkpatrick, R. J.; Tallant, D. R.; Bunker, B. C.; Montez, B. *J. Non-Cryst. Solids* 1988, 99, 418.

[42]   Allen, S. G.; Stephenson, P. C. L.; Strange, J. H. *J. Chem. Phys.* 1997, 106, 7802.

[43]   Mel'nichenko, Y. B.;. Schüller, J.; Richert, R.; Ewen, B.; *J. Chem. Phys.* 1995, 103, 2016.

[44]   Benesi, H. A.; Jones, A. C.; *J. Phys. Chem.* 1959, 63, 179.

[45]   Belfort, G.; *Nature* 1974, 249, 593.

[46]   Crupi, V.; Magazù, S.; Majolino, D.; Maisano, G.; Migliardo, P.; *J. Mol. Liquids* 1999, 80, 133.

[47]   Agamalian, M.; Drake, J. M.; Sinha, S. K.; Axe, J. D.; *Phys. Rev.* E 1997, 55, 3021.

[48]   Armistead, G.; Tayler, A. J.; Hambleton, F. H.; Mitchell, S. A.; Hockey, J. A.; *J. Phys. Chemistry* 1969, 73, 3974.

[49]   Burneau, J. Lapage, G. Maurice, *J. Non-Cryst. Solids* 1997, 217, 1.

[50]   Lapage, J. ; Burneau, A.; Guyot, N. ; Maurice, G.; *J. Non-Cryst. Solids* 1997, 217, 11.

[51]   Rovere, M.; Ricci, M. A.; Vellati, D.; Bruni, F.; *J. Chem. Phys.* 1998, 108, 9859.

[52]   Bruni, M. A. Ricci, A. K. Soper, *J. Chem. Phys.* 1998, 109, 1478.

[53]   Soper, K.; Bruni, F.; Ricci, M. A.; *J. Chem. Phys.* 1998, 109, 1486.

[54]   Krol, M.; van Lierop, J. G. J. *Non-Cryst. Solids* 1984, 63, 131.

[55]   Klein, L. C.; Gallo, T. A.; Garvey, G. J. *J. Non-Cryst. Solids* 1984, 63, 23.

[56]   Gottardi, V.; Guglielmi, M.; Bertoluzza, A.; Fagnano, C.; Morelli, M. A. *J. Non-Cryst. Solids* 1984, 63, 71.

[57]   Kinowski, C.; Bouazaoui, M.; Bechara, R.; Henc, L. L.; Nedelec, J. M.; Turrell, S. *J. Non-Cryst. Solids* 2001, 291, 143.

[58]   Geltech Inc. (US), technical report.

[59]   Anedda, A; Carbonaro, C.M.; Clemente, F.; Corpino, R.; Ricci, P.C.; J. *Phys. Chem. B* 2003, 107, 13661.

[60]   Davis, K. M.; Tomozawa, M. J. Non-Cryst. Solids, 1996, 201, 177, and references therein.

[61]   Morrow, B. A.; McFarlan, A. J. *J. Phys. Chem.* 1992, 96, 1395.

[62]   Benesi, H. A.; Jones, A. C. *J. Phys. Chem.* 1959, 63, 179.

[63]   Glinka, Y. D.; Lin, S. H.; Chen, Y. T. *Phys. Rev.* B 2000, 62, 4733.

[64]   Fan, J.C.; Chen, C.H.; Chen, Y.F.; *Appl. Phys. Lett.* 1998, 72, 1605.

[65]   Anedda, A.; Carbonaro, C.M.; Clemente, F.; Corpino, R.; Grandi, S; Mustarelli, P.; Magistris, A; . *Non-Cryst. Solids* 2003, 322, 68.

[66]   Grandi, S.; PhD thesis, Department of Physical-Chemistry, University of Pavia, 2002.

[67]   Anedda, A.; Carbonaro, C. M.; Clemente, F.; Corpino, R.; Raga, F.; Serpi, A.; *J. Non-Cryst. Solids* 2003, 322, 95.

[68]   Radzig, V. A. in "Defects in SiO2 and related Dielectrics: Science and Technology"; Pacchioni, G., Skuja, L., Griscom, D. L., Eds.; Kluwer Academic Publishers: Dordrecht, 2000; p 339.

[69]   Zimmerer, G. *Nucl. Instrum. Methods* A 1991, 308, 178.

[70]   Anedda, A.; Carbonaro, C.M.; Clemente, F.; Corpino, R.; Ricci, P.C.; *J. Phys. Chem.* B 2005, 109, 1239.

[71] Anedda, A.; Carbonaro, C.M.; Clemente, F.; Corpino, R.; Ricci, P.C.; *Mater. Sci. Eng.* C 2005, 25, 631.

[72] Anedda, A.; Carbonaro, C.M.; Clemente, F.; Corpino, R.; Ricci, P.C.; *Optical Mater.* 2005, 27, 958.

[73] Carbonaro, C.M.; Clemente, F.; Corpino, R.; Ricci, P.C.; Anedda, A.; *J. Phys. Chem.* B 2005, 109, 14441.

[74] Dubois, M. Canva, A. Burn, F. Chaput abd J.P. Boilot, *Appl. Opt.* 1996, 35, 3193.

[75] Yariv, E.; Schultheiss, S.; Saraidarov, T.; Reisfeld, R.; *Opt. Mat.* 2001, 16, 29.

[76] Kobbe, E. T.; Dunn, B.; Fuqua, P. D.; Nishida, F.; *App. Opt.* 1990, 29, 2729.

[77] De Matteis, F.; Proposito, P.; Sarcinelli, F.; Casalboni, M.; Pizzoferrato, R.; Russo, M.V.; Vannucci, A.; Varasi, M.; *J. Non-Cryst. Solids* 1999, 245, 15 .

[78] Suratwala, T.; Gardlund, Z.; Davidson, K.; Uhlmann, D. R.; *Chem. Mater.* 1998, 10, 199.

[79] McKiernan, J.M.; Yamanaka, S.A.; Dunn, B.; Zink, J.I.; *J. Phys. Chem.* 1990, 94, 5652.

[80] Deshpande, A.V.; Namdas, E.B.; *Chem. Phys. Lett.* 1996, 263, 449.

[81] Anedda, A.; Carbonaro, C.M.; Clemente, F.; Corpino, R.; Grandi, S; Mustarelli, P.; Magistris, A; . *Non-Cryst. Solids* 2003, 351, 1850.

[82] Harrison, C.C.; McGiveron, J.K.; Li, X. *J. Sol-Gel Sci. and Technol.* 1994, 2, 855.

[83] Gall, G.J.; Li, X. King, T.A. *Sol-Gel Sci. and Technol.* 1994, 2, 775.

[84] Costa, T.M.H.; Stefani, V.; Balzaretti, N.; Francisco, L.T.S.T.; Gallas, M.R.; de Jornada, J.A.H.; *J. Non-Cryst. Solids* 1997, 221, 157.

[85] Oh, H.T.; Kam, H.-S.; Kwon, T.Y.; Moon, B.K.; Yun, S.I.; *Materials Letters* 1992, 13, 139.

[86] Canham, L.T.; *Appl. Phys. Lett.* 1993, 63, 337.

[87] del Monte, F.; Mackenzie, J.D.; Levy, D.; Langmuir 2000, 16, 7377.

[88] Xu, C.; Xue, Q.; Zhong, Y.; Cui, Y.; Ba, L.; Zhao, B.; Gu, N. *Nanotechnology* 2002, 12, 47.

[89] Reisfeld, R.; Zusman, R.; Cohen, Y.; Eyal, M. *Chem. Phys. Lett.* 1988, 147, 142.

[90] Lu, Y. ; Penzkofer, A.; *Chem. Phys.* 1986, 107, 175.

[91] Selwyn, J.E.; Steinfeld, J.I.; *J. Phys. Chem.* 1972, 76, 762.

[92] Hench, L.L.; West, J.K.; *Chem. Rev.* 1990, 90, 33.

[93] Grandi, S.; Mustarelli, P.; Tomasi, .; Sorarù, G.; Spanò, G. *J. Non Cryst. Solids* 2004, 343, 71.

[94] Del Monte, F.; Levy, D. *J. Phys. Chem.* B 1998, 102, 8036.

[95] Carbonaro, C.M.; Marceddu, M.; Ricci, P.C.; Corpino, R.; Anedda. A.; Grandi, S.; Magistris, A. *Mater. Sci. Eng.* C 2006, 26, 1038.

[96] Weiss, A.M.; Yariv, E.; Reisfeld, R.; *Opt. Mat.* 2003, 24, 31.

In: Atomic, Molecular and Optical Physics...   ISBN: 978-1-60456-907-0
Editor: L.T. Chen, pp. 387-426   © 2009 Nova Science Publishers, Inc.

Chapter 10

PUMP TO SIGNAL RELATIVE INTENSITY NOISE TRANSFER IN OPTICAL AMPLIFIERS AND LASERS

Junhe Zhou

The State Key Laboratory on Fiber-Optic Local Area Communication Networks and Advanced Optical Communication Systems, Shanghai Jian Tong Univ., Shanghai, P.R. China

Abstract

Noise figure analysis is one of the key topics in optical amplifiers design and analysis. The pump to signal RIN transfer is one of the major causes for the noise in optical fiber amplifiers. If there is intensity modulation to the pump power, the relative intense noise (RIN) will transfer from the pump to the signal wavelength and degrade the system performance. There have been corresponding papers published on this issue with analytical expressions, The existing analysis gives deep insight into the problem; however, the analysis mentioned above focused on Raman amplifiers or Brillouin fiber lasers with single pump and single signal channel, which are not the most general case. For the case of multiple pumps case, there has been no model for the pump to signal RIN transfer. Moreover, the current analyses are based on the temporal model, which is very time-consuming and does not give a clear picture of the frequency response. In this chapter, we will propose a novel frequency model to evaluate the pump to signal RIN transfer in optical amplifiers and lasers with arbitrary pumps. Analytical expressions could be derived for one pump specific case based on the model.

1. Introduction

Noise figure analysis is one of the key topics in optical amplifier's design and analysis. The pump to signal RIN transfer is one of the major causes for the noise in optical fiber amplifiers. If there is intensity modulation to the pump power, the relative intense noise (RIN) will transfer from the pump to the signal wavelength and degrade the system performance.

There have been corresponding papers published on this issue with analytical expressions, The existing analysis gives deep insight into the problem; however, the analysis mentioned above focused on Raman amplifiers or Brillouin fiber lasers with single pump and single signal channel, which are not the most general case. For the case of multiple pumps case, there has been no model for the pump to signal RIN transfer. Moreover, the current analyses are based on the temporal model, which is very time-consuming and does not give a clear picture of the frequency response.

In this chapter, we will propose a novel frequency model to evaluate the pump to signal RIN transfer in optical amplifiers and lasers with arbitrary pumps. Analytical expressions could be derived for one pump specific case based on the model.

2. Pump to Signal RIN Transfer in Raman Fiber Amplifiers[1]

2.1. Noise in RFAs

There are three major noise sources in RFAs: the RIN transfer from the pump to the signal, Raleigh scattering and amplified spontaneous emission. The three sources are independent of each and can be simply added to compute the noise figure. We will prove it in the following.

Considering the single pump and single signal case, the waves power evolution in the fiber obeys the following:

$$
\frac{\partial P_p}{\partial z} + \frac{1}{v_{g,p}} \frac{\partial P_p}{\partial t} = g(v_p, v_s)\left(P_s^+ + P_s^-\right)P_p - \alpha_p P_p
$$

$$
\frac{\partial P_s^+}{\partial z} + \frac{1}{v_{g,s}} \frac{\partial P_s^+}{\partial t} = g(v_s, v_p)P_p P_s^+ - \alpha_s P_s^+
$$

$$
+ g(v_s, v_p)P_p \frac{h v_s d\nu}{1 - \exp\left(-h\left(v_p - v_s\right)/kT\right)} + \gamma P_s^- \qquad (0.1)
$$

$$
-\frac{\partial P_s^-}{\partial z} + \frac{1}{v_{g,s}} \frac{\partial P_s^-}{\partial t} = g(v_s, v_p)P_p P_s^- - \alpha_s P_s^-
$$

$$
+ g(v_s, v_p)P_p \frac{h v_s d\nu}{1 - \exp\left(-h\left(v_p - v_s\right)/kT\right)} + \gamma P_s^+
$$

If pump depletion is neglected and the noise is a small term as compared with the steady state, we can separate the power as an average power P(whose derivative versus t is zero) and ΔP, neglecting the higher order term, we have:

$$\frac{\partial P_p}{\partial z} = -\alpha_p P_p$$

$$\frac{\partial \Delta P_p}{\partial z} + \frac{1}{v_{g,p}} \frac{\partial \Delta P_p}{\partial t} = -\alpha_p \Delta P_p$$

$$\frac{\partial P_s^+}{\partial z} + \frac{1}{v_{g,s}} \frac{\partial P_s^+}{\partial t} = g(v_s, v_p) P_p P_s^+ - \alpha_s P_s^+$$

$$\frac{\partial \Delta P_s^+}{\partial z} + \frac{1}{v_{g,s}} \frac{\partial \Delta P_s^+}{\partial t} = g(v_s, v_p) P_p \Delta P_s^+ - \alpha_s \Delta P_s^+$$

$$+ g(v_s, v_p) \Delta P_p P_s^+ + g(v_s, v_p) P_p \frac{hv_s dv}{1 - \exp\left(-h\left(v_p - v_s\right)/kT\right)} + \gamma P_s^-$$

$$P_s^- = \Delta P_s^-$$

$$-\frac{\partial P_s^-}{\partial z} + \frac{1}{v_{g,s}} \frac{\partial P_s^-}{\partial t} = g(v_s, v_p) P_p P_s^- - \alpha_s P_s^-$$

$$+ g(v_s, v_p) P_p \frac{hv_s dv}{1 - \exp\left(-h\left(v_p - v_s\right)/kT\right)} + \gamma P_s^+ \tag{0.2}$$

We can see from the right hand side of the equation of ΔP_s^+, the final three terms stand for the RIN transfer, ASE and Raleigh scattering. We can simply prove that the solution of the equation is the addition of formulas derived separately for the three phenomena. So they do not affect each other.

2.2. Introduction to RIN Transfer in RFAs[1]

The Optical Society of America considers the below requested use of its copyrighted material to be allowed under the OSA Author Agreement submitted by the requestor on acceptance for publication of his/her manuscript.

There have been studies on this issue with analytical expressions [2-4], and the results showed that counter-pumped RFAs have lower RIN transfer than co-pumped RFAs at high modulation frequency. This is caused by the different walk-off effect between the counter-propagated pump and the co-propagated signal wave, i.e., the group velocity difference between the pump and signal wave is more significant in counter-pumping scheme. This discovery has brought more attention to the counter-pumped RFAs. It was not until the very low RIN (less than −110dB) laser diodes were available that the co-pumped RFAs received comparable attention. As mentioned before, the existing analysis gives deep insight into the problem; however, it is focused on RFAs with single pump and single signal channel, which are not the most commonly used RFAs. To achieve the broadband amplification, usually WDM pumps are required to balance the gain spectrum and the noise figure. Meanwhile, WDM transmission requires simultaneous transmission of a bunch of signal channels. For the case of multiple pumps and multiple signal channels, there has been no frequency domain model for the pump to signal RIN transfer.

In this section, we proposed a novel frequency model to evaluate the pump to signal RIN transfer in WDM pumped RFAs. The model is derived without undepleted pump assumption. And the complicated interactions between the pump-to-pump, pump-to-signal and signal-to-signal are taken into account. It is capable of evaluating the RIN transfer in RFAs with multiple pumps and multiple signals. The calculation of the RIN transfer can be accomplished by evaluating a matrix, which can be computed via Picard method or Runge-Kutta method. With some simplification, the analytical formulas for RFAs with single pump and single signal channel are derived and show exact agreement with the published results. After the demonstration of the validity of our model in the specific case, the detailed simulation results are presented.

2.3. Frequency Domain Model for RIN Transfer in RFAs with Multiple Pumps

The power evolution in RFAs is governed by the following equations [5]:

$$s(i)\frac{\partial P_i}{\partial z}+\frac{1}{v_{g,i}}\frac{\partial P_i}{\partial t}=\sum_{j=1,j\neq i}^{n+m}g(v_i,v_j)P_iP_j-\alpha_iP_i \tag{0.3}$$

$$i=1......n+m$$

where s(i) is the sign function that indicates the direction of the wave transmission, 1 indicates the forward propagation and -1 indicates the backward propagation, respectively, m is the pump number, n is the signal channel number, $v_{g,i}$ is the group velocity at frequency v_i, α_i is the attenuation coefficient at frequency v_i, $g(v_i,v_j)$ is the Raman gain coefficient between the frequencies v_i and v_j:

$$g(v_i,v_j)=\begin{cases} gr(v_i-v_j)/2A_{eff} & (v_i>v_j) \\ (-v_j/v_i)gr(v_j-v_i)/2A_{eff} & (v_i<v_j) \end{cases} \tag{0.4}$$

To derive the equations governing the RIN transfer in frequency domain in WDM pumped RFAs, we assume that the amplitude of the pump modulation is relatively small compared with the steady state pump power $P_i(z)$, whose derivative of t is zero. The RIN power on each pump as well as the consequent transferred RIN power on each signal channel is denoted as $\Delta P_i(z,t)$, which has the non-zero derivative with respect to t, $\Delta P_i(z,t)+P_i(z)$ satisfies equation(0.3), and if the second order term is omitted, we have:

$$s(i)\frac{\partial \Delta P_i(z,t)}{\partial z}+\frac{1}{v_{g,i}}\frac{\partial \Delta P_i(z,t)}{\partial t}=\sum_{j=1,j\neq i}^{n+m}g(v_i,v_j)P_i(z)\Delta P_j(z,t)$$

$$+\sum_{j=1,j\neq i}^{n+m}g(v_i,v_j)P_j(z)\Delta P_i(z,t)-\alpha_i\Delta P_i(z,t)\tag{0.5}$$

$$i=1......n+m$$

Taking Fourier transform on Eq. (0.5) and rewriting it in matrix form, one has:

$$\frac{\partial \Delta \mathbf{P}(z,\omega)}{\partial z}=\mathbf{A}\Delta \mathbf{P}(z,\omega)\tag{0.6}$$

Where:

$$\Delta \mathbf{P}(z,\omega)=\begin{pmatrix}\Delta P_1(z,\omega)\\ \vdots\\ \Delta P_m(z,\omega)\end{pmatrix}$$

$$\mathbf{A}=\begin{pmatrix}s(1)\left(\displaystyle\sum_{j=1,j\neq 1}^{m+n}g(v_1,v_j)P_j-\alpha_1-\frac{j\omega}{v_{g1}}\right) & \cdots & s(1)g(v_1,v_m)P_1\\ \vdots & & \vdots\\ s(n+m)g(v_{n+m},v_1)P_{n+m} & \cdots & s(n+m)\left(\displaystyle\sum_{j=1,j\neq n+m}^{n+m}g(v_{n+m},v_j)P_j-\alpha_{n+m}-\frac{j\omega}{v_{g,m+n}}\right)\end{pmatrix}$$

The solution of the equation(0.6) has the following form:

$$\Delta \mathbf{P}(L,\omega)=\mathbf{M}_{RIN}\Delta \mathbf{P}(0,\omega)$$

$$\mathbf{M}_{RIN}=\lim_{\Delta z\to 0}\prod_{k=1}^{L/\Delta z}\left(\mathbf{I}+\mathbf{A}\left(k\Delta z\right)\Delta z\right)\tag{0.7}$$

$\mathbf{M}_{RIN}$ can be evaluated numerically via forward Euler method, Runge-Kutta method or Picard method. For the co-pumped RFAs, $\Delta \mathbf{P}(0,\omega)$ is known and the RIN transfer $\Delta \mathbf{P}_s(L,\omega)$ can be simply calculated by multiplying the matrix. For the counter-pumping scheme, the RIN transfer can also be evaluated. We achieve this by separating the vector $\Delta \mathbf{P}(0,\omega)$ into $\begin{bmatrix}\Delta \mathbf{P}_p(0,\omega)\\ \Delta \mathbf{P}_s(0,\omega)\end{bmatrix}$, $\Delta \mathbf{P}(L,\omega)$ into $\begin{bmatrix}\Delta \mathbf{P}_p(L,\omega)\\ \Delta \mathbf{P}_s(L,\omega)\end{bmatrix}$ and the matrix $\mathbf{M}_{RIN}$ into $\begin{pmatrix}\mathbf{M}_{11} & \mathbf{M}_{12}\\ \mathbf{M}_{21} & \mathbf{M}_{22}\end{pmatrix}$ respectively. Since the RIN on the signal at the input end of the fiber is zero, $\Delta \mathbf{P}_s(0,\omega)=0$, therefore the Eq. (0.7) becomes:

$$\begin{pmatrix} \Delta \mathbf{p}_{p\_out} \\ \Delta \mathbf{p}_{s\_out} \end{pmatrix} = \begin{pmatrix} \mathbf{M}_{11} & \mathbf{M}_{12} \\ \mathbf{M}_{21} & \mathbf{M}_{22} \end{pmatrix} \begin{pmatrix} \Delta \mathbf{p}_{p\_in} \\ 0 \end{pmatrix} \tag{0.8}$$

and the signal RIN at the output end is:

$$\Delta \mathbf{p}_{s\_out} = \mathbf{M}_{21}\mathbf{M}_{11}^{-1}\Delta \mathbf{p}_{p\_out} \tag{0.9}$$

2.4. Analytical Formula for RIN Transfer in RFAs with Single Pump and Single Signal Channel

Before investigating the RIN transfer in WDM pumped RFAs, a specific case of RFA, i.e, one pump and one signal wavelength RFA, will be studied using the model proposed above. The case has already been studied by C. Fludger, et. al. with analytical formulas. With the undepleted pump assumption, the same expressions can be derived using Eq. (0.5).

With the undepleted pump assumption, Eq. (0.5) for the co-pumped RFA becomes:

$$\frac{\partial \Delta P_p(z,\omega)}{\partial z} + \frac{j\omega}{v_{g,p}}\Delta P_p(z,\omega) = -\alpha_p \Delta P_p(z,\omega)$$

$$\frac{\partial \Delta P_s(z,\omega)}{\partial z} + \frac{j\omega}{v_{g,s}}\Delta P_s(z,\omega) = gP_p(z)\Delta P_s(z,\omega) + gP_s(z)\Delta P_p(z,\omega) - \alpha_s \Delta P_s(z,\omega)$$

$$P_p(z) = P_{p0}\exp(-\alpha_p z)$$

$$P_s(z) = P_{s0}\exp\left(\int_0^z gP_{p0}\exp(-\alpha_p l) - \alpha_s dl \right) \tag{0.10}$$

the solution of the first equation in (0.10) is:

$$\Delta P_p(z,\omega) = \exp\left(\left(-\frac{j\omega}{v_{g,p}} - \alpha_p \right)z \right)\Delta P_p(0,\omega) \tag{0.11}$$

Substituting (0.11) into the second equation of (0.10), one has:

$$\frac{\partial \Delta P_s(z,\omega)}{\partial z} = \left(-\frac{j\omega}{v_{g,s}} - \alpha_s + gP_{p0}\exp(-\alpha_p z) \right)\Delta P_s(z,\omega)$$

$$+ gP_{s0}\exp\left(\int_0^z gP_{p0}\exp(-\alpha_p l) - \alpha_s dl \right)\exp\left(-\left(\frac{j\omega}{v_{g,p}} + \alpha_p \right)z \right)\Delta P_p(0,\omega) \tag{0.12}$$

The solution of (0.12) is:

$$\Delta P_s(L,\omega) = \exp\left(\int_0^L \left[-\frac{j\omega}{v_{g,s}} - \alpha_s + gP_{p0}\exp(-\alpha_p z)\right]dz\right)$$

$$*gP_{s0}\Delta P_p(0,\omega)\int_0^L \exp\left(\left(-\left(\frac{j\omega}{v_{g,p}}+\alpha_p\right)+\frac{j\omega}{v_{g,s}}z\right)z\right)dz \qquad (0.13)$$

$$= gP_{s0}\Delta P_p(0,\omega)\exp(-\frac{j\omega}{v_{g,s}}L)G\frac{v_{g,s}\left(1-\exp\left((\frac{j2\pi fb}{v_{g,s}}-\alpha_p)L\right)\right)}{-\alpha_p v_{g,s}+j2\pi fb}$$

Where:

$$G = \exp\left(\int_0^L gP_{p0}\exp(-\alpha_p l)-\alpha_s dl\right) = \exp\left(gP_{p0}L_{eff}-\alpha_s L\right)$$

$$L_{eff} = \frac{1-\exp(-\alpha_p L)}{\alpha_p}$$

$$\qquad (0.14)$$

$$b = j2\pi(1-\frac{v_{g,s}}{v_{g,p}})$$

$$j\omega\left(-\frac{1}{v_{g,p}}+\frac{1}{v_{g,s}}\right) = \frac{j2\pi fb}{v_{g,s}}$$

Hence, one has:

$$\frac{\Delta P_s(L,\omega)}{P_s(L)} = \frac{\Delta P_s(L,\omega)}{P_{s0}G}$$

$$\qquad (0.15)$$

$$= \frac{\Delta P_p(0,\omega)}{P_{p0}}\frac{gP_{p0}L_{eff}}{L_{eff}}\exp(-\frac{j\omega}{v_{g,s}}L)\frac{v_{g,s}\left(1-\exp\left((\frac{j2\pi fb}{v_{g,s}}-\alpha_p)L\right)\right)}{-\alpha_p v_{g,s}+j2\pi fb}$$

By defining:

$$r_s = \left|\frac{\Delta P_s(L,\omega)}{P_s(L)}\right|^2$$

$$\qquad (0.16)$$

$$r_p = \left|\frac{\Delta P_p(0,\omega)}{P_p(0)}\right|^2$$

One has:

$$r_s = r_p \left(g P_{p0} L_{eff} \right)^2 \frac{\left(\dfrac{v_{g,s}}{L_{eff}} \right)^2}{\left(\alpha_p v_{g,s} \right)^2 + \left(2\pi fb \right)^2}$$
$$* \left(1 - 2\cos\left(\frac{2\pi fb}{v_{g,s}} L \right) \exp\left(-\alpha_p L\right) + \exp\left(-2\alpha_p L\right) \right) \tag{0.17}$$

Eq. (0.17) is exactly the same as the equation derived in Ref. [2].

Similarly, the equation for counter-pumping scheme can also be derived, which also exactly agrees with the corresponding equation in Ref. [2].

2.5. Numerical Results and Discussions

In our simulation, a standard single mode fiber with the length of 50km is used as the gain media. The second order dispersion coefficient β_2 and the third order dispersion coefficient β_3 are -20.4071711919 ps^2 / km and 0.17348694743 ps^3 / km at the wavelength of 1550nm. 80 channels of signals are launched into the fiber with 100GHz channel spacing. The input signal power for each channel is −10dbm. Different pumping schemes are investigated upon this fiber sharing the same pumping wavelength, i.e. 1425nm, 1440nm, 1450nm, 1465nm and 1490nm. The data of Raman gain spectrum comes from the published literature [13]. The gain profile of the RFA is optimized using the matrix-based algorithms [9], so that the net gains at the different signal channels are almost the same.

Since the RIN transfer in multi-pump RFAs is determined by the matrix, as derived in Eq. (0.7) and (0.9), the total RIN transfer for one signal channel is the linear combination of the RIN transfer induced by different pumps. Hence, we present our simulation results of the RIN transfer assuming that only one pump is regarded as the noise source and other pumps have no RIN. The RIN transfer is defined as $10\log\dfrac{r_s}{r_p}$ as Ref [2]suggested, where r_s and r_p are defined in Eq. (0.16). Since r_s is proportional to r_p, the RIN transfer value $10\log\dfrac{r_s}{r_p}$ will always be the same despite different pump noise level. However, as mentioned in part 2, the pump noise power should be relatively weaker than the pump power. Usually this assumption is valid.

Co-pumping

The Raman gain profile of the co-pumping scheme is illustrated in Fig. 1. The gain profile has been equalized and the corresponding pump powers are 440.8mW, 312.9mW, 116.7mW, 180mW, and 39.1mW. The maximum gain ripple is 0.9dB. For comparison, we also calculated the Raman gain profile in a wider range, i.e. from 1515nm-1605nm with 110 channels. It can be seen that the gain difference between the two is quite slight. In fact, during simulation, we discover that RIN transfer difference between the different signal configurations is also slight. Without loss of generality, 80 channels of signal will be used in the later simulations.

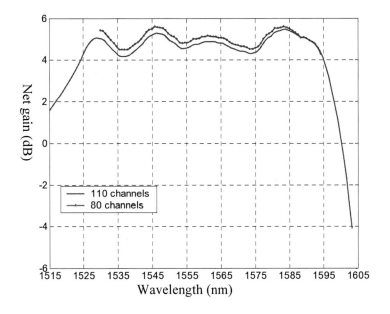

Figure 1. Net gain spectrum of the co-pumped 50km RFA.

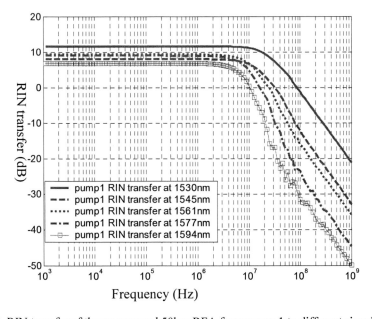

Figure 2. RIN transfer of the co-pumped 50km RFA from pump 1 to different signal channels.

Although the gain spectrum has been flattened, it is not the case for RIN transfer. Different signal wavelength suffers different RIN transfer. As illustrated in Fig. 2, the RIN transfer induced by pump 1 shows that longer wavelength suffers less RIN transfer, this can be explained by the fact that signal channel at longer wavelength has larger gain from this pump, and also by the fact that signal channel at longer wavelength has larger velocity difference to the pump than the shorter wavelength signal channel.

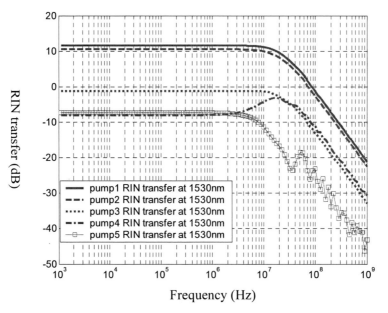

Figure 3. RIN transfer of the co-pumped 50km RFA at 1530nm.

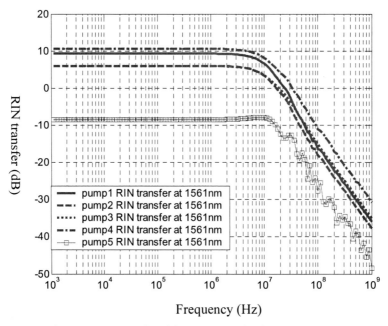

Figure 4. RIN transfer of the co-pumped 50km RFA at 1561nm.

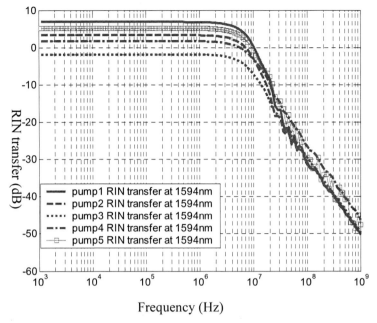

Figure 5. RIN transfer of the co-pumped 50km RFA at 1594nm.

In Fig. 3, Fig. 4 and Fig. 5, the RIN transfer from the pumps to the signal channels at the wavelengths of 1530nm, 1561nm and 1594nm is demonstrated.   Different pump induce different RIN transfer on the signal channels. For signal channel at 1530nm, pump 1 causes the most significant RIN transfer while for signal channel at 1561nm, pump 4 plays the major role. From the figures, it can be inferred that the pump providing more gain on the signal will cause more RIN transfer. One more interesting phenomenon is that the RIN transfer from pump 4 to the signal channel at 1530nm reaches maximum at the frequency of about 200MHz. This has not been observed in single pump case and it might be caused by the complex coupling between the pumps and the signals, which transfer the RIN from pump 4 to pump 3.

Counter-pumping

For the counter-pumping scheme, the Raman gain spectrum is illustrated in Fig. 6.   The gain profile has also been equalized and the corresponding pump powers are 397.1mW, 288.2mW, 111.4mW, 180.1mW, and 47.4mW. The maximum gain ripple is about 1dB.

In Fig. 7, RIN transfer from pump 1 to different signal wavelength is illustrated. Since the difference of the group velocity between the pump and the signal wavelength is larger than that of the co-pumping scheme, the decrease of the RIN transfer versus frequency is much faster. This is consistent with the already published results. Like co-pumping scheme, flat net gain profile does not guarantee flat RIN transfer, i.e. different signal wavelength experiences different RIN transfer.

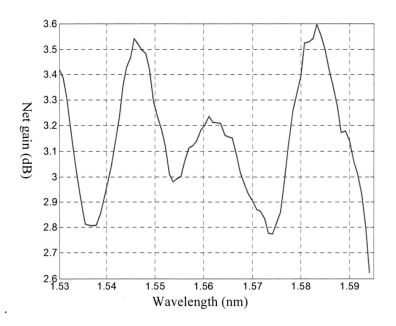

Figure 6. Net gain spectrum of the co-pumped 50km RFA.

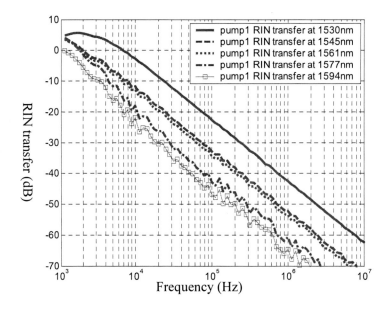

Figure 7. RIN transfer of the counter-pumped 50km RFA from pump 1 to different signal channels.

The RIN transfer from pump 1 to pump 5 to the signal channel at 1530nm, 1561nm and 1594nm is plotted in Fig. 8, Fig. 9 and Fig. 10. Similarly, it can also be inferred that in counter-pumping scheme the pump providing more gain on the signal will cause more RIN transfer.

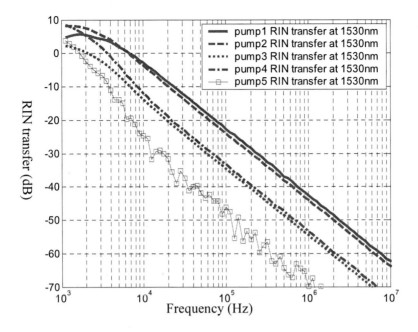

Figure 8. RIN transfer of the counter-pumped 50km RFA at 1530nm.

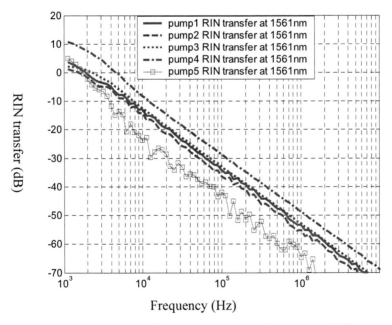

Figure 9. RIN transfer of the counter-pumped 50km RFA at 1561nm.

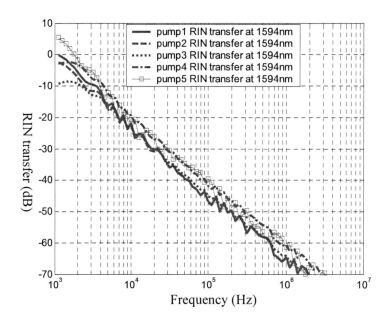

Figure 10. RIN transfer of the counter-pumped 50km RFA at 1594nm.

We have proposed a novel frequency domain model to compute the pump to signal RIN transfer in WDM pumped RFAs without undepleted pump assumption. When the pump and signal numbers are reduced to one and the undepleted pump assumption is used, the analytical expressions derived from the proposed model exactly agree with the published results. The verified model is used to compute the RIN transfer in multi-pump RFAs. The results show that the forward pumping scheme suffers more severe pump to signal RIN transfer than counter-pumping scheme. This is consistent with the published results. The cutoff frequency of RIN transfer is around 10 MHz for the forward pumping scheme and around 1kHz for the backward pumping scheme respectively. RIN on different pump wavelengths result in different RIN transfer on signals. In the results of Fig. 2-5 and Fig. 6-10, it can be roughly seen that the more gain the pump produces, the more severe RIN transfer is. However, as shown in Fig. 3, complex pump interactions may affect the RIN transfer and result in the shift of the peak RIN transfer frequency. Different signal wavelengths also suffers different RIN transfer, it also roughly depends on the gain of the pumps.   By adjusting the pump RIN, one may balance the noise figure in RFAs, and this is a rather interesting topic worth studying.

3. RIN Transfer in Parametric Amplifiers

3.1. Introduction

OPAs with one or two pumps have received considerable attention due to their broad amplification range. There have been great interests in accurately modeling the gain properties, polarization effect and random drift of the zero dispersion wavelength. Meanwhile, noise characterization is also one of the key problems in OPAs' design and analysis, which have been widely addressed. One major noise source is Raman scattering [6], whereas pump to signal RIN transfer is the other. The later is caused by pump power fluctuations, which result from two causes: pump phase modulation and ASE accumulation during the amplification of the pumps. In order to achieve the amplification range over 200 nm, high pump power is required, which will result in high stimulated Brillion scattering (SBS). To reduce SBS, pump phase modulation is required and therefore RIN transfer occurs below the modulation frequency (usually several GHz). Even without pump phase modulation, there is intensity modulation on the pump laser source since the pump is often amplified by an EDFA and the residual ASE noise becomes the RIN on the pump, even at very high frequency. Pump to signal RIN transfer will degrade the system performance, which makes it an important problem to be studied in OPAs, and there have been literatures on this issue.

Marhic et al [7] have proposed a model for low intensity modulation frequency case. The model ignores velocity difference between the pump and the signal. However, in fiber with high third order dispersion coefficient and long distance, for high intensity modulation frequency, the wave walk-off effect between the different wavelengths is not negligible. A much more complete model was proposed by F. Yaman et al [8], considering the wave walk-off effect in the OPAs. The model shows that, at the frequency of 100GHz, the wave walk-off effect will reduce the RIN transfer to 0dB. The complete model brings more insight into the RIN transfer in OPAs than the previous one at high modulation frequency. However, extensive simulations in the time domain are required. Moreover, since the noise is treated as a random process, one has to do the simulations many times in order to compute the RIN transfer. Afterwards, the obtained results must be transformed into frequency domain. Therefore, the computation is rather time-consuming. Meanwhile, the model proposed in Ref. [8] did not consider the pump depletion, which might be significant if the signal power is beyond −10dBm. If a frequency domain model can be used to evaluate the RIN transfer in OPAs, it will greatly reduce the computational time. In fact, it has already been used to calculate the RIN transfer in Raman amplifiers and showed good consistency with the well-known literature.

In this section, based on perturbation theory, we propose a novel frequency domain model for two pump OPAs as well as the one pump OPAs, and use it to compute the RIN transfer in OPAs. With the consideration of the pump power assumption on signals, the model would be able to compute the RIN transfer in the pump depletion region. The section is organized as follows: in part 2, we present the frequency domain model which takes pump depletion into account both for the two pump and the one pump OPAs. Also, for simplification, the model without pump depletion is proposed in order to be compared with the results in Ref. [8]. In part 3, the numerical results by our method are compared with the results obtained by the time domain method, which show good consistency despite some oscillation structures at high frequency. In part 4, the model is used to calculate several examples with the alteration of the OPA parameters and thorough discussions are provided afterwards.

3.2. Mathematical Modeling

For the analysis of RIN transfer in OPAs, we begin with the equations governing the amplitudes evolution in two pump OPAs [5]:

$$\frac{\partial A_p}{\partial z} + \frac{1}{v_{g,p}} \frac{\partial A_p}{\partial t} = j\beta_p A_p$$

$$+ j\gamma \left(\left|A_p^2\right| + 2\left|A_{3-p}^2\right| + 2\left|A_3^2\right| + 2\left|A_4^2\right| \right) A_p + j\gamma 2 A_3 A_4 A_{3-p}^*$$

$$\frac{\partial A_l}{\partial z} + \frac{1}{v_{g,l}} \frac{\partial A_l}{\partial t} = j\beta_l A_l$$

$$+ j\gamma \left(\left|A_l^2\right| + 2\left|A_{7-l}^2\right| + 2\left|A_1^2\right| + 2\left|A_2^2\right| \right) A_l + j\gamma 2 A_1 A_2 A_{7-l}^*$$

(0.18)

Where, A is the wave amplitude and β is the propagation constant, v_g is the group velocity, α is the loss coefficient, p=1,2, stand for the pump waves, and l=3,4, stand for the signal and idler waves. Higher order dispersions are important for OPA gain, and we did include them while calculating the phase mismatch. But higher order dispersion terms with the derivative of t on the left side of nonlinear Schrödinger equation are not important during RIN calculation (we did compare that numerically, and the error is less than 0.1%).

The steady state amplitude is denoted as $A_i(z)$, whose derivative of t is zero. Assuming the pump noise transfer induced amplitude disturbance $\Delta A_i(z,t)$ is rather smaller than the corresponding steady state amplitude $A_i(z)$ and substituting $A_i(z) + \Delta A_i(z,t)$ into Eq. (0.18), we have (0.19) by neglecting the higher order terms:

$$\frac{\partial \Delta A_p}{\partial z} + \frac{1}{v_{g,p}} \frac{\partial \Delta A_p}{\partial t} = j\beta_p \Delta A_p$$

$$+ j\gamma \left(2\left|A_p^2\right| + 2\left|A_{3-p}^2\right| + 2\left|A_3^2\right| + 2\left|A_4^2\right| \right) \Delta A_p$$

$$+ j\gamma \left(A_p \Delta A_p^* + 2 \sum_{i=3-p,3,4} A_i \Delta A_i^* + \Delta A_i A_i^* \right) A_p$$

$$+ j\gamma 2 A_3 A_4 \Delta A_{3-p}^* + j\gamma 2 \Delta A_3 A_4 A_{3-p}^* + j\gamma 2 A_3 \Delta A_4 A_{3-p}^*$$

$$\frac{\partial \Delta A_l}{\partial z} + \frac{1}{v_{g,l}} \frac{\partial \Delta A_l}{\partial t} = j\beta_l \Delta A_l$$

$$+ j\gamma \left(2\left|A_l^2\right| + 2\left|A_{7-l}^2\right| + 2\left|A_1^2\right| + 2\left|A_2^2\right| \right) \Delta A_l$$

$$+ j\gamma \left(A_l \Delta A_l^* + 2 \sum_{i=1,2,7-l} A_i \Delta A_i^* + \Delta A_i A_i^* \right) A_l$$

$$+ j\gamma 2 A_1 A_2 \Delta A_{7-l}^* + j\gamma 2 \Delta A_1 A_2 A_{7-l}^* + j\gamma 2 A_1 \Delta A_2 A_{7-l}^*$$

(0.19)

Taking Fourier transform on (0.19), we have:

$$\frac{\partial \Delta \tilde{A}_p}{\partial z} - \frac{j\omega}{v_{g,p}} \Delta \tilde{A}_p = j\beta_p \Delta \tilde{A}_p + j\gamma(2|A_p^2| + 2|A_{3-p}^2| + 2|A_3^2| + 2|A_4^2|)\Delta \tilde{A}_p$$

$$+ j\gamma(A_p \Delta \tilde{A}_p^* + 2 \sum_{i=3-p,3,4} A_i \Delta \tilde{A}_i^* + \Delta \tilde{A}_i A_i^*)A_p$$

$$+ j\gamma 2A_3 A_4 \Delta \tilde{A}_{3-p}^* + j\gamma 2\Delta \tilde{A}_3 A_4 A_{3-p}^* + j\gamma 2A_3 \Delta \tilde{A}_4 A_{3-p}^*$$

$$\frac{\partial \Delta \tilde{A}_l}{\partial z} - \frac{j\omega}{v_{gl}} \Delta \tilde{A}_l = j\beta_l \Delta \tilde{A}_l + j\gamma(2|A_l^2| + 2|A_{7-l}^2| + 2|A_1^2| + 2|A_2^2|)\Delta \tilde{A}_l$$

$$+ j\gamma(A_l \Delta \tilde{A}_l^* + 2 \sum_{i=1,2,7-l} A_i \Delta \tilde{A}_i^* + \Delta \tilde{A}_i A_i^*)A_l$$

$$+ j\gamma 2A_1 A_2 \Delta \tilde{A}_{7-l}^* + j\gamma 2\Delta \tilde{A}_1 A_2 A_{7-l}^* + j\gamma 2A_1 \Delta \tilde{A}_2 A_{7-l}^*$$

(0.20)

where $\Delta \tilde{A}_i(z,\omega)$ and $\Delta \tilde{A}_i^*(z,-\omega)$ are the Fourier transformations of ΔA_i and ΔA_i^*. (0.20) is a set of ordinary differential equations for RIN transfer computation in the frequency domain. With (0.20), we can directly obtain the RIN transfer without time-consuming computation and discarding the pump depletion term.

For comparison, as well as to simplify the analysis, we assume that the pump is not depleted and it yields the equations in Ref. [8]:

$$\frac{\partial A_p}{\partial z} + \frac{1}{v_{g,p}} \frac{\partial A_p}{\partial t} = j\beta_p A_p + j\gamma(|A_p^2| + 2|A_{3-p}^2|)A_p$$

$$\frac{\partial A_l}{\partial z} + \frac{1}{v_{g,l}} \frac{\partial A_l}{\partial t} = j\beta_l A_l + j\gamma 2(|A_1^2| + |A_2^2|)A_l + j\gamma 2A_1 A_2 A_{7-l}^*$$

(0.21)

Similarly, we may get the model in the frequency domain:

$$\frac{\partial \Delta \tilde{A}_p}{\partial z} - \frac{j\omega}{v_{g,p}} \frac{\partial \Delta \tilde{A}_p}{\partial t} = j\beta_p \Delta \tilde{A}_p + j\gamma(2|A_p^2| + 2|A_{3-p}^2|)\Delta \tilde{A}_p$$

$$+ j\gamma(A_p \Delta \tilde{A}_p^* + 2A_{3-p} \Delta \tilde{A}_{3-p}^* + \Delta \tilde{A}_{3-p} A_{3-p}^*)A_p$$

$$\frac{\partial \Delta \tilde{A}_l}{\partial z} - \frac{j\omega}{v_{g,l}} \frac{\partial \Delta \tilde{A}_l}{\partial t} = j\beta_l \Delta \tilde{A}_l + j\gamma(2|A_1^2| + 2|A_2^2|)\Delta \tilde{A}_l$$

$$+ j\gamma 2A_l \sum_{i=1,2} \left(A_i \Delta \tilde{A}_i^* + \Delta \tilde{A}_i A_i^* \right)$$

$$+ j\gamma 2A_1 A_2 \Delta \tilde{A}_{7-l}^* + j\gamma 2\Delta \tilde{A}_1 A_2 A_{7-l}^* + j\gamma 2A_1 \Delta \tilde{A}_2 A_{7-l}^*$$

(0.22)

3.3. Detailed Numerical Results and Discussion

First of all, we demonstrate the difference between the results obtained by the models with depletion and without depletion. The parameters of the OPAs are exactly the same as the previous example. Since the group velocity difference is different between different pumps and the signal, unlike Ref. [8], we consider the RIN transfer from different pumps separately. Without loss of generality, we consider the RIN transfer from the pump 1, i. e. the pump at 1525.28 nm. The signal input power is –10 dBm.

In Fig. 11, we can see the pump depletion will cause the reduction of the RIN transfer level. The model with pump depletion is more accurate if the input signal power is higher than –10 dBm, and the error is about 2 dB. However, the results for the –20 dBm input signal indicate that the pump power depletion can be neglected in the low signal power case (the error is less than 0.2 dB, and we did not print the curve because the difference between the curves can not be distinguished). Hence, we may draw the conclusion that when the input signal power is lower than –20 dBm, we may neglect the pump depletion, but when the signal input power is higher than –10 dBm, the pump depletion is not negligible. Since we only consider RIN transfer from one pump, the amount of RIN transfer is reduced by 6 dB comparing with the data in Fig. 1.

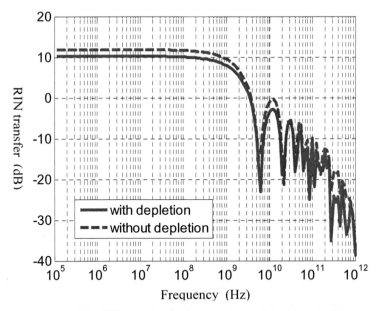

Figure 11. Comparison of the RIN transfer obtained by the model with and without pump depletion.

In Fig. 12 and Fig. 13, we illustrate the RIN transfer from pump 1and pump 2 to the signal wavelength separately. The parameters of the OPAs do not vary. The input signal power is set to be –20 dBm. The signal wavelength varies by 1550 nm and 1536nm. From the figures, it can be seen that signal at different wavelength experiences different RIN transfer, however the difference is quite slight if the gain of the signal remains the same. The cutoff frequency is a little different due to the different walk-off effect on between the pump and

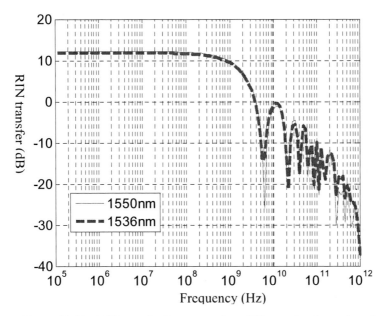

Figure 12. The RIN transfer from pump 1 to different signal wavelength.

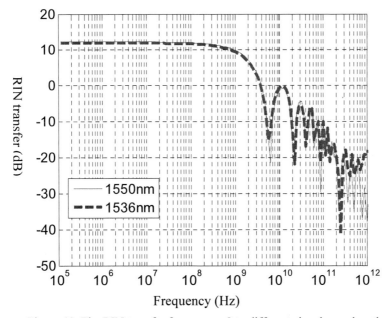

Figure 13. The RIN transfer from pump 2 to different signal wavelength.

different signal channels. The different walk-off effects make the RIN cut-off frequency for 1550 nm higher than 1536 nm both for RIN transfer from pump 1 and pump 2. This can be explained by the fact that the cut-off frequency of RIN transfer is determined by the difference of the group velocity between the signal wavelength and the center wavelength of the pumps. Actually, during computation, we have found that the RIN will transfer from one pump to the other via coupling. The slight difference of RIN transfer between the adjacent signal channels is consistent with the results in Ref. [8]. Moreover, from the figures, we may see that the difference between the RIN transfer from different pumps is also slight. We may study the RIN transfer from one pump during the research.

In the following figures, we are going to illustrate how parameters of OPAs, such as dispersion, nonlinear coefficient, fiber length, and pump power, will affect the RIN transfer function. Other parameters of the OPA remain unchanged. The signal wavelength is 1550 nm and the signal input power is –20 dBm. The RIN transfer results from pump 1 at 1525.28 nm.

First of all, we demonstrate how dispersion affects the RIN transfer in OPAs. The different dispersion coefficient β_3 results in different walk-off effect. The cut-off frequency and the RIN transfer level at low frequency depend on the dispersion coefficient, and the RIN transfer function varies accordingly in the figure. But the RIN transfer level change is slight comparing with the shift of the cut off frequency.

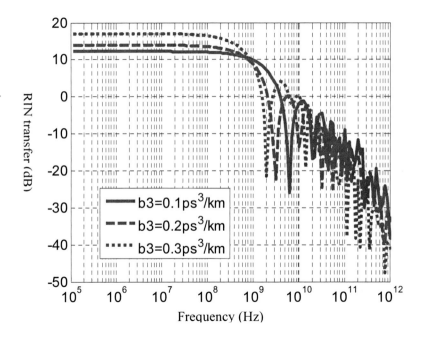

Figure 14. The impact of dispersion coefficient on RIN transfer.

In Fig. 15, we studied the variation of the RIN transfer function with the variation of the nonlinear coefficient. The nonlinear coefficient γ is associated with the gain, so it will affect the RIN transfer level at low frequency as expected.

In Fig. 16, impact of fiber length on RIN transfer is studied. Since the fiber length is associated with the gain and walk-off effect, we expect the change of RIN transfer level and cut-off frequency simultaneously and it appears as expected. Longer fiber introduces more gain and therefore more RIN transfer at low frequency, but it results in more wave walk-off and reduces RIN transfer at high frequency.

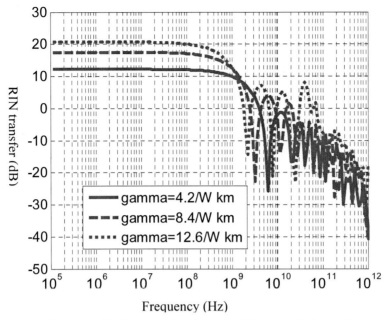

Figure 15. The impact of nonlinear coefficient on RIN transfer.

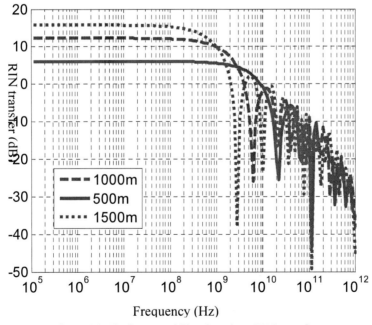

Figure 16. The impact of fiber length on RIN transfer.

Finally, the pump power is varied during the simulation. The results are similar to the impact of nonlinear coefficient on RIN transfer. The level of RIN transfer at low frequency is increased with the increase of pump power. The shift of the cut-off frequency is not significant, which is the same as Fig. 17.

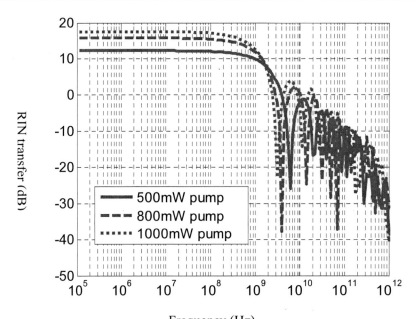

Frequency (Hz)
Figure 17. The impact of pump power on RIN transfer.

We have proposed a novel frequency domain model to compute the RIN transfer in OPAs. The method greatly saves the time for evaluating the RIN transfer in OPAs and the results are in good agreement with the published literature. The computed RIN transfer is constant at low frequency; however, it drops dramatically after the frequency reaches the order of 100 GHz. Meanwhile, at the frequency of 10 GHz order, the RIN transfer also drops comparing with the low frequency case. This means that if high frequency modulation is used to suppress SBS, the noise performance of OPA will be improved. Finally, thorough discussions on impact of OPAs' parameters on RIN transfer are provided. The dispersion will mainly affect the cut-off frequency; the nonlinear coefficient and the pump power will affect the gain and therefore change the RIN transfer level at low frequency; the fiber length will affect the RIN transfer level and the cut-off frequency simultaneously.

4. RIN Transfer in Brillouin Fiber Amplifiers[9, 10]

Portions reprinted, with permission, from (Junhe Zhou, Jianping Chen, Yves Jaouën, Lilin Yi, Xinwan Li, Hervé Petit, and Philippe Gallion, "A New Frequency Model for Pump to Signal RIN Transfer in Brillouin fiber amplifiers," IEEE Photonics Technology Letters VOL. 19, NO. 13, JULY 1, pp, 978-980, 2007). © [2007] IEEE.

4.1. Introduction

Like Raman scattering, Brilllouin scattering is a process that transfers the energy from the shorter wavelength to the longer wavelength. There have been extensive researches on SBS since 1964. The most significant differences between Raman scattering and Brillouin scattering are the following:

> ➢ Brillouin scattering only occurs when the pump and the Stokes waves propagate in the counter direction, while Raman scattering can happen in either co-pumping scheme or counter-pumping scheme.
> ➢ Brillouin gain shift is rather small comparing with Raman scattering. It is only about 10GHz, a lot less than the Raman gain shift 13THz.
> ➢ Brillouin gain spectrum is also narrower, which is about 40MHz.
> ➢ Brillouin gain coefficient is three orders larger than Raman gain coefficient.

Brillouin scattering can be used in optical sensors[11-13], narrow-band filter realizations[14, 15], fiber lasers, and slow light applications [16-18]. The two major sources of noise in SBS process are: the spontaneous amplified emission (ASE) and the pump-to-Stokes relative intensity noise (RIN) transfer.

RIN transfer has been thoroughly studied in Raman amplifiers [2], whereas in Brillouin amplification, most of the research on noise performance is up to now focused on ASE. For RIN transfer in SBS, there have been literatures on RIN transfer in single stage Brillouin lasers[19, 20]. However, there have paid little attention on the impact of RIN transfer on the performance of SBS amplifiers, which are usually used to induce slow light propagation. In this chapter, we will propose a novel frequency domain model to study RIN transfer in Brillouin amplifiers. Moreover, in practical applications, cascaded Brillouin fiber lasers are more of concern. Therefore, in section 4, the RIN transfer in cascaded Brillouin fiber lasers will be discussed.

4.2. The Mathematical Model for SBS

The amplitudes of the pump and the Stokes waves obey the following equations during the propagation in fibers[5]:

$$-\frac{\partial E_p}{\partial z} + \frac{n_{fg}}{c}\frac{\partial E_p}{\partial t} = -\frac{\alpha}{2}E_p - \frac{g_B}{2A}E_s Q$$

$$\frac{\partial E_s}{\partial z} + \frac{n_{fg}}{c}\frac{\partial E_s}{\partial t} = -\frac{\alpha}{2}E_s + \frac{g_B}{2A}E_p Q^* \qquad (0.23)$$

$$\frac{\partial Q}{\partial t} + \left(\frac{\Gamma_B}{2} - i\Delta\omega\right)Q = \frac{\Gamma_B}{2}E_p E_s^*$$

where E_p and E_s are the amplitudes of the pump and the Stokes waves, Q is the amplitude of the acoustic wave, g_B is the Brillouin peak gain coefficient, $\Delta\omega$ is the detuning of the Stokes wave from the Brillouin peak gain wavelength, A is the effective area, Γ_p is the phonon decay rate, α is the attenuation coefficient, and n_{fg} is the group refractive index.

4.3. RIN Transfer in SBS

RIN transfer is weak at high frequency, especially for counter-pumping scheme. In SBS process, the pump wave counter-propagates with the signal wave. However, the Brillouin gain coefficient is three orders larger than the Raman one and requires shorter fiber as the gain medium, which results in the reduction of the walk-off effect. Moreover, since the pump is usually broadened to achieve wider amplification bandwidth, the RIN transfer spectrum will also be broadened. The RIN transfer has been investigated in the case of Brillouin fiber lasers [19]. However, no complete expression of RIN transfer function has been proposed in the case of Brillouin fiber amplifiers.

Theoretical studies on the dynamics and the noise of Brillouin lasers are usually performed in the time domain because the fiber lengths are generally short[19]. However, the method is very computer time-consuming if the fiber length is long (which is usually the case in SBS induced slow light experiments) and the pump modulation frequency is high. In this paper, we propose a frequency model, which is based on perturbation theory, to study the pump to signal RIN transfer in Brillouin amplifiers. Comparison between the results obtained by the classical time domain model and our method illustrates its validity and efficiency. Analytical expressions can be derived from this model neglecting the pump depletion. Detailed numerical simulations are provided along with experimental measurements afterwards.

4.4. RIN Transfer in Brillouin Fiber Amplifiers [9]

We treat the noise term on the pump as a small perturbation and the resulting amplitudes fluctuations are denoted as $\Delta E_p(z,t)$, $\Delta E_s(z,t)$ and $\Delta Q(z,t)$. The fluctuations are small as compared with the corresponding steady state amplitudes $E_p(z)$, $E_s(z)$ and $Q(z)$. Substituting them into (0.23) and taking Fourier transform on the both sides, we have:

$$-\frac{\partial \Delta E_p}{\partial z} + \frac{n_{fg}}{c}\frac{\partial \Delta E_p}{\partial t} = -\frac{\alpha}{2}\Delta E_p + \frac{g_B}{2A}\Delta E_s Q + \frac{g_B}{2A}E_s\Delta Q$$

$$\frac{\partial \Delta E_s}{\partial z} + \frac{n_{fg}}{c}\frac{\partial \Delta E_s}{\partial t} = -\frac{\alpha}{2}\Delta E_s + \frac{g_B}{2A}\Delta E_p Q^* + \frac{g_B}{2A}E_p\Delta Q^* \qquad (0.24)$$

$$\frac{\partial \Delta Q}{\partial t} + \left(\frac{\Gamma_B}{2} - i\Delta\omega\right)\Delta Q = \frac{\Gamma_B}{2}\Delta E_p E_s^* + \frac{\Gamma_B}{2}E_p\Delta E_s^*$$

taking Fourier transform on the both sides of (0.24), we have:

$$-\frac{\partial \Delta \tilde{E}_p}{\partial z} - i\omega\frac{n_{fg}}{c}\Delta \tilde{E}_p = -\frac{\alpha}{2}\Delta \tilde{E}_p - \frac{g_B}{2A}\Delta \tilde{E}_s Q - \frac{g_B}{2A}\tilde{E}_s\Delta \tilde{Q}$$

$$\frac{\partial \Delta \tilde{E}_s}{\partial z} - i\omega\frac{n_{fg}}{c}\Delta \tilde{E}_s = -\frac{\alpha}{2}\Delta \tilde{E}_s + \frac{g_B}{2A}\Delta \tilde{E}_p Q^* + \frac{g_B}{2A}E_p\Delta \tilde{Q}^* \qquad (0.25)$$

$$\Delta \tilde{Q} = \frac{\dfrac{\Gamma_B}{2}\Delta \tilde{E}_p E_s^* + \dfrac{\Gamma_B}{2}E_p\Delta \tilde{E}_s^*}{\dfrac{\Gamma_B}{2} - i(\omega + \Delta\omega)}$$

where $\tilde{E}(z,\omega)$ and $\tilde{E}^*(z,-\omega)$ are the Fourier transformations of $E(z,t)$ and $E^*(z,t)$. Eq. (2) is a set of ordinary differential equations. It can be integrated by the numerical methods such as Runge-Kutta method.

Now we are going to derive the analytical formula for RIN transfer by neglecting the pump depletion. By substituting the following formula into (0.25), and after some simple mathematical calculation, the second equation in (0.25) can be rewritten as:

$$\frac{\partial \Delta \tilde{E}_s}{\partial z} = \left(i\omega \frac{n_{fg}}{c} - \frac{\alpha}{2} + k\left|E_p\right|^2 \right) \Delta \tilde{E}_s + kE_s \left(\frac{k'}{k} \Delta \tilde{E}_p E_P^* + E_p \Delta \tilde{E}_p^* \right)$$

$$k = \frac{g_B}{2A} \frac{\dfrac{\Gamma_B}{2}}{\dfrac{\Gamma_B}{2} + i(\omega + \Delta\omega)} \tag{0.26}$$

$$k' = \frac{g_R}{2A} \frac{\dfrac{\Gamma_B}{2}}{\dfrac{\Gamma_B}{2} + i\Delta\omega}$$

We can get the solution for (0.26):

$$\Delta \tilde{E}_s = \exp\left(\int_0^L \left(i\omega \frac{n_{fg}}{c} - \frac{\alpha}{2} + k\left|E_p\right|^2 \right) dz \right)$$

$$* \int_0^L kE_s \left(\frac{k'}{k} \Delta \tilde{E}_p E_P^* + E_p \Delta \tilde{E}_p^* \right) \exp\left(-\int_0^z \left(i\omega \frac{n_{fg}}{c} - \frac{\alpha}{2} + k\left|E_p\right|^2 \right) dl \right) dz \tag{0.27}$$

If we neglect the pump depletion, (0.27) becomes:

$$\frac{\Delta \tilde{E}_s}{E_s(L)} = kG(L) \int_0^L \left(\frac{k'}{k} \Delta \tilde{E}_p E_P^* + E_p \Delta \tilde{E}_p^* \right) G(z)^{-1} \exp\left(i\omega \frac{n_{fg}}{c}(L-z) \right) dz \tag{0.28}$$

where:

$$G(z) = \exp\left((k-k') P_p Z_{eff} \right)$$

$$Z_{eff} = \frac{1 - \exp(-\alpha z)}{\alpha}$$

$$E_p(z) = E_p(L) \exp\left(\frac{\alpha}{2}(z-L) \right)$$

$$E_s(z) = E_s(0) \exp\left(k' \int_0^z \left|E_p(l)\right|^2 dl - \frac{\alpha z}{2} \right)$$

From(0.28), we can see that the RIN transfer function is a pseudo-periodical function, and the local minimal frequencies for the RIN transfer function are given by:

$$f_N = \frac{N V_g}{2L} \tag{0.29}$$

where N is a integer and V_g is the group velocity. At low frequency, i.e. $\omega = 0$, $G(z) = 1$, (0.28) can be further simplified:

$$\frac{\Delta \tilde{E}_s}{E_s(L)} = \frac{g_B}{2A} \frac{\dfrac{\Gamma_B}{2}}{\dfrac{\Gamma_B}{2} + i\Delta\omega} \Delta \tilde{P}(L,0) L_{eff} \tag{0.30}$$

$$\Delta \tilde{P}(L,0) = \Delta \tilde{E}_p(L,0) E_p^*(L,0) + E_p(L,0) \Delta \tilde{E}_p^*(L,0)$$

$$r_s = \left(\frac{\Delta \tilde{P}_s}{P_s}\right)^2 = r_p \left(\frac{\dfrac{g_B}{A}}{1 + \left(\Delta\omega / \dfrac{\Gamma_B}{2}\right)^2} P_p L_{eff}\right)^2 \tag{0.31}$$

The RIN transfer function is defined as $10\log\dfrac{r_s}{r_p}$, just like Raman and parametric amplifiers

$$r_s = \left|\frac{\Delta P_s(L,\omega)}{P_s(L)}\right|^2$$

$$r_p = \left|\frac{\Delta P_p(0,\omega)}{P_p(0)}\right|^2 \tag{0.32}$$

4.5. Numerical Results of RIN Transfer in Brillouin Amplifiers

In our simulation, the gain media is a dispersion-shifted fiber (DSF) with the effective area of $50\,\mu m^2$, the peak Brillouin gain of $5 \times 10^{-11} m/W$ with the pump wavelength at 1550nm, the Brillouin bandwidth $\Gamma_B / 2\pi = 40 MHz$. For sake of simplicity, the signal detuning from the central gain peak $\Delta\omega$ is 0. Since the disturbance power should be negligible as compared with the steady state power, the pump RIN r_p is 10^{-10}, and the input signal power is –20dBm.

In Fig. 18, the pump power is set to be 100mW. RIN transfer in SBS with the fiber length of 12m, and 500m are calculated based time domain model (0.24) and time domain model (0.25) are compared for 12m and 500m cases, which show good consistency. The discrepancy between the curves for RIN transfer for 500m fiber comes from the fact that the time step is not small enough, which limits the accuracy of the spectrum at high frequency. It takes about 6 hours to calculate the RIN transfer in time domain with the fiber length of 500m, whereas only 2 minutes are required to finish the same job in the frequency domain. The efficiency of the frequency domain model results from the following reasons: the two-dimensional partial differential equations with the derivatives of t and z are transformed into one-dimensional ordinary differential equations; and one is able is adjust the frequency step with the requirement of accuracy at different frequency ranges.

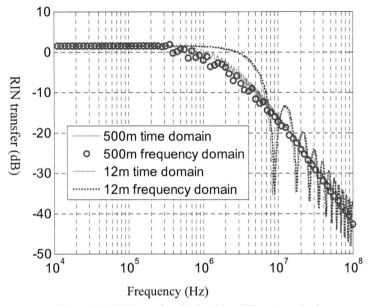

Figure 18. RIN transfer obtained by different method.

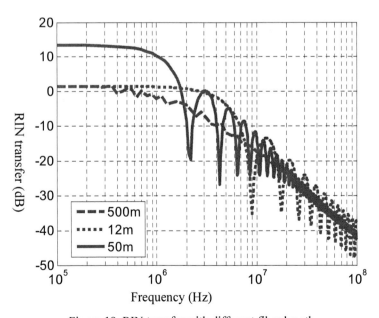

Figure 19. RIN transfer with different fiber length.

In Fig. 19, RIN transfer with different fiber length is illustrated. The pump power is also 100mW. With short fiber, the pump is not depleted; therefore, the fiber length can be viewed as its effective length. Using (0.31), the RIN transfer at low frequency can be directed calculated, which agrees well with the numerical model. With long fiber, i. e., 500m, (0.30) and (0.31) are not valid with the consideration of pump depletion. Gain saturation will cause the reduction of the RIN transfer. From the figure, we may also see the cut-off frequency for RIN transfer is about 5MHz for 50m fiber, which is usually used in Brillion lasers. For 500m fiber, the RIN transfer cut-off frequency is around 1MHz. The walk-off effect is more severe for long fiber and it will reduce RIN transfer at high frequency.

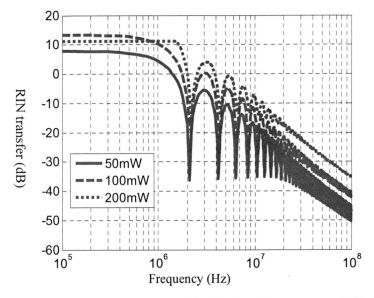

Figure 20. RIN transfer with different fiber length and pump power of 100mW.

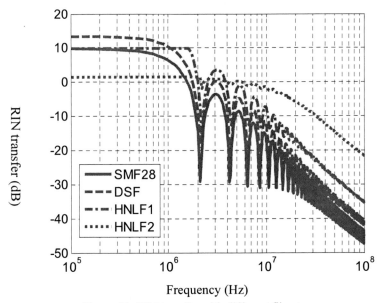

Figure 21. RIN transfer with different fiber type.

In Fig. 20, we demonstrate the results of RIN transfer with different pump power with the fiber length of 50m. The results show that when the gain is not saturated, the RIN transfer rises with the increase of the pump power, which is consistent with (0.31). When the gain is fully saturated, the RIN transfer experiences reduction due to depletion of the pump power. It is comparable with Fig. 2.

In fig. 21, RIN transfer curves for different fiber type is plotted. High nonlinear fiber 1 (HNLF1) has the gain 2 times larger than the DSF, while HNLF2 gain is 10 times larger. The input pump power is 100mW and the fiber length is 50m. It can be seen that high Brillouin gain results in gain saturation which lower the RIN transfer at low frequency while enhancing it at high frequency. The RIN transfer spectrum is broadened under saturation.

In Fig. 22, we compared the numerical results and the analytical results obtained by (0.30). In Fig. 3, the pump power is 100mW and the fiber lengths are 100m and 10m. The analytical formula produced an excellent agreement with the numerical results with the fiber length of 10m. However, when the fiber is longer and the pump depleted, the different curves show that the formula is no more accurate. However, (0.30) and (0.31) are quite useful for evaluating the RIN transfer in SBS process with a broadened pump and broadband signal, whose powers are distributed over the wide spectrum (usually 10G). In this case, the pump operates in the undepleted regime and RIN transfer 3-dB bandwidth is narrow especially for long fibers.

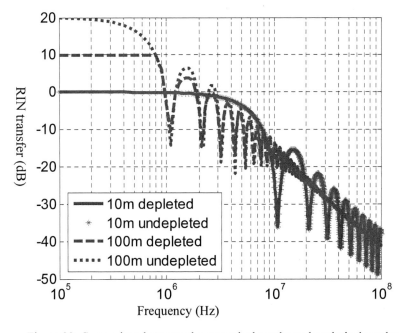

Figure 22. Comparison between the numerical results and analytical results.

4.6. Experimental Measurement of RIN Transfer in Brillouin Amplifiers[21]

The RIN transfer in SBS amplifier pumped by a narrow spectral linewidth diode is measured. The experimental setup is illustrated in Fig. 23. Truewave fibers with the length of 300-m and

20-km are used as the gain media. A monochromatic tunable diode at 1548-nm is used as the light source, whose power is divided by a coupler. Part of the power is used as the pump, which is modulated by using a Mach-Zehnder modulator driven by the electrical sinusoidal generator from a network analyzer (HP 4194A ranging 0.1-Hz~100-MHz). A high power EDFA is used to amplify the pump before its launching into the fiber. The other part of the diode power is modulated by another Mach-Zehnder modulator driven by a signal generator generating a 10.764-GHz sinusoidal wave, corresponding to the Brillouin gain peak in Truewave fiber. The modulated light is then filtered by an optical filter with 0.1nm bandwidth. The filtered side-band is used as the input signal and its power is –20-dBm. An optical circulator is placed at the fiber output to extract the signal counter-propagating with the pump. Two polarization controllers are placed after the pump and signal laser sources to optimize the output power. The output signal is detected by a photo-detector and the RIN transfer function is measured by the network analyzer.

The RIN transfer function of the 300-m Truewave fibre is shown in Fig. 24. The input pump power is 14-dBm. The comparison between the measurements and theoretical results shows excellent agreement. It is also shown that counter propagation averages the noise over the fibre, creating the extinction of 20-dB per decade at high frequency, just as an electrical low pass-filter does. The cut-off frequency results from the Brillouin gain bandwidth and from the walk-off effect. Substituting N=1, L=300-m into the Eq.(0.29), the first dip frequency is about 300-KHz, which exactly agrees with Fig. 5-2. The frequencies of the dips are not dependent on the pump power changes.

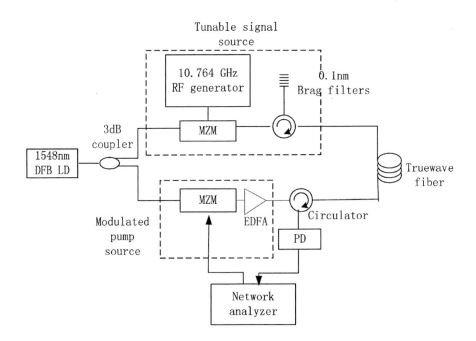

Figure 23. Experimental setup for measurement of the Brillouin amplifier RIN transfer function.

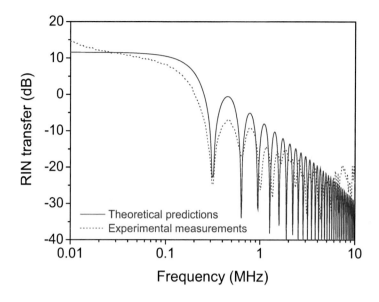

Figure 24. RIN transfer in 300 m Truewave fiber.

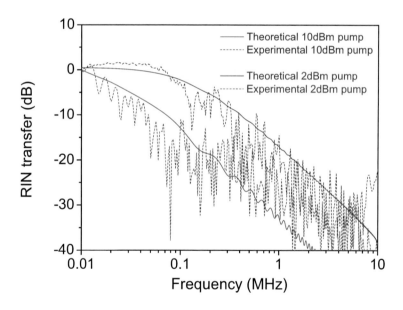

Figure 25. RIN transfer in 20 km Truewave fiber.

In Fig.25, RIN transfer with different pump powers for a fibre length of 20-km is presented. The RIN transfer has been measured for the 2-dBm and 10-dBm pump power respectively. When the pump power is 2-dBm, the RIN transfer decays exponentially when the frequency increases. The overall trends between the theoretical and experimental curves are identical, but the dips in the experimental curves are not resulting of the phase mismatch between the pump and Stokes wave. According to Eq. (0.29), the dips resulting from the walk-off effect are too close to be observed. For longer fiber, Brillouin threshold is rather lower than the short fiber and hence it is easier for the pump to reach the Brillouin threshold and to be saturated. From the figure, we can see that when the input power is 10-dBm, the gain is already saturated and the RIN transfer does increase along with the increase of the pump power, which is consistent with the numerical simulation presented in [21]. When the gain is fully saturated, a RIN transfer reduction due to the depletion of the pump power and the decrease of the effective fiber length can be observed.

5. RIN Transfer in Cascaded Brillouin Fiber Lasers[10]

Portions reprinted, with permission, from (Junhe Zhou, Yves Jaouën, Lilin Yi, and Philippe Gallion, "Pump to Stokes Waves Intensity Noise Transfer in Cascaded Brillouin FiberLasers," IEEE Photonics Technology Letters, 2008 (in press)). © [2008] IEEE.

5.1. Introduction

Brillouin fiber lasers (BFLs) are more of interest than Brillouin amplifiers. Among them, CBFLs are able to provide multiple lasing wavelengths and are very useful in WDM transmission [22, 23]. Generation of over 160 Brillouin lasing wavelengths with the assistance of Erbium doped fiber amplifiers has been reported[24], as well as the researches on the static and dynamic behaviors of CBFLs[25, 26].

It is well known that intensity noise can be a major noise source that degrades the performance of optical communications systems [2], and pump to signal noise transfer in BFLs is quite strong due to high gain coefficient although it can be off-set by the wave walk-off. RIN transfer in BFLs [19, 20] with single lasing wavelength has already drawn attention. However, up to date, no model has been proposed to study the noise transfer in CBFLs, which are more of concern in potential practical applications. Moreover, most of the theoretical works were performed by using a temporal domain model[9]. However, it is less inefficient to solve the two dimensional partial differential equations as compared with the frequency model[21], which transfers the equations into one-dimensional ordinary differential equations. The frequency domain model [21] for RIN transfer in Brillouin amplifiers has been validated both by time domain simulation and experimental demonstration. Hence, we propose to extend the frequency model to study the RIN transfer in Cascaded Brillouin fiber ring lasers (CBFRLs). The Fabry-Perot Brillouin fiber lasers can also be characterized by the similar procedures.

5.2. Theoretical Frame

The typical setup for the CBFRLs is illustrated in Fig. 26. In the fiber, the pump propagates co-directionally with the even order Stokes waves and counter-directionally with odd order Stokes waves. In this following, we will introduce the multiple wave equations and then incorporate the boundary condition into the matrix form solution.

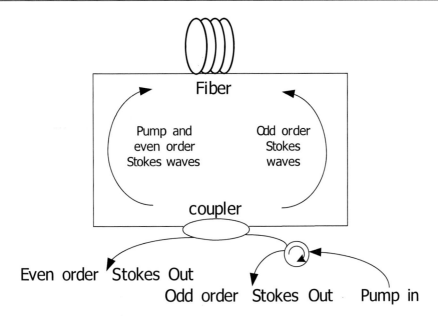

Figure 26. Brillouin fiber ring laser scheme.

A. Multi-wave Equations

The Brillouin scattering occurs with the interactions between the pump wave, Stokes wave and the acoustic wave. For the CBFRLs, the wave equations are [5]:

$$\frac{\partial E_1}{\partial z} + \frac{1}{V_g}\frac{\partial E_1}{\partial t} = -\frac{\alpha}{2}E_1 - \frac{g_B}{2A}E_2 Q_1$$

$$\ldots$$

$$\pm\frac{\partial E_i}{\partial z} + \frac{1}{V_g}\frac{\partial E_i}{\partial t} = -\frac{\alpha}{2}E_i + \frac{g_B}{2A}E_{i-1}Q_{i-1}{}^* - \frac{g_B}{2A}E_{i+1}Q_i$$

$$\ldots \hspace{4cm} (0.33)$$

$$\pm\frac{\partial E_N}{\partial z} + \frac{1}{V_g}\frac{\partial E_N}{\partial t} = -\frac{\alpha}{2}E_N + \frac{g_B}{2A}E_{N-1}Q_{N-1}{}^*$$

$$\frac{\partial Q_i}{\partial t} + \frac{\Gamma_B}{2}Q_i = \frac{\Gamma_B}{2}E_i E_{i+1}^*$$

Where E_i is the amplitude of the wave, $i=1$ stands for the pump wave, $i=2-N$ stands for the Stokes waves, +/- indicates the forward/backward wave propagation, Q is the amplitude of the acoustic wave, g_B is the Brillouin peak gain coefficient, A is the effective area, Γ_B is the phonon decay rate, α is the fiber loss, and v_g is the group velocity.

B. Frequency Domain Model

The amplitude E_i can be separated into its steady state term and its noise fluctuating term. The noise term on the pump can be considered as a small perturbation and the resulting amplitudes fluctuations are respectively denoted as $\Delta E_i(z,t)$ and $\Delta Q_i(z,t)$. The fluctuations are small as compared with the corresponding steady state amplitudes $E_i(z)$ and $Q_i(z)$. Substituting them into (0.33) and taking Fourier transform on the both sides, we have:

$$\frac{\partial \Delta \tilde{E}_1}{\partial z} - \frac{j\omega}{V_g} \frac{\partial \Delta \tilde{E}_1}{\partial t} = -\frac{\alpha}{2} \Delta \tilde{E}_1 - \frac{g_B}{2A} \Delta \tilde{E}_2 Q_1 - \frac{g_B}{2A} E_2 \Delta \tilde{Q}_1$$

$$\cdots$$

$$\pm \frac{\partial \Delta \tilde{E}_i}{\partial z} - \frac{j\omega}{V_g} \frac{\partial \Delta \tilde{E}_i}{\partial t} = -\frac{\alpha}{2} \Delta \tilde{E}_i + \frac{g_B}{2A} \Delta \tilde{E}_{i-1} Q_{i-1}^* + \frac{g_B}{2A} E_{i-1} \Delta \tilde{Q}_{i-1}^*$$

$$-\frac{g_B}{2A} \Delta \tilde{E}_{i+1} Q_i - \frac{g_B}{2A} E_{i+1} \Delta \tilde{Q}_i \qquad (0.34)$$

$$\cdots$$

$$\pm \frac{\partial \tilde{E}_N}{\partial z} - \frac{j\omega}{V_g} \frac{\partial \tilde{E}_N}{\partial t} = -\frac{\alpha}{2} \tilde{E}_N + \frac{g_B}{2A} \Delta \tilde{E}_{N-1} Q_{N-1}^* + \frac{g_B}{2A} E_{N-1} \Delta \tilde{Q}_{N-1}^*$$

$$\left(\frac{\Gamma_B}{2} - j\omega \right) \Delta \tilde{Q}_i = \frac{\Gamma_B}{2} \Delta \tilde{Q}_i E_{i+1}^* + \frac{\Gamma_B}{2} Q_i \Delta \tilde{E}_{i+1}^*$$

where $\Delta \tilde{E}(z, \omega)$ and $\Delta \tilde{E}^*(z, -\omega)$ are the temporal Fourier transformations of $\Delta E(z,t)$ and $\Delta E^*(z,t)$. Eq. (0.34) transforms the partial differential equations (Eq.(0.33)) into the ordinary differential equations. The reduction of the two-dimensional problem into the one-dimensional one greatly saves the computational time because the later can be integrated by the numerical methods such as Runge-Kutta method. The model is valid both in the pump undepleted and depleted regime [9].

C. Solution and Boundary Conditions

The significant difference between the models of BFLs and Brillouin amplifiers is the different boundary conditions. We will propose a semi-analytical solution for Eq. (0.34) and incorporate the boundary condition into the solution thereafter.

Rewriting Eq. (0.34) in the matrix form:

$$\frac{\partial \Delta \tilde{\mathbf{E}}}{\partial z} = A(z) \Delta \tilde{\mathbf{E}} \qquad (0.35)$$

Where $\Delta\tilde{\mathbf{E}} = \left(\Delta\tilde{E}_1, \Delta\tilde{E}_1^*, \cdots, \Delta\tilde{E}_N, \Delta\tilde{E}_N^*\right)^T$, A is the matrix whose components resulting from the corresponding terms in Eq. (0.34). The solution of Eq. (0.35) can be written as:

$$\Delta\tilde{\mathbf{E}}(L, \omega) = \mathbf{M}_{RIN}\Delta\tilde{\mathbf{E}}(0, \omega)$$

$$\mathbf{M}_{RIN} = \lim_{\Delta z \to 0} \prod_{k=1}^{L/\Delta z} \left(\mathbf{I} + \mathbf{A}\left(k\Delta z\right)\Delta z\right) \tag{0.36}$$

Where $\mathbf{I}$ is the identity matrix. For simplicity, we used the forward Euler method to represent $\mathbf{M}_{RIN}$, but the more complicated and more efficient Runge-Kutta method may also be used.

The boundary condition can be written as:

$$\Delta\tilde{E}_1\left(0\right) = \Delta\tilde{E}_1\left(L\right)\sqrt{R_1} + j\Delta\tilde{E}_{p0}\sqrt{1-R_1}$$

$$\Delta\tilde{E}_2\left(0\right) = \Delta\tilde{E}_2\left(L\right)/\sqrt{R_2}$$

$$\Delta\tilde{E}_3\left(0\right) = \Delta\tilde{E}_3\left(L\right)\sqrt{R_3}$$

$$\ldots$$

$$\Delta\tilde{E}_N\left(0\right) = \begin{cases} \Delta\tilde{E}_N\left(L\right)\sqrt{R_N} & \left(\textit{for N odd}\right) \\ \Delta\tilde{E}_N\left(L\right)/\sqrt{R_N} & \left(\textit{for N even}\right) \end{cases} \tag{0.37}$$

Where R_i is the coupling coefficient of the coupler at the ith wavelength. Substituting Eq. (0.37) into (0.36), we have:

$$\Delta\tilde{\mathbf{E}}(L, \omega) = \left(\mathbf{I} - \mathbf{M}_{RIN}\mathbf{R}\right)^{-1}\mathbf{M}_{RIN}\mathbf{b} \tag{0.38}$$

Where $\mathbf{R}$ is a diagonal matrix $dig\left(\sqrt{R_1}, \sqrt{R_1}, 1/\sqrt{R_2}, 1/\sqrt{R_2}, \cdots\right)$, $\mathbf{b}$ is the RIN input vector $\left(j\Delta\tilde{E}_{p0}\sqrt{1-R_1}, -j\Delta\tilde{E}_{p0}^*\sqrt{1-R_1}, 0, 0\cdots\right)^T$.

When $\Delta\tilde{\mathbf{E}}(L, \omega)$ is obtained, we have:

$$\Delta\tilde{\mathbf{E}}(0, \omega) = \mathbf{M}_{RIN}^{-1}\Delta\tilde{\mathbf{E}}(L, \omega) \tag{7}$$

The output RIN on even order Stokes waves can be obtain in the vector $\Delta\tilde{\mathbf{E}}(L,\omega)$, whereas the RIN on odd order Stokes waves can be found in $\Delta\tilde{\mathbf{E}}(0,\omega)$ respectively. The RIN transfer function is defined as $10\ log\ (r_s/r_p)$, where $r = (\Delta P/P)^2$

5.3. Results and Discussions

In our simulations, we consider RIN transfer in a 3 order CBFRLs. The gain medium is a dispersion shifted fiber, with the following parameters: $A = 50\ \mu m^2$, $g_B = 5\times10^{-11}$ m/W and $\Gamma_B/2\pi = 40$ MHz. The coupling coefficients R_i are assumed to be wavelength independent and the pump power coupled into the cavity is $(1-R)P_{p0}$. RIN transfer in 3 order Brillouin scattering can be characterized by setting $N=4$ (including the pump wave) in Eq. (2). The pump RIN is -100 dB. It is worth mentioning that the model shows excellent agreement with the time model and the experimental measurement for the 1 order BFRLs[19, 20].

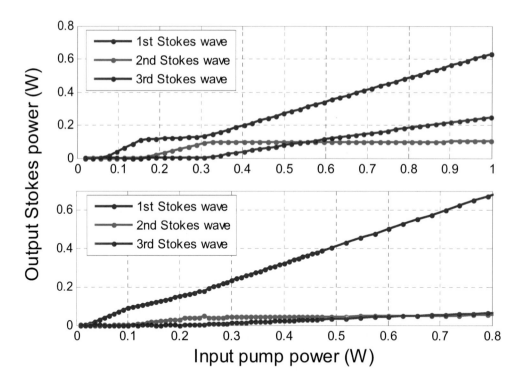

Figure 27. The output power of Stokes waves as a function of input pump power (a) (The fiber length is 10m and the coupling coefficient is 0.36.) (b) (The fiber length is 100m and the coupling coefficient is 0.09.)

Fig. 27 illustrates the output power of Stokes waves as a function of input pump power (the pump power refers to the power before it is coupled into the cavity). In the upper subfigure, the fiber length is 10 m and the coupling coefficient is 0.36. The curves shown in the figure are in agreement with the ones in Ref.[25]. The 1st order Stokes wave first reaches its threshold, and then the 2nd and the 3rd order Stokes waves. When the input pump power is far beyond the threshold, the 1st and the 3rd Stokes waves are much stronger than the 2nd Stokes wave. We can also see that due to the high gain coefficient, the pump power is fully transferred to the Stokes waves meaning that the pump works in the depleted regime. In the second subfigure, the fiber length is 100 m and the coupling coefficient is 0.09. Due to the increase of fiber length and the pump coupling efficiency, the lasing threshold is much lower than the previous case. We can expect the 1st and the 3rd Stokes waves become stronger than the 2nd Stokes wave as the input pump power grows.

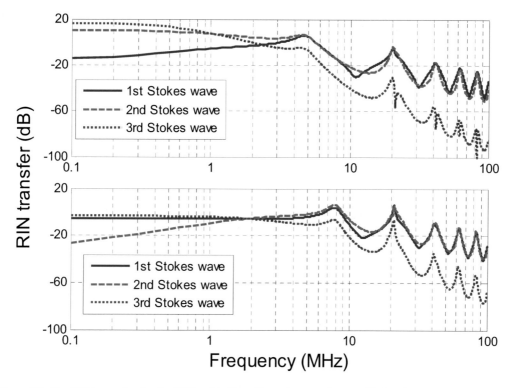

Figure 28. The RIN transfer for the BFL with 10m fiber, the coupling coefficient is 0.36. (a) The input pump power is 400mW. (b) The input power is 1W.

Thereafter, we demonstrated the calculated RIN transfer function for each Stokes wave in Fig. 29 and Fig. 30. In Fig. 29, the fiber length is 10m and the coupling coefficient is 0.36. We choose the pump power to be 0.4W (for the upper subfigure) and 1W (for the second subfigure). At low frequency, the walk-off effect is negligible and therefore, the RIN transfer remains constant. As the frequency increases, the RIN transfer shows oscillations due to the walk-off and the resonant effects. The resonant effect creates very high RIN transfer at resonance frequencies, at which phase of the wave changes multiple π during one round trip of propagation in the cavity. The frequencies can be calculated by:

$$f_N = \frac{N V_g}{L} \tag{0.39}$$

Substituting L=10m, V_g=2*10^8, N=1 into Eq.(0.39), the first peak frequency can be roughly calculated as 2*10^7. From Fig. 29, we can see the frequency well agrees with the calculated one and it is independent of input pump power. The shape of the RIN transfer curves on 1st order Stokes wave shown in Fig. 28 agrees with the experimental results in Ref.[19], which decays 20dB/decade in the high frequency regime. The RIN transfer also decays 20dB/decade on the 2nd Stokes wave whereas it decays 40dB/decade on the 3rd Stokes wave in high frequency regime. This is because the pump, the 1st and 2nd Stokes couples with each other directly in Eq.(0.34), however, the 3rd order Stokes wave couples with 1st and 2nd Stokes wave only. We can also see that when the gain saturation is larger, i.e. when the input pump power is far beyond the threshold (second figure), the RIN transfer is reduced.

In Fig. 29, the fiber length is changed to 100m and the coupling coefficient is 0.09. The input pump power is 0.4W (for the first subfigure) and 0.8W (for the second subfigure). The 2nd Stokes wave has the strongest RIN transfer function; this is because the 1st order Stokes wave has the strongest power. The RIN transfer shows similar resonant effect as Fig. 29 as well. According to Eq.(0.39), since the fiber length is 10 times longer than the previous one, the resonant frequencies are expected to be 1/10 as the ones shown in Fig. 29. It is illustrated as expected. Also, similarly, the walk-off effect strongly reduces the RIN transfer on the 3rd Stokes wave at high frequency.

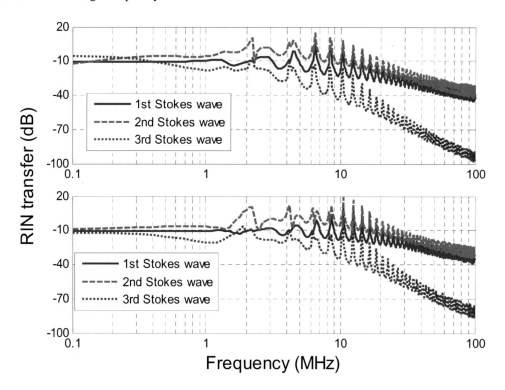

Figure 29. The RIN transfer for the BFL with 100m fiber, the coupling coefficient is 0.09. (a) The input pump power is 400mW. (b) The input power is 800W.

We have proposed a frequency model to address the RIN transfer in CBFLs. The ordinary differential equations are derived in the frequency domain and the boundary conditions are incorporated into the matrix form solution. The method is capable of computing RIN transfer in CBFLs with arbitrary order Sokes waves with higher efficiency than the time domain model. As an example, numerical simulations are performed to study the RIN transfer in a 3-order CBFRL. Simple formula is found to calculate the RIN transfer peak at resonant frequencies.

6. Summary

We have studied the pump-to-signal RIN transfer in Raman amplifiers, parametric amplifiers, Brillouin fiber amplifiers and Brillouin fiber lasers. Based on the frequency domain model, analytical expression is derived in the pump undepleted regime for Raman and Brillouin amplifiers with monochromatic pump. Experimental demonstrations have been performed with a 300-m and a 20-km fibre for Brillouin amplifiers and illustrated very good agreement with the theoretical predictions. The frequency domain model proposed in this chapter can be applied to more general case of Optical amplifiers and lasers which possess arbitrary pumps. Moreover, this model is efficient in computation.

References

[1]   J. Zhou, J. Chen, X. Li, G. Wu, Y. Wang, and W. Jiang, "A novel frequency domain model to calculate the pump to signal RIN transfer in multi-pump Raman fiber amplifiers," *Opt. Express*, vol. 14, pp. 11024-11035, 2006.

[2]   C. R. S. Fludger, V. Handerek, and R. J. Mears, "Pump to signal RIN transfer in Raman fiber amplifiers," *IEEE J. Lightwave Tech.*, vol. 19, pp. 1140-1148, 2001.

[3]   M. D. Mermelstein, C. Headley, and J.-C. Bouteillier, "RIN transfer analysis in pump depletion regime for Raman fiber amplifiers," *IEE Electron. Lett.*, vol. 38, pp. 403-405, 2002.

[4]   B. Bristiel, S. Jiang, P. Gallion, and E. Pincemin, "New Model of Noise Figure and RIN Transfer in Fiber Raman Amplifiers," *IEEE Photon. Technol. Lett.*, vol. 18, pp. 980-982, 2006.

[5]   G. P. Agrawal, *Nonlinear Fiber Optics*, 3rd ed. New York: Academic, 2001.

[6]   P. L. Voss and P. Kumar, "Raman-noise-induced noise-figure limit for parametric amplifiers," *Opt. Lett.*, vol. 29, pp. 445-447, 2004.

[7]   M. E. Marhic, G. Kalogerakis, K. K. Y. Wong, and L. G. Kazovsky, "Pump-to-Signal Transfer of Low-Frequency Intensity Modulation in Fiber Optical Parametric Amplifiers," *IEEE J. Lightwave Technol.*, vol. 23, pp. 1049-1055, 2005.

[8]   F. Yaman, Q. Lin, and G. P. Agrawal, "Pump-noise transfer in dual-pump fiber-optic parametric amplifiers: walk-off effects," *Opt. Lett.*, vol. 30, pp. 1048-1050, 2005.

[9]   J. Zhou, J. Chen, Y. Jaouën, L. Yi, X. Li, H. Petit, and P. Gallion, "A New Frequency Model for Pump to Signal RIN Transfer in Brillouin fiber amplifiers," *IEEE Photon. Tech. Lett.*, vol. 19, pp. 978-980, 2007.

[10] J. Zhou, Y. Jaouën, L. Yi, and P. Gallion, "Pump to Stokes Waves Intensity Noise Transfer in Cascaded Brillouin Fiber Lasers," *IEEE Photon. Tech. Lett.*, vol. (in press), 2008.

[11] T. Horiguchi, K. Shimizu, T. Kurashima, M. Tateda, and Y. Koyamada, "Development of a distributed sensing technique using brillouin scattering," *IEEE J. Lightwave Technol.*, vol. 13, pp. 1296-1302, 1995.

[12] T. R. Parker, M. Farhadiroushan, V. A. Handerek, and A. J. Rogers, "A fully distributed simultaneous strain and temperature sensor using spontaneous Brillouin backscatter," *IEEE Photon. Technol. Lett.*, vol. 9, pp. 979-981, 1997.

[13] X. Zeng, X. Bao, C. Y. Chhoa, T. W. Bremner, A. W. Brown, M. D. DeMerchant, G. Ferrier, A. L. Kalamkarov, and A. V. Georgiades, "Strain measurement in a concrete beam by use of the Brillouin-scattering-based distributed fiber sensor with single-mode fibers embedded in glass fiber reinforced polymer rods and bonded to steel reinforcing bars," *Appl. Opt.*, vol. 41, pp. 5105-5114, 2002.

[14] C. G. Atkins, D. Cotter, D. W. Smith, and R. Wyatt, "Application of Brillouin amplification in coherent optical transmission," *IEE Electron. Lett.*, vol. 22, pp. 556, 1986.

[15] T. Tanemura, Y. Takushima, and K. Kikuchi, "Narrowband optical filter, with a variable transmission spectrum, using stimulated Brillouin scattering in optical fiber," *Opt. Lett.*, vol. 27, pp. 1552-1554, 2002.

[16] Z. Zhu, D. J. Gauthier, Y. Okawachi, J. E. Sharping, A. L. Gaeta, R. W. Boyd, and A. E. Willner, "Numerical study of all-optical slow-light delays via stimulated Brillouin scattering in an optical fiber," *J. Opt. Soc. Am. B*, vol. 22, pp. 2378-2384, 2005.

[17] S. Cui, J. Liu, and X. Ma, "A novel efficient optimal design method for gain-flattened multiwavelength pumped fiber Raman amplifier," *IEEE Photon. Tech. Lett.*, vol. 16, pp. 2451 - 2453, 2004.

[18] Y. Okawachi, M. S. Bigelow, J. E. Sharping, Z. Zhu, A. Schweinsberg, D. J. Gauthier, R. W. Boyd, and A. L. Gaeta, "Tunable All-Optical Delays via Brillouin Slow Light in an Optical Fiber," *Phys. Rev. Lett.*, vol. 94, pp. 153902, 2005.

[19] L. Stepien, S. Randoux, and J. Zemmouri, "Intensity noise in Brillouin fiber ring laser," *J. Opt. Soc. Amer. B*, vol. 19, pp. 1055-1066, 2002.

[20] J. Geng and S. Jiang, "Pump-to-Stokes transfer of relative intensity noise in Brillouin fiber ring lasers," *Opt. Lett.*, vol. 32, pp. 11-13, 2007.

[21] J. Zhou, L. Yi, Y. Jaouën, J. Chen, and P. Gallion, "Pump-to-Stokes relative intensity noise transfer in Brillouin amplifiers," in *ECOC 2007*. Berlin, Germany, 2007.

[22] K. O. Hill, B. S. Kawasaki, and D. C. Johnson, "cw Brillouin laser," *Appl. Phys. Lett.*, vol. 28, pp. 608-609, 1976.

[23] L. F. Stokes, M. Chodorow, and H. J. Show, "All-fiber stimulated Brillouin ring laser with submilliwatt pump threshold," *Opt. Lett.*, vol. 7, pp. 509-511, 1982.

[24] W. Zhang, X. Feng, J. Peng, and X. Liu, "A Simple Algorithm for Gain Spectrum Adjustment of Backward-Pumped Distributed Fiber Raman Amplifiers," *IEEE Photon. Tech. Lett.*, vol. 16, pp. 69-71, 2004.

[25] K. Ogusu, "Analysis of steady-state cascaded stimulated Brillouin scattering in a fiber Fabry-Pérot resonator," *IEEE Photon. Tech. Lett.*, vol. 14, pp. 947-949, 2002.

[26] S. Randoux, V. Lecoeuche, B. Ségard, and J. Zemmouri, "Dynamical behavior of a Brillouin fiber ring laser emitting two Stokes components," *Phys. Rev. A*, vol. 52, pp. 2327-2334, 1995.

In: Atomic, Molecular and Optical Physics... ISBN 978-1-60456-907-0
Editor: L.T. Chen, pp. 427-461 © 2009 Nova Science Publishers, Inc.

Chapter 11

QUANTUM INFORMATION PROCESSING BY CAVITY QED UNDER DISSIPATION

M. Feng, Z.J. Deng,† W.L. Yang‡ and Hua Wei§*
State Key Laboratory of Magnetic Resonance and Atomic and Molecular Physics,
Wuhan Institute of Physics and Mathematics, Chinese Academy of Sciences,
Wuhan 430071, China

Abstract

Among various candidates considered to implement quantum information processing (QIP), cavity quantum electrodynamics (QED) has attracted much attention over past years due to the availability to demonstrate few-qubit quantum gates experimentally and the possibility to construct future quantum network. We review recent work of QIP using cavity QED in weak dissipation. The concrete work we review includes W-state preparation, Toffoli gating, Grover search implementation, and QIP by geometric phase. Under the idea of quantum trajectory, we could present analytical expressions to show the detrimental influence of cavity decay in QIP.

PACS: 03.67.Lx, 42.50.Dv

1. Introduction

Since the publication of Shor's astonishing algorithm for quantum factoring within the polynomial time [1], quantum information processing (QIP) has drawn more and more attention. It has been proven [2] that a universal QIP task could be carried out by a series of single-qubit rotations and two-qubit conditional operations. A candidate system qualified for accomplishing QIP tasks should satisfy DiVincenzo's criteria [3]: Well characterized qubits with easy initialization, availability of universal quantum gates, long decoherence time, and high efficiency of readout. The pedagogical introduction of QIP could be found in [4].

---

*E-mail address: mangfeng@wipm.ac.cn
†E-mail address: dengzhijiao926@163.com
‡E-mail address: yanglingfeng1980@yahoo.com.cn
§E-mail address: huawei.hw@gmail.com

The chapter is focused on QIP based on atom-cavity system, where in most places we mean Rydberg atoms in microwave cavities. For Rydberg atoms, the circular Rydberg states [5] correspond to large principal and maximum orbital and magnetic quantum numbers. The valence electron orbital is a thin torus centered on the atom's core, revealing quantum position fluctuations around the classical Bohr orbit. According to the correspondence principle, the properties of these states can be understood in classical terms. The transitions between neighboring circular states fall in the millimeter-wave domain for principal quantum numbers of the order of 50. In the discussion part of the chapter, we will adopt the experimental data from Haroche's group [6], where the three circular levels with principal quantum numbers 51, 50, and 49 are usually considered, for example called e, g, and i, respectively. The e, g and g, i transitions are at 51.1 and 54.3 GHz, respectively. The radiative lifetimes of e, g, and i—of the order of 30 ms—are much longer than those for noncircular Rydberg states. In free space, the atoms would propagate a few meters at thermal velocity before decaying. Radiative decay is thus negligible along the 20-cm path inside the apparatus. In our discussion, we consider the atoms flying through a microwave cavity with a single quantized mode. The quantization of the electromagnetic field in quantum electrodynamics (QED) is in general achieved by starting with a finite volume defined by conducting walls where the field appears in a discrete, though infinite, set of modes [7]. Introducing canonical variables casts the Hamiltonian into a form similar to that for the harmonic oscillator which can easily be quantized by imposing canonical commutation relations. For simplicity, we will only consider a single mode in our treatment.

The chapter is to review our recent research work on QIP by using cavity-atom system. Our treatment, basically leading to analytical results, is made by quantum-trajectory approach [8], which is widely employed in solving dissipative dynamics of quantum systems, with the main idea the time evolution governed by a non-Hermitian operator and interrupted by instantaneous jumps by the detection of a photon. Normally, a solution by quantum-trajectory approach is resorted to numerical calculations. In our case, however, we suppose that no photon leakage actually happens during our implementation of QIP. Therefore, we only focus on the time evolution of the system governed by the non-Hermitian Hamiltonian. An obvious advantage of our treatment over conventional methods is the availability of analytical solution clearly demonstrating the detrimental influence from the cavity decay.

In section II, we will concentrate on implementation of different QIP tasks by resonant interaction of atoms with cavity mode. The three subsections correspond to three different but correlated schemes. In contrast, section III is only for an accomplishment of two-qubit conditional quantum gates by geometric method, which, different from the implementation in section II, is robust to parametric fluctuation. We will conclude in section IV.

2. Resonant Interaction of Atoms with Cavity QED

In this section, we will consider some atoms resonantly interacting with the cavity, which implies that the cavity mode or one of the cavity modes is of the same frequency as the transition frequency of two atomic levels. To describe such a case, we may simply use Jaynes-Cummings model. But with consideration of physics, the mathematics for this situation should be more complicated because of the atomic dissipation and the decay regarding the cavity mode.

2.1. W-state Preparation

Entanglement is one of the most striking features of quantum mechanics. Entangled states not only help quantum mechanics win over local hidden theory [9], but also have applications in quantum teleportation [10], quantum cryptography [11], quantum dense coding [12], high-precision frequency measurement [13] and so on. Recently, much interest has been devoted to multi-particle entangled states, the classification and properties of which are more complex than that of the bipartite entangled states. In Ref. [14], authors show that there exist two inequivalent classes of tripartite entangled states, i.e., the Greenberger-Horne-Zeilinger (GHZ) class [15] and the W class. They cannot be converted to each other even under stochastic local operations and classical communication. One of the interesting properties of the W state (such as $\frac{1}{\sqrt{3}}(|001\rangle + |010\rangle + |100\rangle)$) is that if one particle is traced out, there remains entanglement of the remaining two particles, or if one particle is measured in basis $\{|0\rangle, |1\rangle\}$, then either the state of remained two particles is in a maximally entangled state or in a product state.

Four-particle entanglement in ion trap [16] and three-particle entanglement [17] in cavity QED have been demonstrated experimentally. More recently, five-photon entanglement has also been reported [18]. As far as we know, three-particle W state in ion trap [19] has been realized, but there is no report of experimental realization of the W state in cavity QED. While there have been several papers discussing the preparation of W state in cavity QED [20, 21, 22, 23]. In Ref. [20], atoms, interacting with nonresonant cavity, can be entangled through virtual excitation of the cavity mode, which can loosen the quality requirement on the cavity. However, the scheme requires that the atom-cavity coupling strength be much smaller than the detuning between atomic transition frequency and cavity frequency, which restricts the operation speed. Ref. [21] proposed how to generate both three-atom W state and three-cavity W state. In the case of the three-atom W state (or three-cavity W state), each atom (or atom) interacts with the cavity (or each cavity) sequentially with different interaction time. In Ref. [23], W state is investigated by adiabatic passages and decoherence-free subspace.

The general form of W state for n particles is $W_n = \frac{1}{\sqrt{n}}|n-1, 1\rangle$, where $|n-1, 1\rangle$ denotes all the totally symmetric states involving $n-1$ zeros and 1 one. In this paper, we present an alternative scheme to realize n-qubit W state via cavity QED. The scheme only requires that n identical atoms interact simultaneously and resonantly with a single-mode high Q microwave cavity. After an appropriate interaction time, the atoms will be entangled in W state provided that there is no photon leakage from the cavity. Compared with Ref. [20], our scheme is much faster due to the resonant interaction instead of virtual excitation of the cavity mode. This is important experimentally in view of decoherence. Moreover, after all atoms being prepared in ground states and cavity in one-photon state, one step of our implementation can achieve the W state, which is much simpler and more straightforward than in Ref. [21], in which the preparation of three-atom W state needs the three atoms going through the cavity one by one, and each atom interacting with the cavity for a different period of time. Furthermore, the success probability in our scheme increases with the atom number, in contrast to the decreasing probability of success in Ref. [20].

We first consider an ideal model of n identical two-level atoms interacting resonantly with a single-mode cavity field in the absence of any decay. The Hamiltonian (assuming

$\hbar = 1$) in the interaction picture reads

$$H = \sum_{j=1}^{n} g(a^{\dagger}\sigma_j^- + a\sigma_j^+), \tag{1}$$

where $\sigma_j^+ = |1\rangle_{jj}\langle 0|$, $\sigma_j^- = |0\rangle_{jj}\langle 1|$, with $|1\rangle_j$ ($|0\rangle_j$) being the excited (ground) state of the jth atom. g is the atom-cavity coupling strength. $a^{\dagger}$, a are, respectively, the creation and annihilation operators for the cavity mode. Assume that the cavity is initially in one-photon state $|1\rangle$ and all the n atoms are in the ground states, i.e., $|0\rangle_{1,2,...n}$. The evolution of the system goes as follows:

$$\Psi(t) = \cos(\sqrt{n}gt)|0\rangle_{1,2,...n}|1\rangle - i\sin(\sqrt{n}gt)W_n|0\rangle. \tag{2}$$

By choosing $\sqrt{n}gt = \frac{\pi}{2}$, i.e., $t = \frac{\pi}{2g\sqrt{n}}$, we can get the n-atom W state. This result can be easily understood from the property of Eq. (1): The excitation number is unchanged. The initial one photon is shared by n atoms and each atom has an equal probability to absorb the photon. As only one atom can succeed, the n-atom W state is naturally generated.

We introduce the cavity decay from now on. As long as there is no photon decay from the cavity, the evolution of the system is governed by the non-Hermitian Hamiltonian (assuming $\hbar = 1$)

$$H' = H - i\frac{\kappa}{2}a^{\dagger}a, \tag{3}$$

where κ is the cavity decay rate. If the initial state of the atom-cavity system is $|0\rangle_{1,2,...n}|1\rangle$, then the evolution of the system is

$$\Psi'(t) = \exp(-\frac{\kappa t}{4})[(\cos(\frac{\lambda t}{4}) - \frac{\kappa}{\lambda}\sin(\frac{\lambda t}{4}))|0\rangle_{1,2,...n}|1\rangle - i\frac{4g\sqrt{n}}{\lambda}\sin(\frac{\lambda t}{4})W_n|0\rangle], \tag{4}$$

with $\lambda = \sqrt{16ng^2 - \kappa^2}$. The probability that no photon has leaked out of the cavity during the evolution is

$$P_{suc} = \exp(-\frac{\kappa t}{2})\{[\cos(\frac{\lambda t}{4}) - \frac{\kappa}{\lambda}\sin(\frac{\lambda t}{4})]^2 + \frac{16ng^2}{\lambda^2}\sin^2(\frac{\lambda t}{4})\}. \tag{5}$$

If the interaction time τ is chosen to satisfy $\tan(\frac{\lambda t}{4}) = \frac{\lambda}{\kappa}$, we can obtain

$$\Psi'(\tau) = -i\frac{4g\sqrt{n}}{\lambda}\exp(-\frac{\kappa\tau}{4})\sin(\frac{\lambda\tau}{4})W_n|0\rangle, \tag{6}$$

and the corresponding success probability of getting the n-qubit W state is

$$P_{suc} = \frac{16ng^2}{\lambda^2}\exp(-\frac{\kappa\tau}{2})\sin^2(\frac{\lambda\tau}{4}). \tag{7}$$

To carry out our scheme, we first need to prepare the single-photon cavity state, which can be done by sending an auxiliary atom being in state $|1\rangle_{aux}$ to go through an initially vacuum cavity state, after an appropriate interaction time, the resonant interaction between the atom and the cavity will leave the cavity in state $|1\rangle$. Then, all the atoms initially prepared in the ground states (i.e., $|0\rangle_{1,2,...n}$), are sent into the single-mode high Q microwave cavity

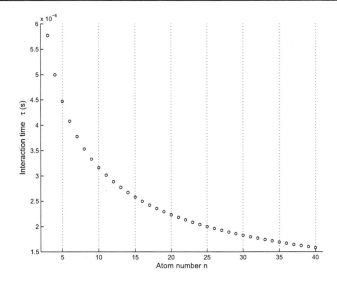

Figure 1. The interaction time τ versus the number of atoms n with n ranging from 3 to 40, where $g = 25 \times 2\pi$ kHz, $\kappa = 170.3 \times 2\pi$ Hz.

simultaneously with the same velocity. The desired interaction time τ can be achieved by choosing an appropriate velocity of the atoms.

We briefly discuss the experimental possibility of our proposal. Considering Rydberg atoms with principal quantum numbers 50 and 51 and the radiative lifetime $T_r = 3 \times 10^{-2}$ s, we assume that the atom-cavity coupling strength g is $25 \times 2\pi$ kHz and the cavity quality factor Q is 3×10^8 [24]. For the present case, the cavity frequency ν is 51.1 GHz. So we have $\kappa = \frac{2\pi\nu}{Q} \simeq 170.3 \times 2\pi$ Hz. In Fig. 1, we plot interaction time τ versus the number of atoms n, where the interaction time is of the order of 10^{-6} s (much shorter than T_r, so we can safely neglect atomic spontaneous decay) and is sharply decreased with the increase of the atom number. In Fig. 2, we show the dependence of success probability P_{suc} on cavity decay rate κ in different cases from 3 to 20 atoms. For a certain atom number, the success probability decreases with the increase of κ. For a certain κ, P_{suc} increases with the atom number n, which may be understood from Fig. 1 that the interaction time is sharply decreased with the increase of the atom number. Given $\kappa \simeq 170.3 \times 2\pi$ Hz, the success probability is above 99.6%.

As mentioned above, there have been some proposals for preparing W- state in cavity QED. For example, atoms in Ref. [20] get entangled in a non-resonant cavity through virtually exciting the cavity mode, in which the implementation time is very long due to the required weak atom-cavity coupling. Moreover, to prepare n-atom W state, Ref. [20] needs (n+1) atoms when n>4, and the success rate decreases with the increase of the atom number. In contrast, our scheme does not include any auxiliary measurement, and can be carried out with much higher success probability, which could lower the repeated time in experiments. In Ref. [21], it is obvious that the interaction time increases with the atom number for a series of atoms with one-by-one interaction with the cavity; thus if n is too large, the cavity decoherence (harmful to fidelity) should be seriously considered. Particularly, due to the different interaction time for different atoms with the cavity, it is very difficult to control

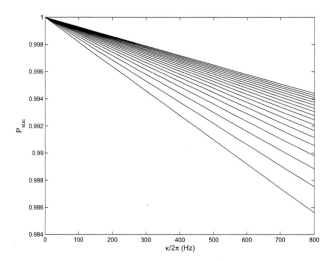

Figure 2. The dependence of success probability P_{suc} on cavity decay rate κ, where $g = 25 \times 2\pi$ kHz, and from bottom to top, the curves correspond to the atom number varying from 3 to 20.

the delay time between nearest-neighbor atoms. In contrast, our scheme includes only one step of resonant interaction. The above mentioned problems does not exist in our scheme.

The main experimental challenge for our scheme is to control N atoms going through a cavity simultaneously and particularly to keep identical coupling strength for each atom to the cavity. In many of previous proposals in this respect [25], the identical coupling strength was employed with the excuse of treatment simplicity. However, this is not easily achieved in experiments. Nevertheless, a recent work [26] gives us a hope that the these two requirements might be achievable in the near future. Therefore, we don't think the requirement of simultaneous control of many atoms in our scheme is more stringent experimentally than in the proposal with a series of atoms [21]. While we have not yet found any experimental report for more than two atoms going through a cavity simultaneously. So to carry out our scheme, we expect more advanced cavity QED technique.

2.2. Toffoli Gating

In the field of QIP, although the building blocks of quantum computers are one- and two-qubit logic gates [2], multi-qubit operations $C^n - NOT$ ($n > 2$), also universal, are also very useful for quantum computation [27] because a direct implementation of $C^n - NOT$ gates is more efficient than implementations consisting of a series of one- and two-qubit operations. In NMR system, some general methods for creating $C^n - NOT$ gates have been proposed, e.g., [28]. Recently, a scheme for realization of conditional quantum gates is proposed, including an N–atom Toffoli gate and some other nonlocal gates on remote atoms, through cavity-assisted photon scattering [29], in which the single photon detectors and a series of linear optical elements are necessary. An alternative scheme for the three-qubit Toffoli gate could be found in Ref. [30], based on vacuum-induced Stark shifts using a collective system in a high-quality dispersive cavity. Nevertheless, the Toffoli gate in cavity

QED has not yet been achieved experimentally so far.

In this subsection, we propose a potentially practical scheme to implement a three-qubit Toffoli gate by only a single resonant interaction in cavity QED. The scheme does not require two-qubit controlled-NOT gates but uses a three-qubit phase gate and two Hadamard gates to construct a Toffoli gate. We will focus on a phase gate implemented by sending atoms through a microwave cavity simultaneously. Our idea is quite straightforward because no classical laser field is involved, and the required interaction time is very short due to the resonant interaction. Moreover, our scheme is close to be reached with current cavity QED techniques and can be in principle generalized to an ion trap system.

First, the resonant interaction of N three-level atoms with a single-mode microwave cavity is considered. The atomic internal states can be expressed by $|i_j\rangle$, $|g_j\rangle$, and $|e_j\rangle$, with $|g_j\rangle$ and $|i_j\rangle$ being states lower than $|e_j\rangle$. Throughout our scheme, the states $|i_j\rangle$ are not involved in the interaction with the cavity mode. So the Hamiltonian in units of $\hbar = 1$ is

$$H_i = \sum_{j=1}^{N} g_j(a^\dagger S_j^- + a S_j^\dagger), \tag{8}$$

where g_j is the coupling constant of the jth atom to the cavity mode, and $S_j^\dagger = |e_j\rangle \langle g_j|$ and $S_j^- = |g_j\rangle \langle e_j|$ are the atomic spin operators for raising and lowering, respectively. $a^\dagger$ (a) is the creation (annihilation) operator for the cavity mode. We first assume that the atoms and the cavity mode are initially in the state $\prod_{j=2}^{N} |e_1\rangle |g_j\rangle$ and in the vacuum state $|0\rangle$, respectively. The evolution of the system is obtained straightforwardly by,

$$\begin{aligned}
|\psi(t)\rangle &= U(t) \prod_{j=2}^{N} |e_1\rangle |g_j\rangle |0\rangle \\
&= \exp[-it \sum_{j=1}^{N} g_j(a^+ S_j^- + a S_j^+)] \prod_{j=2}^{N} |e_1\rangle |g_j\rangle |0\rangle \\
&= [(g_1^2/G^2)\cos(Gt) + (G^2 - g_1^2)/G^2] \prod_{j=2}^{N} |e_1\rangle |g_j\rangle |0\rangle \\
&\quad + (g_1/G^2)[\cos(Gt) - 1] \sum_{k=2}^{N} g_k |g_1\rangle |e_k\rangle \prod_{j=2,j\neq k}^{N} |g_j\rangle |0\rangle \\
&\quad - (ig_1/G)\sin(Gt) \prod_{j=1}^{N} |g_j\rangle |1\rangle,
\end{aligned} \tag{9}$$

where $G = \sqrt{\sum_{j=1}^{N} g_j^2}$. Before going ahead, let us first consider two other simple cases. If the N atoms are initially in the state $\prod_{j=2}^{N} |e_1\rangle |i_j\rangle$, then the atoms, except the first one, do not interact with the cavity mode. So we acquire the corresponding time evolution

$$|\psi_1(t)\rangle = U(t)\prod_{j=2}^{N}|e_1\rangle\,|i_j\rangle\,|0\rangle$$

$$= [\cos(g_1 t)\,|e_1\rangle\,|0\rangle - i\sin(g_1 t)\,|g_1\rangle\,|1\rangle]\prod_{j=2}^{N}|i_j\rangle. \tag{10}$$

While if the N atoms are initially in the state $\prod_{j=2,j\neq k}^{N}|e_1\rangle\,|g_k\rangle\,|i_j\rangle$, then we only have the first and the kth atoms interacting with the cavity mode, as shown below,

$$|\psi_k(t)\rangle = U(t)\prod_{j=2,j\neq k}^{N}|e_1\rangle\,|g_k\rangle\,|i_j\rangle\,|0\rangle$$

$$= [(g_1^2/G_k^2)\cos(G_k t) + (G_k^2 - g_1^2)/G_k^2]\prod_{j=2,j\neq k}^{N}|e_1\rangle\,|g_k\rangle\,|i_j\rangle\,|0\rangle$$

$$+ (g_1 g_k/G_k^2)[\cos(G_k t) - 1]\,|g_1\rangle\,|e_k\rangle\prod_{j=2,j\neq k}^{N}|i_j\rangle\,|0\rangle$$

$$- (ig_1/G_k)\sin(G_k t)\,|g_1\rangle\,|g_k\rangle\prod_{j=2}^{N}|i_j\rangle\,|1\rangle, \tag{11}$$

where $G_k = \sqrt{g_1^2 + g_k^2}$. Based on above four equations, we now try to construct a three-qubit phase gate. By considering quantum information encoded in a subspace spanned by the atomic states $\{|g_1\rangle, |e_1\rangle, |g_2\rangle, |i_2\rangle, |g_3\rangle, |i_3\rangle\}$, we assume $Gt = \sqrt{g_1^2 + g_2^2 + g_3^2}\,t = 2l\pi$ $(l = 1, 2, ...)$. Then we have from Eq. (9),

$$|e_1\rangle\,|g_2\rangle\,|g_3\rangle\,|0\rangle \rightarrow |e_1\rangle\,|g_2\rangle\,|g_3\rangle\,|0\rangle, \tag{12}$$

which means unchanged. From Eq.(10), by setting $g_1 t = \pi$, we have

$$|e_1\rangle\,|i_2\rangle\,|i_3\rangle\,|0\rangle \rightarrow -\,|e_1\rangle\,|i_2\rangle\,|i_3\rangle\,|0\rangle, \tag{13}$$

which corresponds to a phase flip. For Eq. (11), if we consider $k = 2$, and assume $G_2 t = \sqrt{g_1^2 + g_2^2}\,t = 2m\pi$ $(m = 1, 2, ...)$, the evolution of the state is

$$|e_1\rangle\,|g_2\rangle\,|i_3\rangle\,|0\rangle \rightarrow |e_1\rangle\,|g_2\rangle\,|i_3\rangle\,|0\rangle, \tag{14}$$

which is also unchanged. By using the above conditions for G, g_1 and G_2, we obtain $\sqrt{g_1^2 + g_3^2}\,t = \sqrt{4(l^2 - m^2) + 1}\,\pi$, which yields the evolution from Eq. (11) in the case of $k = 3$,

$$|\psi_3(t)\rangle = \{[g_1^2/(g_1^2 + g_3^2)]\cos(\sqrt{4(l^2 - m^2) + 1}\,\pi) + g_3^2/(g_1^2 + g_3^2)\}\,|e_1\rangle\,|i_2\rangle\,|g_3\rangle\,|0\rangle$$

$$+ [g_1 g_3/(g_1^2 + g_3^2)][\cos(\sqrt{4(l^2 - m^2) + 1}\,\pi) - 1]\,|g_1\rangle\,|i_2\rangle\,|e_3\rangle\,|0\rangle$$

$$- (ig_1/\sqrt{g_1^2 + g_3^2})\sin(\sqrt{4(l^2 - m^2) + 1}\,\pi)\,|g_1\rangle\,|i_2\rangle\,|g_3\rangle\,|1\rangle. \tag{15}$$

We expect to have $|e_1\rangle |i_2\rangle |g_3\rangle |0\rangle \rightarrow |e_1\rangle |i_2\rangle |g_3\rangle |0\rangle$ from above equation, which implies $\cos(\sqrt{4(l^2 - m^2) + 1}\pi) = 1$. However, it is obvious that this condition cannot be exactly met. To make the condition approximately satisfied to the best, we assume $l = 5$ and $m = 3$, which yields $\cos(\sqrt{4(l^2 - m^2) + 1}\pi) = \cos(\sqrt{65}\pi) = 0.9810$. Then Eq. (15) reduces to

$$\left|\psi_3'(t)\right\rangle = 0.9997 |e_1\rangle |i_2\rangle |g_3\rangle |0\rangle - 0.0023 |g_1\rangle |i_2\rangle |e_3\rangle |0\rangle - i0.024 |g_1\rangle |i_2\rangle |g_3\rangle |1\rangle$$

$$\simeq 0.9997 |e_1\rangle |i_2\rangle |g_3\rangle |0\rangle. \tag{16}$$

This approximation will be seriously checked later. As other states under consideration, including $|g_1\rangle |g_2\rangle |g_3\rangle |0\rangle$, $|g_1\rangle |g_2\rangle |i_3\rangle |0\rangle$, $|g_1\rangle |i_2\rangle |g_3\rangle |0\rangle$, and $|g_1\rangle |i_2\rangle |i_3\rangle |0\rangle$, remain unchanged in the evolution, a three-qubit phase gate can be reached as follows:

$$
\begin{aligned}
|g_1\rangle |g_2\rangle |g_3\rangle |0\rangle &\rightarrow |g_1\rangle |g_2\rangle |g_3\rangle |0\rangle, \\
|g_1\rangle |g_2\rangle |i_3\rangle |0\rangle &\rightarrow |g_1\rangle |g_2\rangle |i_3\rangle |0\rangle, \\
|g_1\rangle |i_2\rangle |g_3\rangle |0\rangle &\rightarrow |g_1\rangle |i_2\rangle |g_3\rangle |0\rangle, \\
|g_1\rangle |i_2\rangle |i_3\rangle |0\rangle &\rightarrow |g_1\rangle |i_2\rangle |i_3\rangle |0\rangle, \\
|e_1\rangle |g_2\rangle |g_3\rangle |0\rangle &\rightarrow \alpha_0 |e_1\rangle |g_2\rangle |g_3\rangle |0\rangle, \\
|e_1\rangle |g_2\rangle |i_3\rangle |0\rangle &\rightarrow \beta_0 |e_1\rangle |g_2\rangle |i_3\rangle |0\rangle, \\
|e_1\rangle |i_2\rangle |g_3\rangle |0\rangle &\rightarrow \gamma_0 |e_1\rangle |i_2\rangle |g_3\rangle |0\rangle, \\
|e_1\rangle |i_2\rangle |i_3\rangle |0\rangle &\rightarrow -\delta_0 |e_1\rangle |i_2\rangle |i_3\rangle |0\rangle,
\end{aligned} \tag{17}
$$

where $\alpha_0 = \beta_0 = \delta_0 = 1$ and $\gamma_0 = \frac{g_1^2}{g_1^2+g_3^2} \times \cos(\sqrt{65}\pi) + \frac{g_3^2}{g_1^2+g_3^2} = 0.9997$, and from which we can approximately obtain a Toffoli gate in our computational subspace,

$$T = H_3^\dagger T_P H_3,$$

$$
= \begin{bmatrix}
1 & 0 & 0 & 0 & 0 & 0 & 0 & 0 \\
0 & 1 & 0 & 0 & 0 & 0 & 0 & 0 \\
0 & 0 & 1 & 0 & 0 & 0 & 0 & 0 \\
0 & 0 & 0 & 1 & 0 & 0 & 0 & 0 \\
0 & 0 & 0 & 0 & (\alpha_0+\beta_0)/2 & (\alpha_0-\beta_0)/2 & 0 & 0 \\
0 & 0 & 0 & 0 & (\alpha_0-\beta_0)/2 & (\alpha_0+\beta_0)/2 & 0 & 0 \\
0 & 0 & 0 & 0 & 0 & 0 & (\gamma_0-\delta_0)/2 & (\gamma_0+\delta_0)/2 \\
0 & 0 & 0 & 0 & 0 & 0 & (\gamma_0+\delta_0)/2 & (\gamma_0-\delta_0)/2
\end{bmatrix}
$$

$$
= \begin{bmatrix}
1 & 0 & 0 & 0 & 0 & 0 & 0 & 0 \\
0 & 1 & 0 & 0 & 0 & 0 & 0 & 0 \\
0 & 0 & 1 & 0 & 0 & 0 & 0 & 0 \\
0 & 0 & 0 & 1 & 0 & 0 & 0 & 0 \\
0 & 0 & 0 & 0 & 1 & 0 & 0 & 0 \\
0 & 0 & 0 & 0 & 0 & 1 & 0 & 0 \\
0 & 0 & 0 & 0 & 0 & 0 & 0.00015 & 0.99985 \\
0 & 0 & 0 & 0 & 0 & 0 & 0.99985 & 0.00015
\end{bmatrix}
\simeq
\begin{bmatrix}
1 & 0 & 0 & 0 & 0 & 0 & 0 & 0 \\
0 & 1 & 0 & 0 & 0 & 0 & 0 & 0 \\
0 & 0 & 1 & 0 & 0 & 0 & 0 & 0 \\
0 & 0 & 0 & 1 & 0 & 0 & 0 & 0 \\
0 & 0 & 0 & 0 & 1 & 0 & 0 & 0 \\
0 & 0 & 0 & 0 & 0 & 1 & 0 & 0 \\
0 & 0 & 0 & 0 & 0 & 0 & 0 & 1 \\
0 & 0 & 0 & 0 & 0 & 0 & 1 & 0
\end{bmatrix},
$$

$$\tag{18}$$

where H_3 is the Hadamard gate on the third atom with $|g_3\rangle \to (|g_3\rangle + |i_3\rangle)/\sqrt{2}, |i_3\rangle \to (|g_3\rangle - |i_3\rangle)/\sqrt{2}$, and T_P is the operator for the evolution in Eq. (17). To achieve Eq. (18), we should have $g_1 : g_2 : g_3 = 1 : \sqrt{35} : 8$ and $t = \pi/g_1$ from above conditions for G, g_1 and G_2.

We now turn to study the influence from the cavity decay on a three-qubit phase gate (or say, three-qubit Toffoli gate). The corresponding Hamiltonian is,

$$H_{di} = \sum_{j=1}^{N} g_j(a^\dagger S_j^- + a S_j^\dagger) - i\frac{\kappa}{2}a^\dagger a, \tag{19}$$

where κ is the cavity decay rate. Provided that the atoms and the cavity mode are initially in the state $\prod_{j=2}^{N} |e_1\rangle |g_j\rangle$ and in the vacuum state $|0\rangle$, respectively, we obtain the evolution of the system before the leakage of a photon from the microwave cavity happens,

$$\begin{aligned}
|\psi_{decay}(t)\rangle &= U_d(t) \prod_{j=2}^{N} |e_1\rangle |g_j\rangle |0\rangle \\
&= \exp\{-it[\sum_{j=1}^{N} g_j(a^\dagger S_j^- + a S_j^+) - i\frac{\kappa}{2}a^\dagger a]\} \prod_{j=2}^{N} |e_1\rangle |g_j\rangle |0\rangle \\
&= \{(g_1^2/G^2)\exp(-\kappa t/4)[\cos(A_\kappa t) + \frac{\kappa}{4A_\kappa}\sin(A_\kappa t)] \\
&\quad + (G^2 - g_1^2)/G^2\} \prod_{j=2}^{N} |e_1\rangle |g_j\rangle |0\rangle \\
&\quad + (g_1/G^2) \left\{ \exp(-\kappa t/4)[\cos(A_\kappa t) + \frac{\kappa}{4A_\kappa}\sin(A_\kappa t)] - 1 \right\} \\
&\quad \times \sum_{k=2}^{N} g_k |g_1\rangle |e_k\rangle \prod_{j=2, j\neq k}^{N} |g_j\rangle |0\rangle - (ig_1/A_\kappa)\sin(A_\kappa t) \prod_{j=1}^{N} |g_j\rangle |1\rangle , \tag{20}
\end{aligned}$$

where $G = \sqrt{\sum_{j=1}^{N} g_j^2}$, and $A_\kappa = \sqrt{\sum_{j=1}^{N} g_j^2 - \kappa^2/16}$. As done in the ideal case, we consider other two initial conditions. If the N atoms are initially in the state $\prod_{j=2}^{N} |e_1\rangle |i_j\rangle$, and the cavity mode in vacuum state $|0\rangle$, we have the corresponding evolution

$$\begin{aligned}
|\psi_{decay1}(t)\rangle &= U_d(t) \prod_{j=2}^{N} |e_1\rangle |i_j\rangle |0\rangle \\
&= \exp(-\kappa t/4)\{[\cos(A_{1\kappa}t) + \frac{\kappa}{4A_{1\kappa}}\sin(A_{1\kappa}t)] \prod_{j=2}^{N} |e_1\rangle |i_j\rangle |0\rangle \\
&\quad - (ig_1/A_{1\kappa})\sin(A_{1\kappa}t) \prod_{j=2}^{N} |g_1\rangle |i_j\rangle |1\rangle]\}, \tag{21}
\end{aligned}$$

where $A_{1\kappa} = \sqrt{g_1^2 - \kappa^2/16}$. If the total system is initially in the state

$\prod_{j=2,j\neq k}^{N} |e_1\rangle |g_k\rangle |i_j\rangle |0\rangle$, the evolution of the system is

$$|\psi_{decayk}(t)\rangle = U_d(t) \prod_{j=2,j\neq k}^{N} |e_1\rangle |g_k\rangle |i_j\rangle |0\rangle$$

$$= \{(g_1^2/G_k^2)\exp(-\kappa t/4)[\cos(A_{k\kappa}t) + \frac{\kappa}{4A_{k\kappa}}\sin(A_{k\kappa}t)]$$

$$+ (G_k^2 - g_1^2)/G_k^2\} \prod_{j=2,j\neq k}^{N} |e_1\rangle |g_k\rangle |i_j\rangle |0\rangle$$

$$+ (g_1 g_k/G_k^2) \left\{ \exp(-\kappa t/4)[\cos(A_{k\kappa}t) + \frac{\kappa}{4A_{k\kappa}}\sin(A_{k\kappa}t)] - 1 \right\}$$

$$\times \prod_{j=2,j\neq k}^{N} |g_1\rangle |e_k\rangle |i_j\rangle |0\rangle$$

$$- (ig_1/A_{k\kappa})\sin(A_{k\kappa}t) \prod_{j=2,j\neq k}^{N} |g_1\rangle |g_k\rangle |i_j\rangle |1\rangle , \qquad (22)$$

with $G_k = \sqrt{g_1^2 + g_k^2}$ and $A_{k\kappa} = \sqrt{g_1^2 + g_k^2 - \kappa^2/16}$. Assuming that $\sqrt{g_1^2 - \kappa^2/16}t = \pi$, $\sqrt{g_1^2 + g_2^2 - \kappa^2/16}t = 6\pi$, and $\sqrt{g_1^2 + g_2^2 + g_3^2 - \kappa^2/16}t = 10\pi$, resulting in $\cos(\sqrt{g_1^2 + g_3^2 - \kappa^2/16}t) = \cos(\sqrt{65}\pi) = 0.9810$, we acquire a three-qubit phase gate T_P'

$$|g_1\rangle |g_2\rangle |g_3\rangle |0\rangle \to |g_1\rangle |g_2\rangle |g_3\rangle |0\rangle ,$$
$$|g_1\rangle |g_2\rangle |i_3\rangle |0\rangle \to |g_1\rangle |g_2\rangle |i_3\rangle |0\rangle ,$$
$$|g_1\rangle |i_2\rangle |g_3\rangle |0\rangle \to |g_1\rangle |i_2\rangle |g_3\rangle |0\rangle ,$$
$$|g_1\rangle |i_2\rangle |i_3\rangle |0\rangle \to |g_1\rangle |i_2\rangle |i_3\rangle |0\rangle ,$$
$$|e_1\rangle |g_2\rangle |g_3\rangle |0\rangle \to \alpha_1 |e_1\rangle |g_2\rangle |g_3\rangle |0\rangle ,$$
$$|e_1\rangle |g_2\rangle |i_3\rangle |0\rangle \to \beta_1 |e_1\rangle |g_2\rangle |i_3\rangle |0\rangle ,$$
$$|e_1\rangle |i_2\rangle |g_3\rangle |0\rangle \to \gamma_1 |e_1\rangle |i_2\rangle |g_3\rangle |0\rangle ,$$
$$|e_1\rangle |i_2\rangle |i_3\rangle |0\rangle \to -\delta_1 |e_1\rangle |i_2\rangle |i_3\rangle |0\rangle , \qquad (23)$$

where $\alpha_1 = 1 - \frac{g_1^2}{g_1^2+g_2^2+g_3^2}(1 - e^{-\kappa t/4})$, $\beta_1 = 1 - \frac{g_1^2}{g_1^2+g_2^2}(1 - e^{-\kappa t/4})$, $\delta_1 = e^{-\kappa t/4}$, and $\gamma_1 = 1 - \frac{g_1^2}{g_1^2+g_3^2}[1 - e^{-\kappa t/4}\cos(\sqrt{65}\pi)]$ with the term $\frac{\kappa}{4A_{1\kappa}}\sin(A_{1\kappa}t)$ ($\sim 10^{-4}$) omitted. To achieve our gating, we require $g_1 : g_2 : g_3 = 1 : \sqrt{35} : 8$ and $t = \pi/A_{1\kappa}$. When $\kappa = g_1/50$ (or $g_1/10$), α_1, β_1, γ_1, and δ_1 are equal to 0.9994 (0.9973), 0.9983 (0.9925), 0.9988 (0.9956), and 0.9391 (0.7304), respectively, leading to an approximate three-qubit phase gate. Therefore, by combining the phase gate T_P' with the Hadamard transform H_3 on the third qubit, we can obtain an approximate Toffoli gate for $\kappa/g_1 = 1/50$ (1/10), which is of the similar form to Eq. (18), but replaces α_0, β_0, γ_0, and δ_0 by α_1, β_1, γ_1, and δ_1, respectively. The approximation is from our omission of the non-zero but very small terms regarding $(\alpha_1 - \beta_1)/2$ and $(\gamma_1 - \delta_1)/2$. On the other hand, as $(\alpha_1 + \beta_1)/2$ and $(\gamma_1 + \delta_1)/2$ are not exactly equal to unity, we have to check how well our scheme works. To this end, we utilize Eqs. (17) and (23) to prepare an entangled state from an initial state $|\varphi_0\rangle = \frac{1}{2\sqrt{2}}(|g_1\rangle + |e_1\rangle)(|g_2\rangle + |i_2\rangle)(|g_3\rangle + |i_3\rangle)$. So we have $|\varphi_{s\tan dard}\rangle = \frac{1}{2\sqrt{2}}[|g_1\rangle(|g_2\rangle|g_3\rangle + |g_2\rangle|i_3\rangle + |i_2\rangle|g_3\rangle + |i_2\rangle|i_3\rangle) + |e_1\rangle(|g_2\rangle|g_3\rangle + |g_2\rangle|i_3\rangle + |i_2\rangle|g_3\rangle - |i_2\rangle|i_3\rangle)]$, $|\varphi_{ideal}\rangle = \frac{1}{\sqrt{7.9994}}[|g_1\rangle(|g_2\rangle|g_3\rangle + |g_2\rangle|i_3\rangle + |i_2\rangle|g_3\rangle + |i_2\rangle|i_3\rangle) + |e_1\rangle(|g_2\rangle|g_3\rangle + |g_2\rangle|i_3\rangle + 0.9997|i_2\rangle|g_3\rangle - |i_2\rangle|i_3\rangle)]$, $|\varphi_{decay}\rangle_{\kappa=g_1/50} = \frac{1}{\sqrt{7.8764}}[|g_1\rangle(|g_2\rangle|g_3\rangle + |g_2\rangle|i_3\rangle + |i_2\rangle|g_3\rangle + |i_2\rangle|i_3\rangle) + |e_1\rangle(0.9994|g_2\rangle|g_3\rangle + 0.9983|g_2\rangle|i_3\rangle + 0.9988|i_2\rangle|g_3\rangle - 0.9391|i_2\rangle|i_3\rangle)]$, and $|\varphi_{decay}\rangle_{\kappa=g_1/10} = \frac{1}{\sqrt{7.505}}[|g_1\rangle(|g_2\rangle|g_3\rangle + |g_2\rangle|i_3\rangle + |i_2\rangle|g_3\rangle + |i_2\rangle|i_3\rangle) + |e_1\rangle(0.9973|g_2\rangle|g_3\rangle + 0.9925|g_2\rangle|i_3\rangle + 0.9959|i_2\rangle|g_3\rangle - 0.7304|i_2\rangle|i_3\rangle)]$. So the fidelities and the corresponding success probabilities are

$$
\begin{aligned}
F_{ideal} &= |\langle\varphi_{s\tan dard}|\varphi_{ideal}\rangle|^2 \\
&= \frac{(4 + \alpha_0 + \beta_0 + \gamma_0 + \delta_0)^2}{8(4 + \alpha_0^2 + \beta_0^2 + \gamma_0^2 + \delta_0^2)} = 0.9999,
\end{aligned}
\tag{24}
$$

and

$$
P_{ideal} = \frac{4 + \alpha_0^2 + \beta_0^2 + \gamma_0^2 + \delta_0^2}{8} = 0.9999,
\tag{25}
$$

for the ideal case, and

$$
\begin{aligned}
F_{decay} &= |\langle\varphi_{s\tan dard}|\varphi_{decay}\rangle|^2, \\
&= \frac{(4 + \alpha_1 + \beta_1 + \gamma_1 + \delta_1)^2}{8(4 + \alpha_1^2 + \beta_1^2 + \gamma_1^2 + \delta_1^2)} = 0.9995(0.9916),
\end{aligned}
\tag{26}
$$

and

$$
P_{decay} = \frac{4 + \alpha_1^2 + \beta_1^2 + \gamma_1^2 + \delta_1^2}{8} = 0.9924(0.9381),
\tag{27}
$$

for the decay case $\kappa = \frac{g_1}{50}(\frac{g_1}{10})$. Note that Eqs. (24) $\sim$ (27) are actually for averaging over all qubit states, which gives the general assessment for our gating implementation. We can see from Eqs. (26) and (27) how F_{decay} and P_{decay} change with the cavity decay rate κ. Since $\alpha_1, \beta_1, \gamma_1$ and δ_1 decrease with κ, both F_{decay} and P_{decay} become smaller when κ is bigger. Above calculations also present us that both F_{ideal} and P_{ideal} are approaching unity, which implies that our proposed three-qubit gate is of high-fidelity and high-success-probability as long as the cavity decay is small enough.

We may employ Rydberg atoms for our proposal with $|i_j\rangle$, $|g_j\rangle$, and $|e_j\rangle$ being states of principal quantum numbers 49, 50 and 51, respectively [31], as shown in Fig. 3(a). In current microwave cavity experiments [31], we describe the interaction between atoms and the cavity mode by $g = g_0 \cos(2\pi z/\lambda_0)\exp(-r^2/w^2) \simeq g_0 \cos(2\pi z/\lambda_0)$, where g_0 is the coupling strength at the cavity center, r is the distance of the atom away from the cavity center, and λ_0 and w are the wavelength and the waist of the cavity mode, respectively. To meet the condition $g_1 : g_2 : g_3 = 1 : \sqrt{35} : 8$, we send the third atom going through the center of the microwave cavity along y axis, that is, $z_3 = 0$. The first and the second atoms should be sent away from the third one by distances $|z_1| = (\lambda_0/2\pi)\arccos(1/8)$ and $|z_2| = (\lambda_0/2\pi)\arccos(\sqrt{35}/8)$, respectively, as shown in Fig. 3(b). Due to $g_0 = 2\pi \times 49$ kHz, the gating time t $(= \pi/g_1 = 8\pi/g_0)$ is about 80 μs in our scheme, much shorter than the cavity decay time, i.e., 1 ms [31]. As the Rydberg atomic lifetime is 30 ms, much longer

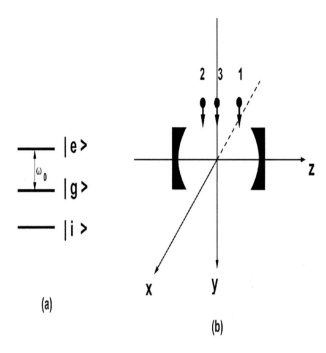

Figure 3. (a) The Rydberg atomic level structure under our consideration, where ω_0 is the resonant frequency of the cavity. (b) Schematic of our scheme, where the three atoms are sent along y axis, experiencing different couplings to the standing wave of the cavity.

than both the gating time and the cavity decay time, we have completely neglected it in our model. This is different from the treatment in optical cavity [32] in which both cavity decay and atomic spontaneous emission are necessarily considered..

However, as mentioned above, simultaneously sending three Rydberg atoms through a cavity with precise tracks and velocities is still very challenging experimentally although two Rydberg atoms going through a microwave cavity simultaneously have been achieved [26]. Here we simply assess the infidelity due to the imperfection in atomic velocity and position by two examples below. First of all, we assume that the atoms are sent to the cavity simultaneously and with precise atomic trajectories. But the times of the atoms passing the cavity, due to the different atomic velocities, to be, respectively, $t_1 = 1.10t$ ($t = \pi/A_{1\kappa}$ being the desired interaction time), and $t_2 = t_3 = t$. Straightforward algebra may yield the fidelity and the success probability for $\kappa = g_1/50$ to be $F_{impv} = 0.9980$ and $P_{impv} = 0.9254$, where we have $\alpha'_1 = \eta\alpha_1 = 0.9460, \beta'_1 = \eta\beta_1 = 0.9450,$ $\gamma'_1 = \eta\gamma_1 - g_1^2/(A_{3\kappa}A_{1\kappa})\sin(\sqrt{65}\pi)\sin(0.1A_{1\kappa}t) = 0.9081,$ and $\delta'_1 = \eta\delta_1 = 0.8890,$ where $\eta = e^{-0.1\kappa t/4}[\cos(0.1A_{1\kappa}t) + \frac{\kappa}{4A_{1\kappa}}\sin(0.1A_{1\kappa}t)] = 0.9446.$ On the other hand, supposing perfect atomic velocity, but one of the three atoms deviating from the desired track, i.e., $g'_1 = 0.9g_1, g'_2 = g_2,$ and $g'_3 = g_3,$ we can obtain that for $\kappa = g_1/50, F_{impp} = 0.9654$ and $P_{impp} = 0.8838$ with $\alpha''_1 = 0.9848, \beta''_1 = 0.9669, \gamma''_1 = 0.9759,$ and $\delta''_1 = 0.4266.$ Compared to Eqs. (26) and (27), we may conclude that our proposed gate is sensitive to the small imperfection, particularly for the imperfection in atomic velocities. If we consider more complex or more serious situation of imperfection, the gating would be worse.

Therefore, to achieve our scheme with high fidelity, we require the implementation to be as perfect as possible. Our scheme will be useful for quantum information processing, for example, for a three-qubit Grover search to label the target qubit, as discussed in the next subsection.

2.3. Three-qubit Grover Search

In the field of QIP, Shor's discovery of a polynomial-time quantum factoring algorithm [1] and Grover search [33] for an item with quantum computer from a disordered system are two most frequently cited algorithms, which could work in quantum computers to solve intractable computational problems more efficiently than present classical computers. Many efforts have been devoted to achievement of these quantum algorithms theoretically and experimentally by using trapped ions [34, 35, 36], NMR system [37, 38], superconducting mesocircuits [39], cavity quantum electrodynamics (QED) [26, 40, 41], and linear optical elements [42, 43].

In this subsection, we propose a potentially practical scheme for implementing three-qubit Grover search with cavity QED by the Toffoli gate in last subsection. The rapid development in relevant experimental technologies has enabled us to achieve entanglement between two atoms in a microwave cavity [24], based on which there have been some proposals for two-qubit Grover search with cavity QED [26, 40]. We have also noticed a very recent publication for three-qubit Grover search with three four-level atoms going through a three-mode cavity [41]. Actually, the important difference of the three-qubit Grover search from the two-qubit case is the probabilistic achievement. To reach a case with high success probability, we have to implement the basic searching step (also called iteration) for several times. So implementation of a three-qubit Grover search is much more complex than that of a two-qubit case. In contrast to Ref. [41], we will design a Grover search scheme by three identical Rydberg atoms sent through a single-mode microwave cavity. We will store quantum information in long-lived internal levels of the Rydberg atoms, and consider the resonant interaction between the atoms and the cavity mode, which yields a very fast implementation of the search. As the cavity decay is the main dissipative factor of our design, we will seriously consider its detrimental effect on our scheme.

Generally speaking, the Grover search algorithm consists of three kinds of operations [34]. The first one is to prepare a superposition state $|\Psi_0\rangle = (\frac{1}{\sqrt{N}}) \sum_{i=1}^{N-1} |i\rangle$ using Hadamard gates. The second is for an iteration Q including following two operations: (a) Inverting the amplitude of the marked state $|\tau\rangle$ using a quantum phase gate $I_\tau = I - 2|\tau\rangle\langle\tau|$, with I the identity matrix; (b) Inversion about average of the amplitude of all states using the diffusion transform $\hat{D}$, with $\hat{D}_{ij} = \frac{2}{N} - \delta_{ij}$ and $N = 2^q$ (q being the number of the qubits). This step should be carried out by at least $\pi\sqrt{N}/4$ times to maximize the probability for finding the marked state. Finally, a measurement of the whole system is done to get the marked state. In other word, the Grover search consists in a repetition of the transformation $Q = H I_{000} H I_\tau$ with $I_{000} = I - 2|000\rangle\langle000|$ (defined later).

In three-qubit case, the number of possible quantum states is $2^3 = 8$, and the operation to label a marked state by conditional quantum phase gate is I_τ with τ one of the states $\{|000\rangle, |001\rangle, |010\rangle, |011\rangle, |100\rangle, |101\rangle, |110\rangle, |111\rangle\}$. For clarity of description, we first consider an ideal situation. For the three identical atoms, the atomic internal states under our consideration are denoted by $|i_j\rangle$, $|g_j\rangle$, and $|e_j\rangle$, with $|g_j\rangle$ and $|i_j\rangle$ being states lower than $|e_j\rangle$. Because the resonant transition happens between $|g_j\rangle$ and $|e_j\rangle$ by the cavity mode, $|i_j\rangle$ is not involved in the interaction throughout our scheme. So the Hamiltonian in units of $\hbar = 1$ reads,

$$H = \sum_{j=1}^{3} \Omega_{jc}(a^\dagger S_j^- + a S_j^\dagger),\qquad(28)$$

where Ω_{jc} is the coupling constant of the jth atom to the cavity mode, $S_j^\dagger = |e_j\rangle\langle g_j|$ and $S_j^- = |g_j\rangle\langle e_j|$ are the atomic spin operators for raising and lowering, respectively, and $a^\dagger$ (a) is the creation (annihilation) operator for the cavity mode. Following the idea in last subsection, to achieve three-qubit conditional phase gate, we require that the three atoms couple to the cavity mode by $\Omega_{1c} : \Omega_{2c} : \Omega_{3c} = 1 : \sqrt{35} : 8$ and the gating time be $\dfrac{\pi}{\Omega_{1c}}$. We would like to point out that the qubit definitions are not the same for each atom in our proposal. The logic state $|1\rangle$ ($|0\rangle$) of the qubit 1 is denoted by $|g_1\rangle$ ($|e_1\rangle$) of the atom 1; $|g_2\rangle$ and $|i_2\rangle$ of the atom 2 encode the logic state $|1\rangle$ ($|0\rangle$) of the qubit 2; The logic state $|1\rangle$ ($|0\rangle$) of the qubit 3 is represented by $|g_3\rangle$ ($|i_3\rangle$) of the atom 3. Eq. (17) has shown us the possibility to achieve an approximate three-qubit quantum phase gate $I_{e_1 i_2 i_3}$ in our computational subspace spanned by $|e_1\rangle|i_2\rangle|i_3\rangle$, $|e_1\rangle|i_2\rangle|g_3\rangle$, $|e_1\rangle|g_2\rangle|i_3\rangle$, $|e_1\rangle|g_2\rangle|g_3\rangle$, $|g_1\rangle|i_2\rangle|i_3\rangle$, $|g_1\rangle|i_2\rangle|g_3\rangle$, $|g_1\rangle|g_2\rangle|i_3\rangle$, $|g_1\rangle|g_2\rangle|g_3\rangle$,

$$I_{e_1 i_2 i_3} = \begin{bmatrix} -1 & 0 & 0 & 0 & 0 & 0 & 0 & 0 \\ 0 & \gamma_0 & 0 & 0 & 0 & 0 & 0 & 0 \\ 0 & 0 & 1 & 0 & 0 & 0 & 0 & 0 \\ 0 & 0 & 0 & 1 & 0 & 0 & 0 & 0 \\ 0 & 0 & 0 & 0 & 1 & 0 & 0 & 0 \\ 0 & 0 & 0 & 0 & 0 & 1 & 0 & 0 \\ 0 & 0 & 0 & 0 & 0 & 0 & 1 & 0 \\ 0 & 0 & 0 & 0 & 0 & 0 & 0 & 1 \end{bmatrix},\qquad(29)$$

where $\gamma_0 = \dfrac{\Omega_{1c}^2}{\Omega_{1c}^2 + \Omega_{3c}^2}\cos(\sqrt{65}\pi) + \dfrac{\Omega_{3c}^2}{\Omega_{1c}^2 + \Omega_{3c}^2} = 0.9997$. To carry out the Grover search, we define the three-qubit Hadamard gate

$$H^{\otimes 3} = \prod_{i=1}^{3} H_i = \left(\frac{1}{\sqrt{2}}\right)^3 \begin{bmatrix} 1 & 1 \\ 1 & -1 \end{bmatrix} \otimes \begin{bmatrix} 1 & 1 \\ 1 & -1 \end{bmatrix} \otimes \begin{bmatrix} 1 & 1 \\ 1 & -1 \end{bmatrix},\qquad(30)$$

where H_i is the Hadamard gate acting on the ith atom, transforming states as $|e_1\rangle \to (1/\sqrt{2})(|e_1\rangle + |g_1\rangle)$, $|g_1\rangle \to (1/\sqrt{2})(|e_1\rangle - |g_1\rangle)$, $|i_2\rangle \to (1/\sqrt{2})(|i_2\rangle + |g_2\rangle)$, $|g_2\rangle \to (1/\sqrt{2})(|i_2\rangle - |g_2\rangle)$, $|i_3\rangle \to (1/\sqrt{2})(|i_3\rangle + |g_3\rangle)$, $|g_3\rangle \to (1/\sqrt{2})(|i_3\rangle - |g_3\rangle)$. These gatings could be performed by external microwave pulses.

It is easy to find that the transformation $Q = H^{\otimes 3} I_{000} H^{\otimes 3} I_\tau = H^{\otimes 3} I_{e_1 i_2 i_3} H^{\otimes 3} I_\tau = -\hat{D} I_\tau$, which implies that the diffusion transform $\hat{D}$ is always unchanged, no matter which state is to be searched. The only change is the phase gate I_τ for different marked state. Based on Eq. (29), i.e., the gate I_{000} to mark the state $|e_1 i_2 i_3\rangle$, we could construct other seven gates for the marking job: $I_{e_1 i_2 g_3} = I_{001} = \sigma_{x,3} I_{000} \sigma_{x,3}$, $I_{e_1 g_2 i_3} =$

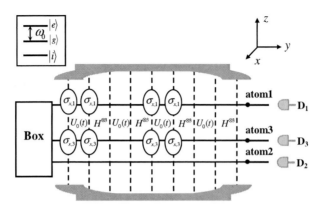

Figure 4. Schematic setup for finding the marked state $|101\rangle$ in a three-qubit Grover search. Three atoms initially prepared in the superposition state $|\Psi_0\rangle$ go through the cavity with the identical velocity from the box B. We send the atom 3 through the center of the microwave cavity along the y axis and other two atoms away from the y axis. We consider twice searching steps in the setup, which yields the largest success rate in the presence of dissipation. The operations $H^{\otimes 3}$, $\sigma_{x,1}$, $\sigma_{x,3}$ and $U_0(t)$ are defined in the text. Only in the case that the marked state is $|011\rangle$ or $|111\rangle$, should additional imhomogeneous electric fields be applied on the regions for single-qubit operation to distinguish the atoms 2 and 3.

$I_{010} = \sigma_{x,2}I_{000}\sigma_{x,2}$, $I_{e_1g_2g_3} = I_{011} = \sigma_{x,3}\sigma_{x,2}I_{000}\sigma_{x,2}\sigma_{x,3}$, $I_{g_1i_2i_3} = I_{100} = \sigma_{x,1}I_{000}\sigma_{x,1}$, $I_{g_1i_2g_3} = I_{101} = \sigma_{x,3}\sigma_{x,1}I_{000}\sigma_{x,1}\sigma_{x,3}$, $I_{g_1g_2i_3} = I_{110} = \sigma_{x,2}\sigma_{x,1}I_{000}\sigma_{x,1}\sigma_{x,2}$, and $I_{g_1g_2g_3} = I_{111} = \sigma_{x,3}\sigma_{x,2}\sigma_{x,1}I_{000}\sigma_{x,1}\sigma_{x,2}\sigma_{x,3}$. Moreover, by combining two Hadamard gates $H^{\otimes 3}$ with the quantum phase gate I_{000}, we can generate a three-qubit diffusion transform $\hat{D}$. So a full Grover search for three qubits is available.

Taking the marked state $|101\rangle$ as an example, we design a three-qubit Grover search setup in Fig. 4. The cavity is a microwave cavity sustaining a single mode with a standing-wave pattern along the z-axis. The atoms 1, 2 and 3 prepared in high-lying circular Reyberg states are sent through the cavity with appropriate speed, resonantly interacting with the cavity mode. Single-qubit rotations are made at certain times by external microwave pulses, and the state-selective field-ionization detectors D_1, D_2, D_3 are settled at the end of the passage for checking the states of the atoms 1, 2 and 3, respectively. One point to mention is that, in searching the state $|011\rangle$ or $|111\rangle$, imhomogeneous electric fields are needed to tune the atomic transitions through the Stark effect [26], which make the single-qubit operations completed individually. But these imhomogeneous electric fields are unnecessary in searching other states.

As the resonant interaction actually excites the cavity mode, although we could carry out the scheme very fast, we should consider the cavity decay seriously. Under the assumption of weak cavity decay that no photon actually leaks out of the microwave cavity during our implementation time, we employ the quantum trajectory method from the Hamiltonian,

$$H = \sum_{j=1}^{3} \Omega_{j_c}(a^\dagger S_j^- + a S_j^\dagger) - i\frac{\kappa}{2}a^\dagger a, \tag{31}$$

where κ is the cavity decay rate. As shown in Eq. (23), under the weak decay condition, the cavity dissipation only affects the diagonal elements of the matrix for the phase gate. For example, by choosing the interaction time $t_\tau = \pi/A_{1\kappa}$ with $A_{1\kappa} = \sqrt{\Omega_{1c}^2 - \kappa^2/16}$ and the condition $\Omega_{1c} : \Omega_{2c} : \Omega_{3c} = 1 : \sqrt{35} : 8$, we generate the three-qubit phase gate I'_{000} in the decay case,

$$I'_{e_1 i_2 i_3} = \begin{bmatrix} -\mu_1 & 0 & 0 & 0 & 0 & 0 & 0 & 0 \\ 0 & \gamma_1 & 0 & 0 & 0 & 0 & 0 & 0 \\ 0 & 0 & \beta_1 & 0 & 0 & 0 & 0 & 0 \\ 0 & 0 & 0 & \alpha_1 & 0 & 0 & 0 & 0 \\ 0 & 0 & 0 & 0 & 1 & 0 & 0 & 0 \\ 0 & 0 & 0 & 0 & 0 & 1 & 0 & 0 \\ 0 & 0 & 0 & 0 & 0 & 0 & 1 & 0 \\ 0 & 0 & 0 & 0 & 0 & 0 & 0 & 1 \end{bmatrix} = U_0(t), \tag{32}$$

where $\alpha_1 = 1 - \frac{\Omega_{1c}^2}{\Omega_{1c}^2 + \Omega_{2c}^2 + \Omega_{3c}^2}(1 - e^{-\kappa t/4})$, $\beta_1 = 1 - \frac{\Omega_{1c}^2}{\Omega_{1c}^2 + \Omega_{2c}^2}(1 - e^{-\kappa t/4})$, $\delta_1 = e^{-\kappa t/4}$, and $\gamma_1 = 1 - \frac{\Omega_{1c}^2}{\Omega_{1c}^2 + \Omega_{3c}^2}[1 - e^{-\kappa t/4}\cos(\sqrt{65}\pi)]$ after the negligible term $\frac{\kappa}{4A_{1\kappa}}\sin(A_{1\kappa})$ is omitted. So for a state $|\Psi\rangle = \frac{1}{2\sqrt{2}}(\bar{A}_j |e_1 i_2 i_3\rangle + B_j |e_1 i_2 g_3\rangle + C_j |e_1 g_2 i_3\rangle + D_j |e_1 g_2 g_3\rangle + E_j |g_1 i_2 i_3\rangle + F_j |g_1 i_2 g_3\rangle + G_j |g_1 g_2 i_3\rangle + H_j |g_1 g_2 g_3\rangle)$, the success probability of the phase gate is defined as

$$P_j = (|D_j|^2 \alpha_i^2 + |C_j|^2 \beta_i^2 + |B_j|^2 \gamma_i^2 + |\bar{A}_j|^2 \mu_i^2 + |E_j|^2 + |F_j|^2 + |G_j|^2 + |H_j|^2)/8, \tag{33}$$

where $j = 0, 1$ correspond to the ideal and decay cases, respectively, with $\alpha_0 = \beta_0 = \mu_0 = 1$. In our case, the atomic sysytem is initially prepared to be $|\Psi_0\rangle = \frac{1}{2\sqrt{2}}(|g_1\rangle + |e_1\rangle)(|g_2\rangle + |i_2\rangle)(|g_3\rangle + |i_3\rangle)$, which corresponds to a success probability of the three-qubit phase gate $P_j = (4 + \alpha_j^2 + \beta_j^2 + \gamma_j^2 + \mu_j^2)/8$.

As mentioned previously, the three-qubit Grover search is carried out only probabilistically. So how to obtain a high success rate of the search is the problem of much interest, particularly in the presence of weak cavity decay. We have numerically simulated the Grover search for finding different marked states in the cases of $\kappa = 0$ (the ideal case), $\kappa = \Omega_{1c}/50$, and $\kappa = \Omega_{1c}/10$. Due to the similarity, we only demonstrate the search for a marked state $|e_1\rangle |i_2\rangle |i_3\rangle$ in Fig. 5 as an example. The result in Fig. 5(a) looks a little bit amazing that the cavity decay helps finding the marked state because higher success probability could be reached with larger decay involved. In fact, this is resulted from the change of the weight of different components in the computational subspace. If we consider the success rate of the three-qubit phase gating (i.e., Eq. (33)), a complete treatment of the dissipative case, the success probability would be smaller and smaller with the increase of the decay rate and the iteration number, as shown in Fig. 5(b). So although the sixth iteration could reach the largest success rate in the ideal consideration, we prefer the second iteration in the presence of dissipation. The detrimental effect from the cavity decay is also reflected in the estimate of fidelity in Fig. 5(c).

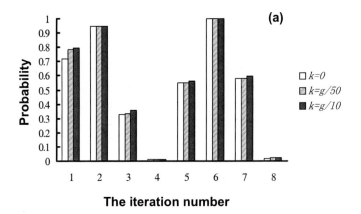

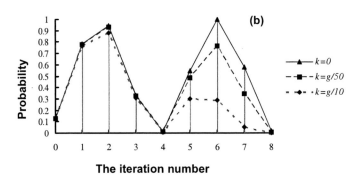

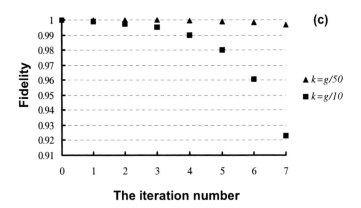

Figure 5. Numerical results for a three-qubit Grover search for the marked state $|e_1\rangle |i_2\rangle |i_3\rangle$, where $k = \kappa$, $g = \Omega_{1c}$. (a) Probability for finding the marked state without considering the success probability of the quantum phase gate (i.e., Eq. (33)); (b) Probability for finding the marked state involving the success probability of the quantum phase gate; (c) Fidelity of the searched state in the case of $\kappa = \Omega_{1c}/50$ and $\Omega_{1c}/10$.

We briefly address the experimental feasibility of our scheme with current microwave cavity technology [6] by considering three Rydberg atoms with principal quantum numbers 49, 50 and 51 to be levels $|i\rangle$, $|g\rangle$ and $|e\rangle$, respectively. Based on the experimental numbers reported [17], the coupling strength at the cavity center could be $\Omega_0 = 2\pi \times 49 kHz$, and the Rydberg atomic lifetime is 30 ms. Since the single-qubit operation takes negligible time in comparison with that for the three-qubit phase gating, an iteration of our proposed Grover search would take $t_0 = 2\pi/\sqrt{\Omega_{1c}^2 - \kappa^2/16}$. Direct calculation shows that the time for one iteration is about $160\mu s$, much shorter than the cavity decay time for both cases of $\kappa = \Omega_{1c}/50$, and $\Omega_{1c}/10$. So our treatment with quantum trajectory method is physically reasonable. We have also noticed that the three atoms going through a microwave cavity have not yet been achieved experimentally, and the two-atom entanglement in a microwave cavity was done by using van der Waals collision between the atoms [31]. However, compared to [41] with four-level atoms sent through a three-mode cavity , our proposal involving a single-mode cavity is much simpler and is closer to the reach with the current cavity QED technology. Considering the intra-atom interaction occurs in the central region of the cavity, we have $\Omega_{jc} \simeq \Omega_0 \cos(2\pi z/\lambda_0)$. So the three atoms should be sent through the cavity with the atom 3 going along the y-axis ($x_3=z_3=0$) and atoms 1 and 2 away from the atom 3 by $|z_1|/|z_2| = \arccos(0.125)/\arccos(\sqrt{35}/8) \approx 1.957$. With the manipulation designed in Fig. 4, a three-qubit Grover search for the marked state $|101\rangle$ is achievable.

We have noticed that four-qubit Grover search with linear optical elements has been achieved [43]. While as the photons are always flying, they are actually unsuitable for a practical quantum computing. In contrast, the atoms could be localized if we consider an ion-trap-cavity combinatory setup [44] or employ optical lattices to confine the atoms [45], in which the model employed here also works. In this sense, our scheme could be straightforwardly applied to those two systems. For these considerations, however, the cavity should be optical one, for which we have to consider both the cavity decay and the atomic spontaneous emission. Based on a previous treatment [46], as long as these dissipations are weak, the three-qubit phase gating would also be available, and thereby our scheme is in principle workable in optical regime.

In an actual experiment, any imperfection, such as non-simultaneous movement of the atoms, diversity in atomic velocities, deflected atomic trajectories, classical pulse imperfection and so on, would affect our scheme. For the clarity and convenience of our discussion, let us take two examples to assess the influence from imperfection. First, as it is still a challenge to simultaneously send three Rydberg atoms through a cavity with precise velocities in experimental performance, we consider an imperfection in this respect. We simply consider a situation that the atom 1 moves a little bit slower than the atoms 2 and 3, i.e., the times of the atoms passing the cavity $t_1 = t_0 + \delta t$ and $t_2 = t_3 = t_0$, with t_0 the desired interaction time for the three-qubit phase gate I_{000}'. Direct calculation yields the infidelity to be

$$I = 1 - \frac{[4 + \xi\alpha_1 + \xi\beta_1 + \xi\mu_1 + \xi\gamma_1 - \Omega_{1c}^2/(A_{1\kappa}A_{3\kappa})\exp(-\kappa\delta t/4)\sin(A_{1\kappa}\delta t)\sin(\sqrt{65}\pi)]^2}{8[4 + \xi\alpha_1 + \xi\beta_1 + +\xi\mu_1 + (\xi\gamma_1 - \Omega_{1c}^2/(A_{1\kappa}A_{3\kappa})\exp(-\kappa\delta t/4)\sin(A_{1\kappa}\delta t)\sin(\sqrt{65}\pi))^2]},$$
(34)

where $\quad \xi \quad = \quad \exp(-\kappa\delta t/4)[\cos(A_{1\kappa}\delta t) \quad + \quad \frac{\kappa}{4A_{1\kappa}}\sin(A_{1\kappa}\delta t)], \quad A_{3\kappa} \quad =$

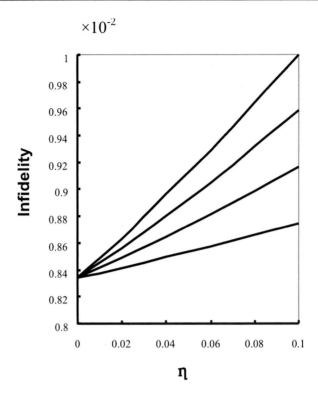

Figure 6. Infidelity in a Grover search versus offset constant η, where the four solid curves from bottom to top correspond to the number of imperfect cavities varying from 1 to 4 in the case of $\kappa = \Omega_{1c}/10$.

$2\pi/\sqrt{\Omega_{1c}^2 + \Omega_{3c}^2 - \kappa^2/16}$, and $A_{1\kappa}$ is defined before under Eq. (31). Secondly, With current cavity QED techniques, the design in Fig. 4 should be realized by four separate microwave cavities with each Ramsey zone located by a cavity. Since each microwave cavity is employed to carry out a three-qubit phase gate I_{000}, the four cavities should be identical. Here we consider the unfavorable influence from the coupling strength Ω'_{jc} in some cavities with the offset $\eta\Omega_{jc}$ from the ideal number. We find the infidelity due to these offsets for a Grover search implementation to be $I' = 1 - \frac{(4+\alpha'_\chi+\beta'_\chi+\gamma'_\chi+\mu'_\chi)^2}{8(4+\alpha_\chi'^2+\beta_\chi'^2+\gamma_\chi'^2+\mu_\chi'^2)}$, where

$\alpha'_\chi = [1 - \frac{\Omega_{1c}'^2}{\Omega_{1c}'^2+\Omega_{2c}'^2+\Omega_{3c}'^2}(1 - e^{-\kappa t_0/4})]^\chi \alpha_1^{4-\chi}$, $\beta'_\chi = [1 - \frac{\Omega_{1c}'^2}{\Omega_{1c}'^2+\Omega_{2c}'^2}(1 - e^{-\kappa t_0/4})]^\chi \beta_1^{4-\chi}$,

$\gamma'_\chi = \{1 - \frac{\Omega_{1c}'^2}{\Omega_{1c}'^2+\Omega_{3c}'^2}[1 - e^{-\kappa t_0/4}\cos(\sqrt{65}\pi)]\}^\chi \gamma_1^{4-\chi}$, and $\mu'_\chi = e^{-\chi\kappa t_0/4}\mu_1^{4-\chi}$, with χ ($= 1, 2, 3, 4$) the number of the cavities with the coupling strength offsets. We plot the dependence of the infidelity on different η and χ in the case of $\kappa = \Omega_{1c}/10$ in Fig. 7. As shown in Figs. 6 and 7, it implies an enlarging infidelity with respect to the time difference δt, the decay rate κ, and the offset constant η. So to carry out our scheme efficiently and with high fidelity, we have to suppress the imperfect factors to the minimum.

In summary, we have analytically demonstrated the preparation of W state, implementation of Toffoli gate and accomplishment of Grover search in a single-mode cavity under weak cavity decay. As the detrimental influence from the dissipation could be explicitly presented in analytical expressions, our treatment would be helpful for understanding physics in real implementation of experiments.

3.   QIP by Geometric Phase Gating

As discussed in section II, decoherence severely damages quantum process and thereby results in imperfection in quantum gating. To accomplish a quantum gate as well as we could, we generally employ the idea of decoherence-free subspace [47] or the geometric phase [48, 49, 50, 51, 52, 53, 54, 55] for achieving built-in fault-tolerant quantum gates. They are different from quantum logic gates [56, 57, 58, 59] by dynamical evolution, which are very sensitive to the parameter fluctuations in the operation. Geometric phase gates are classified into two kinds: the conventional geometric quantum gates (GQG) [48, 49, 50, 51, 52, 53, 54] and the unconventional ones [55, 32]. Although the unconventional GQG [55] also creates the geometric phase with a nonzero dynamical one, it is unnecessary to eliminate the dynamic phase because the dynamic phase is proportional to the geometric phase by a constant independent of the parameters regarding the qubit subspace. We have noticed some quantum computing schemes based on the ideas of conventional GQGs by using super-conducting nanocircuits [49], NMR [51], semiconductor nanstructure [54], and trapped ions [50]. These schemes are focused on the construction of GQGs, without specific consideration about the influence from dissipation. Recently, geometric quantum computation in the presence of dissipation or decoherence has been attracted some attention, where Pachos and Walther [32] addressed specifically quantum computation with trapped $^{40}Ca^+$ ions in an optical cavity by employing adiabatic transitions and the quantum Zeno effect, involving the consideration of ionic spontaneous emission and the cavity decay. Cen and Zanardi [60] proposed double loop scheme (i.e., refocusing scheme) to get rid of the negative influence of dissipation in the no-jump trajectory and to realize the geometric quantum computation. Fuentes-Guridi *et al* [61] studied systematically non-Abelian adiabatic holonomies for an open Markovian system and analyzed the robustness of holonomic quantum computation against decoherence. Carollo *et al* [62] have calculated the geometric phases associated with the evolution of an open system subject to decoherence by a quantum-jump approach, and obtained some approximate results. Nevertheless, there has been actually no experiment achieved so far for the conventional GQGs. In contrast, for the unconventional GQGs, besides the theoretical proposals [55, 63], an experiment has been done with trapped ions [64].

In this section, we will develop a general displacement operator to construct the unconventional GQG in a system with many identical three-level atoms confined in a cavity under decay, driven by a classical field. We will analytically investigate the influence from the cavity decay on the unconventional geometric quantum computation. Our scheme is a further study of Ref. [65], which is based on a large-detuning interaction of the atoms with an ideal cavity mode. We will present conditions for a high-fidelity unconventional GQG, which would be helpful for relevant experimental work.

The section will be separated into several parts. We first define a general displacement operator and the corresponding unconventional geometric phase shift, which are actually extension of the definitions in Ref. [66]. The shift is caused by the general displacement operator along an approximately closed path in the phase space. Then the dynamics of many identical three-level atoms is investigated in the interaction with a quantized cavity mode, driven by a laser field. Additional phase shift and the corresponding amplitude damping factor due to general displacement operator acting on the vacuum state are presented in a separate part. As an example, we will carry out a non-ideal two-qubit unconventional GQG and make some discussion based on experimental feasibility.

3.1. Definitions of a General Displacement Operator and the Corresponding Unconventional Geometric Phases

We first introduce the definitions of a general displacement operator and the corresponding unconventional geometric phases, which are the generalization of the standard displacement operator and unconventional geometric phase shift, respectively. The general displacement operator is defined as follows

$$D_g[f_1(t), f_2(t)] = e^{f_1(t)a^\dagger - f_2(t)a}, \tag{35}$$

where a and $a^\dagger$ are annihilation and creation operators of a harmonic oscillator (for example, the cavity mode in the case of a cavity QED system), respectively. $f_1(t)$ and $f_2(t)$ are time-dependent parameters in this work. When $f_1(t) = f_2^*(t)$, Eq.(35) reduces to the standard definition of the displacement, and if $f_1(t) = f_2(-t)$, we return to the displacement-like operator in [65]. Since the general displacement operator satisfies

$$D_g[f_1(t_2), f_2(t_2)]D_g[f_1(t_1), f_2(t_1)] =$$
$$D_g[f_1(t_1) + f_1(t_2), f_2(t_1) + f_2(t_2)]e^{\frac{1}{2}[f_1(t_2)f_2(t_1) - f_1(t_1)f_2(t_2)]}, \tag{36}$$

we have the total operator for a path consisting of N short straight sections Δf_j

$$D_{gt}^L = D_g^L[\Delta f_1(t_N), \Delta f_2(t_N)] \cdots D_g^L[\Delta f_1(t_1), \Delta f_2(t_1)]$$
$$= D_g^L[\sum_{j=1}^N \Delta f_1(t_j), \sum_{j=1}^N \Delta f_2(t_j)]$$
$$\exp\{\frac{1}{2}[\sum_{j=2}^N \Delta f_1(t_j) \sum_{k=1}^{j-1} \Delta f_2(t_k) - \sum_{j=2}^N \Delta f_2(t_j) \sum_{k=1}^{j-1} \Delta f_1(t_k)]\}. \tag{37}$$

In the limit $N \to \infty$ for an arbitrary path γ and for an initial Fock state $|n\rangle$, we have

$$D_{gt}^L = D_g[(\int_\gamma df_1(t)), (\int_\gamma df_2(t))]e^{-A_{g\kappa 1}}e^{i\Theta_{g\kappa 1}}, \tag{38}$$

where

$$\Theta_{g\kappa 1} = \frac{1}{2}\text{Im}[\int_\gamma f_2(t)df_1(t) - \int_\gamma f_1(t)df_2(t)], \tag{39}$$

and

$$A_{g\kappa 1} = -\frac{1}{2}\text{Re}[\int_\gamma f_2(t)df_1(t) - \int_\gamma f_1(t)df_2(t)], \tag{40}$$

where $\Theta_{g\kappa 1}$ is the definition of a general unconventional geometric phase and $A_{g\kappa 1}$ is for amplitude damping.

For a closed path we have

$$D_{gt}^L = D_g^L[\oint df_1, \oint df_2]e^{-A_{g\kappa 1}}e^{i\Theta_{g\kappa 1}},$$

$$\Theta_{g\kappa 1} = \frac{1}{2}\text{Im}[\oint f_2(t)df_1(t) - \oint f_1(t)df_2(t)],$$

$$A_{g\kappa 1} = -\frac{1}{2}\text{Re}[\oint f_2(t)df_1(t) - \oint f_1(t)df_2(t)], \tag{41}$$

where $\oint df_1, \oint df_2 \neq 0$, and thereby $D_g^L[\oint df_1, \oint df_2] \neq 1$ in the case of dissipation, which is different from the ideal case. Eqs.(38)-(41) will be used below for constructing unconventional GQGs under dissipation.

3.2.   Dynamic Model of a Cavity QED System Assisted by a Driving Field

We consider N identical three-level atoms, each of which has one excited state $|i\rangle$ and two ground states $|e\rangle$ and $|g\rangle$. The qubits are encoded in the states $|e\rangle$ and $|g\rangle$, and the state $|i\rangle$ is an auxiliary state. The two levels $|e\rangle$ and $|i\rangle$ couple to the cavity mode with the coupling constant g and detuning $\Delta = \omega_0 - \omega_c$, assisted by a classical laser field with Rabi frequency Ω and detuning $\Delta - \delta = \omega_0 - \omega_L$, where $\delta \ll \Delta$ and $\delta = \omega_L - \omega_c$. As $|g\rangle$ is not involved in the interaction, under the consideration of the cavity decay and under the condition that no photon leaking out of the cavity is detected, the Hamiltonian can be expressed as (assuming $\hbar = 1$)

$$H_s = \omega_0 \sum_{j=1}^{N} S_{z,j} + \omega_c a^\dagger a + g\sum_{j=1}^{N}(a^\dagger S_j^- + aS_j^\dagger)$$
$$+ \Omega \sum_{j=1}^{N}(e^{-i\omega_L t}S_j^\dagger + e^{i\omega_L t}S_j^-) - i\frac{\kappa}{2}a^\dagger a, \tag{42}$$

where $\omega_0, \omega_c,$and ω_L are the frequencies of the resonant transition between $|e\rangle$ and $|i\rangle$, the cavity mode, and the classical laser field, respectively, with $S_{z,j} = \frac{1}{2}(|i\rangle\langle i| - |e\rangle\langle e|), S_j^\dagger = |i\rangle\langle e|$, and $S_j^- = |e\rangle\langle i|$, $a^\dagger$ and a are the creation and annihilation operators of the cavity mode, respectively, and κ is the cavity decay rate. In the rotating frame with respect to the cavity frequency ω_c, making a transformation

$$u = u_3 u_2 u_1 = e^{-\frac{\kappa}{2}a^\dagger a t}e^{-i\omega_c \sum_{j=1}^{N} S_{z,j}t}e^{-i\omega_c a^\dagger a t}, \tag{43}$$

and by using $H_c = u^{-1}H_s u - u^{-1}(i\frac{\partial}{\partial t}u)$, we obtain

$$H_c = \sum_{j=1}^{N}\Delta S_{z,j} + \sum_{j=1}^{N}(gae^{-\kappa t/2} + \Omega e^{-i\delta t})S_j^\dagger + \sum_{j=1}^{N}(ga^\dagger e^{\kappa t/2} + \Omega e^{i\delta t})S_j^-. \tag{44}$$

In the case that $\Delta \gg \Omega, g$, and $\kappa t \ll \delta t$ (to make sure that no quantum jump happens during the time evolution under our consideration), the Hamiltonian of Eq. (44) can be replaced by an effective Hamiltonian,

$$H_{ceff} = \sum_{j=1}^{N} \frac{1}{\Delta} \{ [g^2 a^\dagger a + \Omega^2 + \Omega g a e^{-\kappa t/2} e^{i\delta t} + \Omega g a^\dagger e^{\kappa t/2} e^{-i\delta t}]$$

$$\times (|i_j\rangle \langle i_j| - |e_j\rangle \langle e_j|)] + g^2 |i_j\rangle \langle i_j| \} + \frac{1}{\Delta} g^2 \sum_{i,j,i\neq j}^{N} (S_i^\dagger S_j^- + S_j^\dagger S_i^-). \quad (45)$$

For further unitary transformation, we assume $H_{ceff} = H_{ce0} + H_{cei}$, where

$$H_{ce0} = \sum_{j=1}^{N} \frac{1}{\Delta} \{ (g^2 a^\dagger a + \Omega^2)(|i_j\rangle \langle i_j| - |e_j\rangle \langle e_j|) + g^2 |i_j\rangle \langle i_j| \}, \quad (46)$$

$$H_{cei} = \sum_{j=1}^{N} \frac{1}{\Delta} \{ [\Omega g a e^{-\kappa t/2} e^{i\delta t} + \Omega g a^\dagger e^{\kappa t/2} e^{-i\delta t}](|i_j\rangle \langle i_j| - |e_j\rangle \langle e_j|)]$$

$$+ \frac{g^2}{\Delta} \sum_{i,j,i\neq j}^{N} (S_i^\dagger S_j^- + S_j^\dagger S_i^-). \quad (47)$$

Performing the unitary transformation $|\psi(t)\rangle = e^{-iH_{ce0}t} |\psi'(t)\rangle$, from the Schrödinger equations $id |\psi(t)\rangle /dt = H_{cei} |\psi(t)\rangle$ and $id |\psi'(t)\rangle /dt = H'_{cei} |\psi'(t)\rangle$, we obtain

$$H'_{cei} = e^{iH_{ce0}t} H_{cei} e^{-iH_{ce0}t}$$

$$= \sum_{j=1}^{N} \frac{\Omega g}{\Delta} \{ [a e^{i(\delta+\frac{g^2}{\Delta})t} e^{-\kappa t/2} + a^\dagger e^{-i(\delta+\frac{g^2}{\Delta})t} e^{\kappa t/2}] |i_j\rangle \langle i_j|$$

$$- [a e^{i(\delta+\frac{g^2}{\Delta})t} e^{-\kappa t/2} + a^\dagger e^{-i(\delta+\frac{g^2}{\Delta})t} e^{\kappa t/2}] |e_j\rangle \langle e_j| \}$$

$$+ \frac{g^2}{\Delta} \sum_{i,j,i\neq j}^{N} (S_i^\dagger S_j^- + S_j^\dagger S_i^-), \quad (48)$$

in which the only term that contributes an unconventional geometric phase shift to the evolution of the encoded qubit states $|g_j\rangle$ and $|e_j\rangle$ is

$$H'_{ceicg} = -\sum_{j=1}^{N} \frac{\Omega g}{\Delta} [a e^{i(\delta+\frac{g^2}{\Delta})t} e^{-\kappa t/2} + a^\dagger e^{-i(\delta+\frac{g^2}{\Delta})t} e^{\kappa t/2}] |e_j\rangle \langle e_j|. \quad (49)$$

According to the definition of the general displacement operator, during the infinitesimal interval $[t, t + dt]$, the corresponding evolution in our qubit subspace is decided by

$$U(dt) = D_{gt}^{L}[df_1(t), df_2(t)] = e^{-iH'_{ceicg}dt}, \quad (50)$$

where

$$df_1(t) = i\frac{\Omega g}{\Delta} e^{-i(\delta+\frac{g^2}{\Delta})t} e^{\kappa t/2} dt, \quad (51)$$

and

$$df_2(t) = -i\frac{\Omega g}{\Delta} e^{i(\delta+\frac{g^2}{\Delta})t} e^{-\kappa t/2} dt. \quad (52)$$

Therefore, the qubit states including a single state $|e_j\rangle$ will evolve according to $D^L_{gt}[df_1, df_2]$, and the one with two $|e_j\rangle$ will evolve according to $D^L_{gt}[2df_1, 2df_2]$. Other states will remain unchanged. For simplicity, we assume from now on the cavity mode to be initially in a vacuum state $|0\rangle$. So after an interaction time t', the general displacement parameters f_1 and f_2 can be, respectively, expressed as

$$
\begin{aligned}
f_1(t') &= i\frac{\Omega g}{\Delta}\int_0^{t'} e^{-i(\delta+\frac{g^2}{\Delta})t} e^{\kappa t/2} dt \\
&= \frac{\Omega g}{\Delta}\frac{(\delta+g^2/\Delta)-i\kappa/2}{(\delta+g^2/\Delta)^2+\kappa^2/4}[1-e^{-i(\delta+\frac{g^2}{\Delta})t'} e^{\kappa t'/2}],
\end{aligned}
\tag{53}
$$

and

$$
\begin{aligned}
f_2(t') &= -i\frac{\Omega g}{\Delta}\int_0^{t'} e^{i(\delta+\frac{g^2}{\Delta})t} e^{-\kappa t/2} dt \\
&= \frac{\Omega g}{\Delta}\frac{(\delta+g^2/\Delta)-i\kappa/2}{(\delta+g^2/\Delta)^2+\kappa^2/4}[1-e^{i(\delta+\frac{g^2}{\Delta})t'} e^{-\kappa t'/2}].
\end{aligned}
\tag{54}
$$

Then the geometric phase shifts $\Theta_{g\kappa 1}$ and $\Theta'_{g\kappa 1}$, regarding $|e_j\rangle|g_k\rangle$ $(|g_j\rangle|e_k\rangle)$ and $|e_j\rangle|e_k\rangle$ $(j \neq k)$, respectively, are

$$
\begin{aligned}
\Theta_{g\kappa 1} &= \frac{1}{2}\,\mathrm{Im}[\int_\gamma f_2(t)df_1(t) - \int_\gamma f_1(t)df_2(t)] \\
&= (\frac{\Omega g}{\Delta})^2\frac{1}{[(\delta+g^2/\Delta)^2+\kappa^2/4]^2}\times \\
&\quad \{[(\delta+g^2/\Delta)^2-\kappa^2/4]\sin[(\delta+g^2/\Delta)t]\cosh(\frac{\kappa t}{2}) \\
&\quad + \kappa(\delta+g^2/\Delta)\cos[(\delta+g^2/\Delta)t]\sinh(\frac{\kappa t}{2}) \\
&\quad + [(\delta+g^2/\Delta)^2+\kappa^2/4](\delta+g^2/\Delta)t\},
\end{aligned}
\tag{55}
$$

and

$$
\Theta'_{g\kappa 1} = \frac{1}{2}\,\mathrm{Im}[\int_\gamma 2f_2(t)d2f_1(t) - \int_\gamma 2f_1(t)d2f_2(t)] = 4\Theta_{g\kappa 1}.
\tag{56}
$$

Moreover, we have terms $A_{g\kappa 1}$ and $A'_{g\kappa 1}$ related to the decay rates, respectively,

$$
\begin{aligned}
A_{g\kappa 1} &= -\frac{1}{2}\,\mathrm{Re}[\int_\gamma f_2(t)df_1(t) - \int_\gamma f_1(t)df_2(t)] \\
&= (\frac{\Omega g}{\Delta})^2\frac{1}{[(\delta+g^2/\Delta)^2+\kappa^2/4]^2}\times \\
&\quad \{[(\delta+g^2/\Delta)^2-\kappa^2/4]\cos[(\delta+g^2/\Delta)t]\sinh(\frac{\kappa t}{2}) \\
&\quad - \kappa(\delta+g^2/\Delta)\sin[(\delta+g^2/\Delta)t]\cosh(\frac{\kappa t}{2}) \\
&\quad + [(\delta+g^2/\Delta)^2+\kappa^2/4]\frac{\kappa t}{2}\},
\end{aligned}
\tag{57}
$$

and

$$A'_{g\kappa 1} = -\frac{1}{2}\text{Re}[\int_\gamma 2f_2(t)d2f_1(t) - \int_\gamma 2f_1(t)d2f_2(t)] = 4A_{g\kappa 1}. \tag{58}$$

It is evident that the cavity decay κ results in impossibility of an exactly closed path movement of the general displacement in the parameter phase space. So we will study in the next section an approximately closed path in a geometric phase gating.

3.3. Additional Phase Shift from the General Displacement Operator

In this subsection, we consider the additional phase shift $\Theta_{g\kappa 2}$ and the corresponding amplitude damping factor $A_{g\kappa 2}$ originated from the general displacement operator acting on a cavity vacuum state. We have

$$D_g[f_1(t), f_2(t)]\,|0\rangle = e^{f_1(t)a^\dagger - f_2(t)a}\,|0\rangle = e^{-\frac{1}{2}f_1(t)[f_2(t)-f_1^*(t)]}\,|f_1(t)\rangle, \tag{59}$$

where $|f_1(t)\rangle$ represents a usual coherent state. It is noted that the prefactor $e^{-\frac{1}{2}f_1(t)[f_2(t)-f_1^*(t)]}$ contributes to the phase shift and the amplitude damping, which is different from that case without dissipation. In our model, the prefactor can be expressed as

$$e^{-\frac{1}{2}f_1(t)[f_2(t)-f_1^*(t)]} = e^{-A_{g\kappa 2}}\cdot e^{i\Theta_{g\kappa 2}}, \tag{60}$$

where

$$\begin{aligned}
\Theta_{g\kappa 2} &= -\frac{1}{2}\text{Im}\,\{f_1(t)[f_2(t)-f_1^*(t)]\} \\
&= (\frac{\Omega g}{\Delta})^2 \frac{1}{[(\delta + g^2/\Delta)^2 + \kappa^2/4]^2} \times \\
&\quad \{[\frac{\kappa^2}{4} - (\delta + g^2/\Delta)^2]\sin[(\delta + g^2/\Delta)t]\sinh(\frac{\kappa t}{2}) \\
&\quad + \kappa(\delta + g^2/\Delta)[1 - \cos[(\delta + g^2/\Delta)t]\cosh(\frac{\kappa t}{2})]\},
\end{aligned} \tag{61}$$

and

$$\begin{aligned}
A_{g\kappa 2} &= \frac{1}{2}\text{Re}\{f_1(t)[f_2(t)-f_1^*(t)]\} \\
&= (\frac{\Omega g}{\Delta})^2 \frac{1}{[(\delta + g^2/\Delta)^2 + \kappa^2/4]^2} \times \\
&\quad \{[(\delta + g^2/\Delta)^2 + \frac{\kappa^2}{4}]e^{\frac{\kappa t}{2}}[\cos[(\delta + g^2/\Delta)t] - \cosh(\frac{\kappa t}{2})] \\
&\quad + \kappa(\delta + g^2/\Delta)\sin[(\delta + g^2/\Delta)t]\sinh(\frac{\kappa t}{2}) \\
&\quad + [(\delta + g^2/\Delta)^2 - \kappa^2/4][1 - \cos[(\delta + g^2/\Delta)t]\cosh(\frac{\kappa t}{2})]\}.
\end{aligned} \tag{62}$$

Then a combination of Eqs. (55)-(58) and Eqs. (61)-(62) would construct the two-qubit unconventional geometric phase gate as below.

3.4.   The Influence of the Cavity Mode Decay on Two-Qubit Unconventional GQG

In the case that $N = 2$ and the cavity is initially in the vacuum state $|0\rangle$, the required condition for the system periodically evolving along an approximately closed path is

$$(\delta + g^2/\Delta)t = 2m\pi, \tag{63}$$

where m is a positive integer. Therefore, we give below the evolution of the system with similar steps to [65],

$$|g\rangle_1 |g_2\rangle |0\rangle \rightarrow |g\rangle_1 |g_2\rangle |0\rangle \rightarrow |g\rangle_1 |g_2\rangle |0\rangle \,,$$

$$|g\rangle_1 |e\rangle_2 |0\rangle \rightarrow e^{-i\Omega^2 t/\Delta} e^{-A_{g\kappa 1}} e^{i\Theta_{g\kappa 1}} D_g^L [\oint df_1, \oint df_2] |g\rangle_1 |e\rangle_2 |0\rangle$$

$$\rightarrow e^{-A_{g\kappa}} e^{i(\Theta_{g\kappa} - \Omega^2 t/\Delta)} |g\rangle_1 |e\rangle_2 \left| f_1(\frac{2m\pi}{\delta + g^2/\Delta}) \right\rangle,$$

$$|e\rangle_1 |g\rangle_2 |0\rangle \rightarrow e^{-i\Omega^2 t/\Delta} e^{-A_{g\kappa 1}} e^{i\Theta_{g\kappa 1}} D_g^L [\oint df_1, \oint df_2] |e\rangle_1 |g\rangle_2 |0\rangle$$

$$\rightarrow e^{-A_{g\kappa}} e^{i(\Theta_{g\kappa} - \Omega^2 t/\Delta)} |e\rangle_1 |g\rangle_2 \left| f_1(\frac{2m\pi}{\delta + g^2/\Delta}) \right\rangle,$$

$$|e\rangle_1 |e\rangle_2 |0\rangle \rightarrow e^{-i2\Omega^2 t/\Delta} e^{-A'_{g\kappa 1}} e^{i4\Theta_{g\kappa 1}} D_g^L [2\oint df_1, 2\oint df_2] |e\rangle_1 |e\rangle_2 |0\rangle \tag{64}$$

$$\rightarrow e^{-A'_{g\kappa 2}} e^{i(\Theta'_{g\kappa} - 2\Omega^2 t/\Delta)} |e\rangle_1 |e\rangle_2 \left| 2f_1(\frac{2m\pi}{\delta + g^2/\Delta}) \right\rangle, \tag{65}$$

where

$$\Theta_{g\kappa} = \Theta_{g\kappa 1} + \Theta_{g\kappa 2}$$

$$= (\frac{\Omega g}{\Delta})^2 \frac{1}{[(\delta + g^2/\Delta)^2 + \kappa^2/4]^2} \times$$

$$\{[(\delta + g^2/\Delta)^2 - \frac{\kappa^2}{4}] e^{-\kappa t/2} \sin[(\delta + g^2/\Delta)t]$$

$$+ \kappa(\delta + g^2/\Delta) \cos[(\delta + g^2/\Delta)t] \sinh(\frac{\kappa t}{2})$$

$$+ [(\delta + g^2/\Delta)^2 + \frac{\kappa^2}{4}](\delta + g^2/\Delta)t$$

$$+ \kappa(\delta + g^2/\Delta)[1 - \cos[(\delta + g^2/\Delta)t] \cosh(\frac{\kappa t}{2})]\}, \tag{66}$$

$$\Theta'_{g\kappa} = 4\Theta_{g\kappa}, \tag{67}$$

and

$$A_{g\kappa} = A_{g\kappa 1} + A_{g\kappa 2} = (\frac{\Omega g}{\Delta})^2 \frac{1}{[(\delta + g^2/\Delta)^2 + \kappa^2/4]^2} \times$$
$$\{[(\delta + g^2/\Delta)^2 - \kappa^2/4][1 - e^{-\kappa t/2} \cos[(\delta + g^2/\Delta)t]]$$
$$- \kappa(\delta + g^2/\Delta)t \sin[(\delta + g^2/\Delta)t]e^{-\kappa t/2}$$
$$+ [(\delta + g^2/\Delta)^2 + \kappa^2/4][\frac{\kappa t}{2} - e^{\frac{\kappa t}{2}}[\cosh(\frac{\kappa t}{2}) - \cos(\delta + g^2/\Delta)t]]\}, \quad (68)$$

$$A'_{g\kappa} = 4A_{g\kappa}. \quad (69)$$

In the computational qubit subspace $\{|g\rangle_j, |e\rangle_j, j = 1, 2\}$, with the single qubit operation $|e\rangle_j \rightarrow e^{-i(\Theta_\kappa - \Omega^2 t/\Delta)}|e\rangle_j$ $(j = 1, 2)$, we obtain the $2\Theta_\kappa$-geometric phase gate

$$|g\rangle_1 |g\rangle_2 |0\rangle \rightarrow |g\rangle_1 |g\rangle_2 |0\rangle,$$
$$|g\rangle_1 |e\rangle_2 |0\rangle \rightarrow e^{-A_{g\kappa}} |g\rangle_1 |e\rangle_2 \left|f_1(\frac{2m\pi}{\delta + g^2/\Delta})\right\rangle,$$
$$|e\rangle_1 |g\rangle_2 |0\rangle \rightarrow e^{-A_{g\kappa}} |e\rangle_1 |g\rangle_2 \left|f_1(\frac{2m\pi}{\delta + g^2/\Delta})\right\rangle,$$
$$|e\rangle_1 |e\rangle_2 |0\rangle \rightarrow e^{-4A_{g\kappa}} e^{i2\Theta_{g\kappa}} |e\rangle_1 |e\rangle_2 \left|2f_1(\frac{2m\pi}{\delta + g^2/\Delta})\right\rangle, \quad (70)$$

where $2\Theta_{g\kappa} = \pi$ yields an approximate $\pi-$phase gate for the two atoms (qubits) under the influence from the cavity decay. Different from the ideal case in [67], Eq. (70) includes the amplitude damping factor as well as coherent states in the cavity mode after the gate operation. The required condition for this gating is

$$(\frac{\Omega g}{\Delta})^2 \frac{1}{[(\delta + g^2/\Delta)^2 + \kappa^2/4]^2} \{\kappa(\delta + g^2/\Delta)\sinh(\frac{\kappa m\pi}{\delta + g^2/\Delta})$$
$$+ 2m\pi[(\delta + g^2/\Delta)^2 + \frac{\kappa^2}{4}] + \kappa(\delta + g^2/\Delta)[1 - \cosh(\frac{\kappa m\pi}{\delta + g^2/\Delta})]\} = \frac{\pi}{2}, \quad (71)$$

and the gating time must satisfy both Eqs. (63) and (71).

From Eq. (70), we can construct an approximate two-qubit controlled-NOT gate in the computational subspace $\{|g\rangle_1 |g\rangle_2, |g\rangle_1 |e\rangle_2, |e\rangle_1 |g\rangle_2, |e\rangle_1 |e\rangle_2\}$,

$$C_{apr-not} = H_2^+ P H_2$$
$$= \frac{1}{2}\begin{bmatrix} 1 + e^{-A_{g\kappa}} & 1 - e^{-A_{g\kappa}} & 0 & 0 \\ 1 - e^{-A_{g\kappa}} & 1 + e^{-A_{g\kappa}} & 0 & 0 \\ 0 & 0 & e^{-A_{g\kappa}} - e^{-4A_{g\kappa}} & e^{-A_{g\kappa}} + e^{-4A_{g\kappa}} \\ 0 & 0 & e^{-A_{g\kappa}} + e^{-4A_{g\kappa}} & e^{-A_{g\kappa}} - e^{-4A_{g\kappa}} \end{bmatrix}, \quad (72)$$

where H_2 and P represent the Hadamard gate on the second qubit and the unconventional geometric phase gate in Eq. (70).

To check how well our scheme works, we have in detail studied the error accumulation by repeated applications of the controlled-NOT gate, i.e., Eq. (72) on two qubits (atoms) initially prepared in the product state $|\psi_0\rangle = (|g\rangle_1 + |e\rangle_1)(|g\rangle_2 + |e\rangle_2)/2$. It is noted that the state can be preserved unchanged under many-time actions of the ideal controlled-NOT gate C_{idnot}, that is, $(C_{idnot})^N |\psi_0\rangle = |\psi_0\rangle$. Since the definition of the fidelity is the overlap between the state after the ideal gate operation and that after the real operations with dissipation, we have the fidelity F_{decay} after the repeated $C_{apr-not}$ gatings,

$$F_{decay} = \left| \langle\psi_0| (C_{apr-not})^N |\psi_0\rangle \right|^2 = \frac{1}{4}(1 + e^{-NA_{g\kappa}})^2, \qquad (73)$$

where N is the number of the controlled-NOT gating. In fact, we can also give the fidelity and success probability after each geometric phase gate based on Eq. (70), which are $F_{pdecay} = \frac{1}{4} \cdot \frac{(1+2e^{-A_{g\kappa}}+e^{-4A_{g\kappa}})^2}{(1+2e^{-2A_{g\kappa}}+e^{-8A_{g\kappa}})}$ and $P_{pdecay} = \frac{1}{4}(1 + 2e^{-2A_{g\kappa}} + e^{-8A_{g\kappa}})$, respectively [65], when a phase gate operation on the initial state $|\psi_0\rangle = (|g\rangle_1 + |e\rangle_1)(|g\rangle_2 + |e\rangle_2)/2$ is completed. According to Eq. (70), after the unconventional geometric phase gating on the two qubits, the cavity is excited to be a coherent state, away from the original vacuum state. This is due to quantum fluctuation during the gating. The mean photon number $\langle n \rangle$ is,

$$\begin{aligned}
\langle n \rangle &= \langle f_1(t)| a^\dagger a |f_1(t)\rangle \\
&= (\frac{\Omega g}{\Delta})^2 \frac{1}{(\delta + g^2/\Delta)^2 + \kappa^2/4} \left\{ e^{\frac{m\kappa\pi}{(\delta+g^2/\Delta)}} - 1 \right\}^2 \\
&= (\frac{\Omega g}{\Delta})^2 \frac{1}{(\delta + g^2/\Delta)^2} \frac{1}{(1 + x^2/4)} \{ e^{m\pi x} - 1 \}^2, \qquad (74)
\end{aligned}$$

where $x = \kappa/(\delta + g^2/\Delta)$ and $t = 2m\pi/(\delta + g^2/\Delta)$. Generally speaking, if $\langle n \rangle \geq 0.75$, we consider the cavity mode to be actually excited, and our GQG with large-detuning fails. Since we can analytically describe the influence from the cavity decay, we could demonstrate the validity of an approximate unconventional GQG by means of Eqs. (66), (68), (73), and (74).

3.5.   Discussion

Assuming that $\delta = g^2/\Delta, \Omega = g, \Delta = 10g$, and $m = 1$ and 4, we have numerically calculated Eqs. (66), (68), (73), and (74) versus the cavity decay rate κ. It appears in Fig. 8 that $\Theta_{g\kappa}$ will decrease with the cavity decay, implying that we will spend more time to achieve the gate in the decay case than in the ideal case. Although the validity of our model is restricted by the condition of $\kappa \ll \delta$, we plot the figure to $\kappa/\delta \leq 50$, which could clearly show that the curve corresponding to $m = 4$ drops more sharply than that of $m = 1$. This is because the value m corresponds to the closed path with respect to $(\delta + g^2/\Delta)t$ (See Eq. (63)). So we have to spend longer time if m is bigger, which implies a slower movement in the phase space. With the increase of the cavity decay, the implementation time regarding $m = 4$ is getting closer to the decoherence time than that of $m = 1$. Therefore, the larger impact on the curve of $m = 4$ than that of $m = 1$, when κ is bigger, is physically understandable. In this sense, to have a good phase gating, we prefer a faster implementation. This feature is reflected in Fig. 9 in which the damping rate of the gating in the case of $m = 4$ is bigger than the case of $m = 1$ even in the presence of weak dissipation, and also in Fig. 10 that the fidelities are evidently higher in the case of $m = 1$

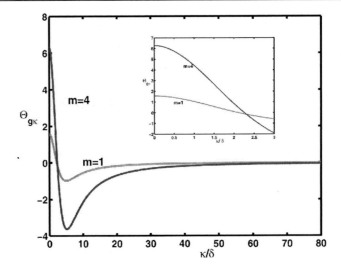

Figure 7. The unconventional geometric phase shift ($\Theta_{g\kappa}$) versus the cavity decay κ/δ based on Eqs. (63) and (66), where we set $\delta = g^2/\Delta$, $\Omega = g$, $\Delta = 10g$, and $m = 1$ or 4.

than $m = 4$. Note that the worst case of F_{decay} is 0.25, corresponding to a completely failed gating, because the number 0.25 comes from the component $|g\rangle_1 |g\rangle_2$.

Fig. 11 demonstrates a larger photon fluctuation in the case of $m = 4$ than $m = 1$, which further confirms that a faster gating works better than a slower one. As the photon fluctuation is closely related to the cavity decay, a slower implementation would yield bigger possibility of cavity mode excitation. This is an important difference of the non-ideal GQG treatment from the ideal one. In the ideal case, our concentration is only on the qubit subspace, so we always consider that a slower implementation would lead to a gating with higher fidelity. Actually, it is correct only in the ideal model that the GQG is related to the area of the closed path in the phase space and the cavity mode keeps constant during an adiabatic gating. While in the presence of cavity decay, as shown in our analytical result Eq. (70), the qubit states are entangled with the cavity mode after the gating. Therefore, an optimal gating time depends on the changes in both the qubit subspace and the cavity mode. Evidently. in the case of dissipation, a faster gating would be preferred.

Here we briefly discuss the experimental feasibility of our scheme in the present microwave regime, in which the lifetime of the Rydberg atomic excited state is about $T_r = 3 \times 10^{-2} s$, the coupling strength between the cavity mode and the atoms is $g = 2\pi \times 49 kHz$, and the photon lifetime in the cavity is $T_c = 3.0 \times 10^{-3} s$ [31], direct algebra from Eq. (63) yields that, our gating time with $m = 1$ is $t_g = \Delta\pi/g^2 \simeq 10^{-4} s$, much shorter than T_c, and T_r. Moreover, with current cavity decay rate, our treatment involving the cavity decay has shown the possibility of high fidelity and high success rate of our scheme. To have a better gating, we need a cavity of higher quality.

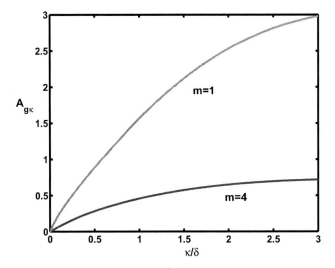

Figure 8. The corresponding amplitude damping factor ($A_{g\kappa}$) versus the cavity decay κ/δ based on Eqs. (63) and (68), where the parameters are the same as in Fig. 8.

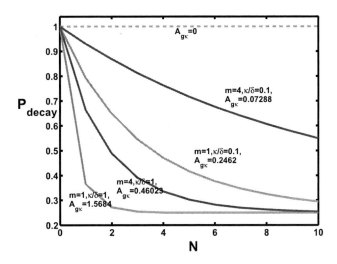

Figure 9. The fidelity F_{decay} between the state after the N ideal controlled-NOT gates and that after application of N approximate controlled-NOT gates based on Eq. (73), where the parameters are the same as in Fig. 8.

4. Conclusion

We have reviewed our recent research work on QIP by cavity QED under cavity decay. By using quantum trajectory method, we have analytically presented the influence from the cavity decay on the quantum gating accomplished by dynamical way or geometric way. What we have discussed in this chapter shows that decoherence is a detrimental factor in implementation of QIP, and to have the QIP working well we have to finish the quantum gates much faster than the decoherence rate.

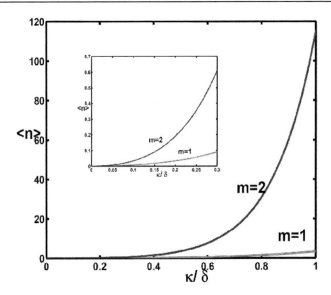

Figure 10. The mean heating photon number $\langle n \rangle$ versus the cavity decay κ/δ based on Eqs. (63) and (74), where the parameters are the same as in Fig. 8. When $\kappa/\delta \leq 0.16$, for $m = 1$ and 4, we have $\langle n \rangle \leq 0.1$ and 0.75, respectively, meaning that photon fluctuation due to the cavity decay is not very big. When $\kappa/\delta \geq 0.16$, we have $\langle n \rangle \geq 0.75$ for $m = 4$, which means that the fluctuation has destroyed our expected quantum logic gate operation.

Ackowledgements

We thank C. Y. Chen and K. L. Gao for helpful discussion. The work is partly supported by National Natural Science Foundation of China under Grants No. 10474118 and No. 10774163, and partly by the National Fundamental Research Program of China under Grants No. 2005CB724502 and No. 2006CB921203.

References

[1] P.W. Shor, *in Proceedings of the 35th Annual Symposium on the Foundations of Computer Science*, edited by S. Goldwasser (IEEE Computer Society, 1994).

[2] A. Barenco, C. H. Bennett, R. Cleve, D. P. DiVincenzo, N. Margolus, P. Shor, T. Sleator, J.A. Smolin, H. Weinfurter, *Phys. Rev. A* **52**, 3457 (1995).

[3] D.P. DiVincenzo, *Fortschr. Phys.* **48**, 771 (2000).

[4] M. A. Nielsen and I. L. Chuang, *Quantum Computation and Quantum Information*, (Cambridge University Press, Cambridge, 2000).

[5] R.G. Hulet, and D. Kleppner, 1983, *Phys. Rev. Lett.* **51**, 1430.

[6] J.M. Raimond, M. Brune and S. Haroche, *Rev. Mod. Phys.* **73**, 565 (2001).

[7]   V.B. Berestetskii, E.M. Lifshitz, and L.P. Pitaevskii, *Quantum Electrodynamics* (Oxford: Pergamon, 1982).

[8]   M.B. Plenio and P.L. Knight, *Rev. Mod. Phys.* **70**, 101 (1998).

[9]   J.S. Bell, Physics (Long Island City, New York) **1**, 195 (1964); S. J. Freedman and J. F. Clauser, *Phys. Rev. Lett.* **28**, 938 (1972); Z.Y. Ou and L. Mandel, *Phys. Rev. Lett.* **61**, 50 (1988); P.G. Kwiat, K. Mattle, H. Weinfurter, A. Zeilinger, A.V. Sergienko and Y. Shih, *Phys. Rev. Lett.* **75**, 4337 (1995).

[10] C.H. Bennett, G. Brassard, C. Crépeau, R. Jozsa, A. Peres, and W.K. Wootters, *Phys. Rev. Lett.* **70**, 1895 (1993).

[11] A.K. Ekert, *Phys. Rev. Lett.* **67**, 661 (1991).

[12] C.H. Bennett and S.J. Wiesner, *Phys. Rev. Lett.* **69**, 2881 (1992) .

[13] J.J. Bollinger, W.M. Itano, D.J. Wineland, and D.J. Heinzen, *Phys. Rev. A* **54**, R4649 (1996).

[14] W. Dür, G. Vidal, and J.I. Cirac, *Phys. Rev. A* **62**, 062314 (2000).

[15] D.M. Greenberger, M. Horne, and A. Zeilinger, in *Bell's Theorem, Quantum Theory, and Conceptions of the Universe*, edited by M. Kafatos (Kluwer, Dordrecht, 1989).

[16] C.A. Sackett, D. Kielpinski, B.E. King, C. Langer, V. Meyer, C.J. Myatt, M. Rowe, Q.A. Turchette, W.M. Itano, D.J. Wineland and C. Monroe, *Nature (London)* **404**, 256 (2000).

[17] A. Rauschenbeutel, G. Nogues, S. Osnaghi, P. Bertet, M. Brune, J.M. Raimond and S. Haroche, *Science* **288**, 2024 (2000).

[18] Z. Zhao, Y.-A. Chen, A.-N. Zhang, T. Yang, H.J. Briegel, J.-W. Pan, *Nature (London)* **430**, 54 (2004).

[19] C.F. Roos, M. Riebe, H. Häffner, W. Hänsel, J. Benhelm, G.P.T. Lancaster, C. Becher, F. Schmidt-Kaler, R. Blatt, *Science* **304**, 1478 (2004).

[20] G.-P. Guo, C.-F. Li, J. Li and G.-C. Guo, *Phys. Rev. A* **65**, 042102 (2002).

[21] G.-C. Guo and Y.-S. Zhang, *Phys. Rev. A* **65**, 054302 (2002).

[22] X. Wang, M. Feng and B.C. Sanders, *Phys. Rev. A* **67**, 022302 (2003).

[23] C. Marr, A. Beige and G. Rempe, *Phys. Rev. A* **68**, 033817 (2003).

[24] A. Rauschenbeutel, G. Nogues, S. Osnaghi, P. Bertet, M. Brune, J.M. Raimond and S. Haroche, *Phys. Rev. Lett.* **83**, 5166 (1999); A. Rauschenbeutel, G. Nogues, S. Osnaghi, P. Bertet, M. Brune, J.M. Raimond and S. Haroche, *Science* **288**, 2024 (2000); P. Bertet, A. Auffeves, P. Maioli, S. Osnaghi, T. Meunier, M. Brune, J.M. Raimond and S. Haroche, *Phys. Rev. Lett.* **89**, 200402 (2002).

[25] For example, in an early paper [G.S. Agarwal, *Phys. Rev. Lett.* **53**, 1732 (1984)] and a recent paper [26], the identical coupling strength is employed.for convenience and simplicity of theoretical treatments. While in the system of cavity-assisted semiconductor quantum dots [e.g., M. Feng et al, *Phys. Rev. A* **67**, 014306 (2003)], the identical coupling strength is used by supposing a good match between the quantum dots spacing and the wave pattern of the cavity mode. Nevertheless, this assumption is hard to apply to our case bacause the atoms under consideration are flying.

[26] F. Yamaguchi, P. Milman, M. Brune, J.M. Raimond and S. Haroche, *Phy. Rev. A* **66**, 010302(R) (2002).

[27] T. Toffoli, in *Automata Languages and Programming, Seventh Colloquium*, edited by J .W. de Bakker and J. van Leeuwen, *Lectures Notes in Computer Science* Vol. 84 (Springer, New York, 1980).

[28] M. D. Price *et. al.*, *Phys. Rev. A* **60**, 2777 (1999).

[29] L.-M. Duan, B. Wang, and H. J. Kimble, *Phys. Rev. A* **72**, 032333 (2005).

[30] A. Gàbris and G. S. Agarwal, *Phys. Rev. A* **71**, 052316 (2005).

[31] S. Osnaghi et al., *Phys. Rev. Lett.* **87**, 037902 (2001).

[32] J. Pachos and H. Walther, *Phys. Rev. Lett.* **89**, 187903 (2002).

[33] L. K. Grover, *Phys. Rev. Lett.* **79**, 325 (1997); ibid. 80, 4329 (1998).

[34] M. Feng, *Phys. Rev. A* **63**, 052308 (2001).

[35] Shingo Fujiwara and Shuichi Hasegawa, *Phys. Rev. A.* **71**, 012337 (2005).

[36] K.-A. Brickman et al, *Phys. Rev. A.* **72**, 050306(R) (2005).

[37] J. A. Jones, M. Mosca and R. H. Hansen, *Nature (Londan)* **393**, 344(1998).

[38] I.L. Chuang, N. Gershenfeld and M. Kubinec, *Phys. Rev. Lett.* **80**, 3408 (1998).

[39] Y. Nakamura, Yu. A. Pashkin and J.S. Tsai, *Nature (Londan)* **398**, 789 (1999); D. Vio et al, *Science* **296**, 886 (2002); M. S. Anwar et al, *Chem. Phys. Lett.* **400**, 94 (2004).

[40] Z. J. Deng, M. Feng and K. L. Gao, *Phys. Rev. A.* **72**, 034306 (2005).

[41] A. Joshi and M. Xiao, *Phys. Rev. A.* **74**, 052318 (2006).

[42] P. G. Kwiat et al, *J. Mod. Opt.* **47**, 257 (2000).

[43] P. Walther et al, *Nature (Londan)* **434**, 169 (2005).

[44] M. Feng and X. Wang, *J. Opt. B: Quantum Semiclass. Opt.,* **4**, 283 (2002).

[45] J. A. Sauer et al, *Phys. Rev. A.* **69**, R051804 (2004).

[46] C. Y. Chen, M. Feng and K.-L. Gao, *J. Phys. A* **39**, 11861 (2006).

[47] P. Zanardi and M. Rasetti, *Phys. Rev. Lett.* **79,** 3306 (1997); L. M. Duan and G. C. Guo, ibid. 79, 1953 (1997).

[48] P. Zanardi and M. Rasetti, *Phys. Lett. A* **264,** 94 (1999); J. Pachos , P. Zanardi and M. Rasetti, *Phys. Rev. A* **61,** 010305 (2000).

[49] G. Falci *et al.*, *Nature (London)* **407,** 355 (2000).

[50] L.-M. Duan, J. I. Cirac, and P. Zoller, *Science* **292,** 1695 (2001).

[51] J. A. Jones, V. Vedral, A. Ekert , and G. Castangnoli, *Nature (London)* **403,** 869 (1999).

[52] J. J. Garcia-Ripoll and J. I. Cirac, *Phys. Rev. Lett.* **90,** 127902 (2003).

[53] X. B. Wang and M. Keiji, *Phys. Rev. Lett.* **87,** 097901 (2001).

[54] P. Solinas *et al.*, *Phys. Rev. B* **67,** 121307 (2003); *Phys. Rev. A* **67,** 052309 (2003).

[55] S. L. Zhu and Z. D. Wang, *Phys. Rev. Lett.* **91,** 187902 (2003).

[56] A. Barenco *et. al.*, *Phys. Rev. Lett.* **74,** 4083 (1995).

[57] P. Domokos *et. al.*, *Phys. Rev. A* **52,** 3554 (1995).

[58] T. Pellizzari *et. al.*, *Phys. Rev. Lett.* **75,** 3788 (1995).

[59] M. Brune, *et. al.*, *Phys. Rev. Lett.* **77,** 4887 (1996).

[60] L. X. Cen and P. Zanardi, *Phys.Rev.A* **70,** 052323 (2004).

[61] I. Fuentes-Guridi, F. Girelli, and E. Livine, *Phys. Rev. Lett.* **94,** 020503 (2005).

[62] A.Carollo et al, *Phys. Rev. Lett.* **92,** 020402 (2004); ibid 90, 160402 (2003).

[63] S. B. Zheng, *Phys. Rev. A* **70,** 052320 (2004).

[64] D. Leibfried et al., *Nature (London)* **422,** 412 (2003).

[65] C.-Y. Chen, M. Feng, X.-L. Zhang, and K.-L. Gao, *Phys. Rev. A* **73,** 032344 (2006).

[66] X. Wang and P. Zanardi, *Phys. Rev. A* **65,** 032327 (2002).

[67] S. B. Zheng, *Phys. Rev. Lett.* **87,** 230404 (2001).

In: Atomic, Molecular and Optical Physics...
Editor: L.T. Chen, pp. 463-489

ISBN 978-1-60456-907-0
© 2009 Nova Science Publishers, Inc.

Chapter 12

HIGH HARMONIC GENERATION FROM MOLECULAR SYSTEMS IN EXCITED ELECTRONIC STATES

M. Kitzler[1] and J. Zanghellini[2]

[1]Photonics Institute, Vienna University of Technology, Austria, EU

[2]Institute of Chemistry, University of Graz, Austria, EU

Abstract

The response of gas atoms or molecules to strong laser fields depends on their internal electronic state. This fact can be exploited to gain insight into bound electron structure, nuclear dynamics, and even electronic dynamics by measuring the emitted photons created by the process of high harmonic generation (HHG). HHG is customarily explained by recombination of a virtually detached electron upon returning to its initial bound state. In this chapter we investigate the emission of high harmonic radiation from molecular systems in excited electronic states.

Specifically we report on numerical results obtained for systems in two types of electronic excited states: (i) states with pronounced net internal angular momentum and (ii) states which are excited by nonlinear plasmon oscillations in highly polarizable molecules by a strong laser field. Our results can be summarized as follows.

If the involved state exhibits pronounced angular momentum both ionization and recombination are influenced and the symmetry of the three-stage process is broken. We show that this can be used to gain access to the phase of the bound state and that recombination to such a bound state leads to creation of circularly polarized, spatially coherent attosecond X-ray pulses.

By solving the time-dependent Schrödinger equation for a model system containing 4 active electrons using the multi-configuration time-dependent Hartree-Fock approximation, we show that the harmonic spectrum exhibits two cut-offs. The first cut-off is in agreement with the well-established, single active electron cut-off law. The second cut-off presents a signature of multi-electron dynamics. Electrons that are ionized from an excited multi-plasmon state and recombine to the ground state gain additional energy, thereby creating the second plateau.

I. INTRODUCTION

The response of gas-phase atoms or molecules to strong laser fields comprises the generation of radiation with photon energies many times higher than those of the incident light. For linearly polarized laser pulses with durations of at least several cycles the emitted radiation consists of clearly separated spectral intensity peaks at odd multiples of the driving wavelength. Consequently the generating process was dubbed high harmonic generation (HHG). The great significance of HHG is based on the fact that it allows the production of coherent XUV/soft-X-ray radiation which is emitted in a directed beam.

HHG can be utilized as a source for sub-femtosecond radiation. Filtering away the low harmonics from the emitted radiation results in a train of attosecond pulses [1] lasting almost over the whole duration of the driving femtosecond laser pulse. Single attosecond pulses can be produced by few-cycle laser pulses, when only the highest energetic region – the so-called cutoff – of the emitted high harmonic (HH) radiation is selected via band-pass filtering [2], or when the emission of all but one of the pulses can be precluded by controlling the time-dependent polarization of the driving field [3, 4].

The process of HHG is usually explained by a semi-classical model [5–8], whereby an electronic wave packet is ejected via tunnel ionization from and directed back onto its parent atom or molecule by the driving electric laser field. Following tunnel ionization, which happens within a fraction of a laser period, the detached electronic wave packet behaves very similarly to a free electron. It is driven by the strong laser field and undergoes wave packet spreading (quantum diffusion) but preserves its sub-femtosecond duration and therefore constitutes an attosecond probe for the parent ion. During propagation in the laser field the electron gains energy from the field. Upon re-collision with the parent ion the electron wave packet may "recombine" to the bound state thereby emitting the excessive kinetic energy as photons. The latter process, "recombination", can be understood in terms of quantum mechanics as a coherent super-position of the re-colliding, quasi-free, electron wave packet with the bound state wavefunction, which leads to the induction of a fast oscillating dipole and consequently to the emission of photons [9].

As HHG preserves the phase of both wave functions – the one of the ion and that of the re-colliding electron – the information about the molecular bound state is imprinted on the temporally varying instantaneous quantities of wavelength, intensity, phase and polarization of the emitted photons. In general, however, not only recombination is sensitive to the bound electronic structure but also ionization, e.g. [10–15], which is the first and fundamental process for any measurement using HHG. It is therefore unavoidable to disentangle the intertwined action of ionization and re-combination on the experimental observable. Thus, measurement of one or some of the above mentioned properties, e.g. the intensity spectra, from aligned or oriented molecules together with several assumptions and additional information about the processes of ionization and re-collision, enables one to map these properties onto a bound-state electronic cloud. Currently this method is heavily used and many different groups have reported sensitivity of their measurements to molecular bound electron structure [16–24] and nuclear dynamics [25]. It is even possible to infer information about the internal attosecond bound dynamics from the HH spectrum [26, 27].

Yet, so far only emission from the electronic ground state has been investigated. However, the response of gas atoms or molecules to strong laser fields depends on their internal state. Emission of radiation with contributions from electronically excited states is therefore expected to be quite different than that from only the electronic ground state. Descriptive examples, where the involvement of excited states may lead to strikingly different behavior, are beating between an resonantly excited state and the ground state [26], population of excited states through laser driven electron dynamics and induced electronic polarization [28, 29], or all types of non-trivial net internal momentum [30].

Here, we present numerical results on HHG in systems where the bound electron state strongly influences the ionization and/or recombination process and therewith leads to thus far unobserved features in the emitted radiation.

In the first part of this chapter we demonstrate the impact of internal net angular momentum on the production of HH radiation. While the effects that we predict happen from states that exhibit pronounced angular momentum in any system that fulfills the symmetry requirements that we develop, we use excited ring-current states [31] in ring-shaped molecules as an incarnation of such states. The main results are that the internal state is transferred to the tunneling wave packets at the time of ionization. Exactly the analogical process happens at the time of recombination, i.e. the internal state is transferred to the polarization state of the emitted photons, which leads to the production of novel, spatially and temporally coherent, nearly circularly polarized attosecond XUV/X-ray pulses.

While the effects of a ring-current state on HHG are sufficiently described by a single active electron (SAE) [30, 31], recent work has shown that for a range of applications the multi-electron (ME) nature of atoms and molecules as well as the dynamics of the bound state cannot be neglected [32, 33]. Thus, in the second part of this chapter, we analyze non-SAE effects with regard to the correlated dynamics of the electronic hull. In particular, we investigate the influence of polarizability on the properties of the emitted radiation. By using multi-configuration time-dependent Hartree-Fock (MCTDHF) [34, 35] we show that field-driven nonlinear collective electron oscillations can populate plasmon states which can be observed by recording the spectrum of the emitted radiation.

The following key results are revealed: In contrast to HHG in noble gases, where the harmonic spectrum exhibits one plateau and cut off, a second cut off is identified in complex, highly polarizable materials. The first cut off is found to be in agreement with the SAE cutoff law [5]. The second cut off originates from the ME nature of the bound electrons. The strong laser field excites nonlinear, collective electron oscillations. This results in a population of multi-plasmon states that oscillate at a multiple of the plasmon frequency. The second plateau is generated by electrons that ionize from the multi-plasmon states and recombine to the ground state. The energy difference between excited and ground state determines the difference between first and second cutoff. The identified plasmon signature presents a novel tool for the investigation of nonlinear, non-perturbative ME dynamics in complex materials.

II. HIGH HARMONIC GENERATION FROM EXCITED STATES EXHIBITING ANGULAR MOMENTUM

In this section we demonstrate that HHG from systems with non-trivial internal momentum states is strongly modified due to so far undescribed effects in both ionization and recombination. We consider a new class of internal states, with a net angular momentum. Specifically we investigate electronic ring-currents which result from the excitation of a suitable molecule or atom by circularly polarized π pulses [31, 36]. We show that the response of ring-currents to strong laser pulses exhibits unique features that can be exploited for measurements and pulse generation: (i) the combined action of ionization and laser-controlled re-scattering can be used to detect the systems' internal symmetry and (ii) spatially and temporally coherent, circularly polarized, XUV/X-ray single attosecond pulses can be produced.

The initial ring-current states $|E_{\pm}\rangle = 1/\sqrt{2}\,(|E_1\rangle \pm i|E_2\rangle)$ that we consider are eigenstates of angular momentum L_z with respect to the axis perpendicular to the (x, y)-plane with eigenvalues $\pm|m|$. The states are constructed as the superposition of two equally populated, degenerate real states of m-fold angular symmetry ($|m| \geq 1$), denoted by $|E_1\rangle$ and $|E_2\rangle$, respectively, with wave functions $\Psi_1(\rho, \varphi) = f(\rho) \cos(|m|\varphi)$ and $\Psi_2(\rho, \varphi) = f(\rho) \sin(|m|\varphi)$, from which one obtains the ring-current wave functions $\Psi_{\pm}(\rho, \varphi) = 1/\sqrt{2}\,f(\rho)\mathrm{e}^{\pm i|m|\varphi}$. Here we use 2D polar coordinates $(x, y) = (\rho \cos \varphi, \rho \sin \varphi)$ and atomic units. Because only the phase of the wave function depends on φ the electron density $|\Psi_{\pm}|^2$ of a ring-current state $|E_{\pm}\rangle$ is constant along the angular direction. It is independent of the current's left-handed (+) or right-handed (-) direction, which can be induced by an anti-clockwise or clockwise rotating circular optical π pulse, respectively [31, 36].

To study ionization dynamics of and HHG from ring-current states we numerically solve the time-dependent Schrödinger equation in two spatial dimensions, x and y, for a single active electron coupled to the laser electric field in dipole approximation. The influence of the magnetic field induced by the ring-current on the free electron's trajectory is by several

orders of magnitude weaker than that of the driving electric field and can therefore be neglected. For our simulations we use a (shielded) ring-shaped molecular potential where
we have adjusted the shielding parameter and introduced some exponential short-range
smoothing to adjust the binding energy of the initial state. The radius of the ring-potential
is chosen $R = 3$ and in our simulations we used different ring-current states with angular
momenta $|m| = \{1 \ldots 6\}$, although we will exclusively demonstrate our results using the
most weakly bound state $|m| = 6$ with a binding energy of 9.5 eV.

A.   Ionization from Ring-Current States

We start by describing how the initial step in HHG, i.e. ionization, is influenced by the
internal electronic state, which in terms of classical physics can be understood as station-
arily rotating electron density. Fig. 1 shows snapshots of the electron probability density
at peak field strength in a linearly polarized laser pulse. The peak intensity of the pulse
was 5.6×10^{13} W/cm^2 at a wavelength of 1600 nm and pulse duration of 10 fs FWHM
(gaussian pulse envelope). The insets of Figs. 1(a) and 1(b) show the electron density for
initial states $|E_1\rangle$ and $|E_2\rangle$, respectively. The nodes in the electron density are imprinted
also in the detached part of the wave function. There are either 3 or 2 lobes of electron
density, depending on the angular nodal distribution perpendicular to the ionizing direction,
as expected for molecular tunnel ionization [14]. When the two states are superimposed to
form the state $|E_+\rangle$, the probability density inside the molecule becomes radially constant;
any nodal structure is absent, cf. inset of Fig. 1(c). Ionization of this ring-current state by
the same linearly polarized field as in panels (a) and (b) results in the emittance of an electronic wave packet strongly rotated off the polarization axis in counter-clockwise direction,
i.e. along the direction of the ring-current, see Fig. 1(c). Correspondingly, ionization from
a state $|E_-\rangle$ results in wave packet emission into the opposite, clockwise, direction (not
shown).

At first glance this *asymmetric* emission from a rotationally *symmetric* electron density distribution could be surprising, as the structures in electron emission reported so far
[10–15] have their counterpart in structures in the molecule's bound electron density. A

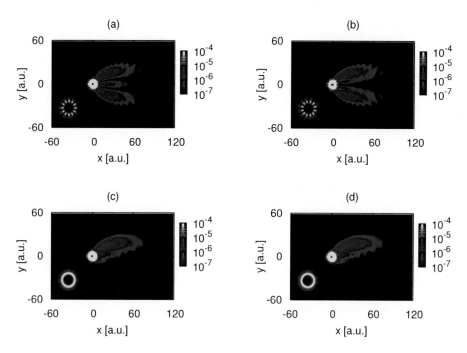

Figure 1. Snapshots of the probability density of an electron wave packet as it is detached from a ring-current state by tunnel ionization in a linearly polarized laser field. Ionization from state (a) $|E_1\rangle$, (b) $|E_2\rangle$, (c) $|E_+\rangle = 1/\sqrt{2}\,(|E_1\rangle + i|E_2\rangle)$. Panel (d) shows the resulting structure when the wave packets from panels (a) and (b) are coherently added.

simple classical picture for the asymmetry observed here is that during ionization internal momentum is imparted to the detached electron. To corroborate this statement we solved the Schrödinger equation for states with different internal electron momentum in the range of $|m| = \{1\ldots6\}$ and calculated the transversal electron momentum distributions immediately after detachment by the numerical probing-technique described in Refs. [37, 38], which shows a monotonic increase of transverse momentum with increasing $|m|$. In the language of quantum mechanics, the effect of asymmetric emission is due to the interference of two tunneling wave functions from the two initial state components, which have opposite parity with respect to reflections about the x-axis. The electron density in Figs. 1(a) has a maximum along $x = 0$, while Figs. 1(b) shows a node. The coherent superposition of the two wave packets produces the density shown in Fig. 1(d), which is identical to the wave packet from the ring-current shown in panel (c): the overall non-zero transverse momentum of the free wave packet is due to constructive interference at $y > 0$ and destructive interference at $y < 0$.

B.   Accessing the Bound State Phase by HHG

Thus far we have shown that the characteristics of the electron emission are determined by the density and momentum distribution of the internal state. Since ionization is the first step in the three-step-process [5, 7] of HHG, we expect the current's direction to influence experimentally accessible observable HHG spectra. This means that the spectra should be sensitive to the direction of the ring-current as given by the sign of the quantum phase $\pm|m|\varphi$ of the bound state. For symmetry reasons, harmonic spectra from ring-current states in a linearly polarized laser pulse are insensitive to the sign of m. By using elliptically polarized laser fields

$$\vec{\mathcal{E}}(t) = \mathcal{E}_0(t)(\cos(\omega t)\vec{e}_x + \varepsilon \sin(\omega t)\vec{e}_y), \tag{1}$$

with $\mathcal{E}_0(t)$ the pulse envelope, ω the laser frequency and ε the ellipticity, respectively, we can break that symmetry, which allows us to probe the transversal momentum distribution of the tunneling wave packets using HH intensity spectra. Fig. 2(a) shows HH spectra from the same ring-current state and with the same pulse parameters as used for Fig. 1, except that the ellipticity of the pulse is varied. One sees some variation of the cut-off frequency and significant variations of intensity. The variations in the cut-off frequency are consistent with the variations of the classical re-collision energies in the simple man model of HHG [5]. The intensity variations reflect the transverse distribution of electron momentum at emission. The field component $\mathcal{E}_\perp$ perpendicular to the field direction at emission time t' can compensate the initial transverse momentum $p_\perp(t')$ of the electrons and guide them back to the nucleus. By varying the ellipticity we steer the re-collision current across the target, thereby imaging the distribution of transverse tunneling electron momenta. The correspondence between $p_\perp(t')$ and ε is established by demanding that the transverse drift between electron release and re-collision is compensated by the weak electric field component in the transverse direction during continuum propagation of the electrons. A similar idea was proposed recently to compensate the lateral birth momentum for a range of harmonics [39]. For small ellipticities we can set $\mathcal{E}_\perp(t) \approx \mathcal{E}_y(t)$, and the instants of release, t', and re-collision, t, can be considered independent of ε. Then the re-collision condition in atomic units can be derived from Newton's equation for the electron's acceleration in perpendicular direction, $\mathcal{E}_y(t) = -\dot{p}_\perp(t)$, by demanding that the electron is steered back to its birth point y_b, i.e. $y(t) - y_b = 0$. The trajectory in perpendicular direction reads

$$y(t) = \int_{t'}^t p_\perp(\tau)\mathrm{d}\tau = -\int_{t'}^t \int_{t'}^\tau \mathcal{E}_y(\tau')\mathrm{d}\tau'\mathrm{d}\tau + p_\perp(t')(t-t') + y_b. \tag{2}$$

Thus, with $\mathcal{E}_y(t) = \mathcal{E}_0(t)\varepsilon\sin(\omega t)$ the re-collision condition becomes

$$p_\perp(t') = \frac{1}{t-t'}\int_{t'}^t \int_{t'}^\tau \mathrm{d}\tau\mathrm{d}\tau'\varepsilon\,\mathcal{E}_0(\tau')\sin(\omega\tau'), \tag{3}$$

i.e. a linear dependence of $p_\perp$ on ε for a given shape of the laser field $\mathcal{E}(t)$. Fig. 2(b) compares harmonic intensity at the cut-off as a function of ε (dotted line, upper abscissa), taken from Fig. 2(a), to the initial transverse momentum (full line, lower abscissa). The transverse momentum distribution at release time, $p_\perp(t')$, was determined from the numerically exact wave function using the probing technique of Refs. [37, 38]. The distribution of harmonic intensity found by varying the ellipticity matches almost exactly the initial momentum distribution, thus confirming the linear relation $p_\perp = \alpha\,\varepsilon$ found above, with $\alpha = 2$ calculated from Eq. (3) for the laser pulse parameters used here. The momentum distribution is asymmetric with respect to the origin and peaks at a positive value as expected from the previously described ionization mechanism for a ring-current state with $\mathrm{sign}(m) = +1$. The method can be used in general to perform all-optical measurements of momentum distributions of tunneling wave packets without the need for more complicated methods based on position- and time-sensitive electron detection which is of crucial importance for development and test of theoretical models for ionization of complex molecules. As the present method offers sub-cycle temporal resolution it can also be applied to probe the effects of fast internal electron dynamics on ionization.

C. Circularly Polarized Isolated Attosecond XUV Pulses

Now we turn to the description of a novel application of $|m| \geq 1$ states in pulse production by HHG. The polarization state of HH radiation created from systems that are not

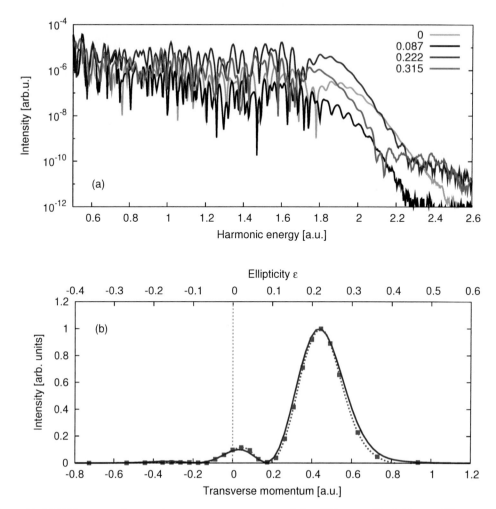

Figure 2. (a) HH spectra from a ring-current state $m = +6$ for different ellipticities ε, (b) comparison of the HH intensity in the cut-off as a function of ellipticity ε (full line) with the distribution of transverse electron momenta $p_\perp$ at the birth time for the highest harmonics (dotted line). Both curves were normalized to peak values of 1.

rotationally symmetric differs, in general, from that of the driving light, as observed recently with molecules [24]. This is also the case for a ring-current state, where the rotational symmetry is not broken in configuration space, but rather in momentum space. We will see that this leads to a very specific modification of the molecule's response. Fig. 3(a) shows the HH intensity spectrum for two orthogonal polarization components created by a slightly elliptical driving pulse, the same as used for maximum HH yield in Fig. 2, emitted once again from the same ring-current state used above. The intensity I_x for harmonic polariza-

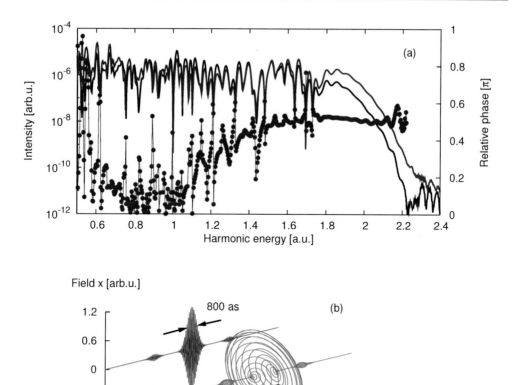

Figure 3. (a) HH spectra of I_x (red line) and I_y (blue line) and their relative phase (gray dots) from a ring-current state with $m = +6$. Pulse parameters as in Fig. 2 with the ellipticity for peak harmonic yield (0.222). (b) Nearly circularly polarized electric field of an attosecond pulse obtained by filtering the highest harmonics of the spectra shown in (a).

tion parallel to the field direction at laser peak and I_y perpendicular to it are comparable, especially in the cut-off around 1.9 a.u. with a ratio in intensity of 1:3. The two components are out of phase by about $\pi/2$ over a broad spectral range (gray dots in Fig. 3). This means that those harmonics correspond to near-circularly polarized XUV radiation, with a ratio of ∼1:1.7 between the largest and the smallest field component. Production of HH radiation with such characteristics, to the best of our knowledge, has never been predicted nor observed before. If we cut out a band around the cut-off, we are left with a circularly polarized single attosecond pulse, cf. Fig. 3(b). This observation might open the door to tabletop-scale imaging and spectroscopy of ferromagnetic materials using magnetic X-ray absorption spectroscopy, e.g. [40], X-ray circular magnetic dichroism (XCMD), e.g. [41], and X-ray resonance magnetic scattering (XRMS), e.g. [42], and at the same time push the temporal resolution of that methods to extreme time scales.

A simple calculation in the frame of the strong field approximation (SFA) elucidates the mechanism of production of circularly polarized HH radiation. Within the SFA, the high frequency dipole response at time t is obtained as a product of the three probability amplitudes for ionization, propagation and recombination please insert new reference: M.Y. Ivanov, T. Brabec, and N. Burnett, Phys. Rev. A 54, 742 (1996)..The polarization state of the harmonic radiation is determined by the recombination step. At frequency $\omega(\vec{k}(t)) = \vec{k}^2(t)/2 + I_p$ it is proportional to the bound-free dipole matrix element $\langle \Psi_m | \vec{r} | \vec{k}(t) \rangle$, where the incident electron wave vector $\vec{k}$ of the plane waves $| \vec{k}(t) \rangle$ points into the direction of the drive field. To evaluate this matrix element we write the ring-current wave function in cylindrical coordinates, $\Psi_m = \psi(\rho)e^{im\varphi}$, with φ the azimuthal angle from the x-axis and ρ the distance from the origin and assume the polarization direction of the laser to be x, such that the vector of re-collision $\vec{k} = k\vec{e}_x$. Then, together with the Bessel expansion $e^{ik\rho\cos\varphi} = \sum_n i^n e^{in\varphi} J_n(k\rho)$, the dipole matrix element can be written as

$$\langle \Psi_m | \vec{r} | e^{i\vec{r}\vec{k}} \rangle = \int d\rho \rho^2 \psi(\rho) \sum_n i^n J_n(k\rho) \int d\varphi e^{-im\varphi} e^{in\varphi} \begin{pmatrix} \cos\varphi \\ \sin\varphi \end{pmatrix}. \tag{4}$$

By making use of the property of the Bessel functions $J_{-m}(z) = J_m(z)(-1)^m$ one easily sees that the x and y components of the polarization at frequency $\omega(\vec{k})$ are

$$\begin{pmatrix} \epsilon_x \\ \epsilon_y \end{pmatrix} \propto \langle \Psi_m | \vec{r} | e^{i\vec{r}\vec{k}} \rangle = \frac{i^{m-1}}{2} \begin{pmatrix} B_{m-1} - B_{m+1} \\ -iB_{m-1} - iB_{m+1} \end{pmatrix}, \tag{5}$$

where we have defined the integral $B_m := \int d\rho \rho^2 \psi(\rho) J_m(k\rho)$. It can be shown that for re-scattering wave lengths comparable to the extension of the initial bound state, $2\pi/k_\rho \sim \Delta\rho$, the integral varies rapidly as a function of m such that the two polarization components are usually dominated by one of the integrals, either B_{m+1} or B_{m-1}. Together with the relative phase of $\pi/2$, indicated by the factor i, which is present for both possibilities in Equ. (5), one obtains strongly elliptical, nearly circular polarization.

This simple analysis shows that creation of circularly (elliptically) polarized harmonics is a generic effect, which appears for any system with $|m| \geq 1$ and harmonic frequencies in the range $\omega \sim 2\pi^2/\Delta\rho^2 + I_p$. As ring-current systems have binding energies below $10\,eV$, one uses drive pulses with lower laser peak intensities but longer wavelengths to avoid excessive ionization. To reach harmonic photon energies beyond the XUV range, i.e. X-rays, it is, however, unavoidable to apply high peak intensities, and ionic systems will have to be used.

In an intuitive picture, production of circularly polarized attosecond pulses via HHG proceeds as follows: First, angular momentum is transferred to the system which can be achieved by any method, for example by a (weak) circularly polarized optical pump pulse [31, 36]. Then the angular momentum is 'read out' by the returning electron wave packet in strong-field rc-collision and the ellipticity of the optical pump pulse is upshifted to the XUV or X-ray domain. For long-lived angular states, the first step is decoupled from the second one, which can happen at any time later. This is reminiscent of the laser principle, where first population inversion is created by any suitable method, which later on can be recovered in a controlled way, e.g. in a short burst.

For experimental applications one must worry about the achievable harmonic intensities. For that we made a comparison with the standard HHG with linearly polarized laser pulses and an $m = 0$ state of a model atom with the same ionization potential, for which we find the same yield of cut-off harmonics within a factor of ~ 2.

Finally, we may note that although for technical reasons we have used the $|m| = 6$ state of our model potential throughout, all our findings are of general nature and equally valid even for $|m| = 1$. Furthermore, thus far we have kept our discussions general and independent of any specific molecular or atomic system. Possible candidates for an experimental implementation of our schemes are derivatives of Benzene, for which the lowest degenerate state $|E_{\pm}\rangle$ has a binding energy of about 6 eV and can be excited from the ground state by the second harmonic of Titanium-Sapphire.

III. HIGH HARMONIC GENERATION IN POLARIZABLE MOLECULES

The theory of HHG is based on the SAE approximation [43]. It is assumed that only the valence electron interacts with the strong laser field. The residual electron core remains static and does not contribute to the interaction. As a result, the valence electron and the core electrons are regarded as uncorrelated. This approach originates in the observation that during HHG essentially single electrons ionize. In fact, this observation has motivated our analysis in the previous section. Experimentally, HHG has been performed with noble gas atoms, noble gas clusters [44], and with small molecules [45, 46]. Many experiments so far were found to be in agreement with SAE theory.

However, experimental [47–49] and theoretical [29] evidence was found that SAE theories cannot describe optical field ionization of highly polarizable systems, such as large molecules and metallic clusters. Due to the high electron mobility and polarizability, a factorization into valence and core electrons is no longer valid and the complete, correlated multi-electron (ME) dynamics has to be taken into account. This raises the question as to which extent the SAE approximation is applicable to non-perturbative phenomena in complex materials [50, 51].

Methodology. Here, HHG in highly polarizable electron systems is investigated by a one dimensional (1D) MCTDHF analysis. Despite various attempts [52, 53], currently MCTDHF still appears to be the most promising method that can properly model non-perturbative, correlated ME quantumdynamics [32, 34, 35, 54–56].

The MCTDHF method makes the ansatz for an n-electron wave function ψ in the form

$$\psi(x_1, ..., x_n; t) = \frac{1}{\sqrt{n}} \sum_{i_1=1}^{m} ... \sum_{i_n=1}^{m} A_{i_1,...,i_n}(t)\varphi_{i_1}(x_1; t)...\varphi_{i_n}(x_n; t). \qquad (6)$$

Here, spatial and spin coordinates are denoted by $x_i = (r_i, s_i)$. Anti-symmetry of the wave function is ensured by imposing anti-symmetry on the expansion coefficients, $A_{i_1,...,i_n}$, i.e. $A_{i_1,..,i_l,...,i_k,...,i_n} = -A_{i_1,..,i_k,...,i_l,...,i_n}$, and $A_{i_1,..,i_l,...,i_k,...,i_n} = 0$ if $i_l = i_k$. Therefor, out of this n^m coefficient only $\binom{m}{n}$ are linearly independent. The sum over all permutations for a given set of $\{i_1, ..., i_n\}$ combines the orbitals into a Slater determinate. This means that MCTDHF consists in approximating the exact wave function as linear combination of

Slater determinants, which are constructed from $m \geq n$ single-particle spin orbitals, φ_i. Note that both, the coefficients, $A_{i_1,...,i_n}$ and the orbitals, φ_i, are time dependent. Without loss of generality we impose ortho-normality on the orbitals in order to obtain a uniquely defined expansion.

The time evolution equations for $A_{i_1,...,i_n}(t)$ and $\varphi_i(x;t)$ are obtained from the Dirac-Frenkel variational principle,

$$\langle \delta\psi | i\partial_t - H | \psi \rangle = 0. \tag{7}$$

Note that it is sufficient to derive the evolution equations for one particle only, since, due to the exchange symmetry, they must be identical for all particles. The resulting equations of propagation form a system of coupled non-linear partial differential equations. Further details on the method my be found elsewhere [34]. In particular, our implementation here is identical to the one in [35]

Modeling. The 1D MCTDHF analysis is based on the solution of the 1D Schrödinger equation for the $n = 4$ electron potential

$$V(r_1, r_2, r_3, r_4) = \sum_{i=1}^{4} \sum_{j>i}^{4} \left(-\frac{Z}{\sqrt{r_i^2 + a_n^2}} + \frac{1}{\sqrt{(r_i - r_j)^2 + a_e^2}} \right). \tag{8}$$

Here, the first term represents the nuclear binding potential, Z is the charge state, and a_n is the shielding parameter of the electron-nucleus interaction. The shielded model potential represents an atom or a small cluster. We believe that it is closer to a small cluster, with a harmonic oscillator potential part close to the center, and a Coulomb far-range potential far away from the center. Further, the second term models the electron-electron interaction potential with shielding parameter a_e. The laser is coupled in velocity gauge and in dipole approximation. Atomic units (a.u.) are used throughout, unless otherwise stated.

Parameters. 1D ME simulations tend to overestimate the polarizability. In order to keep the polarizability at a reasonable level, the ionization potential had to be chosen slightly higher than usual values of complex materials (for instance benzene: $I_p = 0.35$ a.u.). The softening parameter used for the SAE system is $a_n = 1.414$, and the parameters to model highly polarizable atoms are $a_n = 0.80, a_e = 1.0$, and $Z = 4$. The binding energy of the four-electron ground state is $E_0^{(4)} = 8.5$ a.u. and the successive single electron ionization potentials are given by 0.5, 1.07, 3.09 and 3.93 a.u. The static polarizability, α_0, is calculated by using the relation

$$\alpha_0 = \frac{1}{\mathcal{E}(t)} \int \Delta\rho(r;t) r \, \mathrm{d}r, \tag{9}$$

where $\Delta\rho(x;t)$ is the change in electron density caused by the laser field, $\mathcal{E}(t)$. We find a polarizability of $\alpha_0 = 31$ Å^3, which lies between the polarizability of transition metal atoms and clusters; for example, Nb: $\alpha_0 = 15$ Å^3, $I_p = 0.248$ a.u.; C_{60}: $\alpha_0 = 80$ Å^3, $I_p = 0.279$ a.u.. Finally, the laser parameters are as follows: center wavelength $\lambda_0 = 1000$ nm, peak intensity $I = 2 \times 10^{14}$ W/cm^2, Gaussian envelope with FWHM width $\tau = 4T_0$, and oscillation period $T_0 = 3.33$ fs. The evolution of the wave function is calculated between 20 optical cycles before and 80 optical cycles after the laser pulse maximum.

With respect to MCTDHF, the calculations reported here are obtained for $n = 4$ electrons using $m = 8$ expansion functions. The Schrödinger equation is solved in a simulation box with size $l = \pm 360$ a.u. on a uniform 1D grid with 2400 grid points, using a second-order finite difference representation. To avoid reflection at the boundaries, complex absorption potentials (CAP) are used. That is, the total Hamiltonian is modified by adding $i\sum_{i=1}^{4}\{1 + \cos[\pi(x_i - x_{cap})/(|l| - x_{cap})]\}$ for $|x_i| > x_{cap} = 270$ a.u. and 0 otherwise. The time-integration is performed by a self-adaptive, high-order Runge-Kutta integrator with a relative numerical accuracy of 10^{-8}. Convergence was checked with respect to all of these parameters. In particular, increasing the number of expansion functions to $m = 12$ does not change our finding and changes for instance the ionization yield by less than 4%, indicating that our calculations are essentially converged.

A. Appearance of a Second Plateau in High Harmonic Spectra of Polarizable Molecules

In Fig. 4 the harmonic spectrum is shown for a four-electron system (full line) and a SAE system (dashed line) with the same HOMO (highest occupied molecular orbital) ionization potential $I_p = 0.5$ a.u.. The harmonic spectrum is obtained as the modulus of the Fourier transform of the dipole expectation value,

$$d(t) = \left\langle \Psi(t) \left| \sum_{i=1}^{4} r_i \right| \Psi(t) \right\rangle. \tag{10}$$

$d(t)$ was sampled 256 times per optical cycle. Note, however, that the time step in-between these sample points was self-adaptive. The cut-off energy $E_{\text{cut-off}} = I_p + 3.17U_p$ is in agreement with the standard SAE cut-off law [5, 43]. Here $U_p = (\mathcal{E}_0/2\omega_0)^2 = 0.68$ a.u. is the ponderomotive energy, $\mathcal{E}_0$ is the laser peak field strength, and ω_0 denotes the laser circular center frequency. The ME spectrum reveals in addition to the regular, first cut-off a second one. Note, that in the SAE system, ionization is saturated before the peak of the laser pulse, which is not the case for the multi-electron case, where the saturation intensity is increased due to the molecules polarizability [29, 48]. Thus the early saturation of the SAE system reduces the probability of electron trajectories that return with high energy and therewith results in a low HH yield of the plateau.

B. Collective Excitation Cause a New Plateau in High Harmonic Spectra of Polarizable Molecules

To identify the origin of the second plateau we have performed a time-frequency analysis of the ME spectrum, depicted in Fig. 5. The dipole moment is truncated by the window function

$$\sigma_w(t; t_r, T_w) = \frac{1}{\sqrt[4]{\pi T_w}} \exp \left[-\frac{(t - t_r)^2}{2T_w^2} \right], \tag{11}$$

with $T_w = 0.2T_0$ and then Fourier transformed, i.e.

$$d_\sigma(\omega, t_r; T_w) = \mathcal{F}[\sigma_w(t; t_r, T_w)d(t)]. \tag{12}$$

The harmonics corresponding to the first and the second plateau are depicted by orange and violet, respectively. The time-frequency analysis is a way to connect the quantum mechanical result with the classical three-step model [5], model of HHG. It cuts small chunks in time out of the wave function and determines their energy at the time of return to the parent ion. For a SAE harmonic spectrum, the resulting graph of harmonic energy versus return time is very close to the result obtained by the classical three-step analysis.

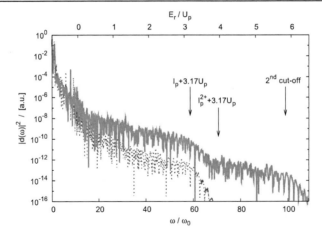

Figure 4. Spectra of the dipole moment squared, $|d(\omega)|^2$, of a highly polarizable ($\alpha_0 = 31\text{Å}^3$), four-electron model system, $I_p = 0.5$ a.u., (full line), and corresponding SAE calculation for the same I_p (dashed line). The lower x-axis gives the harmonic order, the upper x-axis gives the classical return energy, $E_r = E - I_p$ with E the harmonic photon energy in units of the ponderomotive energy, U_p. The standard cut-off harmonic at $(E - I_p)/U_p = 3.17$ and the predicted second cut-off harmonic are marked with an arrow. Additionally we have marked the expected SAE cut-off harmonic if HHG were to occur from the singly charged state of the model system. Laser parameters: $\lambda_0 = 1000$ nm, peak intensity $I = 2 \times 10^{14}$ W/cm^2, FWHM pulse duration $\tau = 4T_0$, optical period $T_0 = 3.33 fs$, Gaussian envelope.

This correspondence allows an interpretation of the time-frequency plot and of HHG in terms of classical trajectories.

The arrow in Fig. 5 denotes the regular, first cut-off. A comparison to the SAE three-step model allows us to determine the importance of ME effects in HHG. The electron return phase of the cut-off trajectory creating the highest harmonic in Fig. 5 is around $60°$ after the pulse maximum. This is shifted with respect to the three-step model cutoff trajectory that is born at $163°$ before and returns at $80°$ after the pulse maximum [5] (see left panel in Fig. 5). The difference in the return times arises from a many-body effect. In ME systems the laser field induces a polarization of the bound electrons that exerts a repelling force on the ionizing electron. This additional potential decreases rapidly with the distance from the parent system and hence affects the electron trajectories only in the vicinity of the nucleus. Therefore, it mostly shifts the birth and return time of the electron trajectories and only

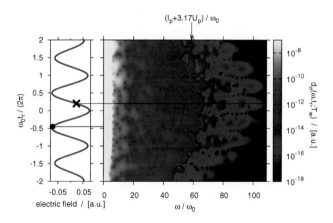

Figure 5. Contour plot of a time-frequency analysis of the four-electron spectrum in Fig. 4 (right panel). The window function is a Gaussian pulse with 0.2 optical cycles FWHM duration. The return time, t_r, is plotted versus the harmonic frequency, ω, normalized to the laser frequency, ω_0. The standard cut-off harmonic has been marked with an arrow. The left panel shows the corresponding laser electric field. The time of birth and the time of return for a classical electron acquiring the maximum kinetic energy during its excursion in the laser field are marked by a dot and a cross, respectively.

weakly affects the highest achievable cut-off energy.

The violet area in Fig. 5 corresponds to the second plateau. Surprisingly, the first and second plateau show similar patterns. The maximum energy in each half cycle occurs at the same return phase for both plateaus. This strongly indicates that the harmonics in both plateaus are generated by the same electron trajectories, starting from different initial states. The strong laser field brings the medium in a coherent superposition of ground and excited states. HHG from excited states can take place as long as the phases of ground and excited states are coherently locked.

In ME systems there exist two types of excitation, collective excitations and individual particle excitations. Single electron excitations can be excluded for the following reasons: (i) the SAE calculation in Fig. 4 does not show a second plateau; (ii) the energy difference between the first and the second cut-off is larger than the HOMO ionization potential; (iii) the absence of doubly ionized states excludes HHG from a deeper bound electron; (iv) while HHG from a coherent superposition of the ground state and an excited single electron state

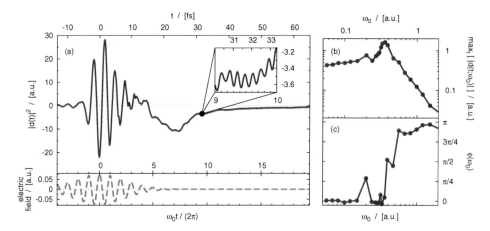

Figure 6. (a) Time dependence of the applied laser electric field (lower panel) and the resulting dipole moment, $d(t)$ (upper panel), of the highly polarizable molecule in Fig. 4. In the inset we show a magnified part of the dipole moment to illustrate the remaining excitations after the end of the laser pulse. (b) Maximum excursion of the center of gravity of the electron density and (c) phase shift, ϕ, between the dipole signal and the laser electric field as a function of the applied laser frequency. In (b) and (c) we have used a continues-wave laser which was linearly switched on over four optical cycles reaching a maximum intensity of 2^{13} W/cm^2. The pulse was propagated over 13 optical cycles.

does indeed produce two plateaus, it does not result in an overall increase of the standard cut-off law [57], because the HOMO electron is born with the same energy after ionization regardless of its initial state.

The excitation mechanism responsible for the second plateau is revealed in Fig. 6. In Fig. 6(a) the dipole moment, $d(t)$, is plotted as a function of time. We find that the dipole moment exhibits oscillations even after the laser pulse, proving a laser induced excitation of the system. The close-up in the inset of Fig. 6(a) shows that the excitation oscillates at a frequency $\omega_p = 0.35$ a.u.. In Fig. 6(b) and (c) the excitation spectrum of the system is determined by probing the response of the system to a plane wave laser signal as function of the laser center frequency ω_0. The maximum excursion of the center of gravity of the electron density and the phase shift ϕ between the dipole signal, $d(t)$, and the laser electric field are plotted. At resonance, light absorption is maximum, and the center of charge motion of the electron is approximately $90°$ out of phase with the laser field.

The resonance in Fig. 6(b) is a collective plasmon resonance. In three-dimensional (3D) plasmas and clusters the collective excitation of the bound electrons is referred to as a plasma wave and as a plasmon, respectively. We define here the term plasmon as the corresponding collective motion of the bound electrons of our 1D model system. The collective frequency depends on the system geometry, explaining the difference between plasma and plasmon frequency. As our model system is neither the 1D limiting case of a bulk nor of a sphere, the usual 3D plasma/plasmon frequency is not applicable. The 1D plasmon frequency is determined by the 1D geometry of our model system. As an analytical expression is currently not known, we use the above determined numerical value.

Single electron excitations cannot explain the resonance in Fig. 6. First because they have a narrow linewidth. The broad width (≈ 0.1 a.u.) of the resonance is a strong indicator of a collective process. Second, the oscillation shown in Fig. 6(a) decays. The decay is also a typical signature of a plasmon, as due to microscopic collisions energy is transferred from the collective electron motion into thermal, single electron motion. In contrast to that, the lifetime of a single electron excitation is infinite. We have tested that the decay is not an artefact of the MCTDHF formalism. Increasing the number of determinants does not change the time-dependence of the dipole signal significantly. Also, this decay does not come from ionization and a decrease of the norm at the absorbing boundaries. We find that the ionization yield is virtually constant (increasing by 0.003 during the last 40 optical cycles of the simulation time), while the amplitude of the plasmon oscillation is reduced by almost a factor of two.

The match between the plasmon frequency of $\omega_p = 0.35$ a.u. and the frequency of the dipole oscillation in Fig. 6(a) proves that the plasmon excitation is responsible for the second cut-off observed in HHG. Moreover, further analysis shows that multi-plasmon states are responsible for the second plateau in the HH signal (Fig. 4). The oscillation in Fig. 6(a) has non-sinusoidal components. A Fourier transform of $d(t)$ shows frequency components at multiples of the plasmon frequency, $k\omega_p, k = 1, 2, 3, ...$ Hence, the non-sinusoidal behavior arises from the interference between these multi-plasmon states, quivering at multiples of the plasma frequency. Consequently, the width of the second harmonic plateau is determined by the highest order of the excited multi-plasmon state, which is $k = 4$ in Fig.

4. The multi-plasmon excitation comes from the nonlinear (anharmonic) part of the binding potential. Whereas, close to the center, the potential has a quadratic (harmonic oscillator) space dependence supporting a single plasmon, the far-range Coulomb part of the potential adds nonlinear terms responsible for the creation of multiple harmonics of the plasmon oscillation.

In contrast to single electron systems, in ME systems the collective excitation energy adds to the harmonic cut-off. The reason is that the collective energy stays in the remaining bound electrons and does not get lost while the valence electron makes its excursion into the continuum. To elucidate this point we define the ground-state energies of the neutral and the singly ionized system, $E_0^{(4)}$ and $E_0^{(3)}$, respectively. The energies of the according plasmon states are given by $E_0^{(4)} + \omega_p$ and $E_0^{(3)} + \omega_p$. Here, we neglect the difference in the plasmon frequencies between the neutral and the singly ionized state, since the difference is of the order of the difference between two adjacent harmonics. As a result, in both cases the HOMO potential is given by $I_p = E_0^{(4)} - E_0^{(3)}$. Although for our ME system the HOMO ionization potential for the plasmon state is slightly smaller, this is a reasonable approximation. In particular, since the difference further decreases for an increasing number of electrons and will eventually disappear in real ME system which usually have considerably more than four electrons.

There are different pathways by which HHG can take place, which are illustrated in figure 5. Before ionization the system is in its ground state, $E_0^{(4)}$, remains in the ground state, $E_0^{(3)}$, after ionization of the valence electron and returns upon recombination to its initial state [figure 5(a)]. This is the standard HHG situation. The system may also start out in a plasmon state, $E_0^{(4)} + \omega_p$, remains in the plasmon state, $E_0^{(3)} + \omega_p$, after ionization, and returns to its four-electron plasmon state upon recombination [figure 5(b)]. For both cases the cut-off law is $3.17U_p + [E_0^{(4)} - E_0^{(3)}] = 3.17U_p + I_p$ since for the latter the plasmon frequency cancels out. However, if ionization starts from the plasmon state, $E_0^{(4)} + \omega_p$, but the electron returns to the ground sate, $E_0^{(4)}$, upon recombination, the plasmon energy is converted into harmonic photon energy, extending the cut-off, i.e. $3.17U_p + I_p + \omega_p$ [figure 5(a)].

So far we have discussed the single system response. At the moment, the observation of

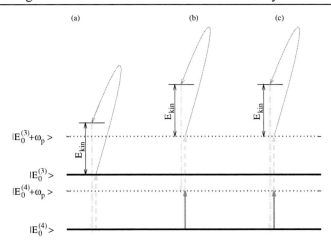

Figure 7. Energy level diagram for different pathways for HHG in polarizable molecules. (a) standard HHG: an electron is set free by tunnel ionization (broken arrow), quivers in the electric laser field (thin full arrow) gaining kinetic energy, E_{kin}, and recombines to the four-electron ground state, $E_0^{(4)}$ (dashed dotted arrow). (b) as (a) but starting from and returning to the plasmonstate $eng4 + \omega_p$. (c) as (b) but returning to the ground state, $E_0^{(4)}$. Here, $E_0^{(3)}$ and $E_0^{(3)} + \omega_p$ denote the singly ionized ground state and singly ionized plasmon state, respectively.

HHG from excited states is experimentally untested. A significant macroscopic signal will only be created, when the individual systems emit harmonic radiation in phase. Hence, the question has to be answered whether the second plateau can be detected in a macroscopic medium consisting of many individual systems. We believe that this is the case for the following reason. The phase of the harmonic signal emitted by a single system is determined by the phase difference between ground state and excited (plasmon) state. In order for coherent emission to occur, the phase difference between ground and excited state has to be the same for all emitters. As the plasmon excitation is driven by the laser field, the phase difference between ground and (plasmon) excited state is exclusively determined by the laser field, and therefore is the same for each system. As a result, the multi-plasmon states are (laser) phase locked to the ground state. The contributions from individual atoms add up coherently and HHG can take place from the ground as well as from excited states.

With respect to experimental observation, we believe that metal clusters with their simple geometry and with plasmon energies of a couple of eV are good candidates. Although atoms also support collective oscillations, which are known as giant (shape) resonances [58], their lifetime is usually very short, three to four times the plasmon oscillation period. As a result the plasmon will likely decay before the active electron can return and create harmonic radiation.

Finally, a three dimensional (3D) numerical analysis of non-perturbative ME dynamics is currently computationally out of reach. Although 3D effects may affect the quantitative structure of the calculated spectra, the essential physics underlying HHG and the nonlinear excitation of multi-plasmon frequencies is contained in our 1D analysis. Ionization potential and polarizability of our model potentials were chosen to be close to values that are representative for highly polarizable, complex materials. Therefore, our 1D analysis of HHG in ME systems gives a reasonable approximation to the "real world" process.

IV. CONCLUSION

In conclusion we have investigated HHG from excited molecular systems. We have demonstrated that the properties of the emitted radiation are strongly influenced by the internal state of the molecules. In particular we have focused on two types of excited states: plasmon states and ring-current states. While the former ones are of multi-electronic nature and can be excited in polarizable systems, the latter ones are single-electron states but result in non-zero net internal electron momentum. The outcome of our studies can be summarized as follows.

Using multi-electron numerical methods the existence of an a additional – second – plateau in the HH intensity spectrum is predicted. While the cutoff of the first plateau agrees with the cutoff law of noble gases, the second plateau presents a signature of electron correlation and is due to the non-linear excitation of collective plasmon oscillations. It arises from electrons that are ionized from an excited plasmon state and recombine to the ground state. The plasmon signatures present a novel tool for the investigation of the non-perturbative multi-electron dynamics in complex materials, a regime that is experimentally very difficult to access otherwise.

By solving the time-dependent Schrödinger equation in two spatial dimensions for initial states with internal momentum we could demonstrate that both ionization and HHG by strong laser pulses proceeds in a non-standard way. Specifically we have used ring-currents as an incarnation of such states. It was shown that the internal state is transferred to the tunneling wave packets, which receive an initial momentum kick; a so far unexplored mechanism. The direction of the kick is determined by the sign of the bound state's phase. To measure this effect we introduced a general method for mapping the transversal momentum distribution of tunneling wave packets by all-optical technology, which uncovers the pronounced asymmetry in the momentum spectrum. It was demonstrated that during the recombination process the internal state is transferred to the polarization state of the emitted photons, which leads to the production of novel, spatially and temporally coherent, nearly circularly polarized attosecond XUV/X-ray pulses. These pulses might open the door to tabletop-scale imaging and spectroscopy of ferromagnetic materials on extreme time-scales.

Acknowledgments

Fruitful discussions with S. Gräfe and O.D. Mücke are gratefully acknowledged. This work was partly financed by the Austrian Science Fund, grants SFB016 and U33-16.

References

[1] P. Paul et al., *Science* **292**, 1689 (2001).

[2] R. Kienberger et al., *Nature* **427**, 817 (2004).

[3] I. Sola et al., *Nature Phys.* **2**, 319 (2006).

[4] G. Sansone et al., *Science* **314**, 443 (2006).

[5] P. B. Corkum, *Phys. Rev. Lett.* **71**, 1994 (1993).

[6] K. C. Kulander, K. J. Schafer, and J. L. Krause, *Proceedings of the Workshop, Super Intense Laser Atom Physics (SILAP) III*, edited by B. Piraux (Plenum, New York) (1993).

[7] P. B. Corkum and F. Krausz, *Nature Phys.* **3**, 381 (2007).

[8] M. Lewenstein et al., *Phys. Rev. A* **49**, 2117 (1994).

[9] A. Scrinzi, M. Y. Ivanov, R. Kienberger, and D. Villeneuve, *jpb* **39**, R1 (2006).

[10] I. V. Litvinyuk et al., *Phys. Rev. Lett.* **90**, 233003 (2003).

[11] A. S. Alnaser et al., *Phys. Rev. Lett.* **93**, 113003 (2004).

[12] D. Pavičić et al., *Phys. Rev. Lett.* **98**, 243001 (2007).

[13] J. Muth-Böhm, A. Becker, and F. H. M. Faisal, *Phys. Rev. Lett.* **85**, 2280 (2000).

[14] X. Tong, Z. Zhao, and C. D. Lin, *Phys. Rev. A* **66**, 033402 (2002).

[15] G. L. Kamta and A. D. Bandrauk, *Phys. Rev. A* **74**, 033415 (2006).

[16] R. Velotta et al., *Phys. Rev. Lett.* **87**, 183901 (2001).

[17] C. Vozzi et al., *Phys. Rev. Lett.* **95**, 153902 (2005).

[18] J. Itatani et al., *Nature* **432**, 867 (2004).

[19] R. Torres et al., *Phys. Rev. Lett.* **98**, 203007 (2007).

[20] J. Levesque et al., *Phys. Rev. Lett.* **98**, 183903 (2007).

[21] T. Kanai, S. Minemoto, and H. Sakai, *Phys. Rev. Lett.* **98**, 053002 (2007).

[22] T. Kanai, S. Minemoto, and H. Sakai, *Nature* **435**, 470 (2005).

[23] J. Itatani et al., *Phys. Rev. Lett.* **94**, 123902 (2005).

[24] J. Levesque et al., *Phys. Rev. Lett.* **99**, 243001 (2007).

[25] S. Baker et al., *Science* **312**, 424 (2006).

[26] H. Niikura, D. M. Villeneuve, and P. B. Corkum, *Phys. Rev. Lett.* **94**, 083003 (2005).

[27] M. Kitzler et al., *Phys. Rev. A* **76**, 011801(R) (2007).

[28] M. Lezius, V. Blanchet, D. Villeneuve, A. Stolow, and M. Y. Ivanov, *Phys. Rev. Lett.* **86**, 51 (2001).

[29] M. Kitzler, J. Zanghellini, C. Jungreuthmayer, M. Smits, A. Scrinzi, and T. Brabec, *Phys. Rev. A* **70**, 041401(R) (2004).

[30] X. Xie et al. *Phys. Rev. Lett.* **101**, 033901 (2008).

[31] I. Barth et al., *J. Am. Chem. Soc.* **128**, 7043 (2006).

[32] G. Jordan and A. Scrinzi, *New. J. Phys.* **10**, 025035 (2008).

[33] A. Gordon, F. X. Kartner, N. Rohringer, and R. Santra, *Phys. Rev. Lett.* **96**, 223902 (2006).

[34] J. Caillat, J. Zanghellini, M. Kitzler, O. Koch, W. Kreuzer, and A. Scrinzi, *Phys. Rev. A* **71**, 012712 (2005).

[35] J. Zanghellini, M. Kitzler, T. Brabec, and A. Scrinzi, *J. Phys. B* **37**, 763 (2004).

[36] I. Barth and J. Manz, *Phys. Rev. A* **75**, 012510 (2007).

[37] X. Xie et al., *Phys. Rev. A* **76**, 023426 (2007).

[38] X. Xie et al., *J. Mod. Optics* **54**, 999 (2007).

[39] N. Dudovich et al., *Phys. Rev. Lett.* **97**, 253903 (2006).

[40] S. Lovesey and S. Collins, *X-Ray Scattering and Absorption by Magnetic Materials* (Oxford Univ. Press, 1996).

[41] F. Nolting et al., *Nature* **405**, 767 (2000).

[42] J. Chakhalian et al., *Nature Phys.* **2**, 244 (2006).

[43] J. L. Krause, K. J. Schafer, and K. C. Kulander, *Phys. Rev. Lett.* **68**, 3535 (1992).

[44] J. W. G. Tisch, T. Ditmire, D. J. Fraser, N. Hay, M. B. Mason, E. Springate, J. P. Marangos, and M. H. R. Hutchinson, *J. Phys. B* **30**, L709 (1997).

[45] Y. Liang, S. August, L. Chin, Y. Beaudoin, and M. Chaker, *J. Phys. B* **27**, 5119 (1994).

[46] R. de Nalda, E. Heesel, M. Lein, N. Hay, R. Velotta, E. Springate, M. Castillejo, and J. P. Marangos, *Phys. Rev. A* **69**, 031804(R) (2004).

[47] M. Lezius, V. Blanchet, M. Y. Ivanov, and A. Stolow, *J. Chem. Phys.* **117**, 1575 (2002).

[48] V. R. Bhardwaj, P. B. Corkum, and D. M. Rayner, *Phys. Rev. Lett.* **91**, 203004 (2003).

[49] A. N. Markevitch, S. M. Smith, D. A. Romanov, H. B. Schlegel, M. Y. Ivanov, and R. J. Levis, *Phys. Rev. A* **68**, 011402(R) (2003).

[50] V. Véniard, R. Taieb, and A. Maquet, *Phys. Rev. A* **65**, 013202 (2001).

[51] H. S. Nguyen, A. D. Bandrauk, and C. A. Ullrich, *Phys. Rev. A* **69**, 063415 (2004).

[52] S. Patchkovskii, Z. Zhao, T. Brabec, and D. M. Villeneuve, *Phys. Rev. Lett.* **97**, 123003 (2006).

[53] R. Santra and A. Gordon, *Phys. Rev. Lett.* **96**, 073906 (2006).

[54] F. Remacle, M. Nest, and R. D. Levine, *Phys. Rev. Lett.* **99**, 183902 (2007).

[55] M. Nest, R. Padmanaban, and P. Saalfrank, *The Journal of Chemical Physics* **126**, 214106 (2007).

[56] J. Zanghellini, C. Jungreuthmayer, and T. Brabec, *Journal of Physics B: Atomic, Molecular and Optical Physics* **39**, 709 (2006).

[57] J. B. Watson, A. Sanpera, X. Chen, and K. Burnett, *Phys. Rev. A* **53**, R1962 (1996).

[58] M. Y. Amusia and J.-P. Connerade, *Rep. Prog. in Phys.* **63**, 41 (2000).

In: Atomic, Molecular and Optical Physics...

Editor: L.T. Chen, pp. 491-522

ISBN 978-1-60456-907-0

© 2009 Nova Science Publishers, Inc.

Chapter 13

UNIVERSAL DYNAMICAL DECOHERENCE CONTROL OF MULTI-PARTITE SYSTEMS

Goren Gordon and Gershon Kurizki

Department of Chemical Physics, Weizmann Institute of Science,

Rehovot 76100, Israel

Abstract

A unified theory is given of dynamically modified decay and decoherence of field-driven multipartite systems. When this universal framework is applied to two-level systems (TLS) or qubits experiencing either amplitude or phase noise (AN or PN) due to their coupling to a thermal bath, it results in completely analogous formulae for the modified decoherence rates in both cases. The spectral representation of the modified decoherence rates underscores the main insight of this approach, namely, the decoherence rate is the spectral overlap of the noise and modulation spectra. This allows us to come up with general recipes for modulation schemes for the optimal reduction of decoherence under realistic constraints. An extension of the treatment to multilevel and multipartite systems exploits intra-system symmetries to dynamically protect multipartite entangled states. Another corollary of this treatment is that entanglement, which is very susceptible to noise and can die, i.e., vanish at finite times, can be resuscitated by appropriate modulations prescribed by our universal formalism. This dynamical decoherence control is also shown to be advantageous in quantum computation setups, where control fields are applied concurrently with the gate operations to increase the gate fidelity.

PACS: 03.65.Yz, 03.65.Ta, 42.25.Kb

Keywords: Decoherence control; dynamical control; quantum computation

I. INTRODUCTION

A quantum system may decohere, under the influence of its environment, in one (or both) of the following fashions: (a) Its population may decay (amplitude noise, AN) to a continuum or a thermal bath, a process that characterizes spontaneous emission of photons by excited atoms [1], vibrational and collisional relaxation of trapped ions [2] and cold atoms in optical lattices [3], as well as the relaxation of current-biased Josephson junctions [4, 5]. (b) It may undergo proper dephasing (phase noise, PN), which randomizes the phases but does not affect the population of quantum states, as in the case of phase interrupting collisions [6].

Most theoretical and experimental methods aimed at assessing and controlling (suppressing) decoherence of qubits (two-level systems, that are the quantum mechanical equivalents of classical bits) have focussed on one of two particular situations: (a) single qubits decohering *independently*; or (b) many qubits *collectively* perturbed by the same environment. Thus, quantum communication protocols based on entangled two-photon states have been studied under collective depolarization conditions, namely, *identical* random fluctuations of the polarization for both photons [7, 8]. Entangled qubits that reside at the same site or at equivalent sites of the system, e.g., atoms in optical lattices, have likewise been assumed to undergo identical decoherence.

For independently decohering qubits, the most powerful approach suggested thus far for the suppression of decoherence appears to be the "dynamical decoupling" (DD) of the system from the bath [9–22]. The standard "bang-bang" DD, i.e. π-phase flips of the coupling via strong and sufficiently frequent resonant pulses driving the qubit [12–14], has been proposed for the suppression of proper dephasing [23]. Several extensions have been suggested to further optimize DD under proper dephasing, such as multipulse control [19], continuous DD [18] and concatenated DD [20]. DD has also been adapted to suppress other types of decoherence couplings such as internal state coupling [21] and heating [14].

Our group has proposed a universal strategy of approximate DD [24–30] for both decay and proper dephasing, by either pulsed or continuous wave (CW) modulation of the system-bath coupling. This strategy allows us to optimally tailor the strength and rate of the modulating pulses to the spectrum of the bath (or continuum) by means of a simple universal formula. In many cases, the standard π-phase "bang-bang" is then found to be inadequate or non-optimal.

Multiqubit, or more generally, multipartite entanglement, is the cornerstone of many quantum information processing applications. However, it is very susceptible to decoherence, decays faster than single-qubit coherence, and can even completely disappear in finite time, an effect dubbed Entanglement Sudden Death (ESD) [31–38].

Control of multiqubit decoherence is of great interest, because it can help protect entanglement of such systems. Entanglement is effectively protected in the collective decoherence situation, by singling out decoherence-free subspaces (DFS) [39], wherein symmetrically degenerate many-qubit states, also known as "dark" or "trapping" states [6], are decoupled from the bath [17, 40–42].

Entangled states of two or more particles, wherein each particle travels along a different channel or is stored at a different site in the system, may present more challenging problems insofar as combatting and controlling decoherence effects are concerned: if their channels or sites are differently coupled to the environment, their entanglement is expected to be more fragile and harder to protect.

To address these fundamental challenges, we have developed a very general treatment. This treatment extends our previously published single-qubit universal strategy [24, 26, 27, 43, 44] to *multiple entangled systems (particles)* which are either coupled to partly correlated (or uncorrelated) finite-temperature baths or undergo locally-varying random dephasing. Furthermore, it applies *to any difference* between the couplings of individual particles to the environment. This difference may range from the large-difference limit of completely independent couplings, which can be treated by the single-particle dynamical control of decoherence via modulation of the system-bath coupling, to the opposite zero-difference limit of completely identical couplings, allowing for multi-particle collective behavior and decoherence-free variables [16, 17, 40–42, 45–48]. The general treatment presented here is valid anywhere between these two limits and allows us to pose and answer the key question: under what conditions, if any, is *local control* by modulation, addressing each particle individually, preferable to *global control*, which does not discriminate between the particles?

We show that in the realistic scenario, where the particles are differently coupled to the bath, it is *advantageous to locally control each particle by individual modulation, even if such modulation is suboptimal* for suppressing the decoherence for the single particle. This local modulation allows synchronizing the phase-relation between the different modulations and eliminates the cross-coupling between the different systems. As a result, it allows us to preserve the multipartite entanglement and reduces the multipartite decoherence problem to the single particle decoherence problem. We show the advantages of local modulation, over global modulation (i.e. identical modulation for all systems and levels), as regards the preservation of arbitrary initial states, preservation of entanglement and the intriguing possibility of entanglement increase compared to its initial value.

The experimental realization of a universal quantum computer is widely recognized to be difficult due to decoherence effects, particularly dephasing [2, 49–51], whose deleterious effects on entanglement of qubits via two-qubit gates [52–54] is crucial.

To help overcome this problem, we put forth a universal dynamical control approach to the dephasing problem during *all the stages of quantum computations*, namely (i) storage, wherein the quantum information is preserved in between gate operations; (ii) single-qubit gates, wherein individual qubits are manipulated, without changing their mutual entanglement; and (iii) two-qubit gates, that introduce controlled entanglement. We show that in terms of reducing the effects of dephasing, it is advantageous to concurrently and specifically control all the qubits of the system, whether they undergo quantum gate operations or not. Our approach consists in specifically tailoring each dynamical quantum gate, with the aim of suppressing the dephasing, thereby greatly increasing the gate fidelity. In the course of two-qubit entangling gates, we show that cross-dephasing can be completely eliminated by introducing additional control fields. Most significantly, we show that one can increase the gate duration, while simultaneously reducing the effects of dephasing, resulting in a total increase in gate fidelity. This is at odds with the conventional approaches, whereby one tries to either reduce the gate duration, or increase the coherence time.

In Sec. II we present a universal formula for the control of single-qubit zero-temperature relaxation and discuss several limits of this formula. In Sec. III the treatment is extended to the control of finite-temperature relaxation and decoherence and culminates in single-particle Bloch equations with dynamically modified decoherence rates that essentially obey the universal formula of Sec. II. We then discuss in Sec. IV the possible modulation arsenal for either AN or PN control. In Sec. V we discuss the extensions of the universal control formula to entangled multipartite systems. In Sec. VI we discuss the implementations of the universal formula to multipartite quantum computation. Sec. VII summarizes our conclusions whereby this universal control can effectively protect complex systems from a variety of decoherence sources.

II. SINGLE-QUBIT ZERO-TEMPERATURE RELAXATION

To gain insight into the requirements of decoherence control, consider first the simplest case of a qubit with states $|e\rangle, |g\rangle$ and energy separation $\hbar\omega_a$ relaxing into a zero-temperature bath via off-diagonal (σ_x) coupling, Fig. 1(a). The Hamiltonian is given by:

$$H = H_S + H_B + H_I \tag{1}$$

$$H_S = \hbar\omega_a|e\rangle\langle e| \tag{2}$$

$$H_B = \hbar \sum_j \omega_j|j\rangle\langle j| \tag{3}$$

$$H_I = \sum_j \mu_{ej}|e,0\rangle\langle g,1_j| + H.c. \tag{4}$$

the sum extending over all bath modes, with $|\{0\}\rangle$ and $|1_j\rangle$ denoting the bath vacuum and jth-mode single excitation, respectively, μ_{ej} being the corresponding transition matrix element and $H.c.$ being Hermitian conjugate. The Schrödinger equation for the evolution of the initially excited state amplitude $\alpha(t) = \langle e,0|\Psi(t)\rangle e^{i\omega_a t}$ with $\alpha(0) = 1$ can be exactly rendered in the form [24]:

$$\partial_t\alpha_e = -\int_0^t dt'\Phi(t-t')e^{i\omega_a(t-t')}\alpha_e(t'), \tag{5}$$

where

$$\Phi = \hbar^{-2}\langle e,0|H_I e^{-iH_0 t/\hbar} H_I|e,0\rangle = \hbar^{-2}\sum_j |\mu_{ej}|^2 e^{-i\omega_j t} \tag{6}$$

is the bath response/correlation function, expressible in terms of a sum over all transition matrix elements squared oscillating at the respective mode frequencies ω_j.

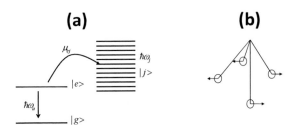

Figure 1. (a) Schematic drawing of system with off-diagonal coupling to a continuum or a bath. (b) Schematic drawing of a bath comprised of many harmonic oscillators with different frequencies, whose temporal dephasing after correlation time t_c renders the system-bath interaction practically irreversible.

It is the spread of oscillation frequencies ω_j that causes the environment response to decohere after a (typically short) *correlation time* t_c (Fig. 1(b)). Hence, the Markovian assumption that the correlation function decays to 0 instantaneously, $\Phi(t) \approx \delta(t)$, is widely used: it is in particular the basis for the venerated Lindblad's master equation describing decoherence [55]. It leads to exponential decay of α_e at the Golden Rule (GR) rate [1, 6]

$$R_{\mathrm{GR}} = 2\pi \sum_j |\mu_{ej}|^2 \delta(\omega_a - \omega_j). \tag{7}$$

We, however, are interested in the extremely non-Markovian time scales, much shorter than t_c, on which all bath modes excitations oscillate in unison and the system-bath exchange is *fully reversible*. How does one probe, or better still, maintain the system in a state corresponding to such time scales?

To this end, we assume modulations of $H_S(t)$ and $H_I(t)$,

$$H_S(t) = \hbar\,(\omega_a + \delta_a(t))\,|e\rangle\langle e| \tag{8}$$

$$H_I(t) = \sum_j \tilde{\epsilon}(t)\mu_{ej}|e,0\rangle\langle g,1_j| + H.c. \tag{9}$$

$$\epsilon(t) = \tilde{\epsilon}(t)e^{i\int_0^t d\tau\,\delta_a(\tau)}, \tag{10}$$

that results in the time-dependent function $\epsilon(t)$, which may pertain to *any intervention* in the system-bath dynamics: (i) measurements that effectively interrupt and completely dephase the evolution, describable by stochastic $\epsilon(t)$ [56]; (ii) coherent perturbations that describe *phase modulations* of the system-bath interactions [26, 57].

For any $\epsilon(t)$, the *exact* Eq. (5) is then rewritten as

$$\partial_t \alpha_e = -\int_0^t dt' \Phi(t-t')\epsilon(t)\epsilon^*(t')e^{i\omega_a(t-t')}\alpha_e(t'). \tag{11}$$

We now resort to the *crucial* approximation that $\alpha_e(t)$ *varies slower* than either $\epsilon(t)$ or $\Phi(t)$. This approximation is justifiable in the weak-coupling regime (to second order in H_I), as discussed below. Under this approximation, Eq. (11) is transformed into a differential equation describing relaxation at a time-dependent rate:

$$\partial_t \alpha_e = -R(t)\alpha_e. \tag{12}$$

The average time-dependent relaxation rate can be rewritten, by using the finite-time Fourier transforms of $\Phi(t)$ and $\epsilon(t)$, in the following form:

$$R(t) = 2\pi \int_{-\infty}^{\infty} d\omega\, G(\omega + \omega_a) F_t(\omega), \tag{13}$$

where

$$G(\omega) = \pi^{-1} \mathrm{Re} \int_0^{\infty} dt\, e^{i\omega t} \Phi(t) \tag{14}$$

is the spectral-response function of the bath, and

$$F_t(\omega) = |\epsilon_t(\omega)|^2/t \tag{15}$$

is the finite-time spectral intensity of the (random or coherent) intervention/modulation function, where the $1/t$ factor comes about from the definition of the decoherence rate *averaged* over the $(0, t)$ interval.

The relaxation rate $R(t)$ described by Eqs. (13)-(15) embodies our *universal recipe for dynamically controlled relaxation* [26, 57], which has the following merits: (a) it holds for any bath and any type of interventions, i.e. coherent modulations and incoherent interruptions/measurements alike; (b) it shows that in order to suppress relaxation we need to minimize the spectral overlap of $G(\omega)$, given to us by nature, and $F_t(\omega)$, which we may design to some extent; (c) most importantly, it shows that in the short-time domain, only broad (coarse-grained) spectral features of $G(\omega)$ and $F_t(\omega)$ are important. The latter implies that, in contrast to the claim that correlations with each individual bath mode must be accounted for, if we are to preserve coherence, we actually only need to characterize and suppress (by means of $F_t(\omega)$) the *broad features* of $G(\omega)$, the bath response function. The universality of Eqs. (13)-(15) will be elucidated in what follows, by focusing on several limits.

A. The Limit of Slow Modulation Rate

If $\epsilon(t)$ corresponds to sufficiently slow rates of interruption/modulation ν_t, the spectrum of $F_t(\omega)$ is much narrower than

$$\xi_a = \left| \frac{dG(\omega)}{d\omega} \right|_{\omega_a} \bigg/ G(\omega_a) \right|, \tag{16}$$

the interval of change of $G(\omega)$ around w_a, the resonance frequency of the system. Then $F_t(\omega)$ can be replaced by $\delta(\omega)$, so that the spectral width of G plays no role in determining R, and we may as well replace $G(\omega + \omega_a)$ by a spectrally finite, flat (*white-noise*) reservoir, i.e., take the *Markovian limit*. The result is that Eq. (13) coincides with the Golden-Rule (GR) rate, Eq. (7) (Fig. 2(a)):

$$R \approx R_{\mathrm{GR}} = 2\pi \int_{-\infty}^{\infty} d\omega \delta(\omega) G(\omega + \omega_a) = 2\pi G(\omega_a) \tag{17}$$

Namely, slow interventions do not affect the onset and rate of exponential decay.

B. The Limit of Frequent Modulation

If $\epsilon(t)$ describes extremely frequent interruptions or measurements $\nu_t \gg 1/t_c$, $F_t(\omega) = (\tau/2\pi)\mathrm{sinc}^2(\omega\tau/2)$ is much broader than $G(\omega + \omega_a)$, $\tau = 1/\nu_t$ being the time-interval between consecutive interruptions. We may then pull $F_t(\omega)$ out of the integral, whereupon Eq. (13) yields

$$R(t) \approx \tau \int_{-\infty}^{\infty} G(\omega + \omega_a) d\omega. \tag{18}$$

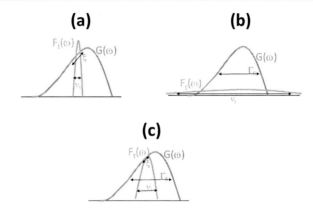

Figure 2. Frequency-domain representation of the dynamically-controlled decoherence rate in various limits (Sec. III). (a) Golden-Rule limit. (b) Quantum Zeno Effect (QZE) limit. (c) Anti-Zeno Effect (AZE) limit.

This limit is that of the quantum Zeno effect (QZE), namely, the suppression of relaxation as the interval between interruptions decreases [58–60]. In this limit the system-bath exchange is reversible and the system coherence is fully maintained (Fig. 2(b)). Namely, the essence of the QZE is that sufficiently rapid interventions prevent the excitation escape to the continuum, by *reversing* the exchange with the bath.

C. Intermediate Modulation Rate

In the intermediate time-scale of interventions, where the width of $F_t(\omega)$ is broader than the width ξ_a (so that the Golden Rule is violated) but narrower than the width of $G(\omega)$, (so that the QZE does not hold) the overlap of $F_t(\omega)$ and $G(\omega)$ *grows* as the rate of interruptions, or modulations increases. This brings about the *increase of relaxation rates $R(t)$ with the rate of interruptions*, marking the anti-Zeno effect (AZE) [15, 29, 61] (Fig. 2(c)). On such time-scales more frequent interventions (in particular interrupting measurements) *enhance the departure of the evolution from reversibility*. Namely, the essence of the AZE is that if you do not *intervene in time* to prevent the excitation escape to the continuum, then any intervention only drives the system further from its initial state.

We note that the AZE can only come about when the peaks of $F_t(\omega)$ and $G(\omega)$ *do not* overlap: its condition can be expressed as $\xi_a \neq 0$, i.e. the resonant coupling is shifted from the maximum of $G(\omega)$. If, by contrast, the peaks of $F_t(\omega)$ and $G(\omega)$ do coincide, any rate of interruptions will result in QZE (Fig. 2(b)).

III. FINITE-TEMPERATURE RELAXATION AND DECOHERENCE CONTROL

So far we have treated the case of an empty (zero-temperature) bath. In order to account for finite-temperature situations, where the bath state is *close* to a thermal (Gibbs) state, we resort to a master equation (ME) for any dynamically-controlled reduced density matrix of the system [27, 57], that we have derived using the Nakajima-Zwanzig formalism [62–65]. This ME becomes manageable and transparent under the following assumptions: (i) The weak-coupling limit of the system-bath interaction H_I prevails, corresponding to the neglect of $O(H_I^3)$ terms. This is equivalent to the Born approximation, whereby the back-effect of the system on the bath and their resulting entanglement are ignored. (ii) The system and the bath states are initially factorizable. (iii) The initial mean-value of H_I vanishes.

We adopt these assumptions for interaction Hamiltonians of the form

$$H_I(t) = \hat{S}(t) \cdot \hat{B} \tag{19}$$

$\hat{S}(t)$ and $\hat{B}$ being any operators of the system and bath, respectively.

We focus on two regimes: a two-level system coupled to either an amplitude- or phase-noise (AN or PN) thermal bath. The bath Hamiltonian (in either regime) will be explicitly taken to consist of harmonic oscillators and be linearly coupled to the system (generalizations to other baths and couplings are obvious):

$$H_B = \sum_\lambda \omega_\lambda a_\lambda^\dagger a_\lambda \tag{20}$$

$$B = \sum_\lambda (\kappa_\lambda a_\lambda + \kappa_\lambda^* a_\lambda^\dagger). \tag{21}$$

Here $a_\lambda, a_\lambda^\dagger$ are the annihilation and creation operators of mode λ, respectively, and κ_λ is the coupling amplitude to mode λ.

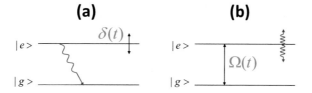

Figure 3. Schematic drawing of system and bath. (a) Amplitude noise (red) and AC-Stark shift modulation (green). (b) Phase noise (red) and resonant-field modulation (green).

A.   Amplitude-Noise Regime

We first consider the AN regime of a two-level system coupled to a thermal bath. We will use off-resonant dynamic modulations, resulting in AC-Stark shifts. The Hamiltonians then assume the following form:

$$H_S(t) = (\omega_a + \delta_a(t))|e\rangle\langle e| \tag{22}$$

$$\hat{S}(t) = \tilde{\epsilon}(t)\sigma_x \tag{23}$$

where $\delta_a(t)$ is the dynamical AC-Stark shifts, $\tilde{\epsilon}(t)$ is the time-dependent modulation of the interaction strength, and the Pauli matrix $\sigma_x = |e\rangle\langle g| + |g\rangle\langle e|$.

B.   Phase-Noise Regime

Next, we consider the PN regime of a two-level system coupled to thermal bath, where we will use near-resonant fields with time-varying amplitude as our control. The Hamiltonians then assume the following forms:

$$H_S(t) = \omega_a|e\rangle\langle e| + V(t)\sigma_x \tag{24}$$

$$\hat{S}(t) = \tilde{\epsilon}(t)\sigma_z \tag{25}$$

where $V(t) = \Omega(t)e^{-i\omega_a t} + c.c$ is the time-dependent resonant field, with real envelope $\Omega(t)$, $\tilde{\epsilon}(t)$ is the time-dependent modulation of the interaction strength, $\sigma_z = |e\rangle\langle e| - |g\rangle\langle g|$.

Since we are interested in dephasing, phases due to the (unperturbed) energy difference between the levels are immaterial.

C. Universal Master-Equation

To derive a universal ME for both amplitude- and phase-noise scenarios, we move to the interaction picture and rotate to the appropriate diagonalizing basis, where the appropriate basis for the AN case of Eq. (22) is

$$|\uparrow, \downarrow\rangle = |e, g\rangle, \tag{26}$$

while for the PN case of Eq. (24) the basis is

$$|\uparrow, \downarrow\rangle = \frac{1}{\sqrt{2}} \left(e^{-i\omega_a t} |e\rangle \pm |g\rangle \right). \tag{27}$$

Allowance for *arbitrary* time-dependent intervention in the system and interaction dynamics $H_S(t)$, $H_I(t)$, respectively, yields the following *universal* ME for dynamically controlled decohering systems [27, 57]:

$$\dot{\rho}(t) = \int_0^t dt' \{ \Phi(t - t')[\tilde{S}(t')\rho(t), \tilde{S}(t)] + H.c. \} \tag{28}$$

Here $\tilde{S}(t) = \epsilon(t)\sigma_x$ is the modulated interaction operator and ρ is a function of t *only* (*not of t'*): this *convolutionless* form of the ME is *fully non-Markovian* to second-order in H_I, as proven exactly in Ref.[57].

D. Universal Modified Bloch Equations

The resulting modified Bloch equations, in the appropriate diagonalizing basis (Eq. (26) for AN, Eq. (27) for PN), are given by:

$$\dot{\rho}_{\uparrow\uparrow} = -\dot{\rho}_{\downarrow\downarrow} = -R_\uparrow(t)\rho_{\uparrow\uparrow} + R_\downarrow(t)\rho_{\downarrow\downarrow}. \tag{29}$$

The corresponding average relaxation rates of the upper and lower states have the form

$$R_\uparrow = \int_{-\infty}^{\infty} d\omega G(+\omega) F_t(\omega), \tag{30}$$

$$R_\downarrow = \int_{-\infty}^{\infty} d\omega G(-\omega) F_t(\omega). \tag{31}$$

For AN (Eq. (22))

$$G(\omega) = \begin{cases} \rho(\tilde{\omega})\mu_{eg}^2(\tilde{\omega}) \left(n_T(\tilde{\omega}) + 1 \right) & \tilde{\omega} \geq 0 \\ \rho(-\tilde{\omega})\mu_{eg}^2(-\tilde{\omega})n_T(-\tilde{\omega}) & \tilde{\omega} < 0 \end{cases}, \tag{32}$$

where $\tilde{\omega} = \omega + \omega_a$, $\rho(\omega)$ and $\mu_{eg}(\omega)$ are the frequency-dependent density of bath modes and the $e \leftrightarrow g$ transition matrix element, respectively, and $n_T(\omega)$ is the temperature-dependent bath mode population.

For PN (Eq. (24))

$$G(\pm\omega) = \pi^{-1}\text{Re} \int_0^\infty dt e^{\pm i\omega t}\overline{\delta_r(t)\delta_r(0)} \tag{33}$$

represents the ensemble-average noise spectrum, if we assume a random *stationary* process.

For either AN or PN we may control the decoherence by either off-resonant or near-resonant modulations, respectively. The modulation spectrum has the *same form* for both (see Sec. IV):

$$F_t(\omega) = \frac{1}{t}\left|\int_0^t dt' e^{i\omega t' + i\int_0^{t'} d\tau \delta_a(\tau)}\right|^2. \tag{34}$$

The time-dependent modulation phase factor is obtained for AN in the form of an AC-Stark shift, time-integrated over

$$\delta_a(t) = \int_0^t dt' \frac{|\Omega(t')|^2}{\Delta(t')}, \tag{35}$$

where $\Omega(t)$ is the Rabi-frequency of the control field (off-resonant for AN and resonant for PN) and $\Delta(t')$ is the detuning. The corresponding phase-factor for PN is the integral of the Rabi frequency

$$\delta_a(t') = \Omega(t'), \tag{36}$$

i.e. the pulse-area of the resonant control field.

Hence, upon making the appropriate substitutions the Bloch equations, Eqs. (29), have the *same universal* form for either AN or PN. An arbitrary combination of AN and PN requires an appropriate choice of the diagonalizing basis, and correspondingly of $G(\omega)$ and $F_t(\omega)$, but the universal form is maintained.

E. Optimal Decoherence Control

The merit of this form is that it allows us to *optimally suppress decoherence* for a given form of $G(\omega)$. To this end, we have applied the Euler-Lagrange variational approach to

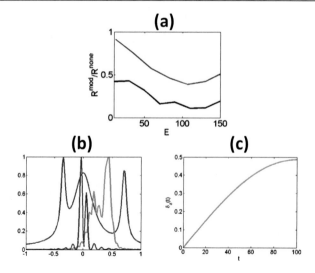

Figure 4. Optimal decoherence control. (a) Decoherence rate (the ratio of these rates with and without modulation control) as a function of the control-pulse energy constraint: Optimal (black) compared to bang-bang (blue) modulations. (b) Schematic representation of system-bath coupling spectrum $G(\omega)$(red), and the corresponding optimal (green) and bang-bang (magenta) modulation spectra, $F_t(\omega)$. (c) Time-domain representation of the optimal modulation $\delta_a(t)$ corresponding tot he optimal modulation spectrum $F_t(\omega)$ in (b).

minimizing $R(t)$, under the constraint that the total energy of the modulating driving control pulse is limited to a chosen value

$$\int_0^{t_{\text{pulse}}} \left| \Omega(t') \right|^2 dt' = E, \tag{37}$$

where t_{pulse} is the pulse duration. This optimized universal pulse control of decoherence has been numerically checked by us to consistently *outperform the standard Bang-Bang* method of periodic π-phase flips [12–14]. Our method yields considerably lower $R(t)$ for the same E (Fig. 4a). This is understandable, since the Bang-Bang π-phase flips at intervals τ corresponds to a modulation spectrum with two peaks at $\pm\pi/\tau$, which is unrelated to the form of $G(\omega)$. By contrast, our method optimally tailors $F_t(\omega)$ to the shape of $G(\omega)$ in such a manner that the peaks of one coincides with the dips of the other, resulting in the *least possible overlap* of $F_t(\omega)$ and $G(\omega)$ for a given pulse E (Fig. 4b).

IV.   MODULATION ARSENAL FOR AN AND PN

Introducing the control-field fluence $Q(t)$, the spectral modulation $F_t(\omega)$ can be normalized to unity:

$$Q(t) = \int_0^t d\tau |\epsilon(\tau)|^2, \tag{38}$$

$$F_t(\omega) = \frac{|\epsilon_t(\omega)|^2}{Q(t)}, \tag{39}$$

where

$$\epsilon_t(\omega) = \frac{1}{\sqrt{2\pi}} \int_0^t d\tau \epsilon(\tau) e^{i\omega\tau} \tag{40}$$

is the finite-time Fourier transform of $\epsilon(t)$.

Any modulation with quasi-discrete, finite spectrum is deemed quasiperiodic, implying that it can be expanded as

$$\epsilon(t) = \sum_k \epsilon_k e^{-i\nu_k t} \tag{41}$$

where ν_k $(k = 0, \pm 1, ...)$ are arbitrary discrete frequencies such that

$$|\nu_k - \nu_{k'}| \geq \Omega \quad \forall k \neq k', \tag{42}$$

where Ω is the minimal spectral interval.

One can define the long-time limit of the quasi-periodic modulation, when

$$\Omega t \gg 1 \quad \text{and} \quad t \gg t_c, \tag{43}$$

where t_c is the bath-memory (correlation) time, defined as the inverse of the largest spectral interval over which $G(\omega)$ and $G(-\omega)$ change appreciably near the relevant frequencies $\omega_a + \nu_k$. In this limit, the fluence is given by

$$Q(t) \approx \epsilon_c t \quad \epsilon_c = \sum_k |\epsilon_k|^2, \tag{44}$$

resulting in the average decay rate:

$$R_\uparrow = 2\pi \sum_k |\lambda_k|^2 G(\omega_a + \nu_k), \tag{45}$$

$$\lambda_k = \epsilon_k/\epsilon_c. \tag{46}$$

A.   Phase Modulation (PM) of the Coupling

1.   Monochromatic Perturbation

Let

$$\epsilon(t) = \epsilon_0 e^{-i\Delta t}. \tag{47}$$

Then

$$R_\uparrow = 2\pi G(\omega_a + \Delta), \tag{48}$$

where $\Delta = $ const. is a frequency shift, induced by the AC Stark effect (in the case of atoms) or by the Zeeman effect (in the case of spins). In principle, such a shift may drastically enhance or suppress R relative to the Golden - Rule decay rate, i.e. the decay rate without any perturbation

$$R_{\mathrm{GR}} = 2\pi G(\omega_a). \tag{49}$$

Equation (48) provides the *maximal change* of R achievable by an external perturbation, since it does not involve any averaging (smoothing) of $G(\omega)$ incurred by the width of $F_t(\omega)$: the modified R can even *vanish*, if the shifted frequency $\omega_a + \Delta$ is beyond the cutoff frequency of the coupling, where $G(\omega) = 0$ (Figure 5a). This would accomplish the goal of dynamical decoupling [11–17, 66, 67]. Conversely, the increase of R due to a shift can be much greater than that achievable by repeated measurements, i.e. the anti-Zeno effect [24, 25, 28, 29]. In practice, however, AC Stark shifts are usually small for (cw) monochromatic perturbations, whence pulsed perturbations should often be used, resulting in multiple ν_k shifts, as per Eq. (45).

2.   Impulsive Phase Modulation

Let the phase of the modulation function periodically jump by an amount ϕ at times $\tau, 2\tau, \ldots$. Such modulation can be achieved by a train of identical, equidistant, narrow pulses of nonresonant radiation, which produce pulsed AC Stark shifts of ω_a. Now

$$\epsilon(t) = e^{i[t/\tau]\phi}, \tag{50}$$

where $[\ldots]$ is the integer part. One then obtains that

$$Q(t) = t, \quad \epsilon_c = 1, \tag{51}$$

$$F_{n\tau}(\omega) = \frac{2 \sin^2(\omega\tau/2) \sin^2[n(\phi + \omega\tau)/2]}{\pi n \tau \omega^2 \sin^2[(\phi + \omega\tau)/2]}. \tag{52}$$

For sufficiently long times (Eq. (43)) one can use Eq. (45), with

$$\nu_k = \frac{2k\pi}{\tau} - \frac{\phi}{\tau}, \quad |\lambda_k|^2 = \frac{4 \sin^2(\phi/2)}{(2k\pi - \phi)^2} \tag{53}$$

For *small phase shifts*, $\phi \ll 1$, the $k = 0$ peak dominates,

$$|\lambda_0|^2 \approx 1 - \frac{\phi^2}{12}, \tag{54}$$

whereas

$$|\lambda_k|^2 \approx \frac{\phi^2}{4\pi^2 k^2} \quad (k \neq 0). \tag{55}$$

In this case one can retain only the $k = 0$ term in Eq. (45), unless $G(\omega)$ is changing very fast with frequency. Then the modulation acts as a constant shift, (Fig. 5a)

$$\Delta = -\phi/\tau. \tag{56}$$

As $|\phi|$ increases, the difference between the $k = 0$ and $k = 1$ peak heights diminishes, *vanishing* for $\phi = \pm\pi$. Then

$$|\lambda_0|^2 = |\lambda_1|^2 = 4/\pi^2, \tag{57}$$

i.e., $F_t(\omega)$ for $\phi = \pm\pi$ contains *two identical peaks symmetrically shifted in opposite directions* (Figure 5b) [the other peaks $|\lambda_k|^2$ decrease with k as $(2k - 1)^{-2}$, totaling 0.19].

The foregoing features allow one to adjust the modulation parameters for a given scenario to obtain an *optimal* decrease or increase of R. Thus, the phase-modulation (PM) scheme with a small ϕ is preferable near a continuum edge (Figure 5a,b), since it yields a spectral shift in the required direction (positive or negative). The adverse effect of $k \neq 0$ peaks in $F_t(\omega)$ then scales as ϕ^2 and hence can be significantly reduced by decreasing $|\phi|$. On the other hand, if ω_a is near a *symmetric* peak of $G(\omega)$, R is reduced more effectively for $\phi \simeq \pi$, as in Refs. [10, 11], since the main peaks of $F_t(\omega)$ at ω_0 and ω_1 then shift stronger with τ^{-1} than the peak at $\omega_0 = -\phi/\tau$ for $\phi \ll 1$.

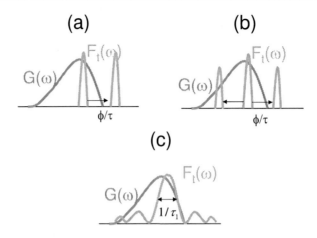

Figure 5. Spectral representation of the bath coupling, $G(\omega)$, and the modulation, $F_t(\omega)$. (a) Monochromatic modulation, or impulsive phase modulation, with small phase shifts, $\phi \ll 1$, and $1/\tau$ repetition rate. (b) Impulsive phase modulation, (π-pulses), $\phi = \pi$. (c) On-off modulation, with $1/\tau_1$ repetition rate for $\tau_1 \ll \tau_0$.

B.   Amplitude Modulation (AM) of the Coupling

Amplitude modulation (AM) of the coupling may be applicable to certain AN or PN scenarios. It arises, e.g., for radiative-decay modulation due to atomic motion through a high-Q cavity or a photonic crystal [68, 69] or for atomic tunneling in optical lattices with time-varying lattice acceleration [70, 71].

1.   On-off Modulation

The simplest form of AM is to let the coupling be turned on and off periodically, for the time τ_1 and $\tau_0 - \tau_1$, respectively, i.e.,

$$\epsilon(t) = \begin{cases} 1 & \text{for } n\tau_0 < t < n\tau_0 + \tau_1, \\ 0 & \text{for } n\tau_0 + \tau_1 < t < (n+1)\tau_0 \end{cases} \tag{58}$$

($n = 0, 1, \dots$). Now $Q(t)$ in (38) is the total time during which the coupling is switched on, whereas

$$F_{n\tau_0}(\omega) = \frac{2\sin^2(\omega\tau_1/2)\sin^2(n\omega\tau_0/2)}{\pi n\tau_1\omega^2\sin^2(\omega\tau_0/2)}. \tag{59}$$

This case is also covered by (45), where the parameters are now found to be

$$\epsilon_c^2 = \frac{\tau_1}{\tau_0}, \quad \nu_k = \frac{2k\pi}{\tau_0}, \quad |\lambda_k|^2 = \frac{\tau_1}{\tau_0}\text{sinc}^2\left(\frac{k\pi\tau_1}{\tau_0}\right). \tag{60}$$

It is instructive to consider the limit wherein $\tau_1 \ll \tau_0$ and τ_0 is much greater than the correlation time of the continuum, i.e., $G(\omega)$ does not change significantly over the spectral intervals $(2\pi k/\tau_0, 2\pi(k+1)/\tau_0)$. In this case one can approximate the sum (45) by the integral (30) with

$$F_t(\omega) \approx (\tau_1/2\pi)\text{sinc}^2(\omega\tau_1/2), \tag{61}$$

characterized by the spectral broadening $\sim 1/\tau_1$ (figure 5c). Then equation (30) for R reduces to that obtained when ideal projective measurements are performed at intervals τ_1 [24]. Thus the AM on-off coupling scheme *can imitate measurement-induced (dephasing) effects* on quantum dynamics, if the interruption intervals τ_0 *exceed the correlation time of the continuum.*

V.   MULTIPARTITE DECOHERENCE CONTROL

Multipartite decoherence control, for many qubits coupled to thermal baths, is a much more challenging task than single-qubit control since: (i) entanglement between the qubits is typically more vulnerable and more rapidly destroyed by the environment than single qubit coherence [31–38]; (ii) the possibility of cross-decoherence, whereby qubits are coupled to each other through the baths, considerably complicates the control. We have recently analyzed this situation and extended [72, 73] the decoherence control approach to multipartite scenarios, where the qubits are either coupled to zero- or finite- temperature baths. The phase-noise scenario is discussed in detail in Sec. VI.

A.   Multipartite Coupling to Zero-Temperature Baths

The decay of a singly excited multi-qubit system (under amplitude noise) to the ground state, in the presence of off-resonant modulating fields is described by the following relaxation matrix [72, 73]:

$$R_{jj'}(t) = 2\pi \int_{-\infty}^{\infty} d\omega \, G_{jj'}(\omega) F_{t,jj'}(\omega) \tag{62}$$

$$G_{jj'}(\omega) = \hbar^{-2} \sum_k \mu_{k,j} \mu_{k,j'}^* \delta(\omega - \omega_k) \tag{63}$$

$$F_{t,jj'}(\omega) = \epsilon_{t,j}(\omega) \epsilon_{t,j'}^*(\omega) \tag{64}$$

where jj' are the different particle indices.

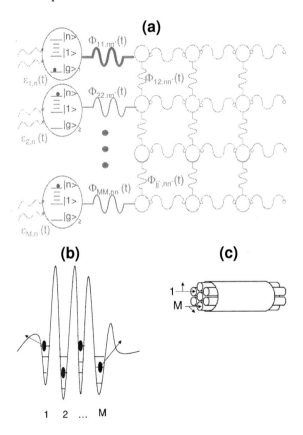

Figure 6. (a) Entangled multilevel systems with different couplings to a phonon bath or different proper dephasings, via $\Phi_{jj,nn'}(t)$. Their cross-coupling is through $\Phi_{jj',nn'}(t)$. The systems are modulated by $\epsilon_{j_n}(t)$. (b) Several polarization-entangled photons propagating through adjacent (coupled) fibers that exhibit fluctuating birefringence. (c) Entangled systems in tunnel-coupled multilevel wells of a washboard potential. There is no direct coupling between the wells at $t > 0$, only different relaxation of each well to the continuum.

The diagonal elements of this matrix, $R_{jj}(t)$, express single-particle decoherence rates that depend on *local* (j-site) *spectral* intensity $|\epsilon_{t,j}(\omega)|^2$ and the corresponding local coupling spectrum $G_{jj}(\omega) = G_j(\omega)$. By contrast, its off-diagonal elements $R_{jj'}(t)$ depend on the *cross-decoherence* or inter-particle noise correlations $G_{jj'}(\omega)$ as well as on the inter-particle (relative) *modulation phases* $\epsilon_j(\omega)\epsilon_{j'}^*(\omega)$.

Without any modulations, decoherence in this scenario has no inherent symmetry. Our point is that one can symmetrize the decoherence by appropriate modulations. The key is that different, "local", phase-locked modulations applied to the individual particles, according to Eq. (64), can be chosen to cause *controlled interference* and/or spectral shifts between the particles' couplings to the bath. The $F_{t,jj'}(\omega)$ matrices (cf.(64)) can then satisfy $2N$ requirements at all times and be tailored to impose the advantageous symmetries. By contrast, a "global" (identical) modulation, characterized by $F_{t,jj'}(\omega) = |\epsilon_t(\omega)|^2$, is not guaranteed to satisfy $N \gg 1$ symmetrizing requirements at all times (Fig. 7).

The most desirable symmetry is that of *identically coupled particles* (ICP), which would emerge if all the modulated particles could acquire the *same* dynamically modified decoherence and cross-decoherence yielding the following $N \times N$ fully symmetrized decoherence matrix

$$R_{jj'}^{\mathrm{ICP}}(t) = r(t) \quad \forall j, j'. \tag{65}$$

ICP would then give rise to a $(N-1)$-dimensional decoherence-free subspace: the entire single-excitation sector less the totally symmetric entangled state. An initial state in this DFS [40] would neither lose its population nor its initial correlations (or entanglement).

Unfortunately, it is generally impossible to ensure this symmetry, since it amounts to satisfying $N(N-1)/2$ conditions using N modulating fields. Even if we accidentally succeed with N particles, the success is not scalable to $N+1$ or more particles. Moreover, the ability to impose the ICP symmetry by local modulation fails completely if not all particles are coupled to all other particles through the bath, i.e. if some $G_{jj'}(\omega)$ elements vanish.

A more limited symmetry that we may *ensure* for N qubits is that of *independent identical particles* (IIP). This symmetry is formed when spectral shifts and/or interferences imposed by N modulations cause the N different qubits to acquire the *same* single-qubit decoherence $r(t)$ and experience no cross-decoherence. To this end, we may choose $\epsilon_{t,j}(\omega) \simeq \epsilon_{t,j}\delta(\omega - \Delta_j)$. We require that at any chosen time $t = T$, the AC Stark shifts satisfy $\int_0^T d\tau \delta_j(\tau) = 2\pi m$, where $m = 0, \pm 1, \ldots$. This requirement ensures that modulations only affect the decoherence matrix (62), but do not change the relative phases of the entangled qubits when their state is probed or manipulated by logic operations at $t = T$.

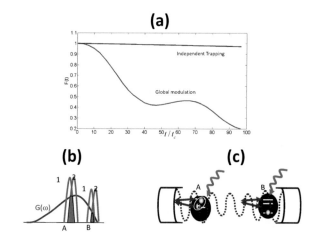

Figure 7. (a) Fidelity as a function of time for two particles driven into independent trapping (blue) as compared to particles driven by global (identival) modulation (red). Global modulation precludes the formation of independent-trapping states since it does not eliminate cross-decoherence that couples the particles effectively. (b) System-bath coupling (red) and modulation (green) spectra for two destructively interfering decoherence channels $1, 2$ in eac of the particles A, B. It is clear that to decouple the particles, i.e. suppress cross-decoherence, we must impose modulation spectra that are far apart. Then independent-trapping may occur, resulting in cancellation of the overlap between $G(\omega)$ and the modulation spectra of the destructive interfering channels $1, 2$ ($R_{A(B)}$ are proportional to the blue overlap areas). (c) Schematic drawing of the independent trapping scenario. The two decoherence/relaxation channels in each particle interfere destructively and the particles are decoupled.

The spectral shifts Δ_j can be different enough to couple each particle to a different spectral range of bath modes so that their cross-coupling vanishes:

$$R_{jj'}(t) = 2\pi \epsilon_{t,j}^* \epsilon_{t,j'} \int d\omega G_{jj'}(\omega_0 + \omega)\delta(\omega - \Delta_j)\delta(\omega - \Delta_{j'}) \to 0. \qquad (66)$$

Here, the vanishing of $G_{jj'}(\omega)$ for some j, j' is not a limitation. The N single-particle decoherence rates can be equated by an appropriate choice of N parameters $\{\Delta_j\}$:

$$R_{jj'}^{\mathrm{IIP}}(t) = 2\pi |\epsilon_{t,j}|^2 G_{jj}(\omega_0 + \Delta_j) = \delta_{jj'} r(t), \qquad (67)$$

where $\delta_{jj'}$ is Kronecker's delta (Fig. 5b). The IIP symmetry results in reducing the multipartite decoherence problem to that of a single decohering particle. If the single-particle $r(t)$ may be dynamically suppressed, i.e. if the spectrally shifted bath response $G_{jj}(\omega_j + \Delta_j)$ is small enough, then the fidelity of any initial state will be kept close to 1.

B. Multipartite Coupling to Finite-Temperature Baths

Multipartite entanglement, which is very susceptible to decoherence, decays faster than single-qubit coherence, and can even completely disappear in finite time, an effect dubbed Entanglement Sudden Death (ESD) [31–38]. While it is known that an initial (pure) Bell state's entanglement decays exponentially for zero-temperature baths, it does not vanish at finite times. Conversely, the coupling of two qubits to finite-temperature baths does result in ESD [34, 37].

To analyze this disentanglement and its control, this section describes the extension of the universal ME given in Sec. III to multipartite systems. The total Hamiltonian is given by:

$$H(t) = H_S(t) + H_B + H_I(t) \quad H_B = \sum_\lambda \omega_\lambda a_\lambda^\dagger a_\lambda \tag{68}$$

$$H_S(t) = \sum_{j=1}^{N} (\omega_a + \delta_{a,j}(t)) |e\rangle_{jj} \langle e| \bigotimes_{j' \neq j} I_{j'} \tag{69}$$

$$H_I(t) = \sum_{j=1}^{N} \sum_\lambda \tilde{\epsilon}_j(t) \sigma_{x,j} (\kappa_{\lambda,j} a_\lambda + \kappa_{\lambda,j}^* a_\lambda^\dagger) \bigotimes_{j' \neq j} I_{j'} \tag{70}$$

where $\sigma_{x,j} = |e\rangle_{jj}\langle g| + |g\rangle_{jj}\langle e|$ is the X-Pauli matrix of qubit j, I is the identity matrix and $\kappa_{\lambda,j}$ is the off-diagonal coupling coefficient of TLS j to the bath oscillator, λ. Note that we did *not invoke the rotating-wave approximation*.

Using the same techniques as in Sec. III, namely, Zwanzig's projection-operator technique to trace out the bath in Liouvilles equation of motion, and moving to the interaction picture, where $\hat\rho(t) = e^{i\int_0^t d\tau H_S(\tau)} \rho(t) e^{-i\int_0^t d\tau H_S(\tau)}$, we have derived, to second order in the system-bath coupling, the non-Markovian master equation:

$$\dot{\hat\rho} = \sum_{j,j'=1}^{N} \int_0^t d\tau \left\{ \Phi_{T,jj'}(t-\tau) \left[\tilde{S}_{j'}(\tau)\hat\rho, \tilde{S}_j(t) \right] + H.c. \right\} \tag{71}$$

where $\Phi_{T,jj'(jj)}(t)$ is the finite-temperature system-bath (cross-)correlation function for systems j, j'. Also, $\tilde{S}_j(t) = \epsilon_j(t)|e\rangle\langle g| + H.c.$, where $\epsilon_j(t) = \tilde{\epsilon}_j(t)e^{i\omega_a t + i\int_0^t dt' \delta_{a,j}(t')}$ is the time-dependent modulation function of qubit j.

The multipartite Bloch equations are also governed by the modified average relaxation rates given in Eq. (62), only with the finite-temperature coupling spectrum given in Eq. (32). Thus, the same symmetries described in the previous section also apply here, with the added effect of the temperature of the baths.

Figure 8 shows the finite-time in which concurrence of bipartite systems [36, 38, 74] vanishes, which we shall dub time-of-death (TOD), t^{TOD}, as a function of temperature for different modulation schemes.

Imposing symmetry, as in the zero-temperature scenario, results in a drastic improvement of disentanglement by incurring destructive interference of the decoherence channels. This can be achieved by equating the cross-decoherence to the individual particle's decoherence rates, i.e. imposing the ICP symmetry and results in the prolongation and delay of ESD.

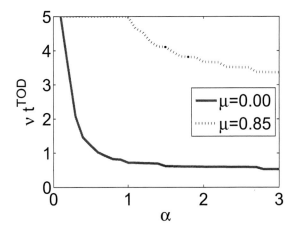

Figure 8. Time-of-death, i.e. the time entanglement vanishes, as a function of thermal-bath temperature, $\alpha = k_B T / \hbar \omega_a$, for an initial singlet. $\nu = R_{11} + R_{22}$ is the sum of individual qubit decoherence rate and $\mu = (R_{12} + R_{21})/\nu$ measures the normalized cross-decoherence. Increasing μ results in imposed symmetry and increased longevity of entanglement.

VI.   DEPHASING CONTROL DURING QUANTUM COMPUTATIONS

We extend the universal phase-noise formalism presented in Sec. II to multipartite systems undergoing quantum computation, namely single- and two-qubit gate operations. One needs a measure of the efficiency of the control schemes during the quantum gate operations. We use fidelity, defined as

$$F(T) = \text{Tr}(\rho_{\text{target}}^{1/2} \overline{\rho}(T) \rho_{\text{target}}^{1/2}), \tag{72}$$

where ρ_{target} is the target density matrix after the quantum computation, e.g. $\rho^{\text{target}} = \rho(0)$ for the storage stage. The error of the gate operation is then

$$E(T) = 1 - F(T). \tag{73}$$

However, since quantum computations presume lack of knowledge of the initial qubits' state, we shall also use the average fidelity,

$$F_{avg}(T) = \langle F(T) \rangle, \tag{74}$$

where $\langle \cdots \rangle$ is the average over all possible initial pure states.

A.   Dynamically-Controlled Fidelity of Single-Qubit Gate Operations

First, we apply local single-qubit gates on each one of the qubits, where the control Hamiltonian is given by:

$$\hat{H}^{(1)}(t) = \hbar \sum_{j=1}^{N} \left(V_j^{(1)}(t)|e\rangle_{jj}\langle g| + H.c. \right) \bigotimes_{k \neq j} I_k \tag{75}$$

Here, $V_j^{(1)}(t) = \Omega_j^{(1)}(t)e^{-i\omega_a t} + c.c.$ is the time-dependent local single-qubit driving field of the j-th qubit. Also, I is the identity matrix and $H.c.$ is Hermitian conjugate.

During the single-qubit gate stage, one can apply several gate operations on different qubits. The gate operations are defined by the accumulated phase of the qubits given by $\phi_j^{(1)}(t) = \int_0^t d\tau \Omega_j^{(1)}(\tau)$, e.g. $\phi_j^{(1)}(T) = \pi/4$ is the Hadamard gate applied to the j-th qubit [75].

However, during these stages, the other qubits that do not participate in the gate operations are in the storage stage, meaning that one can still apply the fields described above,
with the appropriate restrictions.

In quantum computation, the accumulated phase given by $\phi_j^{(1)}(t) = \int_0^t d\tau \Omega_j^{(1)}(\tau)$ determines the gate type, e.g. $\phi_1^{(1)}(T) = 2\pi M$ means storage of the first qubit. The average
fidelity of this scheme is given by:

$$F_{avg}(T) = 1 - \frac{5}{12}t\left(R_{11}^{(1)}(T) + R_{22}^{(1)}(T)\right) \tag{76}$$

where $R_{jj'}^{(1)}(t)$ is again given by Eq. (62), with $\epsilon(t) = e^{i\phi^{(1)}(t)}$.

Equations (76) and (62) reveal that single-qubit gate fields do not cause cross-dephasing,
since Eq. (76) depends only on single-qubit dephasing, $\Phi_{jj}(t)$. This comes about from the
averaging over all initial qubits: for each initial entangled state that loses fidelity from cross-
dephasing there is another entangled state that gains fidelity from cross-dephasing. Thus,
for the triplet, $|\Phi_-\rangle$,

$$F(T, |\Phi_-\rangle) = 1 - t/2(R_{11}^{(1)} + R_{22}^{(1)} + R_{12}^{(1)} + R_{21}^{(1)}), \tag{77}$$

while for the singlet, $|\Psi_-\rangle$,

$$F(T, |\Psi_-\rangle) = 1 - t/2(R_{11}^{(1)} + R_{22}^{(1)} - R_{12}^{(1)} - R_{21}^{(1)}). \tag{78}$$

Equation (76) also shows that the same modified dephasing function appears regardless of
the accumulated phase, meaning that if one applies a gate field on one qubit, one can still
benefit from applying a control field on the other, stored, qubit.

B.   Dynamically-Controlled Fidelity of Two-Qubit Gate Operations

Next, we explore non-local two-qubit gate operations in the presence of control and
dephasing. Here, the control Hamiltonian is given by:

$$\hat{H}^{(2)}(t) = \hbar \sum_{j=1}^N \sum_{k=j+1}^N \left(V_{jk}^{(2)\Psi}(t)|ge\rangle_{jk}\langle eg| \tag{79}$$

$$+V_{jk}^{(2)\Phi}(t)|ee\rangle_{jk}\langle gg| + H.c.\right) \bigotimes_{l\neq j,k} I_l \tag{80}$$

where

$$V_{jk}^{(2)\Psi}(t) = \Omega_{jk}^{(2)\Psi}(t) + c.c \tag{81}$$

$$V_{jk}^{(2)\Phi}(t) = \Omega_{jk}^{(2)\Phi}(t)e^{-i2\omega_a t} + c.c. \tag{82}$$

are two possible time-dependent non-local two-qubit driving fields, acting on qubits j and k, where the notation is derived from their diagonalization basis, i.e. the Bell-states basis,

$$|\Psi_{\pm}\rangle = 1/\sqrt{2}e^{-i\omega_0 t}(|eg\rangle \pm |ge\rangle), \quad |\Phi_{\pm}\rangle = 1/\sqrt{2}(e^{-i2\omega_0 t}|ee\rangle \pm |gg\rangle). \tag{83}$$

During the two-qubit gate stage, one can apply several gate operations on different qubit pairs. These gate operations are also defined by the accumulated phase of the two qubits, given by $\phi_j^{(2)\Psi,\Phi}(t) = \int_0^t d\tau \Omega_j^{(2)\Psi,\Phi}(\tau)$ where the $\Omega_{kk'}^{(2\Psi)}(t)$ change the phase in the $|\Psi_{\pm}\rangle$ plane and $\Omega_{kk'}^{(2\Phi)}(t)$ change the phase in the $|\Phi_{\pm}\rangle$ plane of qubits k and k'. For example, $\phi_{kk'}^{(2\Psi)}(T) = \pi/2$ is a SWAP gate between qubits k and k' [49].

The average fidelity for this scenario is found to be:

$$F_{avg}(T) = 1 - \frac{5}{24}t \sum_{j,k=1,2} \left(R_{jk}^{(2\Phi)}(T) + (-1)^{j+k}R_{jk}^{(2\Psi)}(T) \right) \tag{84}$$

where $R_{jj'}^{(2)\Psi,\Phi}(t)$ is again given by Eq. (62), with $\epsilon^{(2)\Psi,\Phi}(t) = e^{i\phi^{(2)\Psi,\Phi}(t)}$.

Here we see that cross-dephasing does not cancel due to averaging, but has opposite signs for the different two-qubit fields acting on the $|\Phi_+\rangle_{jk} \leftrightarrow |\Phi_-\rangle_{jk}$ transition and $|\Psi_+\rangle_{jk} \leftrightarrow |\Psi_-\rangle_{jk}$ transition, respectively. Hence, for example, a $\sqrt{\text{SWAP}}$ gate [49], which is an entangling gate and in our notation is given by $\phi_{jk}^{(2\Psi)}(T) = \pi/4$, may benefit from cross-dephasing.

Furthermore, we see that applying both two-qubit gate fields can reduce dephasing, even if only one field is needed for the actual gate operation. The other two-qubit field, used for storage, with $\phi_{1,2}^{(2)\Phi}(T) = 2\pi M$, $M = 1, 2, \ldots$, can reduce dephasing, if used along with, e.g., a $\sqrt{\text{SWAP}}$ gate.

This novel approach, consisting in applying an auxiliary (control) field, concurrently with the gate field that performs the logic operation, may require longer gate operations, due to limitations, such as the maximal achievable peak-power. Yet, if the dephasing is reduced by the control fields despite the longer gate duration, the overall gate fidelity may increase, contrary to traditional schemes (Fig. 9).

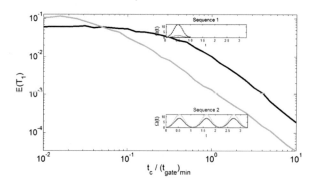

Figure 9. Gate errors at the end of the gate pulse, $E(t^{\text{pulse}})$ as a function of correlation time, t_c, for simple gate pulse (black curve, sequence 1) and for compound gate and control fields (yellow curve, sequence 2). Sequences $1, 2$ involves the following phase shifts $\phi^{(1)}_{1,2,3}(T_1) = \{0, \pi/4, 7\pi/4\}$, $\{4\pi, 17\pi/4, 23\pi/4\}$. The gate durations are chosen such that the peak-power is the same for both sequences. The results were obtained upon averaging over 1000 realizations. Clearly, the combined gate and control sequence outperforms the simple gate-pulse sequence despite its being 3 times longer.

For two-qubit gates, such modulation control requires driving the transitions that are not coupled by the gate field: if the gate field couples Bell states $|\Psi_\pm\rangle$, the modulation field must mix the two orthogonal states $|\Phi_\pm\rangle$, and vice versa.

VII.   CONCLUSIONS

The results presented here attest to the universality and effectiveness of the dynamical control of decoherence embodied by Eqs. (30)-(31): they have essentially the same form for AN and PN and any bath, and their extension to multipartite or multilevel systems, Eq. (62), is very natural.

We can use this universal approach to optimize the performance of quantum gates, storage and communications operations, by *tailoring* the spatiotemporal modulation to the noise spectra acting on *individual* particles (e.g. qubits). Bang-bang is inferior precisely because such tailoring is not involved.

Our general analysis allows one to come up with an optimal choice between global and local control, based on the observation that the maximal suppression of decoherence is not necessarily the best one. Instead, we demand an optimal *phase-relation* between different, but *synchronous* local modulations of each particle.

We have also formulated a universal protocol for dynamical dephasing control during all stages of quantum information processing, namely, storage, single- and two-qubit gate operations. It amounts to controlling all the qubits, whether they participate in the computation or not, and tailoring specific gate and control fields that reduce the dephasing. We have specifically shown that two-qubit entangling control fields can eliminate the cross-dephasing dependence. This counter-intuitive protocol has a great advantage over others and can be of great experimental value in that it increases the fidelity of the operation required, whether storage or computation, despite its longer duration.

References

[1] C. Cohen-Tannoudji, J. Dupont-Roc, and G. Grynberg, *Atom-Photon Interactions* (Wiley, New York, 1992).

[2] C. A. Sackett, D. Kielpinski, B. E. King, C. Langer, V. Meyer, C. Myatt, M. Rowe, Q. A. Turchette, W. M. Itano, D. J. Wineland, et al., Nature (London) **404**, 256 (2000).

[3] M. Greiner, O. Mandel, T. Esslinger, T. W. Haensch, and I. Bloch, Nature **39**, 415 (2002).

[4] J. Clarke, A. N. Cleland, M. H. Devoret, D. Esteve, and J. M. Martinis, Science **239**, 992 (1988).

[5] A. J. Leggett, S. Chakravarty, A. T. Dorsey, M. P. A. Fisher, A. Garg, and W. Zwerger, Rev. Mod. Phys. **59**, 1 (1987).

[6] M. O. Scully and M. S. Zubairy, *Quantum Optics* (Cambridge University Press, Cambridge, 1997).

[7] K. Banaszek, A. Dragan, W. Wasilewski, and C. Radzewicz, Phys. Rev. Lett. **92**, 257901 (2004).

[8] J. L. Ball, A. Dragan, and K. Banaszek, Phys. Rev. A. **69**, 042324 (2004).

[9] G. S. Agarwal, Phys. Rev. A **61**, 013809 (2000).

[10] G. S. Agarwal, M. O. Scully, and H. Walther, Phys. Rev. A **63**, 044101 (2001).

[11] G. S. Agarwal, M. O. Scully, and H. Walther, Phys. Rev. Lett. **86**, 4271 (2001).

[12] L. Viola and S. Lloyd, Phys. Rev. A **58**, 2733 (1998).

[13] K. Shiokawa and D. A. Lidar, Phys. Rev. A **69**, 030302 (2004).

[14] D. Vitali and P. Tombesi, Phys. Rev. A **65**, 012305 (2001).

[15] P. Facchi and S. Pascazio, Prog. in Opt. **42**, 147 (2001).

[16] P. Facchi, D. A. Lidar, and S. Pascazio, Phys. Rev. A **69**, 0302314 (2004).

[17] P. Zanardi and S. Lloyd, Phys. Rev. Lett. **90**, 067902 (2003).

[18] L. Viola and E. Knill, Phys. Rev. Lett. **90**, 037901 (2003).

[19] C. Uchiyama and M. Aihara, Phys. Rev. A **66**, 032313 (2002).

[20] K. Khodjasteh and D. A. Lidar, Phys. Rev. Lett. **95**, 180501 (2005).

[21] M. Stollsteimer and G. Mahler, Phys. Rev. A **64**, 052301 (2001).

[22] L. Faorol and L. Viola, Phys. Rev. Lett. **92**, 117905 (2004).

[23] C. Search and P. R. Berman, Phys. Rev. Lett. **85**, 2272 (2000).

[24] A. G. Kofman and G. Kurizki, Nature (London) **405**, 546 (2000).

[25] A. G. Kofman and G. Kurizki, Z. Naturforsch. A **56**, 83 (2001).

[26] A. G. Kofman and G. Kurizki, Phys. Rev. Lett. **87**, 270405 (2001).

[27] A. G. Kofman and G. Kurizki, Phys. Rev. Lett. **93**, 130406 (2004).

[28] A. G. Kofman, G. Kurizki, and T. Opatrný, Phys. Rev. A **63**, 042108 (2001).

[29] A. G. Kofman and G. Kurizki, Phys. Rev. A **54**, R3750 (1996).

[30] S. Pellegrin and G. Kurizki, Phys. Rev. A **71**, 032328 (2004).

[31] K. Życzkowski, P. Horodecki, M. Horodecki, and R. Horodecki, Phys. Rev. A **65**, 012101 (2001).

[32] S. Bandyopadhyay and D. A. Lidar, Phys. Rev. A **70**, 010301 (2004).

[33] T. Yu and J. H. Eberly, Phys. Rev. Lett. **93**, 140404 (2004).

[34] M. Ban, S. Kitajima, and F. Shibata, J. Phys. A **38**, 4235 (2005).

[35] C. Anastopoulos, S. Shresta, and B. L. Hu (2007), e-print arXiv:quant-ph/0610007.

[36] M. P. Almeida, F. de Melo, M. Hor-Meyll, A. Salles, S. P. Walborn, P. H. S. Ribeiro, and L. Davidovich, Science **316**, 579 (2007).

[37] F. F. Fanchini and R. d. J. Napolitano, Phys. Rev. A. **76**, 062306 (2007).

[38] J. H. Eberly and T. Yu, Science **316**, 555 (2007).

[39] L. Viola, E. Knill, and S. Lloyd, Phys. Rev. Lett. **85**, 3520 (2000).

[40] P. Zanardi and M. Rasetti, Phys. Rev. Lett. **79**, 3306 (1997).

[41] D. A. Lidar, I. L. Chuang, and K. B. Whaley, Phys. Rev. Lett. **81**, 2594 (1998).

[42] L.-A. Wu and D. A. Lidar, Phys. Rev. Lett. **88**, 207902 (2002).

[43] A. Barone, G. Kurizki, and A. G. Kofman, Phys. Rev. Lett. **92**, 200403 (2004).

[44] G. Gordon, G. Kurizki, and A. G. Kofman, J. Opt. B. **7**, 283 (2005).

[45] D. A. Lidar, D. Bacon, and K. B. Whaley, Phys. Rev. Lett. **82**, 4556 (1999).

[46] P. Facchi and S. Pascazio, Phys. Rev. Lett. **89**, 080401 (2002).

[47] R. G. Unanyan and M. Fleischhauer, Phys. Rev. Lett. **90**, 133601 (2003).

[48] E. Brion, V. M. Akulin, D. Comparat, I. Dumer, G. Harel, N. Kèbaili, G. Kurizki, I. Mazets, and P. Pillet, Phys. Rev. A **71**, 052311 (2005).

[49] D. Loss and D. P. DiVincenzo, Phys. Rev. A **57**, 120 (1998).

[50] D. Schrader, I. Dotsenko, M. Khudaverdyan, Y. Miroshnychenko, A. Rauschenbeutel, and D. Meschede, Phys. Rev. Lett. **93**, 150501 (2004).

[51] A. Kreuter, C. Becher, G. P. Lancaster, A. B. Mundt, C. Russo, H. Haffner, C. Roos, J. Eschner, F. Schmidt-Kaler, and R. Blatt, Phys. Rev. Lett. **92**, 203002 (2004).

[52] F. Schmidt-Kaler, H. Haffner, M. Riebe, S. Gulde, G. P. T. Lancaster, T. Deuschle, C. Becher, C. F. Roos, J. Eschner, and R. Blatt, Nature **422**, 408 (2003).

[53] X. Li, Y. Wu, D. Steel, D. Gammon, T. H. Stievater, D. S. Katzer, D. Park, C. Piermarocchi, and L. J. Sham, Science **301**, 809 (2003).

[54] M. Fiorentino, T. Kim, and F. N. C. Wong, Physical Review A (Atomic, Molecular, and Optical Physics) **72**, 012318 (pages 4) (2005), URL http://link.aps.org/abstract/PRA/v72/e012318.

[55] G. Lindblad, *Non-equilibrium Entropy and Irreversibility, volume 5 of Mathematical physics studies* (Dordrecht, Reidel, Dordrecht, 1983).

[56] G. Gordon, G. Kurizki, S. Mancini, D. Vitali, and P. Tombesi, Journal of Physics B **40**, S61 (2007).

[57] G. Gordon, N. Erez, and G. Kurizki, J. Phys. B **40**, S75 (2007).

[58] L. A. Khalfin, JETP Lett. **8**, 65 (1968).

[59] L. Fonda, G. Ghirardi, A. Rimini, and T. Weber, Nuouo Cim. A **15**, 689 (1973).

[60] B. Misra and E. C. G. Sudarshan, J. Math. Phys. **18**, 756 (1977).

[61] A. M. Lane, Phys. Lett. **99A**, 359 (1983).

[62] S. Nakajima, Prog. Theor. Phys **20**, 948 (1958).

[63] R. Zwanzig, Journal of Chemical Physics **33**, 1338 (1964).

[64] H.-P. Breuer, B. Kappler, and F. Petruccione, Annals of Physics **291**, 36 (2001).

[65] H.-P. Breuer and F. Petruccione, *The Theory of Open Quantum Systems* (Oxford University Press, Oxford, 2002).

[66] G. S. Agarwal, Phys. Rev. A **61**, 013809 (1999).

[67] R. A. et al., Physical Review A **70**, 10501 (2004).

[68] B. Sherman, G. Kurizki, and A. Kadyshevitch, Phys. Rev. Lett. **69**, 1927 (1992).

[69] Y. Japha and G. Kurizki, Phys. Rev. Lett. **77**, 2909 (1996).

[70] M. C. Fischer, B. Gutierrez-Medina, and M. G. Raizen, Phys. Rev. Lett. **87**, 040402 (2001).

[71] Q. Niu and M. G. Raizen, Phys. Rev. Lett. **80**, 3491 (1998).

[72] G. Gordon, G. Kurizki, and A. G. Kofman, Opt. Comm. **264**, 398 (2006).

[73] G. Gordon and G. Kurizki, Phys. Rev. Lett. **97**, 110503 (2006).

[74] W. K. Wootters, Phys. Rev. Lett. **80**, 2245 (1998).

[75] M. Nielsen and I. Chuang, *Quantum Computation and Quantum Information* (Cambridge University Press, Cambridge, UK, 2000).

INDEX

D

E

N

O

P

Q

T

W

X

Y

Z